ECOLOGY

ECOLOGY

Sasinath Jha
Kul Prasad Limbu

ANAMIKA PUBLISHERS & DISTRIBUTORS (P) LTD.
4697/3, 21A, Ansari Road, Daryaganj, New Delhi 110002
Phone: 23281655, 23270239; Fax.: 011-27868035
E-mail : anamikapublishers@yahoo.co.in

First Published 2010

ISBN 978-81-7975-269-2

PRINTED IN INDIA

Published by Anamika Publishers & Distributors (P) Ltd.
4697/3, 21A, Ansari Road, Daryaganj, New Delhi 110002.
Typset and printed by Aadil Printographics, Delhi 110092.

CONTENTS

PART TWO
ENVIRONMENTAL FACTORS

PREFACE

The science of ecology is usually defined as the study of the relation of organisms or groups of organisms to their environment. It is concerned with the web of interactions involved in the circulation of matter and flow of energy that makes life possible on the earth, and with the adaptations of organisms to the conditions under which they survive. The term *ecology* refers to the whole environmental *house* in which we live. Resource impoverishment, environmental degradation and soaring population growth in recent years have damaged our natural ecosystems. The United Nations Environment Program (UNEP) has in particular been drawing attention to the need for development without destruction, and the science of ecology can be instrumental in achieving this goal.

This book "Ecology" contains altogether 19 chapters, 17 of which, in general, describe autecology-the responses of populations to their environment, synecology-the community attributes, and the role of the environmental factors (solar radiation, temperature, water, climate, fire and soil) at both the autecological and the synecological levels. Chapter 11 deals with the biodiversity found in Nepal; whereas, basic principles of the environmental impact assessment (EIA) are dealt in the last chapter. This is not a detailed, thorough, unbiased reference work, nor does it take us to a new level of understanding. Rather, it is a basic synthesis which we hope will be useful starting point for both undergraduate and postgraduate students of the South Asian Universities.

We are thankful to all the sources used in synthesizing this work. Thanks are also due to the Anamika Publishers & Distributors (P) Ltd., New Delhi for publishing the book.

Criticisms and suggestions for the improvement of this book are cordially invited.

Sasinath Jha, K.P. Limbu

Part One
BIOTIC/COMMUNITY FACTORS

1. ECOLOGY AND ECOSYSTEMS

1.1 Ecology

Ecology (Greek: *oikos*=home; *logos*= the study of) is the science of the interrelations between living organisms and their environment. **Environment** is the summation of all **biotic** (living) and **abiotic** (non-living) factors which surround and potentially influence an organism. Examples of biotic factors include competition, mutualism, allelopathy and other interactions between organisms, whereas abiotic factors include all chemical and physical aspects of the environment associated with growth and distribution of organisms.

The German zoologist Ernst Haeckel coined the word "ecology" in 1869, and the science of ecology emerged as a distinct field of biology at the beginning of the 20th century. Today, the ecological studies have become indispensable for creating and maintaining the quality of human civilization as the world is crowded with 7 billion people; the gap between rich and poor nations has widened; a large chunk of human population of South Asia, Middle East and Sub-Saharan Africa is undernourished; the fuelwood and potable water have fallen far short of need ; many grasslands and croplands are converting to desert-like conditions; much of the tropical forests and Ramsar wetlands have disappeared ; and thousands of species of plants and animals have become extinct. The burning of fossil fuels is already causing damaging increases in the acidity of rain and snowfall, and it is raising the concentration of carbon dioxide in the earth's atmosphere at the level sufficient to alter the world climate substantially –with possible serious disruption of human activities, especially agriculture.

1.1.1 Sub Divisions of Ecology

There are two main subdivisions of ecology: autecology and synecology (Fig. 1.1).

Autecology (synonyms: physiological ecology, ecophysiology) deals with the life histories /adaptations and behaviour of individual species or populations in relation to their environment. Subdivisions of autecology include **demecology** (speciation), **population ecology** (the regulation of population size), **demography** and **genecology** (genetics). Autecologists try to explain the **why** of a given species' distribution; what phenological, physiological, morphological, behavioural or genetic traits seem most important to the species' continued success in a given habitat?

Synecology (synonyms: community ecology, phytosociology, vegetation ecology) deals with the study of groups of organisms which are associated together as a unit. There are three phases of synecology: (a) plant sociology, (b) community dynamics, and (c) evolutionary ecology.

The description and mapping of vegetation types and communities is called **plant sociology**; wherein the term **community** (**phytocoenose**) denotes any vegetation unit from the regional to the very local. One of the offshoots of plant sociology is **paleoecology** – the description of past vegetation types and association as they have existed through geologic time.

Community dynamics includes processes such as the transfer of nutrients and energy between members, the antagonistic or symbiotic relationships between members, and the process and causes of **succession** (community change over time).Vegetation which does not change with time is in equilibrium with its environment and is said to be **mature**, **stable** or in a **climax state**. In contrast, vegetation which is changing is said to be **seral**. In the process of succession, one seral community replaces another until a climax community is established and perpetuates itself. The path of succession through time is called a **sere**. The study of community dynamics can be abstracted to a mathematical level, where complex formulae and computer programmes summarize, simulate or model the particular dynamic system being examined. This type of research is called **systems ecology**.

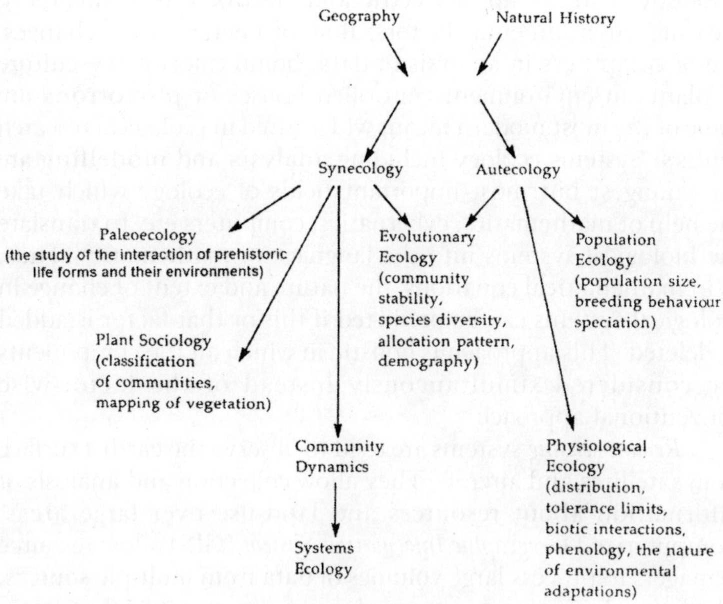

FIGURE 1.1
THE RELATIONSHIP BETWEEN AND ORIGINS OF SPECIALIZATIONS WITHIN THE FIELD OF ECOLOGY

The branch of synecology which tries to discover evolutionary themes that determine the fundamental nature of communities – factors determining the number of species which may coexist in a habitat, stability or fragility of a community, coevolution of plants and animals in the gradual formation of communities which exist today etc., is called **evolutionary ecology** and it overlaps with autecology.

1.1.2 Relation of ecology to other sciences

Four biological disciplines–genetics, evolution, physiology and behaviour are closely related to ecology. Besides these, various tools of other sciences, mathematics and statistics are increasingly applied in the analysis of ecological systems. Application of radioactive materials, use of sophisticated instruments like spectrophotometers, infra-red gas analyzers, flame photometers, elaborate field set up of electric and electronic instruments to measure environmental factors, flow of energy, gas exchanges, use of computers in analysis of data, bomb calorimetry, culture of plants in environment controlled houses or **phytotrons** are some of the most modern means widely used in ecological research centers. Systems ecology including **analysis** and **modelling** are the youngest but most important fields of ecology which take the help of mathematics, cybernatics, computers etc. to translate the biological systems into the language of mathematics. Thus, as in mathematical equations, the nature and extent of change in biological systems can be predicted if this or that factor is added or deleted. This approach is holistic in which all the components are considered simultaneously instead of the factor-wise conventional approach.

Remote sensing systems are used to observe the earth's surface from satellites and aircraft. They allow collection and analysis of information about resources and land-use over large areas. Computerized *Geographic Information System* (GIS) allow resource managers to process large volumes of data from multiple sources, which can be integrated to produce maps, monitor changes in resources and model the impacts of management decisions. Many resource managers have already benefited from these new technologies in fields such as watershed management, agriculture, forestry, conservation, mineral resource development and prevention and mitigation of natural disasters.

1.1.3 History of ecology

The roots of ecology lie in natural history which is as old as humans. The early human race had knowledge of the distribution of wild food and forage plants which sustained them and the prey they hunted, and also about the nature and habitats of the plants which had healing, narcotic or hallucinogenic properties. However, the real development of ecology came through world exploration of the floras and faunas and advances in agriculture, fisheries and medicines in the nineteenth century.

Carl Ludwig Willdenow (1765-1812), who taught plant geography at the University of Gottingen in Germany, noted that similar climates produce similar vegetation types, even in regions thousands of kilometers apart, such as southern Africa and Australia.

The French ecologist Friedrich Heinrich Alexander von Humboldt (1769-1859) - a student of C.L. Willdenow explored Cuba, Venezuela, Peru, Mexico and the Orinoco and Amazon rivers, wrote a 30- volume work "Voyage aux regions equinoxiales", in which he used the term "association", correlated the distribution of vegetation types with environmental factors, and described the synergistic effects of some physical factors (elevation, latitude and temperature).

J.F. Schouw (1789-1852) – a Professor at the University of Copenhagen, emphasized the role of temperature in plant distribution, and also popularized the procedure of naming associations by combining the dominant genus with the suffix-*etum*. Thus *Quercetum* is an oak woodland association (*Quercus* + *etum*) and *Pinetum* is a pine forest association (*Pinus* +*etum*). Some modern schemes of association nomenclature still use this concept.

The Hungarian scientist Anton Kerner von Marilaun (1831-1898) is regarded as one of the first experimental ecologists. He established several transplant gardens at various elevations in the Tyrolean Alps, from Vienna at 180 m to his villa at 1200 m, and up to alpine zone at 2200 m. In each garden he grew alpine and lowland forms of more than 300 species together. He found that some variations within a species were fixed and heritable, whereas other variations were nonheritable modifications induced by the environment.

Jonannes Eugenius Bulow Warming (1841-1924) - a successor to J.F. Schouw at the University of Copenhagen coined the terms *halo-*, *hydro-*, *meso-*, and *xerophyte*, meaning plants of saline, wet, moist and dry habitats, respectively.

The German botanist Andreas Franz Withelm Schimper (1856-1901) explained the causes of regional differences in floras and vegetation.

Jozef Paczoski (1864-1941) is credited as the father of phytosociology. He explained how plants modify the habitat creating their own microenvironment, and discussed the role of competition, the causes of succession, the role of fire, the interdependence of species in a community, the continuum nature of community boundaries, and such physiological adaptations as shade tolerance.

Clinton Hart Merriam (1855-1942) – a founder of the National Geographic Society (USA) believed that temperature, especially warmth in the growing season, was the determining environmental factor for the distribution of vegetation types in the North American Continent.

Henry Chandler Cowles (1869-1939) – a founder of Ecological Society of America investigated plant succession on sand dunes and emphasized the dynamic nature of vegetation.

Frederick Edward Clements (1875-1945) described much of the vegetation of North America, naming regional formations and associations, local variants, and seral stages. He wrote about the causes of succession, the use of certain species as environmental indicators, and the methods one might use in documenting succession or in identifying associations.

In Europe, Christian Raunkiaer who succeeded J.E.B. Warming as Professor of Botany at Copenhagen developed a life form classification in 1934 that is still used widely.

Sir Arthur Tansley, a founder of the British Ecological Society, coined the term **ecosystem** in 1935 and led a conservation movement in England.

In Switzerland and later in France, Josias Braun- Blanquet, following the path laid by Kerner, developed his methods of community sampling and association nomenclature in his book "Plant Sociology: the Study of Plant Communities"(1932) which dominate much of plant ecology today. His approach to plant synecology is called the "Zurich –Montpellier School of Phytosociology".

Among the pioneer animal ecologists, Graunt (1662), who described human population for London in quantitative terms (birth rate, death rate, sex ratio and age structure), is called the father of demography. Leeuwenhoek (1687) made one of the first attempts to calculate theoretical rates of increase for an animal species. Buffon (1756) discussed how the great fertility of every species was counterbalanced by innumerable agents of destruction. In his "Essay on Population" (1798), Malthus calculated that although the numbers of organisms can increase geometrically (1, 2, 4, 8, 16....), their food supply may never increase faster than arithmetically (1, 2, 3, 4....). Later Charles Darwin (1859) used the reasoning of Malthus as one of the bases of his theory of natural selection. Doubleday (1841) stated that whenever a species was threatened, nature made a corresponding effort to preserve it by increasing the fertility of its members. In

1838 Verhulst derived an equation describing the initial rapid growth and eventually levelling off of population overtime. He called the resultant S-shaped curve the *logistic curve*. Edward Forbes in 1844 described the distribution of animals in British coastal waters and part of the Mediterranean Sea, and he wrote of zones of differing depths that were distinguished by the associations of species they contained. Between 1925-1926, A.J. Lotka and Vito Volterra formulated useful *Lotka-Volterra equations* for modeling predator-prey, parasite-host, competition, or other two-species interactions.

During the International Biological Programme (1964-1974), many ecologists worked on the production (biomass and energy) ecology of terrestrial, freshwater and marine ecosystems with emphasis on systems analysis. Some other landmarks in the history of ecology in the last three to four decades are Man and Biosphere (MAB) Programmes of UNESCO, Stockholm Conference (June 1972) on "Human Environment", and the Rio Earth Summit (1992) and Agenda 21 (Programme of Action for Sustainable Development).

The floral and faunal explorations of various parts of the country carried out by the Natural History Museum, London (1952), Kyoto University, Japan (1952), University of Tokyo, Japan (1963), Chiba University, Japan (1963,1971) and others paved the way for ecological studies in Nepal (Rajbhandari 2002). Today, autecology, synecology, conservation ecology and response and tolerance of vegetation to pollution and other disturbances are being actively studied under the aegis of Tribhuvan University, Kathmandu University, NAST (Nepal Academy of Science and Technology), WWF (Worldwide Fund for Nature), ICIMOD (International Center for Integrated Mountain Development), IUCN (The World Conservation Union), NTNC (National Trust for Nature Conservation) and others. Nepal is a signatory to several multilateral environmental agreements including Convention on Wetlands of International Importance especially as waterfowl habitat (Ramsar, 1971), Convention on Biological Diversity (CBD, 1992), and Convention on International Trade in Endangered Species of Wild Fauna and Flora (CITES, 1973). Recently, the Environmental Protection Act 1996 of Government of Nepal has made Initial Environmental Examination (IEE) and Environment Impact Assessment (EIA) mandatory for any sort of developmental project including industry, roads, hydropower, power transmission lines, tourism, solid waste management, etc.

1.2 Ecosystems

Nowhere an organism or a species lives alone. Always there are associates influencing each other and organizing themselves into communities. A biotic community lives in an environment which supplies its material and energy requirements and provides other living conditions. The biotic community, together with the physical environment forms an interacting system called ecosystem. In short,

Ecosystem = community of organisms + environmental conditions

Each ecosystem has a certain spatial extent; together ecosystems form a diverse mosaic in the biosphere. A forest is an ecosystem, as is a meadow, a lake or an ocean. The principles by which all these ecosystems operate hold equally well for the biosphere as a whole, and as well for natural ecosystems as for artificial systems such as aquariums and self-contained manned spacecraft.

1.2.1 Components of an ecosystem

The biosphere and its ecosystems can be separated into two parts: (1) abiotic, or nonliving components: water, air, nutrients, and solar energy; and (2) biotic or living components (plants, animals, and microorganisms, sometimes called *biota*).

1.2.1.1 Abiotic components

The nonliving or abiotic components of an ecosystem are the physical and chemical factors (temperature, humidity, rain and snowfall, soil and substratum, etc.) that influence living organisms in terrestrial and aquatic ecosystems.

Different species thrive under different physical conditions. Some need bright sunlight, and others thrive better in shade. Some need a hot environment and others a cool or cold one. Some do best under wet conditions and others under dry conditions.

Each population in an ecosystem has a *range of tolerance* to variations in its physical and chemical environment. Individuals within a population may also have slightly different tolerance ranges for temperature or other factors because of small differences in genetic makeup, health, and age. In general, there are minimum and maximum limits for physical conditions (such as temperature) and concentrations of chemical substances, called *tolerance limits*, beyond which no members of a particular species can survive. Further, a species may have a wide range of tolerance to some factors and a narrow range of tolerance to others. Most organisms are least tolerant during juvenile or reproductive stages of their life cycles. Highly tolerant species can live in a variety of habitats with widely different conditions.

A variety of factors can affect the number of organisms in a population. However, sometimes one factor, known as a *limiting factor*, is more important in regulating growth than other factors. This ecological principle, related to the law of tolerance, is called the **limiting factor principle**: too much or too little of any abiotic factor can limit or prevent growth of a population, even if all other factors are at or near the optimum range of tolerance. On land, precipitation often is the limiting factor. Lack of water in a desert limits plant growth. Soil nutrients also can act as a limiting factor on land. Suppose a farmer plants corn in phosphorus-poor soil. Even if water, nitrogen, potassium, and other nutrients are at optimum levels, the corn will stop growing when it uses up the available phosphorus. Too much of an abiotic factor can also be limiting. For example, too much water or too much fertilizer can kill plants.

The limiting factor for a particular population can change. For example, at the beginning of a plant's growing season, temperature may be limiting factor; later on, the supply of a particular nutrient may limit growth; if a drought occurs, water may be the limiting factor.

1.2.1.2 Biotic components

Every independently functional ecosystem is composed of at least two biological components: the producers and the decomposers. Between these there may exist a whole chain of consumers.

Primary producers are the autotrophic organisms, which

effect the incorporation of inorganic elements in organic compounds and thus raise them to a higher energy level. The green plants and some bacteria utilize sunlight to form carbohydrates from carbon dioxide and water (photosynthesis), and these become the basis for further syntheses. Various microorganisms accomplish the same thing by using the energy freed by exergonic inorganic reactions (chemosynthesis).

Consumers (or *phagotrophs*) are the heterotrophic organisms that feed directly or indirectly upon the organic substances synthesized by the primary producers. The principal consumers are the herbivores and the plant parasites. The herbivores serve in turn as food for the carnivores, and both are attacked by animal parasites.

Decomposers (or *saprotrophs*) are those organisms that finally reduce plant and animal refuse to the level of its basic inorganic components. This group includes bacteria, fungi, and some soil animals. The decomposers, like the herbivores and other consumers, can serve as food for other organisms. In such cases, they take on the role of **secondary producers**. Thus a single individual, depending on its position in the food chain, can play the role of secondary producer, consumer or decomposer.

1.2.2 Energy flow and mineral cycling

The study of one of the most important aspects of an ecosystem involves the volume and the rate at which various materials circulate and the rate at which energy flows through it. All organisms require energy for their life processes and materials for the formation and maintenance of body structures. Food supplies both energy and materials for the sustenance of life.

The transfer of food energy from the source in plants through a series of organisms with repeated eating and being eaten is referred to as the *food chain*. The net transfer of energy in a chain always proceeds in one direction – from the sun through the primary producers to the decomposers. Matter, on the other hand, is cycled. Each cycling of bioelements takes place at three distinct levels: in the plant, in the ecosystem, and in the biosphere. The individual organism takes up substances, incorporates them in its body or obtains energy from them by metabolism, and excretes them. At the second level, the movement of materials through the ecosystem begins with conversion by the metabolism of the primary producers and leads back to the starting point through intermediate consumers and the decomposers – that is, through the food chain. The food cycle in the ecosystem is associated with geochemical transpositions of matter which extend beyond the ecosystem itself; these exchanges link the metabolism of the organisms with the transformations of inorganic chemicals normally occurring in the biosphere. Because of this complex biogeochemical exchange of matter, each ecosystem itself is actually an *open system*.

1.2.3 Autoregulation

Ecosystems possess intrinsic mechanisms for self-regulation, based both on the pronounced ability of individual organisms, populations and communities to adapt, and on the properties of the closed sequence of producers, consumers and decomposers in the biological nutrient cycle. If the supply of energy is sufficient and the recycling of matter undisturbed, then once a state of equilibrium has been reached with respect to the interrelationships of species and the exchange of materials with the environment, the state is maintained without notable change. Such *stable ecosystems* consist of a large biomass which does not further increase, since production balances consumption in the nutrient cycle over a year. E.P. Odum (1971) termed these ecosystems *protective*. The model of protective ecosystem is the tropical rain forest–the plants produce a great deal of organic substance but a large part of it enters the cycle and is remineralized. When land is newly colonized, or after drastic damage such as that caused by fire in forest, a state of equilibrium in the ecosystem must be gradually established. Young ecosystems are at first susceptible to disturbance–they are *unstable*. Their relatively small biomass does not fully occupy the available space, but their rate of growth is high; they are *productive*. As they mature, their rate of growth is regulated, their productivity approaches a state of equilibrium, and the initially productive ecosystem eventually becomes a protective ecosystem. Autoregulation ensures that the development of mass and the increase in number of members of the system remain subordinate to the functional whole–the population does not increase without bound.

The readiness to adapt and the autoregulatory capacity of a biological community can be overburdened by extraordinary environmental events. Such "stresses" may be transitory though severe–examples are floods, avalanches, frost, fire, drought and outbreaks of parasites– but they can also be long–term disturbances involving air and water pollution or unrestrained exploitation by man. Depending on their species composition, structure and the nature of their equilibrium state, ecosystems are differentially "resilient" in the face of such factors. In general, the "resilience" of an ecosystem is greater the less susceptible it is to disturbance and the more capable it is of recovering rapidly and completely when the disturbance is removed.

1.2.4 Biotic interactions

In addition to the food chain, there are other kinds of interdependence and interaction relationships among the organisms in an ecosystem that may either benefit or hamper the individual. Examples of *beneficial* biotic influences include ecological relationships between plants and animals that promote pollination and distribution; moreover, plants can live in symbiosis with microorganisms, and full–grown plants can shield seedlings from intense radiation, overheating, or cooling. *Inhibitory* influences include the competition among plants for light, water and nutrients, and the repulsion and suppression of organisms by secretions (allelopathy), as well as parasitism in all its manifestations (for example, hemiparasites such as mistletoe interfere with the water balance of the host plant).

1.3 Man and the Biosphere

Like any other organism, man is an environment–dependent member of the living community and an environmental factor with respect to other members of the ecosystem in which he happens to participate. But he is more than that. To the extent that his behaviour is determined by the intelligence, his cultural development has raised him above the narrow confines of limited ecosystems and made him a citizen of the world. As a conscious and purposeful shaper of nature, he intervenes with incomparable effectiveness in natural processes and creates new habitats to his own liking. But he can also misuse these unique abilities, disturbing or destroying every habitat on the planet–especially if his knowledge fails to comprehend environmental relationships and if his will is not tempered by a sense of responsibility.

2. INDIVIDUALS AND SPECIES

2.1 Individuals

Individuals are the fundamental units of populations, communities, ecosystems and biomes. An individual organism exchange materials and information with its surroundings, which helps it to adapt and survive. Individual organisms, either unicellular or multicellular, are always derived from pre-existing ones through the mechanism of reproduction- vegetative, asexual or sexual. The offspring resembles the parents very closely. On the one hand, a particular individual usually has a life span with definite beginning and a definite end. On the other hand, it transmits its characters to the offspring, thus ensuring continuity.

2.1.1 Classification of organisms

Linnaeus placed all organisms in either one of the two large kingdoms, the plant kingdom (kingdom **Plantae**) or the animal kingdom (kingdom **Animalia**). The animal kingdom included the multicellular forms and the unicellular organism (protozoa). These ingest food or particulate matter and often have some means of locomotion. All other organisms, green plant, mosses and multicellular seaweeds, moulds and mushrooms, lichens, minute-coloured or colourless unicellular organism and bacteria, were lumped together in the plant kingdom. A new five-kingdom arrangement of organisms was proposed by Whittaker in 1969 to replace the old, two kingdom groupings. In this arrangement, the separation into five kingdoms is made on the basis of the following three criteria: complexity of cell structure, complexity of the organism's body, and mode of obtaining nutrition (Fig.

2.1). The five kingdoms are Monera, Protista, Plantae (plants), Fungi, and Animalia (animals).

Monera includes all prokaryotic organisms, viz. bacteria, actinomycetes (filamentous bacteria), and photosynthetic cyanobacteria. Moneran cells are microscopic (1 to a few microns in length) and do not contain a nucleus or other membrane-bounded organelles. Most moneran cells have a rigid cell wall. The moneran nutrition is varied: some are heterotrophs, others are autotrophs. Heterotrophic monerans decompose and absorb organic matter in solution. Their wall prevents ingestion of particulate organic matter. The autotrophic monerans prepare their own food by reducing carbon dioxide, using light energy and an electron source different from water. Some of these use organic acids such as acetic or succinic acid as electron sources, others use H_2S and deposit sulphur as a by- product:

$$CO_2+2H_2S+\text{light} \rightarrow (CH_2O) + H_2O+2S$$

Some monerans fix atmospheric nitrogen into ammonia, enriching the soil. Some monerans thrive symbiotically with higher forms of life; others are parasitic. Some monerans are able to thrive under extreme environmental conditions, such as absence of oxygen, high salt concentration, or high temperature (80°C), or acidic pH. They are called **archae-bacteria** (ancient bacteria) because they are probably early forms of life. They continue to live under conditions which resemble the conditions that prevailed on primitive earth. Monerans, in general, are the important decomposers and mineralizers in the biosphere.

The kingdom **Protista** (*protisto*=primary) includes diverse kinds of mostly unicellular and primarily aquatic eukaryotes. They contain typical eukaryotic cell organelles such as nucleus, mitochondria, endoplasmic reticulum, golgi bodies, and plastids (if photosynthetic). Protistan organisms or protists display diverse ways of life. Many are photosynthetic autotrophs and are the chief producers in oceans and many freshwater environments. Included in this category are several types of coloured unicellular algae and diatoms. They are collectively called **phytoplankton** or microscopic, floating photosynthetic organism. Most phytoplanktons have cell wall and several have flagella. Some protists are predatory, feeding on other protists. They are called **protozoa**. Absence of cell wall in protozoans allows ingestion of particulate food, a process termed **holozoic** or animal-like

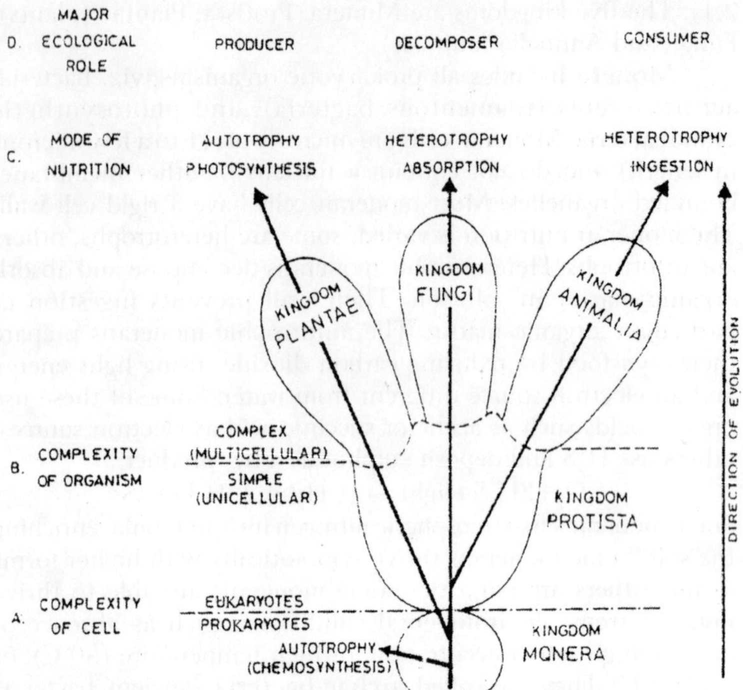

FIGURE 2.1
THE FIVE KINGDOMS OF LIFE. THE KINGDOMS ARE ARRANGED TO SHOW THE PROBABLE PHYLOGENTIC RELATIONSHIPS AMONG THEM BASED ON FOSSIL AND OTHER EVIDENCES (WHITTAKER 1969).

nutritions. Some protozoans live as parasites in other animals. Others are digestive tract (gut) inhabitants in almost all types of animals. Some aid in the decomposition of organic matter like the normally indigestible cellulose of plant cell walls, as in termites and wood-eating cockroaches.

Kingdom **Plantae** includes all the coloured, multicellular photosynthetic plants which we see on land, on sea shores, in lakes and streams. The main groups are the seaweeds consisting of complex red, brown, and green algae, mosses, ferns, and seed plants with or without flowers. Plant cells, owing to their rigid cellulose cell wall, cannot contract and relax like animal cells do. Plants are, therefore, immobile and exhibit no locomotion seen in many unicellular protists or animals. They are primarily producers on land, and along many seashores. They can be the dominant vegetation of shallow lakes and ponds. Plants synthesize all their organic constituents from water, carbon dioxide, and inorganic forms of essential elements, using light energy trapped by chlorophyll and accessory pigments. A few flowering plants have evolved into totally heterotrophic forms, living as partial parasites or full parasites on other plants. Others, called *insectivorous plants* have developed devices to trap and digest extracellularly small animals like insects, and absorb their nitrogenous matter. Still others have developed symbiotic relationship with certain nitrogen-fixing bacteria or with fungi to augment their nitrogen or mineral nutrition.

The kingdom **Fungi** includes diverse kinds of eukaryotic, predominantly multicellular heterotrophic organisms. They have a common mode of nutrition, namely absorption of organic matter made soluble by decay. Fungi are termed *saprobes* when they live on decaying plant or animal matter, and *parasites* when they assimilate tissues of living plants or animals. Fungi liberate digestive enzymes through their cell wall into the immediate environment where complex organic matter is made soluble and absorbed in solution. The visible parts of fungi like many mould, mushrooms, puff balls and bracket fungi on live trees or logs are in fact reproductive structures. These develop from the tangle of filamentous vegetative or assimilative body of the fungus, which grows within the tissues of wood. Parasitic forms of fungi cause diseases such as mildews, rusts, smuts, soft or dry rots, wilts and leaf spots in the host plants. Yeasts are a group of exceptional fungi involved in fermentation. These organisms have single-celled body. Although they resemble protists in being multicellular, their mode of sexual reproduction indicates their relationship with unicellular fungi.

Members of the kingdom **Animalia** are multicellular, holozoic eukaryotes. They are also known as *metazoa*. Their principal mode of nutrition is the pursuit and ingestion of food. Their mobile way of life is made possible by the development of cells with ability to contract (muscle cells) or to transmit impulses (nerve cells). Notable exceptions are the primitive metazoans called *sponges*. They lack nerve cells. Some animal groups have become parasitic. They live on or within the tissues of other metazoans or plants. The animal kingdom exceeds all other kingdoms in diversity. Over a million species have been named. These include diverse forms such as sponges; cnidarians (hydra and jelly fishes); flat, round or segmented worms; snails and other molluscs; arthropods (jointed-leg animals) like insects, scorpions and spiders; starfishes; and vertebrates such as fish, amphibians, reptiles, birds and mammals. Ecologically animals are *consumers*. They form links in different food chains and complex food webs through which part of matter and energy flows from the producer to the decomposer.

2.1.2 Nutrition

All organisms require energy for their life processes and materials for the formation and maintenance of body structures. Food supplies both energy and materials for the sustenance of life.

There are many approaches to classifying the ways in which individuals obtain their food. A useful one is to divide organisms into autotrophs and heterotrophs. **Autotrophs** obtain only the simplest inorganic substance from environment. Green plants are the most obvious autotrophs. These need only visible light, water, CO_2 and inorganic ions such as nitrate to survive, grow and reproduce. The process of photosynthesis enables most plants, the photosynthetic monerans and some protists to synthesize all the complex molecules that they require from these simple building blocks. Animals, fungi, most protists and a few plants cannot synthesize their organic molecules from inorganic ones. They need to take in an external source of organic carbon. These organisms are **heterotrophs**.

Heterotrophic nutrition involves ingestion, egestion, digestion, absorption, assimilation and excretion. *Ingestion* is the process by which heterotrophs take in their food. It completes in two stages: in the first, the food is captured; in the second, it is brought within the alimentary canal. The parts of the food ingested by a heterotroph which have not been digested sufficiently to allow them to be absorbed into the tissues are disposed of by *egestion* (the faeces produced by animals, for example).

Digestion is the process by which heterotrophs breakdown the food they have ingested into particles small enough to be absorbed. In most cases, this means that mastication and hydrolytic enzymes are used to breakdown food particles to their constituent monomers. Proteins are broken down to amino acids; carbohydrates to monosaccharides; fats to tiny fat globules or even to fatty acids, glycerol and other simple molecules. Salts and vitamins can be absorbed without needing to be digested. *Absorption* is the uptake from the gut of vitamins, salts and the products of digestion. In mammals these substances are absorbed by certain of the *villi* of small intestine. The products of absorption are either respired or assimilated. *Assimilation* occurs when the products of absorption are taken up by cells and synthesized into macromolecules. These macromolecules may be stored or used for repair, growth and reproduction. *Excretion* is the expulsion of waste products of metabolism from an organism. Nitrogen is an important excretory product of heterotrophic organisms. When proteins are broken down, ammonia (NH_3) is produced. Ammonia is very toxic and needs to be diluted in a large volume of water. Freshwater fish, which abviously have water in abundance, are able to dilute and then excrete their ammonia directly. Other vertebrates convert the ammonia to compounds which can be concentrated so as to be less wasteful of water, such as trimethylamine oxide, urea, or uric acid.

There are three main types of heterotrophic nutrition: holozoic, parasitic, and saprotrophic. **Holozoans** feed on relatively large pieces of dead organic material. For example, *carnivores* feed on prey, *herbivores* (sheep, cattle and goats) feed on vegetation, and *omnivores* (humans, pigs, etc.) feed on a mixture of plant and animal food. Herbivores include *granivores* (birds) which feed on grains or seeds, *frugivores* which feed on fruits, *folivores* which feed on the leaves of shrubs or trees, *grazers* which feed on forbs and grasses, and *browsers* which feed on the leaves, young shoots and fruits of shrubs and trees. Animals have anatomical, physiological and behavioural adaptations which are associated with their diets. For example, grasses contain very fine particles of silica (SiO_2) which generally wears away the teeth of grazers; hence, the teeth of most of the grazers grow afresh from their roots throughout their life.

Parasites, unlike holozoons, feed of matter that is still alive. Parasites can conveniently be divided into *endoparasites*, such as tapeworms, which live inside their hosts, and *ectoparasites*, such as fleas, which live on the outer surface of their host's body. Whether ectoparasitic or endoparasitic, most parasites obtain their food in liquid form. Parasites of plants may obtain their nourishment from the phloem, as aphids do. Animal parasites may live in gut or tap into the lymph or blood system.

Saprotrophs feed on dead organic matter which they either absorb in solution or ingest as very small pieces. Most saprotrophs obtain their food by extracellular digestion-enzymes are secreted on to the food source and the soluble products are then absorbed. This is the technique used by bacteria, fungi and house fly (*Musca domestica*), for instance. Animals which feed off dead organic matter, but ingest large pieces, may be classified as **scavengers** (for example, vultures).

2.1.3 Metabolic rate

The metabolic rate of an organism is the amount of energy it needs per unit time. This indicates what the demands of that organism are on its environment. For instance, animals with large metabolic rates are generally found near the top of food webs and at quite low population densities. A convenient unit for measuring metabolic rate is the number of joules of energy an organism uses each day. For organisms that rely on aerobic respiration to supply their energy needs, oxygen consumption is directly proportional to energy requirements and metabolic rate may be measured in units of hourly oxygen consumption as this is fairly easy to measure. One of the measurements of metabolic rate is *basal metabolic rate*, the minimum amount of energy an organism can respire. Three conditions need to be fulfilled for the metabolic rate of an organism to be its basal metabolic rate. First, the organism needs to be in a thermoneutral environment (so that no energy has to be used for thermoregulation). Second, the organism must be at rest (so that no energy is needed for locomotion or any other activities). Third, the organism has to be in a post-absorptive state (that is, some time after food has been eaten), because metabolic rate of an organism rises shortly after it eats. An organism's energy expenditure in real life is measured by the daily energy expenditure of organism in the field. The daily energy expenditure includes the energy spent on locomotion, thermoregulation, growth and reproduction. It is usually about 1.5 to 3 times basal metabolic rate.

The metabolic rate of an organism is not directly proportional to its body mass. For example, small birds such as shrews have such a high metabolic rate relative to their body size that they may need to eat more than their body weight in food each day. An elephant, on the other hand, takes about three months to eat its own body weight of food.

The metabolic rate of an organism is also related with its life style. For example, **homoiothermic** ("warm-blooded") animals which maintain a constantly high body temperature, need more energy (25 to 30 times) than **poikilothermic** animals (that is, animals having a body temperature that varies approximately with that of the environment; also called "cold-blooded" animals); and a poikilothermic animal has a metabolic rate 8 to 10 times higher than a unicellular organism of the same size.

What do organisms do with the energy (food)? How much of it can they apportion to growth or to reproduction and how much do they need just to remain alive? These questions can be answered by looking at the energy budgets of organisms.

2.1.4 Energy budgets

When an organism ingests food, only some of it is assimilated. The term assimilation efficiency refers to the percentage of the energy that an organism ingests that is assimilated rather than egested. As one might expect, organisms differ greatly in their assimilation efficiencies depending on the type of food they eat. Carnivores feeding on vertebrates may have assimilation efficiencies in excess of 90%; insectivores have assimilation efficiencies of 30 to 60%, though zooplankton feeding on phytoplankton have assimilation efficiencies of 50 to 90%, and the giant panda (*Ailuropoda melanoleuca*) has the lowest assimilation efficiency of any mammal measured at 20% (Ricklefs 1980; Anon 1982).

Once a heterotroph has assimilated its food, the products of assimilation can either be respired to provide the energy needed to maintain existence and to repair old and damaged body tissues, or they can be diverted to growth and reproduction. Together, growth and reproduction are called **production**.

The percentage of the energy assimilated that an organism

diverts to growth is called its **growth efficiency**. As a general rule, juvenile poikilotherms have higher growth efficiencies than juvenile homoiotherms. However, if one looks at the growth efficiencies of very young organisms, one finds that homoiotherms may have growth efficiencies as high as 50-70 %, almost the same as those of poikilotherms which may be as high as 50-80% (Calow 1977).

Organisms have a limited amount of time and energy with which to complete a life cycle. This means that a fraction of the total time and energy available is budgeted to each activity. An organism's **resource allocation pattern** involves the allocation of time to vegetative growth, seasonal dormancy, and reproduction, and the allocation of energy to growth, maintenance, reproduction, roots, shoots, and leaves. **Reproductive effort** is the percentage of total energy that an organism spends on reproduction. A short-lived annual plant which matures rapidly devotes a large fraction of their photosynthate to produce flowers, fruits, and seeds. On the other hand, a long-lived perennial plant which has a prolonged vegetative stage, devotes only a fraction of their photosynthate to reproduction (that is, low reproductive effort).

2.2 Species and Speciation

A **species** consists of morphologically (structurally) and ecologically (functionally) similar natural populations which may or may not be interbreeding, but which are reproductively isolated from other such groups. *Reproductive isolation* (inability to interbreed) constitutes the most important boundary between different species. This is the main factor which maintains the distinctness of different species. If this were not so, the boundaries of distinctive characters of two species would disappear owing to production of offspring sharing the common characters or having the intermediate features. Reproductive isolating mechanisms are environmental, behavioural, mechanical, and physiological barriers that prevent two individuals of two species from producing viable offspring. These barriers may be grouped as prezygotic and postzygotic barriers. **Prezygotic mechanisms** which prevent fertilization and zygote formation include: (a) *habitat*: the populations live in the same region but occupy different habitats; (b) *seasonal* or *temporal*: The populations exist in the same region but are sexually mature at different times; (c) *ethological* (only in animals): The populations are insolated by different and incompatible behaviour before mating; and (d) *mechanical*: cross-fertilization is prevented or restricted by incompatible differences in reproductive structures (genitalia in animals, flowers in plants). In **postzygotic mechanisms** fertilization takes place and hybrid zygotes are formed but these zygotes are inviable or give rise to weak or sterile hybrids.

In addition to the specific morphology (external appearance) and breeding behaviour, a species always has habitat distinctiveness. It is this third aspect which allows an ecologist to use a species as deductive tools with which to understand ecosystems. Fox example, if the ecological requirements of a plant species *A* are known, and the resource allocation pattern is understood, then the presence and degree of vigour of species *A* anywhere permits us to make many inferences about the environment, such as soil depth, soil nutrient levels, frequency of frosts, light intensity, length of growing season, frequency of distribution, and the presence or absence of other plants and animals which may interact with that species.

Habitat of an organism is the place where it lives, or the place where one could go to find it; it includes the concerned organism of its population, other organisms as well as abiotic environment. (The latin term "habitat" was first used in the eighteenth-century floras and faunas to describe the natural place of growth or occurrence of a species. When floras and faunas began to be written in modern languages, the term "habitat" remained untranslated and began to be used as a technical term).

It is often observed that closely related organisms having similar habits or life forms often do not occur in the same places. If they do occur in the same places, they use different food, are active at different times, or are otherwise occupying somewhat different niches (that is, "one niche, one species"). The world **niche** originally meant a shallow ornamental recess formed in the wall of building, usually for the purpose of containing a statue or other decorative object. In ecology it came to stand for the precise way in which a species fits into its environment. A habitat is a description of where an organism is found, but its niche is a complete description of how the organism relates to its physical and biological environment. Thus, niche involves not only the physical space (habitat) occupied by an organism, but also its place in the community, including its energy source and period of activity. Laboratory works and field observations have shown that minor niche differences exist even between closely related species, and these differences are sufficient to permit coexistence and release from some competition (Ayala 1969; MacArthur 1958).

Hutchinson (1958) introduced the terms fundamental niche and realized niche. He suggested that the niche an organism could occupy in the absence of competitors and predators be called its **fundamental niche**, while the role it actually played in the community should be called its **realized niche**. In real life, therefore, the realized niche of an organism is smaller than its fundamental niche.

MacArthur and Pianka (1966) considered what might happen when several competitors invaded a habitat. They argued that in the presence of greater number of competing species, each species should concentrate its search efforts for food on a smaller part of the total available habitat, the so-called **ecological compression**. On the other hand, within the reduced habitat searched by each species, a full range of food items may still be present. Consequently, the range of food taken by a species should be independent of the number of competing species present. Similarly, when interspecific competition is relaxed, species should increase the range of habitats in which they obtain food, but not the actual range of food taken. In brief, ecological compression means that the species specialize more in terms of habitats they occupy in presence of more species of competitors. Consequently, they might inevitably specialize more in terms of the food they eat. This increased specialization might be sufficient to lead to less niche overlap.

Population of a species is tied together by the sharing of a common gene pool. A population's **gene pool** (genetic makeup) is the set of all genes in the individuals of the population of a species. **Speciations**, or the formation of new species and the development of species diversity, occurs when gene flow (gene exchange) within the common pool is interrupted by an isolating mechanism (Fig. 2.2). The most widely accepted hypothesis explaining the process of multiplication of species is the geographic theory of speciation, called allopatric speciation. However, allopatric speciation is not the only way in which new species can rise. Two additional theories of speciation, namely parapatric speciation and sympatric speciation have been suggested as alternative models.

Allopatric speciation operates in four steps: (a) reproductive isolation occurs because of physical, geographic separation of two populations; (b) the two reproductively and geographically isolated populations undergo independent evolution and become adapted to their separate environments; (c) reproductive isolation evolves, so that mechanisms occur to reduce interbreeding between the two populations; and (d) if geographic isolation stops and the two populations come into contact, and if some reproductive isolating mechanism has evolved, the speciation process is complete, and two species now exist.

Parapatric speciation can occur when a population of a widespread species enters a new habitat. Although no physical barrier separates this new population from other populations, occupancy of the new habitat will result in a barrier to gene flow between the population in the new habitat and the rest of the species. Parapatric speciation may be common in organisms that move very little, such as plants, moles, and flightless insects. Insect herbivores that mate on the host plant could undergo parapatric speciation if different races live on different plant species. Parapatric speciation is probably not as common as allopatric speciation.

Sympatric speciation occurs when reproductive isolation occurs within the range of a population and before any differentiation of the two species can be detected. It was first clearly demonstrated in higher plants in which such genetic isolating mechanisms as *polyploidy* (duplication of chromosome sets that can bring about immediate genetic isolation), hybridization, self-fertilization, and asexual reproduction are more common than animals. It is assumed that behavioural and reproductive patterns tend to break up a species into genetically isolated segments so that gene exchange between or local communities takes place at a much lower-rate than within the colonies, ultimately resulting into "on the spot" or sympatric speciation.

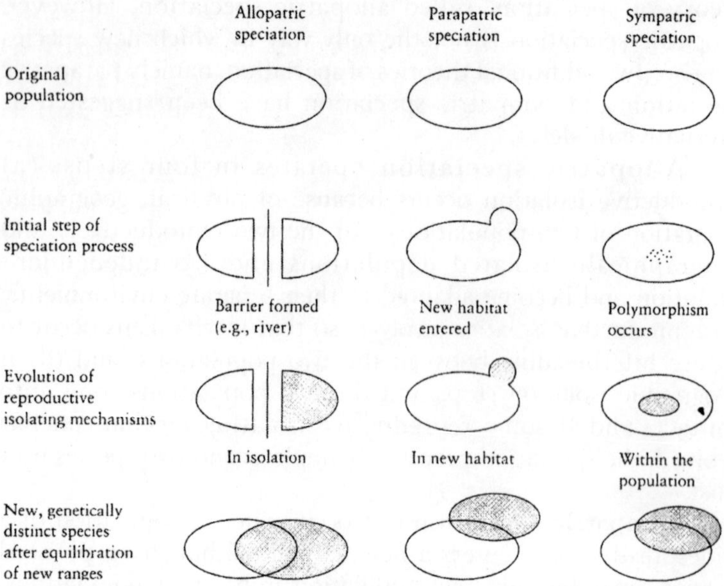

FIGURE 2.2
THREE GENERAL HYPOTHESES OF SPECIATION. ALLOPATRIC SPECIATION IS PROBABLY THE MOST COMMON TYPE

2.3 Ecotypes And Ecoclines

Plant species are found in populations which are not genetically static but dynamic and constantly changing. Variations observed within the populations are determined by various environmental factors (climate, soils, and biotic conditions), breeding systems (outcrossing, selfing, etc.), and internal conditions (polypoidy, mutations, etc.). Within the framework of taxonomic species, Gote Turesson (1922, 1930) developed the concept of ecological races which differ from other types of races in the sense that variation is always associated with some environmental factors. These are principally of two types: ecotypes, and ecads or ecophenes.

There is ample evidence that widespread species are not physiologically, or even morphologicaly, identical in all parts of their range. These ecological races are referred to as **ecotypes**. The most convincing way to demonstrate ecotypic differentiation is by transplantation. Plants of the same species but from different habitats may be grown in a uniform garden (Turesson's approach), or they may be reciprocally transplanted, each into the habitat of the other. A clump may be divided, and its parts planted in different habitats. If the ecological race retains its distinct characters even in the test garden, it is regarded as an ecotype. An example of a plant species with ecotypic differentiation is the harebell (*Campanula rotundifolia*). This species was studied by Turesson (1925), and the results of his study are summarized in Figure 2.3. He found that mountain populations were shorter, flowered earlier and had stronger rosettes of leaves than the lowland forms. This suited their growth in short mountain turf and their adaptation for rapid flowering and seed production in the shorter growing season in high altitude mountain habitats. Were these genetical distinct types of plants different species? Turesson made crosses and found both the types to be interfertile. Therefore, the types were technically part of only a single species.

In brief, an ecotype is the product of genetic response of a population to a habitat. It is a population or group of populations distinguished by morphological and/or physiological characters, interfertile with other ecotypes of the same species, but usually prevented from naturally interbreeding by ecological barriers. In general, most plant species are composed of an assemblage of ecotypes, each ranging in size from a single population to a regional group of many populations; the wider the species' range is, the more ecotypes there are within the species. The terms *race, genecotype,* and *ecological race* are synonyms for ecotype. Random genetic variants (individuals or groups of individuals) within ecotypes are called *biotypes*.

Populations whose uniqueness in nature is due to nongenetic plasticity are called *ecophenes* or *phenecotypes* or *ecads*, to distinguish them from ecotypes. These are plants of the same species which differ in appearance such as size, erect or prostrate nature, reproductive vigour etc. in differing environmental conditions. These variations are not genetically fixed, and when transplanted (usually, seed is collected from different ecophenes of a species for this purpose) to neutral conditions, the variation vanishes. For example, plants of *Euphorbia hirta* growing in disturbed conditions is prostrate and cushion-like, whereas in undisturbed conditions they are erect (Ambasht 1969). Upon transplantation the prostrate form may become erect as the morphological variations are not permanently fixed. *Viola sagittata* has notably long petioles when grown in the shade, and much shorter ones in sunny sites, but the difference is a direct response of the individual to the habitat (Yost 1987). Even some astute field-botanists have been misled to suppose a specific difference here, and the plants of open sites have passed in recent ecological literature as a distinct species under the name *V. fimbriatula* (Cronquist 1988).

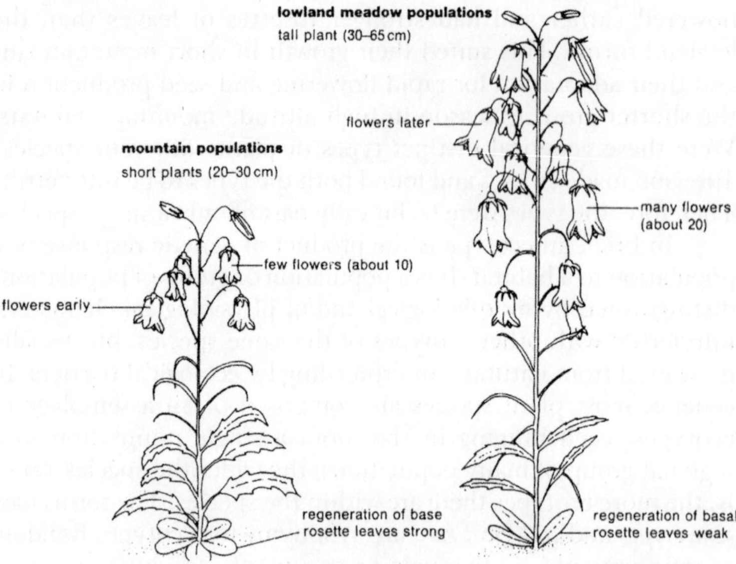

FIGURE 2.3
SOME OF THE DIFFERENCES BETWEEN POPULATIONS OF MOUNTAIN AND LOWLAND HAREBELLS (*CAMPANULA ROTUNDIFOLIA*) AS DESCRIBED BY TURESSON 1925.

Turesson thought of ecotypes as discrete entities, with clear differences separating one ecotype from another, because his sampling method prejudiced his results: plant material was selected from widely separated places and ecotones were ignored. In Scotland, J.W. Gregor (1946) closely examined what at first appeared to be two ecotypes of the coastal plaintain, *Plantago maritima*. One ecotype inhabited salt marshes regularly flooded by high tide with a soil salinity of approximately 2.5%. These plants had short leaves, small seeds, and thick, short, somewhat *decumbent* (lying or growing along the ground but erect at or near the apex) flowering stalks. The other ecotype inhabited nonsaline meadows further inland, and these plants had longer leaves, larger seeds, and thinner, taller, more upright flowering stalks. Gregor collected seeds of each and sowed them in a test garden. In addition, he collected and sowed seeds from plants growing in an *ecotone*, an intermediate habitat. The resulting plants showed that field differences were genetically fixed, but more importantly, they showed a continuous gradation from one extreme, one ecotype, to the other. There were no discrete boundaries between ecotypes or even between ecotypes and plants from the ecotone (Table 2.1). For example, 3.9% of the salt marsh ecotype plants had a habit grade score of 3, which is well within the range of meadow ecotype plants, and 2% of the meadow ecotype plants had a habit grade score of 2, which is well within the range of salt march ecotype plants. Ecotone plants completely overlap the habit grades for both salt marsh and meadow ecotype plants.

Olaf Langlet (1959), a compatriot of Turesson, brought seeds of the pine *Pinus sylvestis* from 580 sites throughout Sweden to a test garden. When he examined saplings for growth rate and morphological features, he found the extremes to be quite different, but a **cline**, a continuum of variation, connected the extremes. There were no sudden breaks in the range of variation, where one could say ecotype A ended and ecotype B began.

The above findings and ecotype research by many other investigators, on many different species throughout the world indicate that ecotypes are no more discrete and distinct from each other than are species. The changes in habitat as well as in ecotypic populations are now regarded gradual and continuous; that is, not like a stair case but gradual slope of a ramp (Fig. 2.4). An **ecoline** is a gradation in the attributes of a species (or community or ecosystem) associated with an environmental gradient. An ecotype, therefore, is only an arbitrary segment of an ecocline, which may be convenient to recognize for reference purposes.

TABLE 2.1

RESULTS OF GREGOR'S CULTIVATION EXPERIMENTS WITH PLAINTAIN (*PLANTAGO MARITIMA*) ECOTYPES. HABIT GRADE 1 INCLUDES THE MOST PROSTRATE, SUCCULENT PLANTS, WITH SHORTEST LEAVES AND SMALLEST SEEDS, HABIT GRADE 5 IS THE OTHER EXTREME, AND HABIT GRADES 2-4 ARE INTERMEDIATE.

Habitat	*Mean scape length (cm)*	*% of plants in habit grades 1-5*				
		1	2	3	4	5
Salt marsh, soil salinity 2.5%	23.0	74.5	21.6	3.9		
Upper marsh edge, ecotone	38.6	10.8	20.6	66.7	2.0	
Nonsaline meadow above marsh, soil salinity 0.25%	48.9	2	2.0	61.6	35.4	1.0

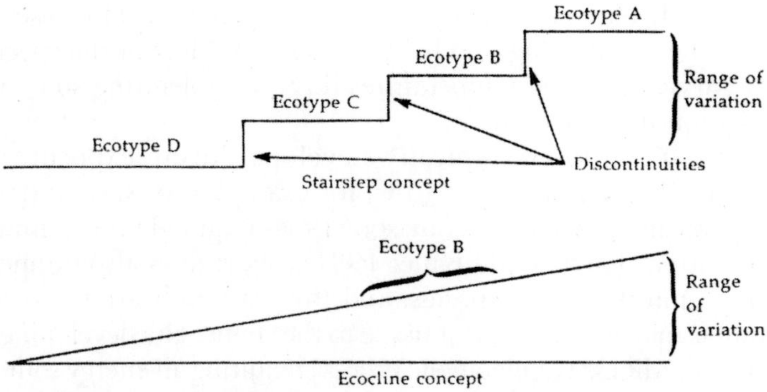

FIGURE 2.4 DIAGRAMS THAT CONTRAST TWO CONCEPTS OF ECOTYPES. IN THE STAIRSTEP VIEW, ECOTYPES ARE DISTINCT AND SEPARATED FROM EACH OTHER BY DISCONTINUITIES IN THEIR MORPHOLOGY OR OTHER ATTRIBUTES (THE VERTICAL LINES INDICATE DISCONTINUITIES). IN THE ECOCLINE VIEW, THERE ARE NO DISCONTINUITIES, THERE IS JUST A CONTINUUM OF VARIATION

2.4 Hardening (Acclimation)

The relationship between plant and environment can be written as:

Phenotype= genotype + prevailing environment + past environment

As such, **hardening** or **acclimation** (also called *acclimatization*) is a plastic, temporary change in an organism caused by an environment to which it has been exposed *in past*. Plants exposed to low water potentials, high light levels, and such other factors as high- phosphorus and low-nitrogen fertilization become drought tolerant or hardy compared to plants of the same species not treated in this way. That is, they become acclimated to drought, a process of considerable importance to agriculture. This is a good example of a conditioning effect.

Actively growing plants, especially herbaceous species, are damaged or killed by temperatures of only -1 to --5°C, but many of these plants can be acclimated to survive winter temperatures of -25°C or lower. In regions where air temperatures drop below this, many plants have underground meristems protected from extreme air temperatures by soil or snow. They avoid or escape cold rather than being cold hardy (tolerant). Most of the species that survive freezing temperatures do so by tolerating some ice formation in their tissues.

Frost hardiness typically develops during exposure to relatively low temperatures (5°C, for example) for several days. Temperatures down to -3°C are sometimes required for maximum acclimation (Larcher and Bauer 1981). Short days also promote acclimation in several species, and there are indications that a stimulus may move from leaf tissue to the stems. The development of frost hardiness is a metabolic process requiring an energy source. Apparently this can be provided by light and photosynthesis. Factors that promote more rapid growth inhibit acclimation: high nitrogen in the soil, pruning, irrigation, and so on. In general, non-growing or slowly growing plants are more resistant to several environmental extremes, including air pollution. Water-stressed plants are more resistant to air pollution, partially because their stomates are closed. Salt hardiness can also be increased somewhat by exposure to saline conditions.

3. POPULATION ECOLOGY

Organisms do not live solitary lives in isolation from other members of their species. Usually they live in groups or, if normally single, they interact with other members of their species at various times in their lives. A **population** may be defined as a group of organisms of the same species occupying a particular space at a particular time. The ultimate constituents of the population are individual organisms that can potentially interbreed. One of the fundamental principles of modern evolutionary theory is that natural selection acts on the individual organism, and through natural selection, population evolve.

The population may be subdivided into local populations, called **demes**, which are groups of interbreeding organisms, the smallest collective unit of a plant or animal population. Several demes linked together through dispersal (Fig.3.1) constitute a large regional population, called **metapopulation**. Over time the regional metapopulation changes less than each local population.

The population has group characteristics, which are statistical measures that can not be applied to individuals. These characters include the arrangement of individuals in space (density and pattern), the arrangement of individuals in time (demography), and the behaviour of individuals or the resource allocation patterns.

3.1 Density and Pattern

Population density refers to the number of individuals present in unit area at a given time. The most direct way to find out how many organisms are living in an area is to count them. One good example of this is the human population census. Usually, however, the investigator must be content to count only a small proportion

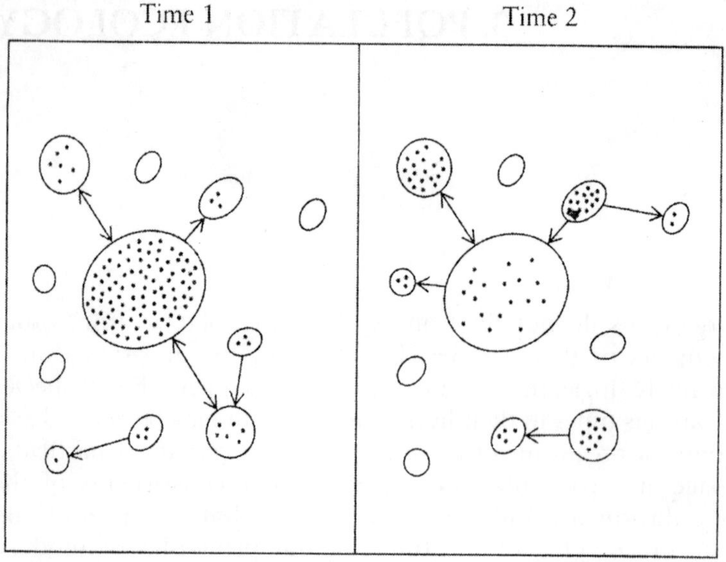

FIGURE 3.1
HYPOTHETICAL METAPOPULATION DYNAMICS. CLOSED CIRCLES REPRESENT HABITAT PATCHES, DOTS REPRESENT INDIVIDUAL PLANTS OR ANIMALS. ARROWS INDICATE DISPERSAL OF INDIVIDUALS BETWEEN PATCHES. OVER TIME THE REGIONAL METAPOPULATION CHANGES LESS THAN EACH LOCAL POPULATION

of the population and to use this *sample* to estimate the total. There are two general ways of sampling, by the use of quadrats or the capture-recapture method. Quadrat sampling methods involve counting and weighing of organisms in plots or transects of appropriate size and number to get an estimate density in the area sampled. In marking-recapture methods (for mobile animals), a sample of population is captured, marked, and released, and the proportion of marked individuals in a later sample is used to determine total populations. For example, if 100 individuals were marked and released and 10 out of a second sample of 100 were found to be marked, then population (N) would be figured as follows:

$$\frac{\text{Marked animals in second sample (R)}}{\text{Total caught in second sample (C)}} = \frac{\text{Marked animals in first sample (R)}}{\text{Total papulation size (N)}}\text{; or}$$

$$\frac{10}{100} = \frac{10}{N}\text{; or, } N = 1000$$

This simple type of population estimation is known as the **Petersen method** (also called Lincoln method). This method is most reliable when turnover is low; however, it does not work well when density is undergoing rapid change.

Individuals in a population may be distributed (arranged) in basically three patterns: random, clumped, or regular (Fig. 3.2). In the **random** pattern, the location of any one individual (plant or animal) has no bearing on the location of another of the same species. This type of distribution is relatively rare in nature, occurring where the environment is very uniform and there is no tendency to aggregate. In a **clumped** pattern (also called aggregated or underdispersed), the presence of one individual (plant or animal) means there is a high probability of finding another of the same species nearby. A **regular** (uniform), or overdispersed, is similar to the pattern of trees in an orchard. The member of a species appear to repel one another so that where one individual is found, there is a lower probability of finding another than would be expected by the assumption of randomness.

Clumping is a the result of the aggregation of individuals: (a) in response to local habitat differences, (b) in response to daily and seasonal weather changes, (c) as the result of reproductive processes, or (d) as the result of social attractions (in higher animals). In the temperate zone, members of most species of plants seem to be clumped, and there are at least two reasons for this: (i) *reproduction*: seeds or fruits tend to fall close to a parent, or runners or rhizomes produce vegetative offspring near a parent, and (ii) *microenvironment*: the habitat is homogeneous at a macroenvironment level, but at a finer level it consists of many different microsites which permit the establishment of a species with varying degrees of success. Those microsites most suitable for a species will tend to become more densely populated with that species.

Many animals are *social*, that is, they associate in a group with other individuals, usually of the same species. Family formation is the characteristic of some animals. Wolves, foxes and swans are monogamous. Once a male and female come together the partnership lasts throughout life. In some others like the sparrow the partnership breaks down at the end of a breeding season. The two birds find new partners in the following years. Seals, walruses and deer are polygamous. A male keeps a harem of many females. Rats, mice and some other animals do not have any fixed partners. Many birds such as ducks, geese and parrots remain in flocks. Antelopes, deer, zebras, elephants and monkeys live in herds. Members of the herd show cooperation for mutual protection and for finding food and water for the herd.

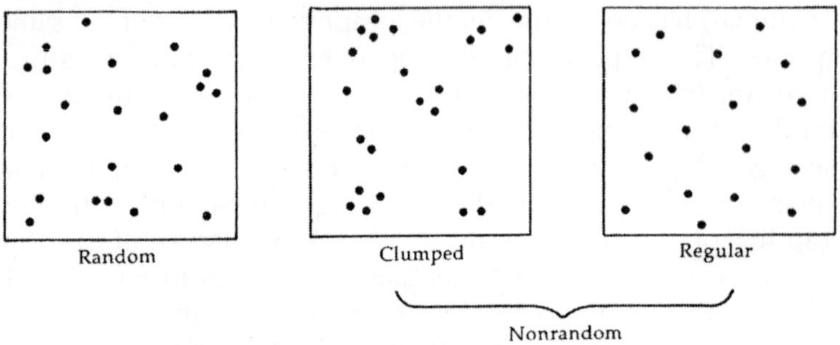

FIGURE 3.2
RANDOM AND NONRANDOM DISTRIBUTION PATTERNS OF INDIVIDUALS. NONRANDOM DISTRIBUTION PATTERNS MAY BE CLUMPED OR REGULAR. DOTS REPRESENT INDIVIDUAL PLANTS AS VIEWED FROM ABOVE.

Within the herd there is a social organization having a hierarchial order and often there is ranking of individuals. Usually the females are subordinate to males and the young are subordinate to the adult. In the red deer the female is the leader of the herd. Dominance of ranking of individuals is established by battles, ranging from a mock show to serious fights. In some species dominance can be established even without physical contact by aggressive posture, display of feathers and different other signals. The exact composition of the herd as well as the relationships between dominant animals and their subordinates also vary from species to species. Irrespective of the species it is found that the individual in the highest rank has the first choice of mate and food followed by individuals of subsequent ranks.

Some animals like the bees, ants and termites form well organized societies with division of labour among its members. In these social insects a large number of individuals of different kinds (*polymorphic*) live together in a colony and work closely together for the benefit of the group. Although the composition and functions vary in different groups, insect societies are formed of different *castes*, such as, workers, males (drones) and queen, which are specialized for different kinds of work. The workers collect and store food, build houses (hills or mounds) of complicated and characteristic architectural design and do many other special tasks. Some species of ants cultivate fungi. The males and the queen are mainly involved in reproduction.

Several members of a species may cover a defined area in search of food and mates. This is called their *home range*. A number of breeding groups (pairs or herds) inhabit a home range. There is an order in the distribution of each group. Each breeding group, or an individual occupies a particular territory. The home range of different groups can overlap but not the *territory* (Fig. 3.3). Each territory centers round the individual's or group's home, nest, burrow or other types of shelter. Territorial boundaries of certain animals are marked by urination (dogs, tigers) or defecation. The special odours of the urine or faeces delineate the territory. The territory of an individual or a group is defended by its members. The home range of different species can, however, overlap.

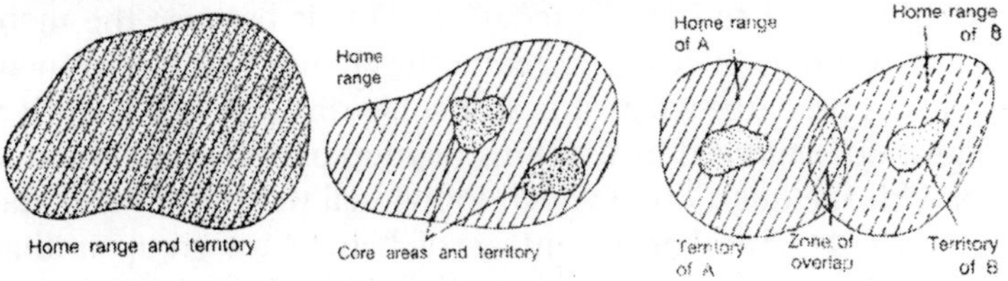

FIGURE 3.3
HOME RANGES AND TERRITORIES OF TWO SPECIES

Almost all species of animals have some mechanism of exchanging information among their members through sight, sound, touch or chemicals. There are numerous visual signals among higher animals. Specific patterns indicate the sex of the individual as in lion, pea fowl and pheasants. The white rump of an antelope serves as a signal by which other animals can follow it. This keeps the frightened herds from scattering. Honey bees perform various dance movements to indicate the distance and direction of the food source.

Different species of animals communicate by sending out sound signals such as croaks, grunts, snorts, squeaks and chirps. Some animals have as many as two dozen different sound signals indicating hunger, fear, threat, annoyance, desire for mating and parental concern. Certain special chemical compounds called *pheromones*, secreted by animals transmit messages to other members of the species. Pheromones are detected by smell or taste. The territory-marking compounds in urine and faeces described above are also pheromones. Among social mammals, the dominant male individuals have large *scent-marking glands* as compared to the others of the group and use them more often. If a dominant male is castrated, his scent-marking glands shrink in size and he gradually loses his advantageous position in the group.

There are several advantages of group living. One such advantage is that it reduces an organism's risk of being eaten. For examples, woodpigeons (*Columba palumbus*) are safer in larger flocks. Kenward (1978) used a trained goshawk (*Accipiter gentilis*) to attack feeding flocks of woodpigeons and found that the predator's attacks were less successful when directed at larger groups of pigeons. Kenward found that the reason for this is that the larger a flock of woodpigeons, the greater the distance at which they tend to detect predators. This is because the more birds there are in a flock, the greater the chance that one of them will spot a predator when it is still some distance off. Even if a predator does succeed in attacking a group of potential prey, it is faced with the question of which individual to attack (Neill and Cullen 1974). Another advantage of being in a group is that together the group may be able to defend itself from predators. For example, a group of musk oxen (*Ovibos moschatus*) is less vulnerable to attacks by grey wolves (*Canis lupus*) than are solitary individuals because a group of musk oxen can gather into a defensive circle when attacked by wolves. They stand facing outwards towards the wolves, their large horns forming a defensive wall to protect their young which hide within the circle of adults.

Group living in animals is often accompanied by instances of helping behaviour or **altruism**. For example, observations in the Bandipur Sanctuary of Karnataka (India) revealed that when a group of spotted deer (cheetals) is under attack by a predator like the wild dog, panther or tiger, the stag with the best antlers is closely surrounded by the other individuals so that it is saved from destruction (Mahadevan *et al.* 1988). This appears as a striking example of the self-sacrifice of a few individual deer allowing themselves to be killed in the interest of the species. There are numerous examples in the natural world of this kind. Some of the most notable ones relate to insects. Among ants, white ants (termites), and honey bees there is a remarkable case of *altruistic behaviour* whereby individuals sacrifice themselves in the interest of the whole group (population). Among honey bees, the workers, which are females, are sterile. The queen bee occupies a large cell in the hive and is the progenitor of future colony. She receives special food and care from the workers. Termites show incomplete metamorphosis; that is, the insects hatching from eggs are very similar to adults and develop into adults through growth and moulting. Only the "king" and the "queen" and sometimes only a few reach adulthood. The worker termites are perpetual adolescents who have lost the capacity of developing into adults.

If individuals huddle together, they effectively decrease their surface area to volume ratio. This may help them to reduce the rate at which they lose water and warmth. Nestling great tits (*Parus major*) huddle together for warmth and lose less weight the more of them there are (Mertens 1969). Woodlice (Isopoda) may huddle together to reduce their chances of drying. Bees provide another example of group survival value; a hive or cluster of bees can generate and retain enough heat in the mass for survival of all the individuals at temperatures low enough to kill all the bees if each were isolated. Allee *et al.* (1949) stated that these types of primitive cooperation (protocooperation) found even in very primitive phyla are the beginning of social organization, which shows varying degrees of development in the animal kingdom, culminating in the intelligent group behaviour of human beings. Allee's principle is diagrammatically illustrated in Fig. 3.4.

Forces which bring about isolation or the spacing of individuals, pairs, or small groups in a population are perhaps not as widespread as those favouring aggregation but they are nevertheless very important, especially in terms of population regulation (intraspecific, interspecific, or both). Isolation usually is the result of (a) interindividual competition for resources in short supply or (b) direct antagonism (allelopathy). In both cases a random or a uniform distribution may result, because close neighbours are eliminated or driven away. Individuals, pairs, or family groups of vertebrates and the higher invertebrates commonly restrict their activities to a definite area, called home range/territory. Territoriality seems to be most pronounced in vertebrates and certain arthropods that have complicated reproductive behaviour patterns involving nest building, egg laying, and care and protection of young. In higher animals the isolating mechanism is likely to be behavioural (neural), while in lower animals and plants it is chemical (allelochemics). In brief, isolation of this sort reduces competition, conserves energy during critical periods, and prevents overcrowding and exhaustion of food supply in the case of animals, or nutrients, water or light in the case of plants.

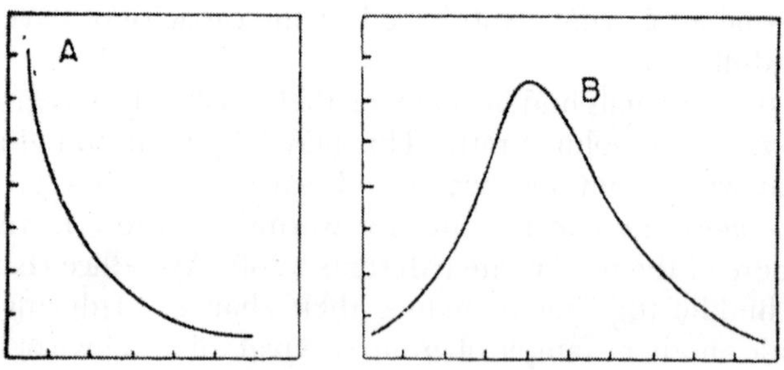

FIGURE 3.4
ILLUSTRATION OF ALLEE'S PRINCIPLE. IN SOME POPULATIONS GROWTH AND SURVIVAL IS GREATEST WHEN POPULATION SIZE IS SMALL (A), WHEREAS IN OTHERS INTRASPECIFIC PROTOCOOPERATION RESULTS IN AN INTERMEDIATE SIZED POPULATION BEING THE MOST FAVOURABLE (B). IN THE LATTER INSTANCE "UNDERCROWDING" IS AS DETRIMENTAL AS "OVERCROWDING" (ALLEE *ET AL.* 1949).

A number of methods have been suggested that may be used to determine the type of spacing and degree of clumping between individuals in a plant population (where it is not self-evident). One method utilizes random quadrats. The number of rooted individuals of species *A* is tallied in each quadrat and summarized in table form (Table 3.1). These are the observed data; the expected data, if members of species *A* were distributed at random, is generated by a rather simple formula, the *Poisson distribution*, which only requires that we know the average number of plants per quadrat, Discrepancy between the observed data and the expected data is evaluated by a *chi-square* calculation. In the example shown in Table 1, the chi-square value is greater than what one would expect by chance, so the conclusion is that members of species *A* are not distributed at random. Are they, then, clumped or regular? Inspection of the table shows that many fewer quadrats with zero individuals were encountered than expected, and many more quadrats with one individual were encountered than expected. This is the sort of result one expected from a regularly distributed population. If the population had been clumped, many quadrats would have had either zero or more than one plant, and very few would have had one plant. By deduction, then we can say that the members of species *A* were distributed regularly.

The variance/mean ratio also determines the type of pattern. A general property of random distribution is that the variance (V) equals the mean (m); variance greater than the mean indicates a clumped distribution, and less than mean a uniform (regular) pattern. Thus, in a random distribution:

$$V/m = 1; \text{ standard error } = \sqrt{2/n-1}$$

If, on the application of standard significance tests, the variance/mean ratio is found to be significantly greater than 1, the distribution is clumped; if it is significantly less, the distribution is regular; if not different from 1, the distribution is random (Table 3.2).

Another method, suggested by Dice (1952), involves actually measuring the distance between individuals in some standardized way. When the square root of the distance is plotted against frequency, the shape of the resulting frequency polygon indicates the distribution pattern. A symmetrical polygon (a normal bell-shaped curve) indicates random distribution, a polygon skewed to the right a uniform distribution, and one skewed to the left a clumped distribution.

TABLE 3.1
POISSON ANALYSIS OF QUADRAT DATA FOR A SPECIES WITH A NONRANDOM PLANT DISTRIBUTION. IN THE FORMULAS, *M* IS THE AVERAGE NUMBER OF PLANTS PER QUADRAT, 1.56 IN THIS CASE, AND E^{-M} IS 0.21. TO EMPLOY THIS TEST, EACH CATEGORY MUST HAVE AN EXPECTED VALUE OF ≥ 5 % OF THE TOTAL QUADRATS; TO ACHIEVE THIS, CATEGORY 5 HAD TO BE LUMPED TOGETHER WITH CATEGORY 4. THE X^2 SUM IS THUS BASED ON ≥5 NUMBERS, AND THE DEGREES OF FREEDOM IS 5-2 = 3. AT THE 99 % LEVEL OF SIGNIFICANCE FOR 3 DEGREES OF FREEDOM, ANY ΣX^2 VALUE > 11.34 SUGGESTS THAT THE NULL HYPOTHESIS BE REJECTED. THE NULL HYPOTHESIS HERE IS THAT THE PLANTS ARE DISTRIBUTED RANDOMLY. THEREFORE, THE PLANTS ARE DISTRIBUTED NONRANDOMLY.

No. of plants per quadrat (x)	Observed no. of quadrats with x plants	Expected no. of quadrats with x plants = $\left(e^{-m}\right)\left(\frac{m^x}{x!}\right)(100)$	$x^2 = \frac{(\text{observed} - \text{expected})^2}{\text{expected}}$
0	13	21.0	3.0
1	51	32.8	10.1
2	23	25.6	0.3
3	3	13.3	8.0
4	0 } 10	1.5 } 6.8	
5	10	16	1.5
Totals	100	99.5	$\Sigma x^2 - 22.9$

TABLE 3.2
MEAN, VARIANCE, AND SPATIAL DISTRIBUTION OF TWO SPECIES OF SMALL CLAMS ON AN INTERTIDAL MUDFLAT IN CONNECTICUT (JACKSON 1968)

Species and age	*Mean*	*Variance*	*Variance/mean ratio*
Mulinia lateralis			
All ages	0.27	0.26	Random
Gemma gemma			
All ages	5.75	11.83	Clumped
1st year	4.43	7.72	Clumped
2nd year	1.41	1.66	Random

3.2 Demography

The study of the population dynamics (that is, the dynamics of the number of individuals in a population) is called demography. It is the science of vital statistics: natality (births), mortality (deaths), age class distribution, and average life span.

Populations gain individuals when young are born in the area or new members join the group from other populations (*immigration*). Immigration is most common in mobile species such as mammals, birds, fish, insects and so on. In plant species immigrants are usually seeds or spores. Populations lose organisms when they die, or when they leave an area to join another population (*emigration*).

Three major variables distinguish individuals in population: *sex, age,* and *size*. The male/female sex ratio of a population affects the potential reproductive rate and may affect social interactions in many vertebrates. Age is a significant variable in human population, and age effects are common to many species. Older individuals are frequently larger, and changes in size may be the main mechanism by which age effects occur. Larger fish lay many more eggs than smaller fish, and larger plants produce more seeds. Young mammals may be more prone to diseases that older animals can resist. Age and size are very significant individual attributes in all animals that have social organization because they help to specify an individual's social position.

Unlike animals, which cease growing when mature, perennial plants possess primary and secondary meristems which theoretically permit continued growth in length and girth forever. Some, in addition, have the ability to reproduce asexually by runners, rhizomes, or layering so that the same genetic "individual" can also live endlessly, theoretically. Indeed, the longest-lived organisms on earth are plants, whether they be simple lichens 4500 years old (Thomson 1974), clones of shrubs 3000-4000 years old (Vasek *et al.* 1975, Cottam *et al.* 1959), or conifer trees nearly 5000 years old (Currey 1965). In addition, the seeds of some species may lie dormant for as long as 1000-10000 years (Harper and White 1974). However, even the record holders do eventually die, falling to disease, physical disasters, excessive herbivore damage, or a changing climate.

Plants represent some unique problems to demographic

studies, compared to animals. Most higher animals are *unitary* organisms with determinate form-cats all have four legs. In unitary organisms an individual is easy to recognize and each individual is a separate genetic individual. By contrast, most plants are *modular* organisms in which the zygote develops into a unit of construction (a module) which then produces more, similar modules (Fig. 3.5). Many of the lower animals such as hydrozoans, corals, and bryozoans, are also modular organisms. Such organisms are always branched, and individuals are composed of variable number of modules. The behaviour of such interconnected units, in terms of mortality and other demographic events, may be different than for true individuals (Hutchings and Barkham 1976).

Another problem is that the timing of germination may be unrelated to the timing of reproduction. The Middle East desert plant *Blepharis persica* retains its seeds in the fruit for 10 or more years until a heavy rain triggers their release, at which time they germinate in *en masse* within 3 hours (Gutterman 1972). The chaparral species of *Ceanothus* produces seeds with a hard coat which delays germination for many years, so that the small number of seeds which do germinate any given year do not describe the enormous seed bank in the soil. Scarification, as by fire, may release them from dormancy *en masse*. Some species produce several kinds of seeds, each with different germination requirements.

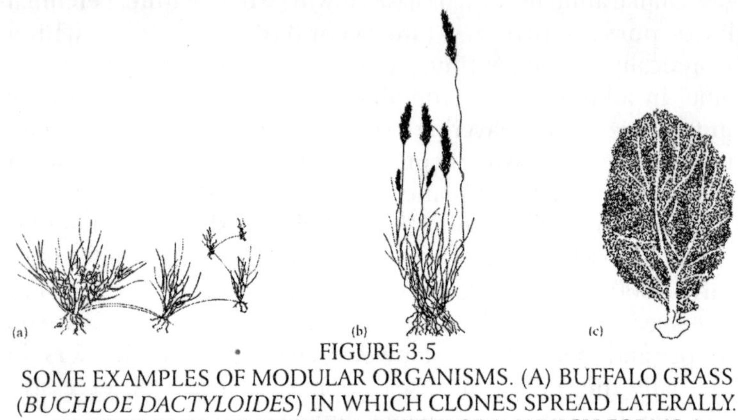

FIGURE 3.5
SOME EXAMPLES OF MODULAR ORGANISMS. (A) BUFFALO GRASS (*BUCHLOE DACTYLOIDES*) IN WHICH CLONES SPREAD LATERALLY. (B) FESCUE GRASS (*FESTUCA OCTOFLORA*) WHICH FORMS A TUSSOCK OF TIGHTLY PACKED MODULES. (C) SEA FAN (*GORGONIA* SP.) A MULTIPLE BRANCHED CORAL (BEGON *ET AL.* 1990).

Finally, the degree of phenotypic plasticity exhibited by plants can be enormous (Bradshaw 1965), so that demographic events can vary for the same species over time or space. Growth rate, age at onset of reproduction, plant size, and life span can all be modified by the environment.

3.2.1 Life spans

There are five characteristic plant life spans, and each has a basic corresponding life form. **Annual plants** live for 1 year or less. Their average life span is 1-8 months, depending on the species and then on the environment (the same desert species may complete its life cycle in 8 months one year, or in 1 month the next, depending on rainfall (Beatley 1967). Annuals with extremely short life cycles, such as *Boerrhavia repens* of the Sahara Desert, which goes from seed to seed in as few as 10 days (Cloudsley-Thompson and Chadwick 1964), deserve the special designation **ephemeral**. Annuals are **herbaceous**, that is, they lack a secondary meristem which would produce lateral, woody tissue. They die following seed production for a number of reasons: nutrient depletion, hormone changes, or inability of the nonwoody tissue to withstand unfavourable environmental conditions following the growing season (Leopold and Kriedemann 1975). If the members of some species are in particularly favourable microhabitats, they may persist for more than 1 year (Barbour 1970).

Biennial plants are also herbaceous, but typically live for 2 years. The first year is spent in vegetative growth, which is generally more below ground than above ground. Reproduction occurs in the second year, followed by plant death. Under poor growing conditions, or by experimental manipulation of the overwintering temperatures, the vegetative stage can be prolonged for more than 1 year.

Herbaceous perennials typically live for 20-30 years (Rabotnov 1969), though exceptional species are estimated to live for 400-800 years (Harper and White 1974). These plants die back to the root system and root crown at the end of each growing season. The root system becomes woody, but the above-ground system is herbaceous. They have a juvenile, vegetative stage for the first 2-8 years, then bloom and reproduce yearly, periodically every 2-3 years, or sometimes only once, at the end of their life span.

Suffrutescent shrubs (synonyms: sub-shrubs, half-shrubs, hemixyles) are half way between herbaceous perennials and true shrubs. They develop perennial, woody tissue only near the base of their stems, and the rest of the shoot system is herbaceous and dies back each year. They are small, nearly 25 cm tall, and short-lived compared to true shrubs.

Woody perennials (trees and shrubs) have the longest life spans: shrubs average 30-50 year, angiosperms (broadleaf) trees average 200-300 years, and conifer (needleleaf) trees average 500-1000 years (Harper and White 1974). Woody perennials spend roughly the first 10 % of their life in a juvenile, totally vegetative state, then enter a combined reproductive and vegetative state, reaching a peak of reproduction several years before death.

The age of some organisms can be determined fairly accurately because they possess some character which leaves a regular pattern in their structure representing some fixed time interval such as a year, lunar month or day. The best known example is **annual growth rings** in temperate trees. Each year as the sap rises in the tree in spring for the growth of new leaves, the tree grows a new layer of wood in a ring just beneath the bark. This spring wood has wide transport cells of tracheids or vessels. The diameter of new cells decreases during the following months until growth of wood stops in late summer. This produces a distinct tree ring for each year of growth and the rings can be counted easily in a felled stump. Tropical trees do not show this distinct growth pattern because of the lack of winter dormancy followed by a spring burst of growth. They may show no variation in wood structure or have rings which correspond with flowering or leaf emergence. However, these rings may be misleading if used to age the tree. For example, some trees flower more than once a year (certain *Ficus* spp., for example) while others may miss one or more years, and leaf production can also vary seasonally with several bursts of growth a year (Creber 1977). In such cases the alternative method is to determine the tree trunk diameter at breast height (dbh) as age correlates closely with dbh. Leaf bases may be useful for dating plants such as palms if they have a regular yearly leaf production as the scars left by the leaves can be counted. The methods used to age herbaceous perennials also depend on counting leaf scars or estimating the rate of spread in tussock (clumped) forms.

Structures which can be used to determine the age of animals include annual growth rings in teeth, horns, in fish scales, in shells of molluscs and in the otoliths (small grains found in the inner ear which are used for balancing) of mammals. In rabbits the weight of the lens of the eye is a good indicator of age.

3.2.2 Age distributions

It is usually impossible to determine the exact age of the organisms being studied, even to the nearest year. This is because most organisms do not have a structure which records yearly cycles or because study of such structures can only be done by killing or harming the organisms. In such cases the organisms are grouped into broad categories based on the state in the life cycle each individual has reached. According to this scheme one can recognize about eight important age states in an individual plant or a population (Rabotnov 1969): (a) viable seed (b) seedling (c) juvenile (d) immature, vegetative (e) mature, vegetative (f) initial reproductive (g) maximum vigour (reproductive and vegetative), and (h) senescent. If a population shows only the first four or five states, it is obviously invading and is part of a seral community. If a population shows all eight states, it is stable and is most likely part of a climax community. If it shows only the last four states, it may not maintain itself and may be part of a seral community.

Knowing the age distribution of a population may permit us to use demography as a predictive tool in community ecology. Figure 3.6 shows an age distribution of longleaf pine (*Pinus palustris*) in a northern Florida forest (trunk dbh correlates closely with age in this case). The pine population has many mature trees, but no seedlings or saplings; it appears to be a senescent population which is not reproducing itself and is part of a seral community. What will replace it? The same graph shows two hardwood populations, water oak (*Quercus nigra*) and wax myrtle (*Myrica cerifera*), which have many seedlings and saplings but no large (mature) individuals. The hardwoods are the invading populations, part of a community that will replace the pine forest.

In contrast, Fig.3.7 shows the age distribution of red spruce

(*Picea rubens*) in virgin forests of the White Mountains of New Hampshire. Here we have a reverse J-shaped curve, with many seedlings and saplings, and fewer larger, older trees. Many ages, and all age states, are represented, and the high density of young trees means there is a high probability of the population maintaining itself as part of a climax community.

Another way to represent age distribution in a population is to arrange the data in the form of a **population age pyramid,** the number of individuals or the percentage in the different age classes being shown by the relative widths of successive horizontal bars. Each horizontal bar consists of two bar graphs back to back, with males to the left and females to the right of a central line (Fig. 3.8). The upper pyramids in Fig. 3.9 illustrate three hypothetical cases (population pyramids): (left) a pyramid with broad base, indicating a high percentage of young individuals; (middle) a bell-shaped polygon, indicating a moderate proportion of young to old; and (right) an urn-shaped figure, indicating a low percentage of young individuals. The latter would generally be characteristic of a senile or declining population. The vole pyramids in Fig. 3.9 show stable age distributions under

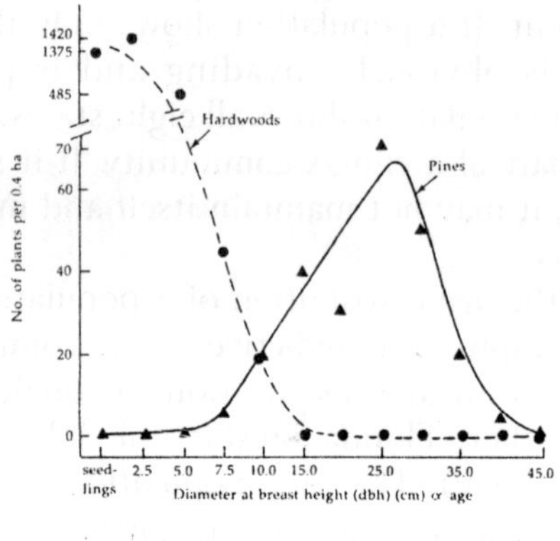

FIGURE 3.6
SUMMARY OF THE AGES OF ALL TREES IN A FOREST NEAR GAINSVILLE, FLORIDA. ACTUALLY, ONLY DIAMETER AT BREAST HEIGHT (DBH) WAS MEASURED, NOT AGE, BUT DBH IS ROUGHLY PROPORTIONAL TO AGE IN YEARS. SUCCESSION IS PROGRESSING FROM A PINE COMMUNITY (SOLID LINE) TO A HARDWOOD COMMUNITY (MAINLY OAK AND WAX MYRTLE) (DASHED LINE) (DAUBENMIRE 1968)

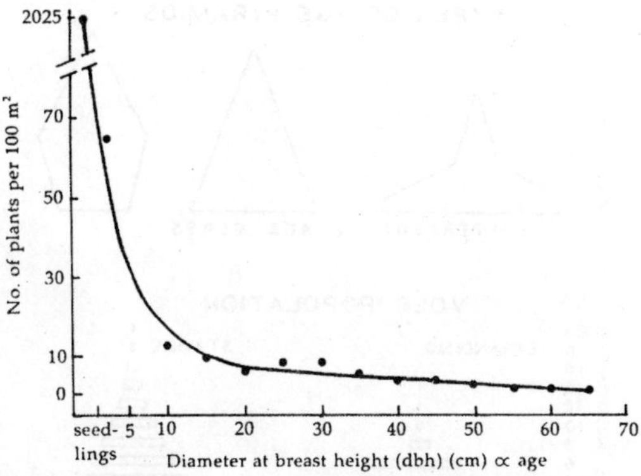

FIGURE 3.7
A REVERSE J-SHAPED AGE DISTRIBUTION CURVE FOR RED SPRUCE (*PICEA RUBENS*) IN FOUR STANDS IN THE WHITE MOUNTAINS OF NEW HAMPSHIRE (OOSTING AND BILLINGS 1951).

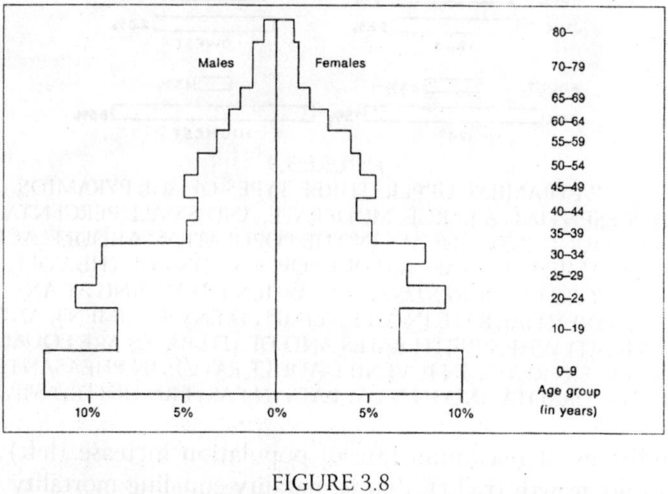

FIGURE 3.8
A POPULATION PYRAMID OF RUSSIAN PEOPLE TAKEN IN 1959. THE DATA ARE DIVIDED INTO MALES AND FEMALES. THIS PYRAMID IS UNUSUALLY IRREGULAR IN OVERALL SHAPE AND THE DATA FOR MALES AND FEMALES ARE NOT SYMMETRICAL (HOLLINGSWORTH 1969)

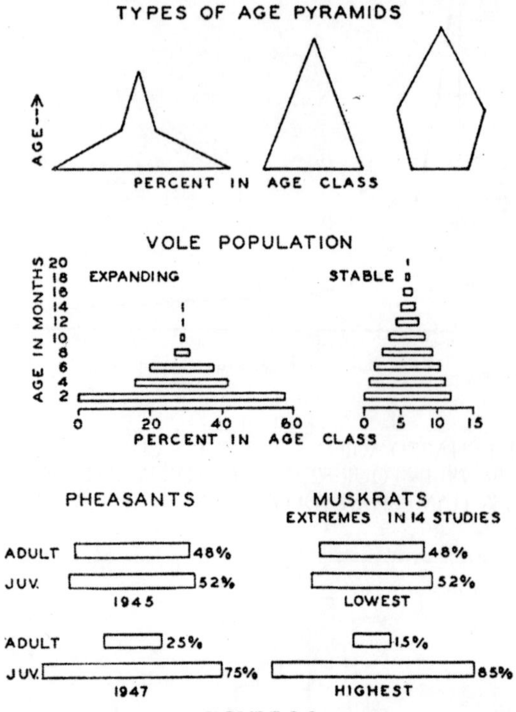

FIGURE 3.9
AGE PYRAMIDS. UPPER: THREE TYPES OF AGE PYRAMIDS REPRESENTING A LARGE, MODERATE, AND SMALL PERCENTAGE OF YOUNG INDIVIDUALS IN THE POPULATION. MIDDLE: AGE PYRAMIDS FOR LABORATORY POPULATIONS OF THE VOLE, *MICROTUS AGRESTIS* (LEFT), WHEN EXPANDING AT AN EXPONENTIAL RATE IN AN UNLIMITED ENVIRONMENT, AND (RIGHT) WHEN BIRTH RATES AND DEATH RATES ARE EQUAL. LOWER: EXTREMES IN JUVENILE-ADULT RATIOS IN PHEASANTS IN NORTH DAKOTA AND IN MUSKRATS IN EASTERN UNITED STATES

conditions of maximum rate of population increase (left) and with no growth (right), that is, natality equaling mortality. The rapidly growing population has the much greater proportion of young individuals.

In game birds and fur-bearing mammals, the ratio of first year animals to older animals, determined during the season of harvest (fall or winter) by examination of samples from the population taken by hunters or trappers, provides an index to the population trends. In general, a high ratio of juveniles to adults, as shown in the bottom diagrams in Fig. 3.9, indicates a highly successful breeding season and likelihood of a larger population the next year, provided juvenile mortality is not excessive. In the muskrat example (Fig. 3.9) the highest percentage of juveniles (85%) occurred in a population which had been heavily trapped for the previous few years; reduction in total population in this manner had apparently resulted in increased natality for those individuals surviving.

3.2.3 Life tables

Plant and animal populations may be composed of several types of individuals which can be classified by sex or age. A life table is an age-specific summary of the mortality rates operating on a cohort of individuals. A *cohort* may include the entire population or it may include only males, or only individuals born in, say, 2001. An example of a life table for song sparrows is given in Table 3.3. The columns of the life table are symbolized by letters:

x = age interval
nx = number of survivors at start of age interval x
lx = proportion of organisms surviving at start of age interval x.
dx = number dying during the age interval x to x+1
qx = rate of mortality during the age interval x to x+1

To set up a life table, one must decide on age intervals to group the data. For humans or trees the age interval may be 5 years, for deer, birds, or perennial plants 1 year, and for annual plants or field mice 1 month. By making the age interval shorter, we increase the detail of the mortality picture shown by the life table.

The first important point to be made is that, given any one of the columns of the life table, one can calculate the rest. To put this another way, there is nothing "new" in each of the three columns lx, dx and qx. They are just different ways of summarizing one set of data. The columns are related as follows:

$$n_x + 1 = n_x - d_x;\ q_x = \frac{d_x}{n_x};\ lx = \frac{n_x}{n_0}$$

TABLE 3.3
LIFE TABLE FOR THE SONG SPARROW ON MANDARTE ISLAND, BRITISH COLUMBIA

Ages (years) (x)	*Observed no. of birds alive* (n_x)	*Proportion surving at start of age interval x* (l_x)	*No. dying within age interval x to x+1* (d_x)	*Rate of mortality* (q_x)
0	115	1.0	90	0.78
1	25	0.217	6	0.24
2	19	0.165	7	0.37
3	12	0.104	10	0.83
4	2	0.017	1	0.50
5	1	0.009	1	1.0
6	0	0.0	-	-

For example, from Table 3.3,
$n_3 = n_2 - d_2 = 19\text{-}7 = 12$

$$q_2 = \frac{d_2}{n_2} = \frac{7}{19} = 0.37$$

$$l_4 = \frac{n_4}{n_0} = \frac{2}{115} = 0.017$$

The rate of mortality qx is expressed as a rate for the time interval between successive census stages of the life table. For example q_0 for the song sparrows in Table 3.3 is 0.78 for the interval between egg and one year, or per year. Thus 78% of the birds are lost in the nest or during their first year of life.

The most frequently used part of the life table is the nx column, the number of survivors at the start of age x. When data from nx column are plotted with the time interval on horizontal coordinate and the number of survivors on the logarithmic vertical coordinate, the resulting curve is called a **survivorship curve**. This is a graph showing the number of individuals which survive per thousand of population through each phase of life. In most studies, of course, there are fewer than a thousand individuals in the population, but this is standardized so that life tables can easily be compared.

There are three basic types of survivorship curves (Fig. 3.10).

Type I populations have low mortality when young. Most individuals die at maturity, within a relatively narrow age span. Many species of large animals, large size plants of perennial habit, and humans tend to have a type I survivorship curve. *Type II* populations have constant mortality at all ages. Many herbaceous perennials, birds, mice, and rabbits have a type II survivorship curve. *Type III* populations have high mortality when young. Those few individuals that reach adulthood have a low death risk and continue living for a long time. Type III curves occur in many short-lived weedy annuals, oak trees, many fishes, marine invertebrates, and parasites.

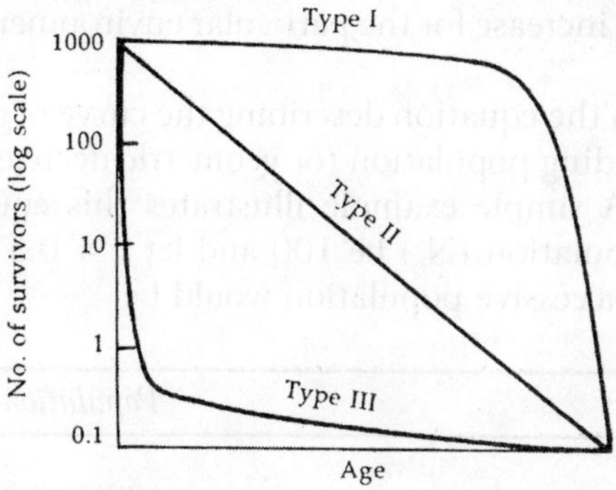

FIGURE 3.10 HYPOTHETICAL SURVIVORSHIP CURVES (DEEVEY 1947)

3.2.4 Population Growth

Any population in a particular environment will have a mean longevity or survival rate, a mean natality or birth rate, and a mean growth rate or speed of development of individuals. The values of these means are determined in part by the environment and in part by the innate qualities of the organisms themselves.

When the environment is unlimited (space, food, other organisms not exerting a limiting effect), the *specific growth rate* (that is, the population growth rate per individual) becomes constant and maximum for the existing microclimatic conditions. Such a population that is subject to a constant schedule of birth and death rates will gradually approach a fixed or *stable age distribution* (a more or less constant proportion of individuals of different ages), whatever the initial age distribution may have been and will then maintain this age distribution indefinitely. When the population has reached this stable age distribution, it will increase in numbers according to the differential equation:

$$\frac{dN}{dt} = rN$$

(Rate of change in numbers per unit time) = (intrinsic capacity for increase) x (population size)

This same equation may be rewritten in integral form:

$N_t = N_0 e^{rt}$

where, N_0= Number of individuals at time 0; N_t = Number of individuals at time t; e =2.71828 (a constant); r =intrinsic capacity for increase for the particular environmental conditions; t = time.

This is the equation describing the curve of *geometric increase* in an expanding population (or geometric decrease to zero if r is negative). A simple example illustrates this equation. Let the starting population (N_0) be 100 and let r = 0.5 per female per year. The successive population would be:

Year	*Population size*
0	100
1	(100) ($e^{0.5}$) = 165
2	(100) ($e^{1.0}$) = 272
3	(100) ($e^{1.5}$) = 448
4	(100) ($e^{2.0}$) = 739
5	(100) ($e^{2.5}$) = 1218

This hypothetical population growth is plotted in Fig.3.11. Note that on a logarithmic scale the increase follows a straight line, but on an arithmetic scale the curve swings upward at an accelerating rate. In brief, any population subject to fixed age schedule of births and deaths will increase in a geometric way, and this geometric increase will dictate a fixed and unchanging age distribution called the stable age distribution.

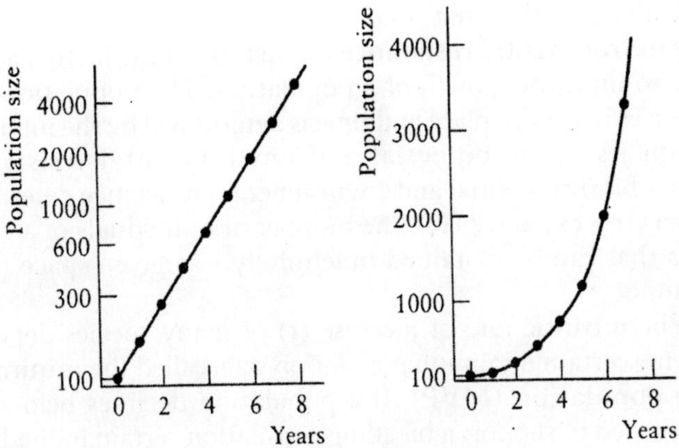

(a) Logarithmic scale (b) Arithmetic scale

FIGURE 3.11 GEOMETRIC GROWTH OF A HYPOTHETICAL POPULATION WHEN N_0 = 100 AND R = 0.5.

When a stationary and stable age distribution exists, the specific growth rate is called the **intrinsic rate of natural increase (r)**. This is the rate at which a population would grow if it had unlimited resource, and calculated as: $r = \frac{\ln N_t - \ln N_0}{t}$. The maximum value of r is often called by the less specific but widely used expression **biotic potential** or **reproductive potential.** Generally, individuals in populations with a high intrinsic rate of increase reproduce early in life, have short generation times (the time between successive generations), can reproduce many times (have a long reproductive life), and have many offspring each time they reproduce.

Some species have an astounding **biotic potential**. For example, without any controls on its population growth, the ancestors of a single female housefly *(Musca domestica)* could total about 5.6 trillion flies within about 13 months, and within a few years there would be enough flies to cover the earth's entire surface. However, this is not a realistic scenario because no population can grow indefinitely. In the real world, a rapidly growing population reaches some size limit imposed by a shortage of one or more limiting factors, such as light, water, space or nutrients or too many competitors or predators. There are always limits to population growth in nature.

Environmental resistance consists of all the factors acting jointly to limit the growth of a population. The population size of a species in a given place and time is determined by the interplay between its biotic potential and environmental resistance. Together biotic potential and environmental resistance determine the **carrying capacity (K)**, the number of individuals of a given species that can be sustained indefinitely in a given space (area or volume).

The intrinsic rate of increase (r) of many species depends on having certain minimum population size, called the **minimum viable population (MVP)**. If a population declines below the MVP needed to support a breeding population certain individuals may not be able to locate mates, genetically related individuals may interbreed and produce weak or malformed offspring, and the genetic diversity may be too low to enable adaptation to new environmental conditions. Then the intrinsic rate of increase falls and extinction is likely.

Populations have characteristic patterns of increase which are called **population growth forms**. For the purpose of comparison we may designate two basic patterns based on the shapes of arithmetic plots of growth curves, namely, the *J-shaped growth form* and the *S-shaped* or *sigmoid growth form* (sigma is the Greek letter S) (Fig. 3.12). A population that has few if any resource limitations grows exponentially. Exponential growth starts out slowly and then proceeds faster and faster as the population increases. If number of individuals is plotted against time, this sequence yields a J-shaped exponential curve. This form of population increase is given by $\frac{dN}{dt} = rN$ (b-d)N, where N = population size, t = time, and r = innate capacity for a population to increase; it is a constant for a given species, being the birth (b) rate minus the death (d) rate in a setting without any resource limitations. But populations do not show continuous geometric increase. When a population is small and growing in a limited space, the density increases slowly at first (establishment or positive acceleration phase), then more rapidly (perhaps approaching a logarithmic phase); but it soon slows down gradually as the environmental resistance increases (the negative acceleration phase), until a more or less equilibrium level is reached and maintained. Thus, the S-shaped curve differs from geometric curve in that it has an upper *asymptote* (that is, the curve does not exceed a certain maximal level), and it approaches this asymptote smoothly, not abruptly. This form may be represented by the simple logistic model:

$$\frac{dN}{dt} = rN\left(\frac{K - N}{K}\right)$$

where dN/dt is the change in population size with time, r is the innate capacity for a population to increase, N is the population size, and K is the highest population density (upper asymptote or maximal value of N) that can be maintained in a constant, but real environment; it is the *carrying capacity* of the habitat. Very early in population growth, there is little difference between the curves for the logistic and geometric equations. As we approach the middle segment of the curve, they begin to diverge more. As we approach the upper limit, the curves diverge much farther, and when we reach the upper limit, the population stops growing because (K-N)/K becomes zero. In brief, ideal, limitless habitats would permit a geometric population increase, but real habitats have limits to growth, and as the population reach K, the growth rate declines.

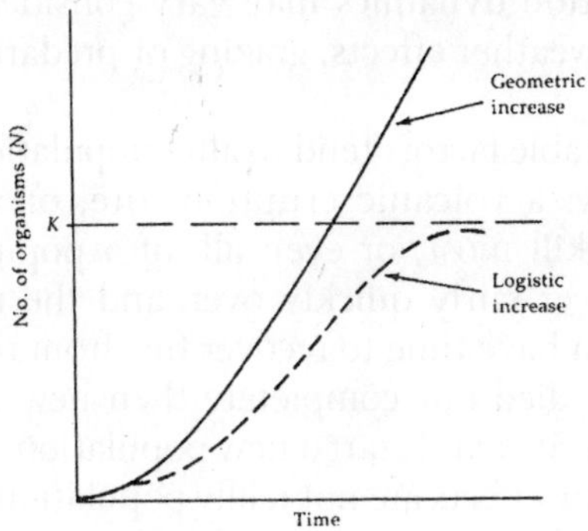

FIGURE 3.12
GEOMETRIC (SOLID LINE) VERSUS LOGISTIC (DASHED LINE) POPULATION GROWTH OVER TIME. K IS THE CARRYING CAPACITY OF THE HABITAT FOR A POPULATION SHOWING DENSITY-DEPENDENT LOGISTIC GROWTH (DASHED LINE)

3.2.5 Factors regulating population size

The environmental factors which are responsible for regulating populations are often the ones that affect the growth of the individuals. This is because populations are made up of individuals. Population increase depends on the reproductive fitness and life span of these individuals. Factors such as energy and nutrient availability, flood, drought, predators and disease will affect population size.

Constantly limiting environmental factors are always in short supply, but are relatively constant so that a population is limited to a certain fairly constant size by the factor. Individuals may have to compete for the resource. For example, plants compete for space and light, birds for nesting territories and many heterotrophs for food. These factors do not usually produce large changes in population. On the other hand, *variably limiting environmental factors* are although variable but predictable, like seasonal drought and cold, or variation in food availability. They are only an influence at certain times of year when they may cause the population to crash. Some species avoid such factors by taking evasive action. For example, some birds migrate and deciduous trees drop their leaves.

The *unpredictable environmental factors* change in an irregular way so population dynamics may vary considerably over time. These include weather effects, grazing or predation pressure and disease levels.

Unpredictable factors tend to alter population size in a fairly haphazard way: a volcanic eruption, fire, or virulent disease epidemic may kill most, or even all, of a population. Such an event is, however fairly quickly over, and the remnants of the population then have time to recover free from the hazard. If the population has died out completely then new individuals may migrate into the area and start a new population. It can be argued that such limiting effects are not really population regulators. This is because such factors are not constantly or repeatedly limiting population size, but merely damaging it in a one-off event, after which the population will recover up to a level where real regulating effects take place. Population regulation, therefore, implies that one or more factors are always involved in controlling population size whether they are constant or temporally variable in operation.

Many factors have an increasing effect on a population the more closely spaced the individuals in the population become. Such factors are said to be *density dependent*. Density-dependent influences often include resources that are in limited supply such as space, water and nutrients. The more individuals there are competing for the resources, the higher the mortality will be in the population. Disease can also be density dependent as epidemics spread more rapidly if organisms are closer together and in frequent contact rather than sparsely distributed. Factors which do not show this effect are said to be *density independent* and include fire, volcanic eruption and other catastrophes, extreme weather conditions like heavy frosts and high winds, and toxic pollution.

3.3 The Behaviour of Individuals: Resource Allocation Patterns

Organisms have a limited amount of time and energy with which to complete a life cycle. Time itself is not allocated, but is important in the gain of photosynthetic energy and in the utilization of energy for maintenance. A fraction of the total energy available is budgeted to each activity in the life cycle: the amount of time spent in a dormant state, in a juvenile, vegetative stage, or in a mature, reproductive stage; the amount of energy used for roots, shoots, leaves, flowers, seeds, or fruits; and the amount of energy used for growth, for maintenance, or for herbivore defense.

Organisms seem to lie on a continuum between two extremes of resource allocation: r and K (the letters r and K come from the logistic equation for population growth rate where r is the innate capacity or maximum rate for a population to increase, and K is the upper asymptote or equilibrium population size). Extreme *r-selection* leads to short-lived plants which mature rapidly, occupy an open habitat in a seral community, and devote a large fraction of their *photosynthate* (their stored energy from photosynthesis) to producing flowers, fruits, and seeds. Their population size is density-independent; that is, population size is regulated by physical factors such as fire, flood, frost, drought, etc. Weeds and other species of pioneer communities are examples of r-selected populations.

Extreme *K-selection* leads to long-lived plants which have a prolonged vegetative stage, occupy a closed, late seral or climax community, and devote a small fraction of their photosynthate to reproduction. Their population size is density-dependent; that is, population size is regulated by biotic interactions such as competition. Population size is near the carrying capacity of the habitat. Forest trees are examples of K-selected populations. Table 3.4 summarizes general traits of r-selected and K-selected populations. It is important to remember that r and K are relative terms. For example, two annual weeds may be compared and one considered to be in more of an extreme r mode than the other (Hickman 1977). Many species lie in the middle of continuum between r and K extremes (Grime 1977, 1979).

Reproductive effort: A convenient way to place a population on the r-K continuum is to measure its reproductive effort, that is, the percentage of its total energy that is spent on reproduction (Harper and Ogden 1970). K-selected plants show relatively low reproductive effort, no matter what year of their life cycle is measured, while r-selected plants show a relatively high reproductive effort. Reproductive effort is partly a fixed, genetic trait, but it can be modified by the environment.

For some species, reproductive effort is relatively constant regardless of the environment, Harper and Ogden (1970) grew common groundsel (*Senecio vulgaris*), an annual weed, in the greenhouse under three levels of rooting stress (pot volume), but found reproductive effort to be roughly 20% regardless of stress. They pointed out that this value was comparable to that of other annuals, including cultivated crop species which had been bred for high yields.

Reproductive effort may be very plastic, changing with stress within the same species. Hickman (1975) observed the annual buckwheat *Polygonum cascadense* at five different sites in the Cascade Mountains of Oregon. The sites ranged from relatively xeric to relatively mesic. In the field, he found reproductive effort to decrease from 58% for a population on the driest site to 38% for a population on the most mesic site (Table 3.5). Were these ecotypic responses? Hickman collected seedlings from a xeric and mesic site early in the season, transplanted them to a uniform greenhouse environment, and measured their reproductive effort (Table 3.5). There was no difference in reproductive effort between

TABLE 3.4
SOME TRAITS WHICH CORRELATE WITH R- AND K-SELECTION (PIANKA 1970).

	r-selection	K-selection
Climate	variable and/or unpredictable; uncertain	fairly constant and/or predictable; more certain
Mortality	often catastrophic; density-independent	density-dependent
Survivorship	often type III (see Fig. 3.10)	usually types I and II (see Fig. 3.10)
Population size	variable in time; not in equilibrium; usually well below carrying capacity of the habitat; recolonization each year	fairly constant in time; in equilibrium; at or near carrying capacity of the habitat; no recolonization necessary
intraspecific and interspecific competition	variable; often lax	usually keen
life span	short, usually less than 1 year	longer, usually more than 1 year
selection favours	rapid development; early reproduction; small body size; single reproduction period in life span	slower development; greater competitive ability; delayed reproduction; larger body size; repeated reproduction periods in life span
overall result	productivity	efficiency

greenhouse grown populations; hence the field differences were not genetic, ecotypic differences, but are examples of phenotypic plasticity. Consequently, reproductive effort may indicate something about the severity of a habitat as well as about a species' resources allocation pattern, and the two components might not be easy to separate under uniform conditions.

Table 3.5 Vegetative growth and reproduction effort in five populations of the annual buckwheat *Polygonum cascadense*. In the field, population 1 was on the most xeric site and population 5 was on the most mesic site. Young plants from two of the sites, 1 and 4, were also grown in a uniform (greenhouse) garden (Hickman 1975)

Trait	*Field*					*Greenhouse*	
	1	*2*	*3*	*4*	*5*	*1*	*4*
Vegetative weight (mg)	13.6	16.0	43.2	49.8	325.3	48.4	73.6
Reproductive weight (mg)	18.9	22.4	46.1	40.1	207.9	37.3	57.4
Reproductive effort %	58.4	57.8	51.6	43.2	38.4	43.8	45.4

Reproductive effort at any given age can be associated with a biological *cost* and a biological *profit*. The biological cost derives from the reduction in growth or survival that occurs as a consequence of using energy to reproduce. The biological profit is measured in the number of descendants left to future generations, which will be affected by the survival rate and the growth rate. The hypothetical organism must ask at each age, should I reproduce this year or would I profit more by waiting until next year? Obviously, if the mortality rate of adults is high, it would be best to reproduce as soon as possible. But if adult mortality is low, it may pay an organism to put its energy into growth and wait until the next year to reproduce.

In perennial plants, reproductive effort can differ from one species to another over their cumulative life span. Some perennial species reproduce only once (*semelparity*), invest a considerable amount of stored energy into the event and become senescent during or following reproduction, for example, century plants (*Agave* species). There are both costs and profits to postponing reproduction in this way. Other perennial species reproduce many times (*iteroparity*), though not necessarily every year nor with equal intensity each time. Each reproductive event uses only a small fraction of the plant's energy store.

3.4 Human Population

Humans (*Homo sapiens*) constitute one of the most abundant species comparable only to certain fishes and insects. Human population has a very wide range of distribution, is adapted to almost all climatic conditions and is reproductively active in all seasons. Although the socio-cultural situations vary considerably in different parts of the globe, the fundamental characteristics and uniqueness are recognizable in all human populations.

The pattern of human population and population growth depend upon several complex factors: climatic and edaphic factors, location of natural resources, transport facilities, and availability of employment and amenities. The size of population, and birth and death rates has great significance on the standard of living of the people, their aspirations, and their economic and social development. Human behaviour and activity cannot be understood by biological principles alone. The birth rate of human population is regulated more by socio-economic factors than biological ones. Social status of women, family structure, age of women at marriage, education, acceptability of contraceptive practices, religious beliefs are some of the factors. Similarly, the death rate is regulated by public hygiene, nutritional status, nature of employment and medical facilities.

Over the last million years, the human population has shown three major bursts in the rate of population increase. The first accompanied the development of Palaeolithic stone tools about 600000 to 14000 years ago when human existence was solely by hunting and gathering, with no cultivation. However, the growth rate for the human population has been estimated as only about 0.001 % up to abut 10,000 years ago. At this time (that is, at the end of Palaeolithic period) the number of the people in the world was approximately 10 million. About this time, the beginning of the Neolithic period, there seems to have been the second burst in the rate of population increase up to about 50 million. This increase in population growth rate up to 0.1% coincides with a changeover from hunter-gatherer societies to more stable agricultural ones (Cohen 1980; Hassan 1980). The third burst in population growth is well documented as it is associated with the recent industrialization of the eighteenth and nineteenth centuries (Deevey 1960).

The world population increased from 0.75 billion to 1.6 billion between 1750-1900 (150 years), from 1.6 billion to 3.3 billion between 1900-1965 (65 years), and from 3.3 billion to 7 billion between 1965-2005 (40 years). At present, the average number of children produced per woman is about 2.6 and falling. To achieve zero population growth the average number of children per woman must drop to 2.1. (The extra tenth of a child, or one child per ten women, compensates for child mortality). If the 2.1 level is reached very soon and held, there will be 7.7 billion people on earth by 2050, levelling off at 8.5 billion in 2150. If a slight negative growth is reached, at 2 children per woman, the world population will peak at 7.8 billion and drop to 5.6 billion, a bit below the current level, also by 2150. To these estimates, unfortunately, must be added an even more worrisome estimate: if the replacement level is held even slightly above 2.1, the world population will continue to soar. Thus, only 2.2 children per woman will yield 12.5 billion by 2050, a potentially catastrophic doubling of present levels. In general, whatever the magnitude of increase, most of it will occur in developing world, which is already economically impoverished.

Why would the current rate of population increase be so bad? After all, average per capita production is rising worldwide, a statistic often cited cheerfully by bottom-line economists. The reason for concern is that the increase is being achieved by an accelerating consumption of the earth's non-renewable capital. Natural-resources experts generally agree that we are running out of arable land and water as a result of rising total demand, and that the finiteness of land and water represents a wall beyond which humanity cannot pass. There are no new continents to colonize, and displacement of other people by conquest and genocide is now generally frowned upon. As a species we are at long last approaching the Malthusian limit. The amount of cropland available per person, having dropped from 0.23 hectares in 1950 to 0.12 hectares in the mid-1990s, is continuing downward. The productivity per hectare has been elevated by fertilizer, Green-Revolution genetic improvement of crop strains, and other technological innovations, but the rate of improvement is diminishing, and its maintenance also depends on water, the other principal declining resource.Meanwhile, the population is shooting upward in most countries.

To the strain on the food and water resources can be added the aftershock of personal economic striving that follows the population explosion. Billions of people are not only producing children in excess of zero population growth but also taking every means available to improve their own and their family's personal security and welfare. The true measure of the human impact on earth's environment is therefore not in pure population numbers but in the "ecological footprint", the amount of productive land needed to supply each person with food, water, energy, living space, commerce, transportation, and waste management. The footprint is far more than just land that each person uses directly. It is also composed of small pieces obtained from many places, some even from other continents - where food is grown, petroleum drawn, clothing manufactured, roads built, waste dumped, and public buildings raised. The ecological footprint of the average person in the developing countries is half a hectare; that of one citizen of the United States is 5 hectares. For the entire current world population to pull itself up to U.S. levels with existing technology would require two more planet Earths.

That amount of bootstrapping, of course, is not going to happen. The greatest challenge of the twenty-first century is to settle humanity down and accommodate 8 to 10 billion people with a decent standard of living before they wreck the planet. Meanwhile, the last remaining natural environments - the world's forests, grasslands, semideserts, and wetlands - are shrinking before the onslaught. Humanity's responsibility to the rest of the life and to future generations is clear: bring with us as much of the environment and biodiversity through the bottleneck as possible (Wilson 1999).

3.4.1 Nepal

Population of Nepal increased from 15 million in 1981 to 18.5 million in 1991 and then to nearly 23 million in 2001 with an annual growth rate of 2.25% in 1991-2001 decade (Table 3.6). The crude birth rate is 31.28 per thousand population, crude death rate 9.22 per thousand population, total fertility rate 3.7

per women, and life expectancy at birth 62.2 years (males 61.78 years, females 62.5 years).

TABLE 3.6
POPULATION CHANGES IN NEPAL DURING 1911-2001

Year	*Population*	*Compound growth rate (%)*
1911	5638749	-
1920	5573788	(-) 0.13
1930	5532574	(-) 0.07
1941	6283649	1.16
1952/54	8256625	2.27
1961	9412996	1.64
1971	11555983	2.05
1981	15022839	2.62
1991	18491097	2.08
2001	22736944	2.25

The census of 2001 counted 7.3, 44.3 and 48.4% of the total population in the Mountain (above 2100 m altitude), Hill (350 to 2100 m altitude), and Terai (60-349 m altitude) regions, respectively. The figure was 9.9, 52.5, and 37.6%, respectively in 1971, and 8.7, 47.7 and 43.6%, respectively in 1981 for the total population of Nepal.

Resources shrink as people multiply and demographic pressures lead to economic pressures. Age-wise, Nepal is a very youthful country (Table 3.7), with about 66% people being children and youth (age group 0-29 years). Such a situation is also very conducive to further population growth. By the year 2021, we are likely to have a population of 34172144 people. Can we afford it? Agricultural land was 0.25 ha per person in 1960. Due to population pressure it declined to 0.2 ha in 1980, 0.13 ha in 2001, and it is estimated to decline further to 0.07 ha by the year 2020. The crux of the matter is that future population growth has to be related to resource base. Our immediate task is to know the carrying capacity of our country. In any case, a zero population growth holds the key to our prosperity, and to achieve it the government must be prepared for a measure of unpopularity. This has to be accomplished in the next few years or so. In turn, it takes us to the birth control strategies, education and vocation for women, which involve considerable scientific and technological inputs including biotechnology and immunology.

There is an increasing trend of population concentration in urban areas. The number of municipalities has increased from 33 in 1981 to 58 in 2001, and nearly 14% people of Nepal live these areas.

Negative impacts of the absence of housing are indeed a cause of a lot of human misery with consequent degradation of physical health, economic, social and cultural environment in urban slum and in rural areas. At present, there is shortage of houses in both rural and urban areas. Commensurate with this is ill-planned or unplanned or even absence of water supply and sanitation systems. Furthermore, in any programme on drinking water, sewerage has been constantly ignored. The refuse, therefore, affect water bodies and in turn causes problems of environmental sanitation and health as 80% of the diseases are water-related. The vulnerable sections of our society, and among them women and children, suffer the most. There is an immediate need for research and development with regard to utilization of traditional, locally available building materials and architecture in relation to local ecology, low cost latrines using minimum water, and parameters for environmentally compatible human settlements.

TABLE 3.7
POPULATION BY FIVE YEAR AGE GROUP FOR NEPAL - 2001.

Age	*Male*	*Female*	*Total*
All ages	11359378	11377566	22736944
0-4	1395715	1359498	2755213
5-9	1633087	1578355	3211442
10-14	1533806	1448126	2981932
15-19	1185826	1203176	2389002
20-24	946742	1070026	2016788
25-29	821014	904464	1725478
30-34	726040	763463	1489503
35-39	651351	659302	1310653
40-44	539993	548051	1088044
45-49	469695	453678	923373
50-54	392659	373395	766054
55-59	318610	283493	602093
60-64	262255	258653	520908
65-69	196053	191170	387223
70-74	141678	132111	273789

Age	*Male*	*Female*	*Total*
75-79	82335	83429	165764
80-84	41192	43063	84255
85-89	13630	14317	27947
90-94	5082	6339	11421
95+	2615	3457	6072

4. SPECIES INTERACTIONS

Table 4.1 catalogues all possible species interactions according to a symbolic scheme developed by Burkholder (1952). Each interaction is described by its effect on two populations or organisms, A and B, when they are in contact (the interaction is "on") and when they are apart (the interaction is "off"). For example, when a herbivore and its food plant are together, the herbivore is stimulated (its growth, reproduction, or general success is improved) and the plant is depressed (its growth, reserves, reproduction, or general success declines); when the two are apart, the herbivore is depressed and the plant remains stable. Mathematically, there are 81 possible interactions with this symbolism, but Burkholder concluded that only the 10 (Table 4.1) are logically possible. Among those 10, three are sufficiently rare or at least unexamined that they have no names. **Neutralism** is included for purposes of comparison and completeness, but it too may be rarity in nature. Among the remaining six interactions some are **negative** (one or another partner is inhibited, as in competition or amensalism), and others are **positive** (one or another partner is stimulated, as in commensalism or mutualism).

In nature, positive interactions produce positive spatial relationships between the partners; where one partner is found, the probability is high that the other will be found nearby. The two populations attract one another and they exist in a *nonrandom*, *clumped* pattern. Similarly, negative interactions produce negative spatial relationship; the two populations appear to repel one another and exist in a *nonrandom*, *regular* pattern. If there is no interaction between the populations, then the location of one individual has no influence on the location of others; the two populations are said to be *randomly* distributed with respect to each other (Fig. 4.1).

TABLE 4.1
A COMPLETE LIST OF ALL BIOLOGICALLY POSSIBLE TYPES OF INTERACTIONS ACCORDING TO BURKHOLDER (1952). STIMULATION IS SYMBOLIZED AS +, NO EFFECT AS O, AND DEPRESSION AS—

Name of interaction	*On*		*Off*	
	A	*B*	*A*	*B*
Neutralism	0	0	0	0
Competition	-	-	0	0
Mutualism	+	+	-	-
Unnamed	+	+	0	-
Protocooperation	+	+	0	0
Commensalism	+	0	-	0
Unnamed	+	0	0	0
Amensalism	0	-	0	0
Parasitism, predation, herbivory	+	-	-	0
Unnamed	+	-	0	0

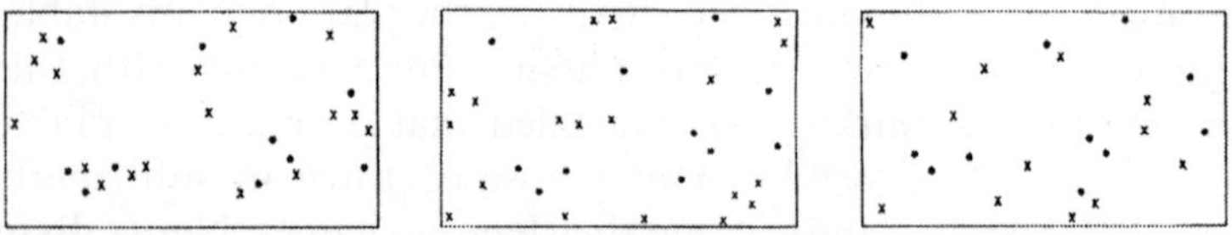

(a) Positive association (b) Negative association (c) Random association

FIGURE 4.1
AN OVERHEAD, DIAGRAMMATIC VIEW OF TWO SPECIES (DOTS AND X'S) WHICH ARE (A) POSITIVELY ASSOCIATED, (B) NEGATIVELY ASSOCIATED, AND (C) RANDOMLY ASSOCIATED.

Chi-square contingency analysis provides initial evidence for an interaction between two species. To determine spatial relationship between two grass species (*Axonopus compressus* and *Chrysopogon aciculatus*) in a round the year grazed pasture, 100 quadrats of 30 cm x 30 cm size were laid randomly. In each quadrat, the presence or absence of each species was noted, then summarized in a contingency table (Table 4.2). This constituted the observed data from which the expected data were derived by formula. Thereafter, observed and expected data were compared by the *chi-square test*. The example in Table 4.2 yielded a chi-square value significantly higher ($P<0.01$ at 1 degree of freedom) than expected by chance, so the two species were not distributed randomly with respect to each other-they were negatively associated.

TABLE 4.2
CONTINGENCY TABLE FOR ANALYSIS OF ASSOCIATION BETWEEN TWO SPECIES, A (*AXONOPUS COMPRESSUS*) AND B (*CHRYSOPOGON ACICULAUS*)

Symbol and description	Observed number of quadrats	Expected number of quadrats	$\chi^2 = \frac{(\text{observed} - \text{expected})^2}{\text{expected}}$
a = A and B present	27	$\frac{(a+b)}{100} x(a+c) = 36$	2.2
b = A present, B absent	29	(a+b)-36= 20	4.0
c = A absent, B present	38	(a+c)-36= 29	2.8
d = A and B absent	6	100-(36+20+29)=15	5.4
Totals	100		$\Sigma\chi^2 = 14.4$

4.1 Negative Interactions

4.1.1 Competition

When individuals interfere one another in order to get more of space to live, or light, or nutrients or water from soil or any other of their requirements, particularly when the resource is short in supply to meet full requirements of all individuals, there develops competition resulting in mutually adverse effects to competing organisms. If the individuals of the same species exert interference on one another, it is called *intraspecific competition* and between different species as *interspecific competition*. Competition is dependent on resource level or carrying capacity of ecosystem and the number of individuals present.

4.1.1.1 Competition for resources

A. Water: The perennial bluebunch wheatgrass (*Agropyron spicatum*) was the dominant plant species in semiarid grasslands in the northern intermountain regions of the United States prior to the mid-nineteenth century. About the mid-nineteenth century, an annual grass called cheatgrass (*Bromus tectorum*) was accidently introduced to the area from Europe which dominates the land previously held by wheatgrass. Both species have similar life cycle. They germinate (if annual) or break dormancy (if perennial) in fall, grow slowly during winter, grow rapidly in spring, form flowers in early summer, and die in June (or begin dormancy in mid-July if perennial).

Harris (1967) observed direct competitive effects in experimental seeding of wheatgrass in the presence of different densities of cheatgrass. For example, average survival of wheatgrass was 39, 69 and 86%, respectively, in dense, moderate, and sparse populations of cheatgrass. The competitive advantage of cheatgrass was the faster root growth than wheatgrass during winter. Consequently, at the start of rapid spring growth in April, cheatgrass roots had penetrated 90 cm, in contrast to 20 cm for wheatgrass. This difference in depth allowed cheat grass to absorb water from a much greater part of the soil profile than wheatgrass could. Thus, when the upper soil became dry in early summer, with both species drawing water from it, only wheatgrass suffered, for water still remained available at greater depths, where only cheat grass roots had penetrated.

B. Space: To avoid overcrowding, plant and animal populations may regulate themselves through isolation or the spacing of individuals. *Isolation* (intraspecific, interspecific, or both) usually is the result of: (i) interindividual competition for resources in short supply, or (ii) direct antagonism. In both cases a random or a uniform distribution may result because close neighbours are eliminated or driven away. In brief, spacing reduces competition, and prevents overcrowding and exhaustion of food supply in the case of animals, or nutrients, water or light in the case of plants. In either case, spacing tends to regulate population at a level below the saturation level.

Forest trees which have reached sufficient height to form a part of the forest crown may show a uniform distribution because competition for sunlight is so great that the trees tend to be spaced at very regular intervals. Similarly, desert shrubs often are very regularly spaced almost as if planted in rows, apparently because of the intense competition in the low moisture environment.

Many animals isolate themselves and restrict their major activities to definite areas or home ranges, which may vary from a few square meters to many square kilometers. Since home ranges often overlap, only partial spacing is achieved. *Territoriality* achieves the ultimate in spacing. In general, a territory is an area which contains a resource of sufficient value to an animal that it defends the area. Nice (1941) classified territories of birds into several basic types : (i) entire mating, feeding, and breeding area defended, (ii) mating and nesting but not feeding area defended, (iii) mating area only defended, (iv) nest only defended, (v) nonbreeding areas defended. In type (i) the defended area may be quite large, larger than needed in terms of food supply for the pair and its young. For example, the tiny gnatcatcher (which weighs about 3 g) establishes a territory averaging 4.6 acres but obtains all the food it needs in a much smaller area around the nest (Root 1969). In most territorial behaviour actual fighting over boundaries is held to a minimum. Owners advertise their land or location in space by song or displays and potential intruders generally avoid entering an established domain. The fact that the area defended by birds is often larger at the beginning of the nesting cycle than later when the demand for food is greatest, and the fact that many territorial species of birds, fish, and reptiles do not defend the feeding area at all give support to the idea that reproductive isolation and control has greater survival value than the isolation of a food supply as such (Odum 1971). Territoriality not only tends to prevent overpopulation, but it also provides for a quick recovery should unusual mortality of breeders occur, because a number of nonbreeding individuals are usually present that are unable to find space; these quickly fill in whenever a territory-holder is lost.

C. Crop-weed interference: Weeds are the unsown and undesired plants of agricultural fields because they compete efficiently with crop species for the vital resources: water, sunlight and nutrients. Crop losses due to weeds are recognized in every cropping system around the globe. The world-wide loss of attainable production of eight principal food and cash crops (paddy, wheat, barley, maize, potatoes, soybeans, cotton and coffee) due to pests has been estimated to be 243.7 billion US dollars annually. Out of this total loss, 76.3 billion US dollars is attributed to weeds (Oerke *et al.* 1994)

Weeds tend to be fast growing and abundant in occurrence. They are difficult to get rid of, because they build up a seed bank which replaces the adult population even if this is destroyed. They produce lots of seed which can remain dormant for many years. Often the adult plant is short-lived, usually an annual.

4.1.1.2 Competition versus coexistence

Since competition involves two organisms utilizing the same resource, it is obvious that competing organisms must have, to some extent, overlapping niches. Intraspecific competition is necessarily more intense than interspecific competition, because niche overlap is greater. Current evolutionary theory holds that selection pressure drives species within a community to partition the environment (utilize different parts of the environment); with the result that competition is minimized.

The notion that no two species can coexist if they occupy the same niche (that is, "one niche one species") is known as **competitive exclusion principle** or **Gause's principle** (after the Russian microbiologist G. F. Gause who first confirmed the principle experimentally). Gause (1934) first grew two closely related ciliate protozoans, *Paramecium caudatum* and *Paramecium aurelia* in separate cultures which contained bacteria on which the *Paramecium* fed. Both the species exhibited typical sigmoid population growth and maintained a constant population level in the culture medium that was maintained constant with a fixed density of food items (bacteria which did not themselves multiply in the media and thus could be added at frequent intervals to keep food density constant).When both protozoans were placed in the same culture, however, *P. aurelia* alone survived after 16 days (Fig. 4.2). In this case neither organism attacked the other or secreted harmful substances; *P. aurelia* population simply had more rapid growth rate (higher intrinsic rate of increase) and thus "out competed" *P. caudatum* for the limited amount of food under the existing conditions. On the other hand, *Paramecium caudatum* and *P. bursaria* were both able to survive and reach a stable equilibrium in the same culture medium because although they were competing for the same food, *P. bursaria* occupied a different part of the culture where it could feed on bacteria without competing with *P. caudatum*. Thus, the habitat feature of the niche of these two species proved to be significantly different, even though their food was identical.

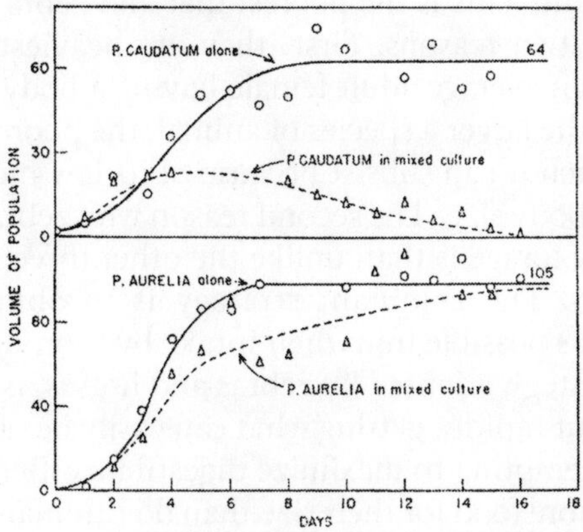

FIGURE 4.2

COMPETITION BETWEEN TWO CLOSELY RELATED SPECIES OF PROTOZOA THAT HAVE SIMILAR NICHES. WHEN SEPARATE, *PARAMECIUM CAUDATUM* AND *P. AURELIA* EXHIBIT NORMAL SIGMOID GROWTH IN CONTROLLED CULTURES WITH CONSTANT FOOD SUPPLY; WHEN TOGETHER, *P. CAUDATUM* IS ELIMINATED (ODUM 1971)

The interspecific competition is minimized through the partition of environment in two basic ways: resource partitioning, and character displacement. *Resource partitioning* occurs when two or more species in a community divide out a resource such as food. On the other hand, *character displacement* occurs when interspecific competition results in natural selection causing the characters (morphological, ecological, behavioural, or physiological) of two closely related species to diverge.

Figure 4.3 shows how the root systems of some common Mojave Desert plants divide up their access to the water (Cody 1986). Some specialize on ephemeral sources of water, which may be trapped by surface roots; others rely on relatively permanent sources of deep water. A classic instance of resource partitioning among animals was provided by Bell (1971) in his studies of grazing mammals on the Serengeti plain in Tanzania. Careful analyses of the stomach contents of most of the principal large herbivores on the Serengeti plain - zebra (*Equus burchelli*), wildebeest (*Connochaetes taurinus*), topi (*Damaliscus korrigum*) and Thomson's gazelle (*Gazella thomsoni*) - revealed considerable resource partitioning. Zebra were able to take the diet highest in cell wall content. This is the poorest diet and zebra were able to live on it for two reasons. First, they are heaviest of the four species, with an average adult female having a body mass of 220 kg. As a rule, the larger a species of animal, the poorer the quality of diet on which it can subsist because of its low metabolic rate, relative to its body size. The second reason why zebra can survive on low quality forage is that, unlike the other three species, they lack a rumen. The ruminant strategy is to obtain as much nourishment as possible from their intake by thorough digestion. The equid strategy, adopted by zebras and horses, is to pass food through the gut rapidly, getting what can easily be obtained from it, but not attempting to maximize digestibility. Because of this, zebras need more food for their size than do ruminants. However, this evolutionary strategy does mean that zebras can survive off the poorest quality herbage, which is unavailable to the other Serengeti herbivores. The species that ate the highest quality diet was the smallest ruminant, the Thomson's gazelle. Adult female Thomson's gazelles weigh about 16 kg. Of the four herbivores, Thomson's gazelles were the only ones to take a large amount of dicotyledonous material: 40%, chiefly fruit. This high-protein high-energy diet is necessary for the small ruminant with its high metabolic rate, relative to its size. Topi (110 kg) and wildebeest (160 kg) took food intermediate in quality between that taken by Thomson's gazelles and zebra.

The phenomenon of character displacement is exhibited by many plants endemic to severe habitats. Restriction of such endemics to severe habitats is mainly because they are poor competitors on less extreme sites. Severe habitats have very low plant cover, and so there is little competition. Plants in severe habitats tolerate the sort of environmental extremes that most other plants cannot, but they would grow better elsewhere, in the absence of competition. Research with serpentine endemics and halophytes supports this conclusion.

Serpentine is a metamorphic, magnesium silicate rock, often green in colour and slippery to touch, which has a number of traits inimical to plant growth. It is low in such essential nutrients as N, Ca, K, and P; its pH may be far from neutrality (either acidic or basic); and it is high in such toxic elements as Ni and Cr. Soils which are derived from serpentine rock are sterile, support unusual, endemic floras, and are covered by vegetation whose physiognomy differs from that of surrounding vegetation on nonserpentine soil. Serpentine outcrops are especially common in the Pacific Coast states.

Arthur Kruckeberg (1954) of the University of Washington experimented with serpentine and nonserpentine ecotypes and species. He reported that herbaceous serpentine endemics became established from seed better and grew faster on nonserpentine soil, provided they were free from interspecific competition. When sown with typical nonserpentine species on nonserpentine soil, they became etiolated and did not survive. On serpentine soil, only the serpentine endemics survived, but there was considerable bare ground and the plants grew slowly.

Figure 4.3: The root system of some common Mojave Desert plants. Species abbreviations: Hc = *Haplopappus cooperi*, As = *Acamptopappus sphaerocephalus*, Gm = *Gutierrezia microcephala*, Or = *Opuntia ramosa*, Ys = *Yucca schidigera*, Hr = *Hilaria rigida*, Ef

= *Eriogonum fasciculatum*, Hs= *Hymenoclea salsola*, Ca = *Cassia armata*, Ee = *Echinocereus engelmannii*, Sd = *Salvia dorrii* (Cody 1986).

Halophytes (literally, salt plants) are capable of growing in soil with more than 0.2% salt concentration, whereas growth of **glycophytes** (literally, sweet plants) is severely reduced above 0.2% salinity. However, like glycophytes most halophytes whether they are mangroves, coastal salt marsh herbs, beach plants which receive salt spray, or salt desert herbs and shrubs, have better germination, growth and reproduction on nonsaline than the saline soil. These species grow in highly saline areas because they do not compete as well in nonsaline wet or dry environments.

The Darwin's finches (*Geospiza* spp.) found on the chain of the Galapagos islands (situated off the west coast of South America) exhibit a classic example of character displacement among the birds. A chain of islands usually form by volcanic activity on the ocean floor gradually building up igneous material until it reaches above sea level. A feature of such island chains is that they are graded in age. A new island is always being added to one end of the chain. Evidences indicate that from a common ancestor a whole group of *Geospiza* species evolved in isolation on the different islands and adaptively radiated so that a variety of potential niches was eventually exploited on reinvasion; species now present include slender-billed insect-eaters and thick-billed seed-eaters, ground - and tree-feeders, large and small bodied finches, and even a wood-pecker-like finch (Fig. 4.4).

4.1.1.3 Measurement of competition

The logistic equation of population growth for species in isolation is:

$$dN/dt = rN - \left(\frac{K-N}{K}\right)$$

where r is the natural, intrinsic rate of population growth, N is population size, and K is population size at saturation (the environment's carrying capacity). Multiplying out, we can see that the equation already contains an intraspecific competition component:

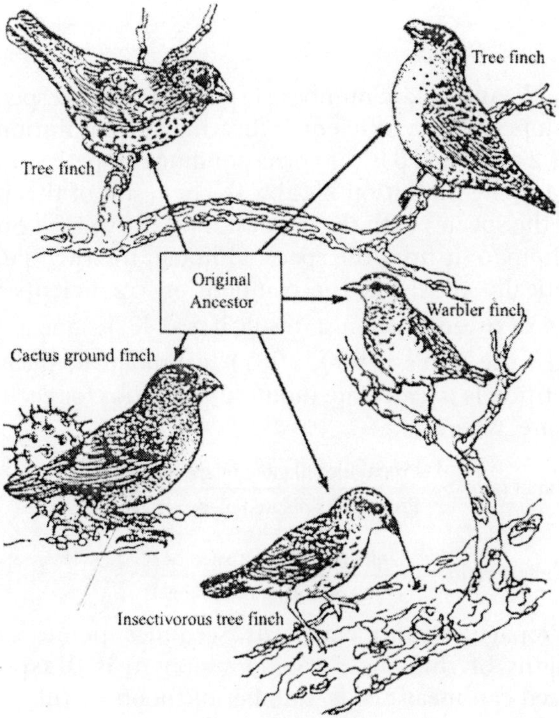

FIGURE 4.4
DARWIN'S FINCHES

$$dN/dt = rN - \frac{rN^2}{K}$$

where rN is the unlimited rate of growth and rN^2/K is the decline caused by crowding (competition).

The *Lotka-Volterra equations*, derived independently by Lotka (1925) in the United States and Volterra (1926) in Italy, describe the *interspecific competition* between two species for food, or space, or any other resources. In terms of competition within a limited space where each population has a definite K, or equilibrium level, the simultaneous growth equations can be written in the following form:

$$dN_1 / dt = r_1 N_1 \frac{(K_1 - N_1 - \alpha N_2)}{K_1}$$

$$dN_2 / dt = r_2N_2 \frac{(K_2 - N_2 - \beta N_1)}{K_2}$$

where, N_1, and N_2 are number of species 1 and 2 respectively, α is the competition coefficient indicating the inhibition effect of species 2 on 1, and β is the corresponding competition coefficient signifying the inhibition of 2 by 1. The result of this interaction is that the species with the greatest inhibitory effect on the other will eliminate it from the space, although the two species might theoretically coexist if the competition coefficients were very small. Coexistence will also result if $\alpha < K_1/K_2$ and $\beta < K_2/K_1$.

C.T. de Wit's (1960, 1961) technique to measure plant competition is to calculate input/output ratios for each species in a mixture, where:

$$\text{Input ratio} = \frac{\text{Propagules planted of species A}}{\text{Propagules planted of species B}}$$

$$\text{Output ratio} = \frac{\text{Units produced of species A}}{\text{Units produced of species B}}$$

Propagules can mean seeds, seedlings, potato eyes, bulbs, or lengths of rhizomes; a good synonym is **diaspore**. Units produced can mean seeds, number of shoots or tillers, length of new stolons, or plant weight. Whatever definition of propagule or unit is chosen, it must be the same for both species. Two species are planted in a replacement series, where the total number of initial propagules is constant, but the ratio of the two species is changed from pure species A, to an even mix of A and B, and on to pure species B.

Figure 4.5 shows the four possible outcomes of any such experiment. First, species A can "win", in which case the input/output line is parallel to but above the equilibrium line. Second, species B can "win" in which case the ratio line is parallel to but below the equilibrium line (Fig. 4.5a). Finally, species A and B can coexist, in which case the ratio line crosses the equilibrium line; that is, the slope is no longer 45° (Fig. 4.5b). If the slope is less than 45°, the equilibrium point is stable because movement away from the equilibrium point tends to be dampened by the populations' competitive qualities, and the populations drift back to equilibrium. If the slope is greater than 45°, the equilibrium is unstable because movement away from the equilibrium point tends to lead to further departure from stability and one species will decline to extinction.

Figure 4.6 (a) - (c) shows actual data rather than hypothetical data. In Fig. 4.6 (a), the annual grasses barely and oats were sown in a replacement series and output measured as seed production. The results show that barley will outcompete oats at any mixture. Figure 6 (b) shows competition between two perennials: sweet vernal grass (*Anthoxanthum odoratum*) and timothy (*Phleum pratense*). The input/output (measured as tillers) line lies so close to the equilibrium line that we may predict the two species can coexist at any mixture. Figure 4.6 (c) shows competition between two annual range plants from California grasslands: wild oat (*Avena fatua*) and slender wild oat (*A. barbata*), sown with a combined density of 1380/m^2. The input/output line, with a slope of less than 45°, crosses the equilibrium line at a point corresponding to a mix of about 20% slender wild oat and 80% wild oat. One can conclude that these two species should be able to coexist in nature at this ratio.

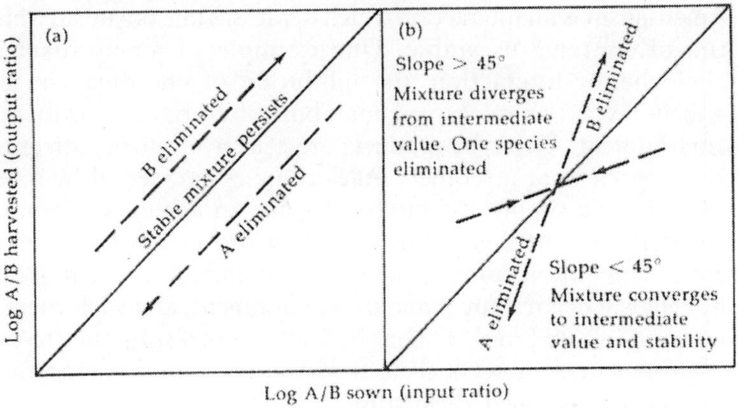

FIGURE 4.5
INPUT/ OUTPUT DIAGRAMS ILLUSTRATING THE POSSIBLE OUTCOMES OF COMPETITION BETWEEN TWO SPECIES, A AND B.

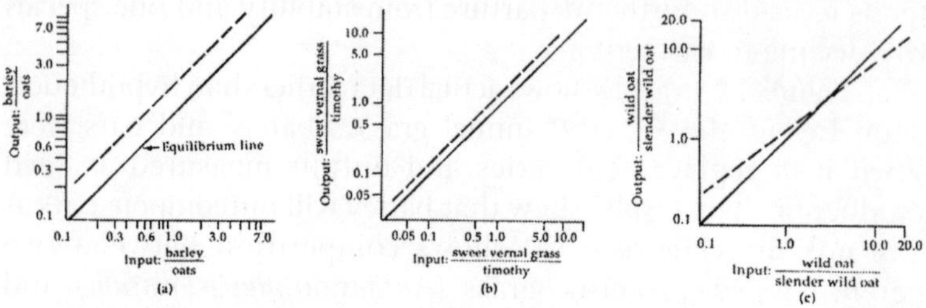

Figure 4.6 Input/putput diagrams from actual experiments: (a) between barley and oats, barley apparently always the "winner", (b) between the perennial grasses sweet vernal grass and timothy, both apparently of equal competitive ability and capable of coexisting at any ratio, and (c) between two species of wild oats, the two capable of coexisting at one ratio only [(a) and (b) from de Wit (1960), (c) Marshall and Jain (1969)]

4.1.2 Amensalism

Amensalism is an interaction which depresses one organism while the other remains stable. One example of amensalism is allelochemic interaction, the inhibition of one organism by another via the release of metabolic by-products into the environment. These by-products are selectively toxic, affecting some species but not others. Allelochemics are viewed by some biologists as simply the mechanism for an aggressive form of competition, but there is an important distinction to be made with competition: competition is an interaction brought about by removal of a resource from the environment, and allelochemic interactions are brought about by addition of a substance to the environment. A subset of allelochemic interactions involving plant species only is called allelopathy.

4.1.3 Herbivory

Herbivory is the consumption of all, or part, of a plant by a consumer including: (a) parasitic microbes or plants, (b) saprophytic microbes which decompose dead tissue, (c) browsing and grazing animals which consume only a part of woody and herbaceous plants, respectively, and (d) animals which consume whole plants (or such a significant part of them that death ensues) or plant propagules (future, potential whole plants). Grazers and browsers are sometimes considered parasites, and whole-plant consumers can be called **predators**. Wiegert and McGinnis (1975) suggested that consumers of living tissues be called **biophages**, and consumers of dead tissue be called **saprophages** (Fig. 4.7). Herbivory is usually thought of as a negative interaction on the part of plants, a one-way flow of benefits to consumers.

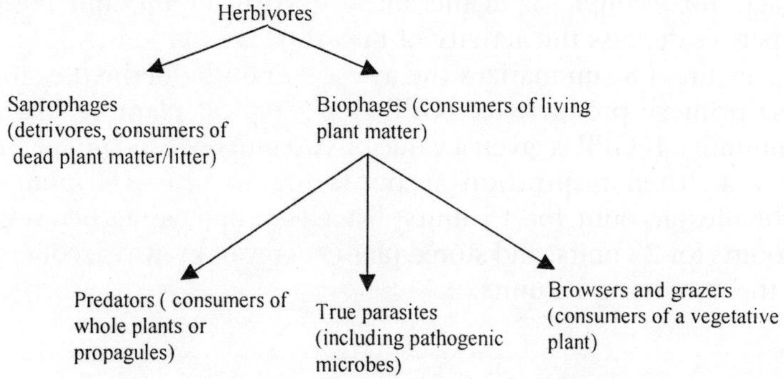

FIGURE 4.7 CATEGORIES OF HERBIVORES

The percentage of net primary production (NPP) consumed by biophage herbivores varies with terrestrial vegetation: 2-3% for desert scrub or arctic/alpine tundra, 4-7% for forest, 10-15% for temperate grasslands with minimal grazing, and 30-60% for African grasslands managed for domesticated animals. In general, vegetation may withstand considerable loss to herbivores without lasting damage. For example, tree wood growth rate is not reduced until more than 50% of the canopy's leaf surface is lost. Certainly, however, death can and does result from some herbivore outbreaks (such as tent caterpillars, locusts, budworms, rabbits, voles, lemmings, or pathogenic microbes), particularly if they persist for more than one growing season.

Intensive seed predation by herbivores in the tropics is considered the single most important factor which regulates tree populations or the distance between two trees of the same species (Janzen 1970).

The annual rain of **litter** (detritus or dead plant material such as bark, twigs, flowers, aborted fruit and leaf in the case of deciduous trees, and virtually entire body in the case of annual plants) is not entirely consumed by the saprophage herbivores each year. Litter **turnover time** (time for complete decay) is two to three years in the temperate deciduous forest and five to fifteen years in the boreal coniferous forest. This is because the fact that the rate of litter decomposition depends on soil temperature, soil moisture, and on the chemical properties of litter. Conifer foliage, for example, is acidic and is high in lignin, and both properties depress the activity of microbial populations.

Figure 4.8 summarizes the average drain by herbivores on gross primary productivity (GPP) for a typical plant or plant community. If GPP is given a value of 100 units per square meter per year, then respiration accounts for 50 units, biophage herbivores account for 15 units, litter (saprophage herbivores) accounts for 23 units, and stored plant energy or growth accounts for the remaining 12 units.

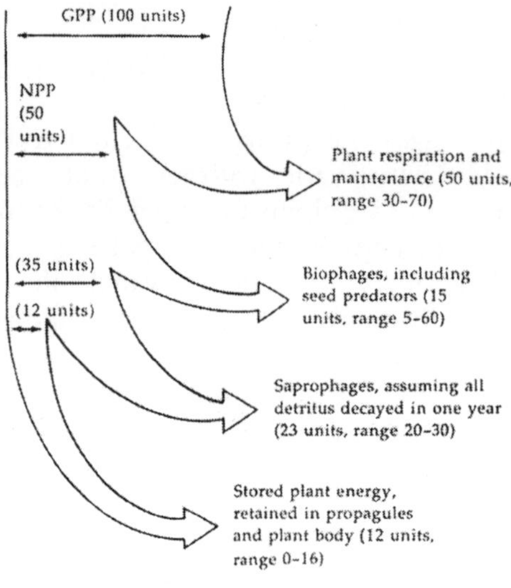

FIGURE 4.8
THE RELATIVE CONSUMPTION OF GROSS PRIMARY PRODUCTIVITY (GPP) AND OF NET PRIMARY PRODUCTIVITY (NPP) OF A PERENNIAL PLANT BY RESPIRATION AND HERBIVORES.

4.1.3.1 Impact of herbivory

For a given herbivore, all species of plants are not equally palatable. Many species are rejected totally, some are eaten preferentially, and others are eaten only when preferential species are absent. One of the reasons for nonpalatability is the allelochemics stored in plant tissue. According to Cates and Orians (1975) pioneer species are more palatable because r-selected taxa invest relatively little energy in herbivore defense, in contrast to K-selecteed taxa. The r- and K-selected taxa also differ in the type of allelochemic substances they contain. Those annuals which do invest energy in chemical herbivore defense appear to do so only with toxins, while woody perennials tend to produce tannins (Rhoades and Cates 1976). **Toxins** are small, relatively inexpensive molecules, of molecular weight 500 g/mole or less, and they are effective in small concentrations. Toxins usually account for less than 2% of leaf dry weight. They interfere with nerve and muscle activity, with hormone activity, or with liver and kidney functions. They are effective against generalist herbivores; specialist herbivores can develop means to detoxify the substances. Woody perennials may contain toxins, but they invest considerably more energy into the manufacture of tannins. **Tannins** are large complex molecules which reduce the digestibility of plant tissue by complexing with starch, proteins, or herbivore enzymes when the tissue is chewed. Tannin content may exceed 6% of leaf dry weight. Tannins are effective against both generalist and specialist hervivores. Only 17% of all annual plants tested have tannins, while more than 80% of all woody perennials have tannins (Barbour *et al.* 1980).

Some temperate or tropical zone trees may lose their leaves as an adaptation preventing the build-up of herbivore population and not for the avoidance of cold or drought. Even while leaves remain on the tree, their chemical composition can change, further limiting their suitability to herbivores. Feeny and Bostock (1968) found such a relationship between larvae of the winter moth *Operophtera brumata* and leaves of the deciduous oak *Quercus robur*. Larvae feed in spring on young, expanding leaves. They are intolerant of high tannin content in leaves. Tannin content in leaves is minimum in spring, about 0.5% on a dry weight basis. By July, tannin reaches 2%, and by the end of September, just prior to leaf drop, it reaches 5%. Thus, larvae of the winter moth will not mature satisfactorily on oak leaves two weeks older than those on which the larvae normally feed. If they hatch too early, they starve before the leaves appear; if they hatch too late, they are defeated by the tannin.

The most sophisticated and specific chemical interaction between plants and herbivores may involve hormones (Fig 4.9). Feeding preferences in some cases may be related to hormonal control rather than simple palatability.

The periodic shedding of cuticle in the course of growth in the preadult stage of insects is called *ecdysis* (moulting). The old cuticle is split and cast off to reveal a new, soft cuticle underneath; the insect increases in size, often by intake of air, and the new cuticle hardens. Each larval stage is referred to as *instar*; so that the first instar is terminated by the first ecdysis, the second instar by the second ecdysis. Ecdysis is initiated by the moulting hormone-**ecdysones**, synthesized by the prothoracic glands of larvae. These hormones form a class of compounds similar in structure to cholesterol. Ecdysone has been isolated from the leaves of several species of ferns, conifers, and angiosperms, the most ubiquitous being β-ecdysone. The ecological significance of β-ecdysone in plants is unclear. It is not toxic when orally ingested (as feeding larvae would obtain it from a food plant), but their is some evidence that it could be a feeding deterrent in concentrations as low as 1 ppb.

Another developmental hormone in insects is the **juvenile** hormone (secreted by *corpus allatum* of the brain), which predominates early in larval life (later, ecdysone predominates). It is a methyl ester of the epoxide of a fatty acid derivative, and there is some evidence that its structure differs in different groups of insects. By a series of coincidences, it was discovered that certain paper toweling prevented the European bug, *Pyrrhocoris apterus*, from developing into sexually mature adults. Instead, an extra one or two larval molts ensued, and all eventually died without being able to complete metamorphosis. The juvenility factor was traced to particular conifers used in American paper pulp, mainly *Abies balsamea, Tsuga canadensis, Taxus brevifolia*, and *Larix laricina*. The active principle was isolated and characterized. It has a structure similar to that of the juvenile hormone, and has been named **juvabione**, although it proved to be effective on only one

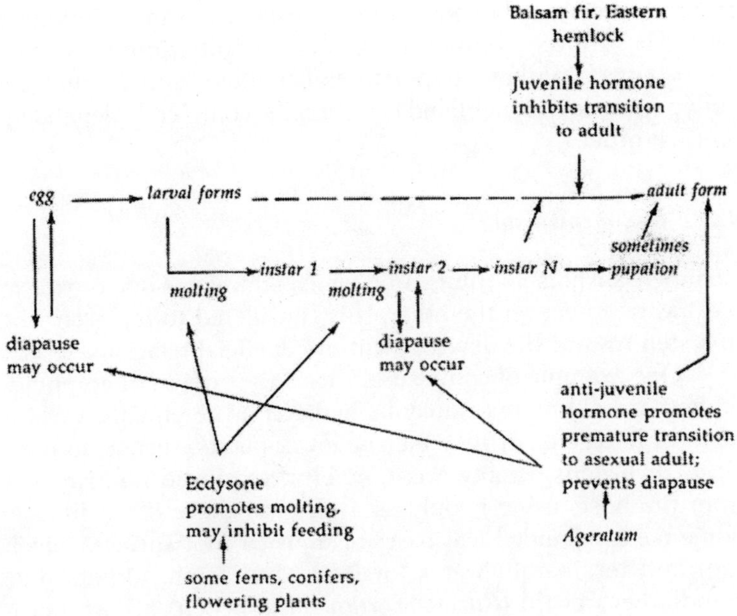

FIGURE 4.9
THE POSSIBLE ROLE OF INSECT HORMONES CONTAINED IN PLANTS, USING AN IDEALIZED INSECT LIFE CYCLE

family of insects, the Pyrrhocoridae. An anti-juvenile hormone has been extracted from an annual garden weed *Ageratum houstonianum*. It has proven capability of inducing precocious molting of immature insects to sterile adults, thus reducing feeding time and reproduction at once (Bowers *et al.* 1976). Other plants may also possess such a hormone, and its effects on other insects could additionally include prevention of *diapause* (critical resting period) and prevention of *sex pheromone* (sexual attractant) production (Fig. 4.9).

In the overall view, herbivory has a negative impact on plant growth, reproduction and survival. The evolution of herbivore defenses, whether chemical or mechanical (prickles, spines, thorns etc.) or phenological, represents energetic costs to the plant.

4.2 Positive Interactions

Associations between two species populations which result in positive effects may be conveniently considered in an evolutionary series as follows: commensalisms-one population benefited; protocooperation-both populations benefited; and mutualism-both populations benefit and have become completely dependent on each other.

4.2.1 *Commensalism*

Commensalism is an interaction which stimulates one organism but has no effect on the other. It is considered to represent the first step toward the development of beneficial relations.

One example of commensalisms is the growth of epiphytes on host trees. Epiphytes may be herbaceous perennials, such as orchids, ferns etc.; or they may be lower plants, such as mosses, algae, or lichens. In any event, epiphytes gain no nourishment from the host, using it only as a physical place of anchorage. Some have expanded leaf bases or unusual root surfaces which trap and retain rain water for later absorption. Others, like grandfather's beard (*Ramalina reticulata*, a lichen) absorb much of their water from humid air.

It is likely that competition with other epiphytes and with the canopy of the host tree for space and light has induced each epiphytic species to specialize its habitat requirements, because censuses of tropical trees (Longman and Jenik 1974) have shown that relatively distinct parts of the canopy can be associated with each epiphyte species (Fig. 4.10). Epiphyte preferences for certain host species seem to reflect such physical attributes as openness to branching and roughness of bark; they avoid pine trees mainly due to the sloughability of pine bark.

Epiphytism may easily grade into other types of interactions, such as mutualism and parasitism. Mutualism can result if the epiphyte produces nutrients (for example, atmospheric nitrogen fixed into organic form by the algal symbiont-*Nøstoc* in lichen eiphytes) which can be leached by rainwater running down the trunk or released during lichen decomposition and entering the soil around the host trees. Parasitism can result if the epiphyte's roots penetrate beneath the bark into phloem and xylem and develop absorbing organs called **haustoria**. Epiphytes can also become parasites if their size or weight creates strain on the host. For example, some *Ficus* species germinate in the canopy of the host and begin life as a typical epiphyte. Then, aerial roots grow towards the soil, reach the ground, and thicken, engulfing the host trunk and preventing further growth. At the same time, *Ficus* canopy enlarges to overtop the host and deprive it of light. Eventually the host dies.

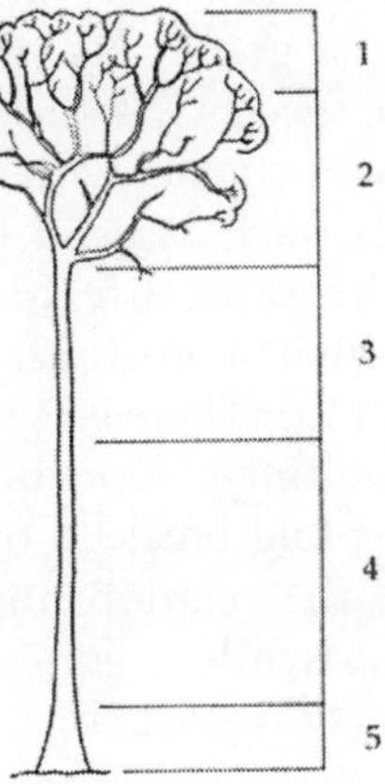

Figure 4.10 Microhabitats of epiphytes within an emergent tree of the tropical rain forest. Small epiphytes are common in zone 1, large epiphytes in zone 2, crustaceous lichen epiphytes in zone 3, and bryophytes in zones 4 and 5 (Longman and Jenik 1974).

Another example of commensalism is the **nurse plant syndrome** shown by saguaro cactus (*Carnegiea gigantea*) whose seedlings are found in close proximity to a shade-producing perennial plant or inanimate object. The positive effects of the nurse plant are shading (reduces temperature and rate of soil drying), hiding the young cactus from rodent herbivores, and some protection from frost.

4.2.2 Protocooperation

Protocooperation is an interaction which stimulates both partners, yet is not obligatory because stable growth continues in the absence of the interaction. For example, in the oceans crabs and coelenterates often associate ("commensal crabs") with mutual benefit. The coelenterates grow on the back of the crabs, providing camouflage and protection (since coelenterates have stinging cells). In turn, the coelenterates are transported about and obtain particles of food when the crab captures and eats another animal. In the above instance the crab is not absolutely dependent on the coelenterate, nor *vice versa*. Another example of protocooperation is root grafts (Fig. 4.11) between members of the same or different species. As the roots of some trees grow through soil and come in contact with each other, a natural graft or union may form. In such cases, if both partners are equally successful and of the same life form, the relationship is protocooperation, with a mutual, balanced exchange of photosynthate (Fig. 4.11a). There is evidence that hormones may also be transferred, resulting in more uniform physiognomy such as synchromous spring bud break. If one partner is smaller and suppressed, however, the relationship may be parasitic, with significantly more photosynthate going into the smaller tree from the larger than *vice versa* (Fig. 4.11b).

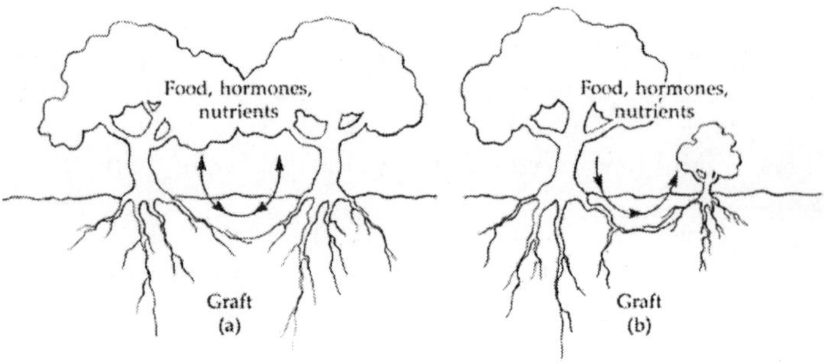

Figure 4.11 Two relationships between trees connected by a root graft, depending on the relative size of the trees. (a) The partners are both overstory trees and the flow of nutrients is equal in both directions, and so the interaction is protocooperation, (b) One partner is an understory tree and it parasitizes the larger tree.

4.2.3 *Mutualism*

Mutualism is an obligate interaction; the absence of the interaction depresses both partners. Closely related organisms do not seem to form such an interaction, for common examples of mutualism are lichens (algae+fungi), mycorrhizae (fungi+higher plants), symbiotic nitrogen-fixation (bacteria or blue-green algae+higher plants), pollination (insects, birds, or mammals + flowering plants), zoochory (animal dispersal of plant propagules), and myrmecophytes (ants+woody plants).

4.2.3.1 Mycorrhizae

Mycorrhizae (singular, mycorrhiza) are associations of nonpathogenic fungus with the roots of higher plants. In some cases, the fungus covers the root exterior near the tip of the root with a thick mantle of hyphae; additional hyphae extend as far as 8 m out in all directions from the root into the soil, and other hyphae penetrate inside the root in the intercellular spaces of the epidermis and cortex. No intracellular penetration into epidermal or cortical cells occurs, but an extensive nutrient absorbing network called the **Hartig net** is formed between these cells (Fig. 4.12). This type of mycorrhizae is classed as *ectomycorrhizae* in which fungi are typically Basidiomycetes. Ectomycorrhizae are common on high altitudes and northern temperate zone trees, including members of the Pinaceae (pine, fir, spruce, larch, hemlock), the Fagaceae (oak, beech, chestnut), the Betulaceae (birch, alder), the Salicaceae (willow, poplar), and a few other families.

Another large group of mycorrhizae are classed as *endomycorrhizae*. The fungal hyphae in this case do not form a sheathing mantle. Hyphae within the cortex do penetrate root cells, and other hyphae extend out into the soil, but not to the extent of ectomycorrhizae. The endomycorrhizae are present on most species of herbaceous angiosperms including grasses (all grasses appear to be endomycorrhizal), orchids, annual and perennial crops, and a variety of tropical and southern hemispheric temperate zone trees. The fungi in endomycorrhizal associations may be Phycomycetes, Ascomycetes, or Basidiomycetes.

The fungal partner of both kinds of mycorrhizae receives

sugars from the host plant, and plants that are grown in shade and deficient in sugars predictably have poor mycorrhizal development. Also, plants growing on fertile soils often have mycorrhizae that are less developed than those of wild plants growing on nonfertile soils. The most well-documented advantage of mycorrhizae to plants is increased phosphate absorption, although absorption of other nutrients and water is often facilitated. The greatest benefit of mycorrhizae is probably to increase absorption of essential ions that normally diffuse slowly toward roots yet are needed in high demand, especially phosphate, NH_4+, K+, and $N0_3$— (Chapin 1980). Mycorrhizae offer great advantages to trees growing on unfertile soils. In fact, without the nutrient-absorbing properties of mycorrhizae, many communities of trees could not exist. A considerable potential exists for revegetation of certain mine-waste areas, landfills, roadsides, and other infertile soils through the introduction of plants inoculated with fungi capable of forming mycorrhizae (Mark and Schenck 1983). Greater contributions of mycorrhizae to agriculture and forestry should occur as we understand them better, and such understanding is developing rapidly.

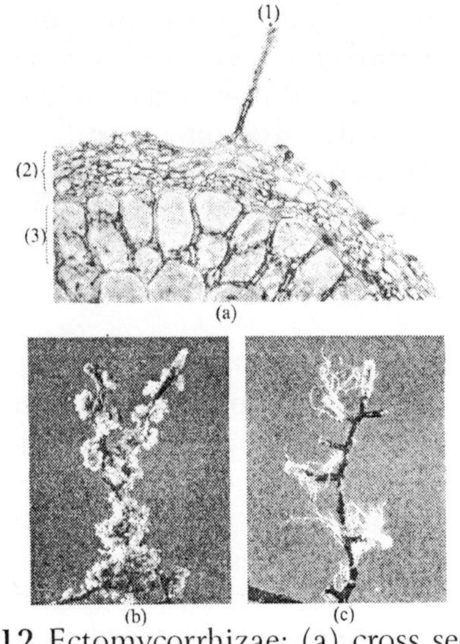

Figure 1.12 Ectomycorrhizae: (a) cross section of a root, showing (1) hyphae branching off into the soil, (2) the fungal mantle coating the outside of the root, and (3) hyphae forming a net between root cortex cells. (b) and (c) show the various shapes

that mycorrhizae may take, depending on the species involved. The fungal mantle covers short, club-shaped lateral roots.

4.2.3.2 Symbiotic nitrogen fixation

Nitrogen fixation is the reduction of atmospheric nitrogen gas into organic ammonium:

$$4N_2 + 6H_2O \xrightarrow[\text{Enzyme Nitrogenase}]{\text{ATP} \rightarrow \text{ADP}} 4NH_3 + 3O_2$$

Only certain prokaryotic organisms are capable of this process. Some of these species are free-living, occurring in soil or in water, while others exist in symbiotic relationships with plants, receiving sugars and energy-rich molecules from them, and supplying nitrates to the hosts in return. The association of *Rhizobium* bacteria with the root nodules of legumes is well known, and many grasses also have nitrogen-fixing bacteria associated with their roots in **rhizosphere** - the soil area immediately surrounding the roots of plants in which their exudations affect the surrounding microbial flora. Many exudations are nutrients which the prokaryotes utilize and include carbohydrates, vitamins and amino acids. Species of the blue-green algae *Nostoc* and *Anabaena* can become associated with bryophyte gametophytes, root nodules of cycads, leaf tissue of the angiosperm salt marsh plant *Gunnera*, and leaf tissue of the aquatic fern *Azolla*. The *Azolla-Anabaena* symbiosis has economic importance in addition to ecological significance. Agricultural trails have shown that three-fourths or more of all nitrogen requirements of rice-paddies can be met by incorporating *Azolla* in cultivation.

4.2.3.3 Pollination

Pollination is a very specialized form of mutualism which has developed in outcrossing species of plants. Even plants that are self-fertile require the transfer of pollen because the male and female functions are not present in the same tissue. There is a physical space or gap between them which must be spanned for pollination to be effected.

A plant species is usually morphologically adapted to the

specific behavioural and morphological characteristics of its pollinator, which is often an insect, bird, or bat. Adapted plants usually exhibit some or all of the following characteristics : (a) petals, sepals or inflorescence attractive, either visually or olfactorily or both; (b) pollen grains often sculptured or sticky, sometimes massed together; (c) nectar or pollen of nutritive value (nectar is rich in amino acids and sugars and pollen gains in starch and proteins) to the pollinator; and (d) attractants available at blooming time (anthesis), which is correlated to activity patterns of the pollinator.

Bees have more blossom intelligence than any other insect group: they have a greater capacity for remembering blossom characteristics, discriminating among them, and remaining faithful pollinators. The significant characteristics of bee pollinated flowers are: (a) bilateral symmetry (zygomorphy); (b) landing platform, mechanically strong flower, often partly closed with sexual organs concealed; (c) bright colours, often blue or yellow, often with nectar guides; (d) moderate quantities of nectar, nectar sometimes partly concealed; and (e) many ovules per ovary and few stamens. The adaptations of the pollinator include: (a) the possession of good colour reception in the blue, yellow, and ultraviolet ranges; (b) a high degree of intelligence and a long memory; and (c) often a long proboscis capable of probing for nectar.

Many tropical plants have flowers that are adapted to bird pollination (*ornithophily*). Such flowers are borne upon a stout stem or stout inflorescence branch to provide a landing place. Table 4.3 presents a summary of the coevolved traits of birds and ornithophilous flowers which are important to their mutualistic interactions.

The obvious advantage conferred upon *entomophilous* (adapted for insect pollination), ornithophilous, or *chiropterophilous* (adapted for bat pollination) flowers is the possibility of pollen dispersal far from the host anther, allowing for outcrossing and therefore genetic variability. By providing a nutritional source for animals, flowers have influenced the evolution of pollinators and have evolved very specific traits to ensure pollination. The result is a series of mutualistic interactions, each with its own set of special adaptations.

TABLE 4.3
A COMPARISON OF FLOWERS POLLINATED BY BIRDS AND BIRDS WHICH POLLINATE FLOWERS (FAEGRI AND VAN DER PIJL 1971).

Flowers pollinated by birds	*Birds which pollinate flowers*
Diurnal anthesis	Diurnal activity
Vivid, highly contrasting colours, often scarlet	Visual sensitivity for red, not for ultraviolet
Lip or margin absent or curved back, flower tubate and/or hanging	Too large to alight on the flower itself
Hard flower wall, filaments stiff or united, projected ovary, nectar stowed away but visible	Hard bill
Absence of odours	Scarcely any sense of smell
Nectar very abundant	Large, requiring a large amount of food
Capillary system bringing nectar up or preventing its flowing out	Long bill and tongue
Usually deep tube or spur, wider than in flowers pollinated by butterflies	Long bill and tongue
Distance between nectar source and sexual organs may be large	Large, long bill; large body
Nectar guide absent or plain	Intelligent in finding entrance

4.2.3.4 *Myrmecophytism*

The classic example of the myrmecophytic mutualism is that between the ant *Psedomyrmex ferruginea* and the tree *Acacia cornigera* (Fig. 4.13). The relationship starts when a queen ant finds a young acacia seedling or sucker which is not already occupied by ants. As they grow, acacias produce swollen stipules at the bases of leaves. These swollen stipules are sharp and look like the horns of a bull, hence the name of these trees. The queen lands on a swollen stipule and cuts a hole into it; she then clears out the soft filling of parenchyma to leave a hollow cavity. Here, she lays her eggs. She wanders over the plant which provides her food by secreting nectar from special glands, called *foliar nectaries*, which are at the base of the leaves. The queen also gathers small swollen nodules from the ends of pinnae of the leaves. These modified leaf tips are called **Beltian bodies** and contain proteins and lipids. With these the queen feeds her developing larvae and a colony of ants develops. After about a month the colony will have developed to about 1200 worker ants which live in various swollen stipules on the tree. After three years or more, there can be as many as 30,000 ants in the colony.

Once the workers become numerous, they swarm all over their acacia attacking insects and other herbivores they find. They also attack any plants which come into contact with their acacia. They even clear the ground of plants under the tree, leaving a bare circle up to a meter in diameter around the trunk. If acacias have their ants removed, they are rapidly overrun by leaf-eating insects and climbing vines. They seldom survive more than one or two years in this state.

4.2.3.5 *Zoochory*

One interesting example of zoochory is that between dodo (Fig. 4.14) and a big tree (*Calveria major*). The dodo (*Raphus cucullatus*) is native to Mauritius, an island in the Indian Ocean. The Portuguese sailors were the first to see the bird, when they landed on the island in 1507. Dodo belongs to the pigeon family but is flightless. What we know about these birds comes from the reports of the Portuguese sailors. They were ungainly in appearance, slow and clumsy in their movements and quite fleshy. A bird weighed

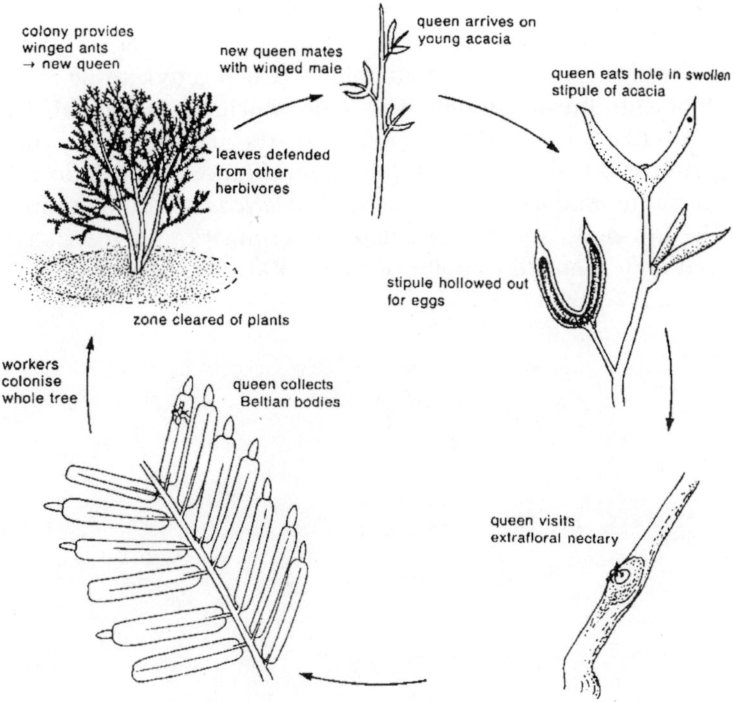

FIGURE 4.13
THE PHASES IN THE ESTABLISHMENT AND LATER DEVELOPMENT OF AN ANT COLONY ON AN ACACIA TREE

anything around 10 to 20 kg. The Portuguese sailors shot them for food. The birds were reported to be so stupid that they simply stared at the guns trained on them. This led to their quick extermination. The last bird was seen in 1681.

FIGURE 4.14 THE DODO

In recent years it was noticed that the tree *Calveria major*, also a native to Mauritius, was getting depleted in numbers. In 1973 there were only 13 *Calveria* trees in Mauritius, all of them around 300 years old and dying. These trees produce big fruits but their seeds do not germinate even in nurseries, because the seed is encased in a stout epicarp which is too strong for the embryo to break through. The dodo which consumed this fruit in plenty softened the epicarp in its gizzard. This made germination possible. With the disappearance of the dodo there was no natural agency left to aid the *Calveria* seed to germinate. The last dodo and the last natural germination of the *Calveria* seed corresponded in point of time - 300 years ago.

5. ALLELOPATHY

Allelopathy concerns the effects of one plant on another (including microorganisms) due to chemicals released by them, or due to the breakdown of their metabolites. Most of the organic compounds (allelochemicals) which are inhibitory at some concentrations are stimulatory to the same processes in very small concentrations. A very important point concerning allelopathy is that its effect depends on a chemical compound being added to the environment. It is thus separated from competition, which involves the removal or reduction of some factor from the environment that is required by some other plant sharing the habitat. Factors that may be reduced include water, minerals, food, and light. The term *interference* refers to the overall influence of one plant on another (Muller 1969). Interference thus includes both allelopathy and competition.

Although many botanists, farmers, and gardeners have observed and suggested many cases of allelopathy for over 2000 years, the controlled scientific experiments on this phenomenon were not conducted until after the year 1900. In 1937, Prof. Dr. Hans Molisch coined 'allelopathy' from the Greek words *allelon* (=mutual), and *pathos* (= harm) to describe the effects that one plant could have on another. Although the etymology of allelopathy suggests that only suppressive interactions should be involved, Molisch did discuss promotive effects as well within his context of allelopathy.

Although Molisch is acknowledged in having coined 'allelopathy', the establishment of the term in the ecological literature is principally due to Grummer (1955). He suggested that special terms be adopted for the chemical agents involved in allelopathy based on the type of plant producing the agent and the type of plant affected. Grummer recommended the commonly used term 'antibiotic' for a chemical produced by a microorganism or macro-fungus which inhibits another microorganism or macro-fungus. A 'phytoncide', is in Grummer's sense a substance produced by a higher plant which suppresses microorganisms and macro-fungi. The term 'marasmin' is applied for any ihhibitory substance produced by a microorganism or macro-fungus which affects a higher plant; and the term 'choline' (or koline) is applied for an inhibitor or toxin produced by a higher plant, which affects another higher plant (Fig. 5.1).

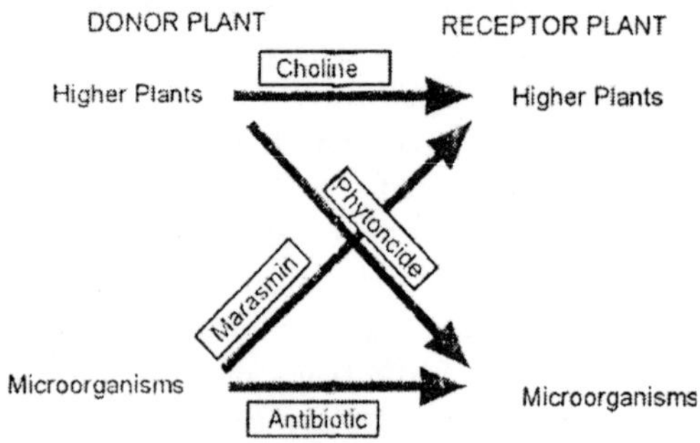

FIGURE 5.1
GRUMMER'S CLASSIFICATION OF ALLELOPATHIC SUBSTANCES.

5.1 Allelochemicals

Plants synthesize a variety of compounds and several of them hmave been implicated in allelopathy. Most of these are secondary metabolites and are identified from soil and plants. Although a wide range of chemical compounds possessing allelochemic action have been isolated, specific chemicals involved in allelopathy remain obscure. Generally, allelochemicals are grouped under the following categories.

Aliphatic compounds: Though these compounds do not comprise a large portion of the allelochemicals identified, a few of them, such as methanol, butanol, oxalic, formic, butyric, and lactic acids, are known to inhibit the germination of seeds and growth of plants.

Unsaturated lactones: Most of the well-known antibiotics, such as patulin and penicillic acid, are simple lactones and are inhibitory to several microorganisms and higher plants.

Fatty acids and lipids: Several fatty acids (nonanoic, decanoic, lauric, myristic, palmitic, steric, linoleic acids) from both terrestrial and aquatic plants are known to exhibit toxicity to plants growing in their vicinity.

Terpenoids: These comprise a large portion of allelochemicals that have been isolated from plants, particularly those growing in arid and semiarid regions. These include α-pinene, camphor, cineole, dipentene and sesquiterpenes.

Cyanogenic glucosides: Various cyanogenic glucosides, such as dhurrin and amygdalin, are widely recognized for their allelopathic potential to various crops. Most of the Brassicaceae members are known to produce a large amount of these glycosides, which on hydrolysis produce isothiocyanates with equal biological potential.

Aromatic compounds: These comprise the largest portion of allelochemicals and include phenols, phenolic acids, cinnamic acid derivatives, quinones, coumarins, flavonoids, and tannins. Among the simple phenols, hydroquinones and arbutin have been isolated from the leachates of *Arctostaphylos* and are known to inhibit the growth of several plants. Benzoic acid and its derivatives, such as hydroxy benzoic acid and vanillic acid, are the most common acids involved in allelopathy. These compounds have been isolated from cucumber, *Avena*, *Rorippa sylvestris* and *Sorghum*. These compounds are also identified from soils. The cinnamic acid and its derivatives are mostly derived from the shikimic acid pathway and are widespread in plants. They have been identified from *Parthenium*, cucumber, sunflower, and guayule. Other cinnamic acid derivatives, such as chlorogenic, caffeic, p-coumaric, and ferulic acids, are widely distributed among the plant kingdom and are inhibitory to a variety of crops and weeds. The toxic effects of these compounds are pronounced due to their long persistence in the soil, and several cinnamic acid derivatives have been identified as germination inhibitors.

Several of the quinones and their derivatives are derived from the shikimic acid. Juglone and related naphthoquinones have been isolated from walnut. Naphthoquinone and α-naphthol are isolated from sunflower. Most of these derivatives are well-known antibiotics such as helminthosporin, tetracyclin, and aureomycin.

Coumarins, the lactones of hydroxy cinnamic acid, are known to occur in several plants. Some of the coumarins, such as methyl esculin, have been identified from *Ruta*, *Avena* and *Imperata*. Among others, scopolin, scopoletin and furanocoumarins are well known for their potential to inhibit growth of several crop plants.

A large variety of flavonoids, such as phlorizin from apple residues and its breakdown products such as glycosides of kaempeferol, quercetin, and myrcetin, are well known allelochemicals.

Tannnins, which include both hydrolyzable and condensed varieties, are well known for their inhibitory effects due to their capacity to bind to the proteins. The common hydrolyzable tannins, such as gallic, ellagic, trigallic, tetragallic, and chebulic acids, are widely distributed in the plant kingdom. Most of these are also identified from the forest soils in concentrations required to inhibit nitrification. The condensed tannins, which arise by the oxidative polymerization of the catechins, are well known for their potential to inhibit nitrifying bacteria in forest soils and to slow the rate of decomposition of organic matter in soil which is very important in mineral cycling.

Alkaloids: Though several alkaloids have been isolated from a variety of plants, only few of them are identified as having allelopathic potential. Some of the alkaloids such as cocaine, physostigmine, caffeine, quinone, strychine and cinchonin, are seed germination inhibitors.

5.1.1 Biosynthesis of allelochemicals

A great majority of allelochemicals are secondary metabolites derived either from acetate or shikimic acid (Fig. 5.2). Some of the allelochemicals derived from acetate are aliphatic aldehydes, ketones, alcohols, simple unsaturated fatty acids, long chain fatty acids, polyacetylenes, naphthoquinones, anthraquinones, complex quinones and phloroglucinol. The allelochemicals derived from shikimic acid include simple phenols, benzoic acid and its derivatives, cinnamic acid and its derivatives, coumarins, sulphides and glycosides, alkaloids and cyanohydrins, some of the quinone derivatives, and hydrolyzable and condensed tannins.

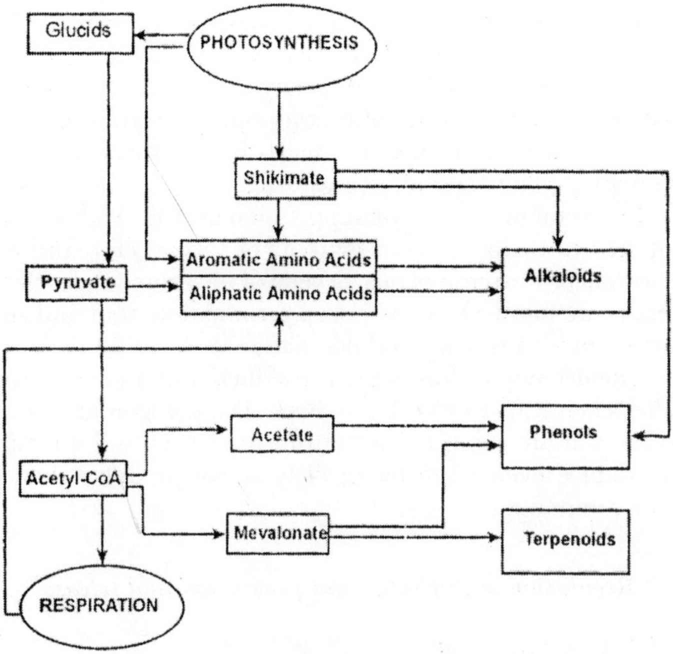

FIGURE 5.2
BIOSYNTHETIC PATHWAYS OF ALLELOCHEMICALS

5.1.2 Mode of release of allelochemicals

Four major routes of entry of allelochemicals into the environment have been recognized: (1) volatilization, (2) leaching, (3) root exudation, and (4) decomposition of plant residue.

The release of allelochemicals by volatilization is often confined to plants producing terpenoids and monoterpenes. The common genera that release volatiles include *Artemisia*, *Salvia*, *Parthenium* and *Eucalyptus*. In addition, a large number of Brassicaceae members (*Brassica juncea*, *B. nigra*, *B. rapa*, *B. oleracea*) are known to release volatiles that inhibit the germination and growth of lettuce and wheat. Apart from their toxic effects on plants, volatile allelochemicals have been shown to have insecticidal and feeding deterrent potential. Furthermore, the toxicity of volatiles is prolonged, due to their adsorption to soil particles; they remain inhibitory to plants for several months.

Leaching is the removal of substances from plants by the action of rain, snow, fog, dew, or mist. The major allelochemicals released *via* leaching include a variety of organic and inorganic substances such as phenolic compounds, terpenoids, and alkaloids, and most of these chemicals have toxic effects on the surrounding plants and microorganisms.

In several instances, the suppression in crop yields has been attributed to toxins released by roots of various crop and weed species adjacent to crop plants. Several of the compounds released via roots are known to reduce seed germination, root and shoot growth, nutrient uptake, and nodulation.

The decomposition of plant residues adds a large quantity of allelochemicals to the rhizosphere. The compounds released into the soil are subjected to transformation by soil microflora and produce even more biologically active products than the parent compounds.

5.1.3 Regulation of allelochemical production and release

5.1.3.1 Abiotic regulation

The synthesis and the release of allelochemicals into the environment are influenced by a variety of abiotic factors (Fig. 5.3), such as light, temperature, soil adsorption, soil texture, soil aeration, mineral deficiencies, soil pH, and soil moisture.

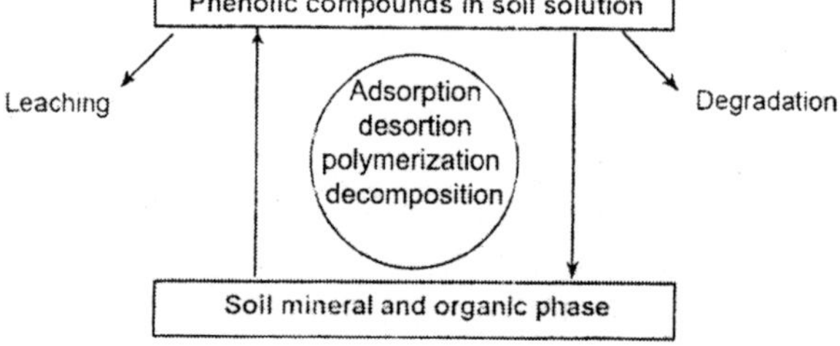

FIGURE 5.3
MAJOR PROCESSES OF REGULATING PHENOLICS
IN PLANT-SOIL SYSTEM

Light plays an important role in the synthesis of allelochemicals, rather than in their release into the environment. Exposure of tobacco and sunflower to ionizing radiation increases the amount of various phenolic compounds (Koeppe *et al.* 1970). Much work has been done on the influence of UV light on the synthesis of many allelochemicals. In several instances, a substantial rise in a variety of flavonoids and phenolic compounds has been noticed in higher plants and in several microorganisms. However, their synthesis in response to UV light vary with the part of the plant, the order of accumulation being young leaves>old leaves> stems > roots (Koeppe *et al.* 1976). An increase in synthesis of ferulic and p-coumaric acids in potato tuber discs was observed upon exposure to red light when compared to far-red light. Short exposure to red light is also known to increase the alkaloid content, whereas plants exposed to far-red light show increased levels of chlorogenic acid and other soluble phenolics (Tso *et al.* 1970). Increased synthesis of chlorogenic acid has been observed in response to short exposure to visible light compared to dark (Zucker 1963). In addition to the intensity of light, day-length has been observed to influence the synthesis. Long days increase the concentration of phenolic acids and terpenes in plants, regardless of the day-length required for flowering.

Temperature influences the release of allelochemicals rather than their synthesis. Increased temperatures favour the exudation of allelochemicals by roots. Chilling increases the concentration of total chlorogenic acid in leaves and stems but decreases it in roots.

Soil adsorption is of considerable importance for volatiles to exert their action, as most of the volatiles released into the environment get adsorbed to soil particles and persist for longer durations.

The soil texture also influences the allelochemical toxicity. Most of the volatile allelochemicals adsorbed to the colloidal particles of the soil remain in an active form for prolonged periods and migrate to sites of action in plants (Muller and del Moral 1966).

The toxicity of phenolic compounds present in the soil is influenced mostly by soil aeration and tillage practices. Under aerobic conditions the decomposition of organic compounds by microbes is faster, and anaerobic conditions favour the production of methane, hydrogen sulphide, ethylene, acetic acid, lactic acid, butyric acid, and various phenolics that have toxic effects under laboratory studies (Chou and Chou 1979).

Inhibition in the uptake of mineral nutrients by plants under allelopathic conditions has often been observed. However, several investigations have demonstrated the decrease of allelochemical effect with increase in nutrient level (Glass 1976). Amelioration of allelochemical-mediated inhibition by increased supplies of nitrogen, phosphorus, and potassium has been observed in *Solidago*, *Festuca*, and *Helianthus* (Fisher *et al.* 1978; Hall *et al.* 1982). In contrast, Bhowmik and Doll (1984) demonstrated the inability of increased levels of nitrogen and phosphorus to reduce the allelopathic potential of five annual weeds. Furthermore, a large number of minerals are known to affect the biosynthesis of allelochemicals. An increase in synthesis of scopolin and caffeic acid chlorogenic acids has been observed in plants grown in boron-deficient soil (Watanabe *et al.* 1961). Similar increases in the levels of scopolin and chlorogenic acid have been observed in plants grown in calcium- and magnesium-deficient soils.

The pH of the soil also influences the release of allelochemicals *via* modifying the microbe-mediated decomposition of litter. Low pH of the soil is known to decrease the decomposition of the litter and decrease the release of allelochemicals.

Interference of allelochemicals with plant water balance is well-known. Reductions in the water potentials are observed with ferulic acid and coumaric acid in sunflower, *Abutilon* and *Kochia* (Colton and Einhellig 1980); and in several instances the allelopathic effects are increased at low water potentials.

5.1.3.2 Biotic regulation of allelochemicals production and release

(a) Plants: In addition to various abiotic factors regulating the level of allelochemicals in the environment, plants have developed a variety of mechanisms that are involved in the dexotification of allelopathic compounds. These include *phase I* reactions like oxidation, reduction, and /or hydrolysis. Oxidation reactions involve the peroxidases, monoxygenases, and other oxygenase enzymes. These reactions are the most common in plants. *Phase II* reactions include conjugation reactions that involve the conjugation of allelochemicals with an endogenous substrate to form a new compound. The structural modification of the allelochemical during the conjugation reduces the toxicity and the higher molecular weights of the conjugates impose a limitation for its free movement. The most important conjugation mechanisms in plants include *glucosylation*, which involves the conjugation of allelochemicals with glucose mediated by an enzyme glucosyl transferase. *Phase III* reactions include further conjugation of phase II conjugates to other endogenous substrates, which results in tertiary conjugates. This mechanism is the most important process in the ultimate reduction of the toxicity of the allelochemicals to considerable extent.

Correlating the age of the plant organs with concentration of allelochemicals, Koeppe *et al.* (1976) showed increased concentration of scopolin and chlorogenic acid in tobacco leaves with increase in the age of the plant. In general, plants during flowering phase are found to be more allelopathic than during vegetative stage.

(b) Pathogens and predators: There is convincing evidence that infection increases the synthesis of various secondary metabolites, among which many are allelopathic in nature, and offers resistance to at least some plants against pathogens.

(c) Microbes: Soil microorganisms play a major role in determining the concentration as well as the potential of allelochemicals. Several microbes and fungi are known to use a variety of allelochemicals as their carbon source and thus reduce their concentration in the soil.

5.1.4 Mode of action of allelochemicals

Though much evidence has been documented on growth inhibition due to allelochemicals, the physiological and molecular mechanisms leading to such inhibition have not been critically investigated. The most probable reason for the lack of understanding of the mechanisms of action is that most of the allelopathic effects under natural conditions are either synergistic or additive, and it is difficult to study the mode of action of a particular allelochemical separately.

In his hypothetical action sequence for effects of allelochemicals on higher plants, Einhellig (1989) suggested that it is the nature of the interaction of the cell membrane with allelopathic substances that determines the type of allelopathic responses (Fig. 5.4).

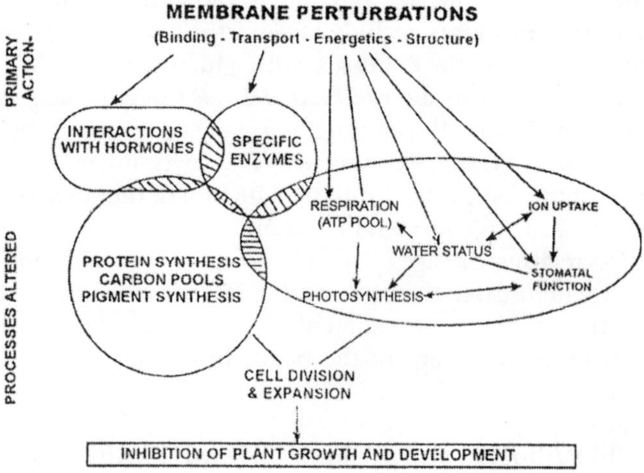

FIGURE 5.4
HYPOTHETICAL ACTION SEQUENCE FOR EFFECTS OF ALLELOCHEMICALS ON HIGHER PLANTS

5.1.5 Effect of allelochemicals on plant physiological processes

5.1.5.1 Growth

Many of the allelochemicals have been found to reduce the growth of root and shoot, reduce leaf expansion, and cause wilting in a variety of crop and vegetable plants. Cell division and elongation are essential phases of growth and development. Thus, alterations in these processes would have a direct effect on growth of plants. The decrease in cell elongation is thought to be due to the accumulation of more lignin, making the cell walls more rigid. In addition to elongation, a few of the allelochemicals (parasorbic acid, coumarin, scopoletin) inhibit root mitosis (Avers and Goodwin 1956).

5.1.5.2 Interaction with hormones

Many of the allelochemicals interact with plant growth regulators, either synergistically or additively, while exerting their function. A variety of allelochemicals have been shown to alter the levels of endogenous indole acetic acid (IAA) by modifying the activity of IAA oxidase, an enzyme that acts on IAA. Depending on their action on IAA, most of the phenolic compounds are divided into two groups: (a) those that promote IAA destruction (p-coumaric, hydroxy benzoic, vanillic, syringic, and phloretic acids), and (b) those that inhibit IAA destruction (chlorogenic, caffeic, ferulic, and protocatechuic acids). Similarly, a variety of tannins inhibit the gibberellic acid-induced growth, accompanied by reduction in amylase and acid phosphatase synthesis in endosperm half seeds of barley. Moreover, the regulation of growth by an allelochemical is dependent on its affinity to abscisic acid. The growth inhibition of cucumber seedlings due to ferulic acid and other phenolic compounds has been correlated to increased abscisic acid levels (Devi *et al.* 1997).

5.1.5.3 Enzymes

Many of the allelochemicals are known to modify either the synthesis and /or activity of various enzymes *in vitro* and *in vivo*. Often most of the allelochemicals exert dual effects in regulating enzyme activities, that is, low concentrations tend to activate while high concentrations inhibit the enzyme activity.

5.1.5.4 Respiration

Allelochemicals have been shown to inhibit or alter respiration depending on the age of the plant. Many of the quinones uncouple the oxidative phosphorylation. Some of the flavonoids are known to inhibit respiration by inhibiting the mitochondrial ATPases and interfering with energization of the membrane.

5.1.5.5 Photosynthesis

Most of the allelochemicals lead to a decrease in the dry matter production accompanied by a decrease in the photosynthetic ability of the plants. In most of the situations, decrease in photosynthesis has been correlated to decrease in chlorophyll content and stomatal conductance. Some of the allelochemicals, namely the substituted phenols, mimic the action of phenolic herbicides in inhibiting photosynthesis, and the possibility of close relation of some of the allelochemicals to potential herbicides cannot be ruled out.

5.1.5.6 Stomatal conductance

Many of the allelochemicals are known to induce stomatal closure and affect the stomatal conductance depending on concentration. Studies with ferulic, p-coumaric, and cinnamic acids show their ability to reverse the abscisic acid-mediated closure of stomata by increasing the K^+ uptake of guard cells when tested at low concentrations. However, at high concentrations, the same phenolics are known to induce stomatal closure by decreasing the turgidity of guard cells, as many of the phenolics are well-known for their potential to decrease water potentials.

5.1.5.7 Water relations

The uptake of water, which is essential to maintain turgor in the cell, becomes adversely affected by a number of allelochemicals (namely, ferulic, p-coumaric, caffeic and gallic acids). A decrease in stomatal conductance, transpiration, and leaf expansion due to allelochemicals have been ascribed to reduction in water potentials, due to reduction in the osmotic potential and turgor pressure.

5.1.5.8 Ion uptake

The uptake of water and nutrients from soil are inhibited under most of the allelopathic situations. At several instances, a decrease in the mineral content or a decrease in the capacity of the plants to absorb minerals from the soil or absorbing solution has been observed. The potential of the allelochemicals depends on their lipid partition coefficient, that is, their capacity to be soluble in membrane lipids. Depending on the partition coefficient, Glass (1975) categorized the potential of cinnamic and benzoic acid derivatives to inhibit phosphate absorption. The higher the lipid partition coefficient, the more inhibitory the allelochemical is to nutrient absorption.

5.1.5.9 Membrane permeability

Many of the allelochemicals, particularly many of the phenolic derivatives, flavonols, and flavonoids are shown to alter membrane properties. In fact, reduction in many of the physiological processes such as ion uptake, respiration, and photosynthesis are thought to be due to alteration in the structural and functional integrity of the membranes. Benzoic acid and cinnamic acid derivatives are known to depolarize the membrane, and the extent of disruption depends on the capacity to be soluble in the membrane lipids. In addition to depolarization, many of the phenolic allelochemicals interfere with membrane ATPases, which provide the necessary energy for their function.

The studies on the allelochemical effect on various plant physiological processes and possible modes of action suggest that these compounds do not act on a single definite process but have multiple sites on which to exert their action. In plants all these processes are interdependent, and any change in one process would certainly have an effect on other processes. In such cases, the relative sensitivity of a particular process is most important when proposing the mode of action of a particular allelochemical.

5.2 Allelopathy in Terrestrial Ecosystems

5.2.1 Agrosystems

Broadly, four types of allelopathic interactions are noticed in agroecosystems, namely: (1) weeds versus crops, (2) crops versus weeds, (3) crops versus crops, and (4) weeds versus weeds (Rice 1984).

5.2.1.1 Weeds versus crops

Under field conditions weed infestation is a major problem, reducing crop yield. In most of the situations, toxic effects of allelochemicals are exerted by a direct inhibition of the crop plant's nitrogen fixation or by an indirect effect on the growth of nitrogen-fixing microorganisms such as *Rhizobium* and *Azotobacter*: Allelochemicals inhibit the nutrient uptake in cereals and legumes. Owing to their greater persistence in soil, weed residues offer a major source of allelochemicals in addition to leachates and root exudates. Weed toxicity is more pronounced in monoculture than in crop rotation and mixed cropping.

5.2.1.2 Crops versus weeds

Compared to weeds, only few crops have been evaluated for their potential to inhibit weeds in various cropping systems. In most cases, allelopathic affects are exerted either by growing crops or by crop residues. The major sources of allelochemicals are root exudates and leachates of growing crops. Crops such as barley, *Sorghum*, and soybean are known to suppress weeds by their phytotoxic exudates. A few crop plants, such as wheat, pea, and sunflower are known to increase weed germination and growth, suggesting the differential sensitivity of weeds to crop allelochemicals. Wild species of crop plants have been shown to have a greater potential to inhibit weeds. Thus the genetic traits that regulate the allelochemical production can be exploited to breed new varieties with natural pesticide or weedicide properties.

5.2.1.3 Crops versus crops

Allelopathic interactions among crop plants are of considerable importance when intercropping or crop-rotation are practiced. Under monoculture practices often autotoxicity of several crops has been noticed, due to accumulation of phytotoxins from the decomposing residue of the previous crop. Allelopathy has been implicated in yield decline in crops when grown after Sorghum, corn, mungbean, and sunflower. In most of the crop rotations, stimulatory as well as inhibitory effects have been observed, depending on the rotating crops. Soybean and corn rotation, for example, has been exploited by farmers for a long time and stimulatory effect of allelochemicals has been mostly observed in such rotation. Allelopathic interactions play a greater role in determining the growth and yield of component crops in intercropping and mixed cropping systems. In these agricultural practices, legume association is often known to increase the yield of cereals by increasing the nutrient availability, improving ion exchange and providing partial weed control due to allelochemicals.

5.2.1.4 Weeds versus weeds

Allelopathic effects of weeds on other weeds have not been extensively studied. Inhibitory effects of some weeds on other weeds have been shown which may be due to competition for water and nutrients.

5.2.2 Forestry systems

Allelopathic interactions have been identified in all types of forests but to a different degree. Most of the time, such interactions have been shown to result in failure of forest trees to regenerate naturally, decrease in tree biomass production, or change in the ability of trees to associate with nitrogen-fixing bacteria and mycorrhizal fungi. Rare are the reports of positive effects of allelopathy in forest ecosystems; hence it is of considerable importance in selection of tree species for various forestry programmes.

Several nitrogen-fixing perennials are being used in tropical agroforestry systems. *Leucaena leucocephala* is one prominent example and is an important multipurpose tree used for forage, fuel, paper, pulp, timber, and for soil improvement. The only problem with *Leucaena* is that its foliage contains potential allelochemicals. Leaf mulches of *Leucaena* suppress the growth of crop plants, weeds, and some forest plant species. Similarly, the suppression of the growth of a number of field crops by several tree species, namely *Eucalyptus, Acacia nilotica, Azadirachta, Dalbergia, Ziziphus, Morus alba* and *Prosopis*, is attributed to allelopathy

5.3 Allelopathy in Aquatic Ecosystems

For many years, the inhibition of phytoplankton growth by macrophytes has been mostly ascribed to competition for nutrients, light and pH changes. However, with recent advances in chemistry, a number of chemicals with allelopathic potential have been identified from aquatic plants that play a greater role in structure and dynamics of aquatic plant communities. The chemistry of aquatic allelochemicals reveals a wide range of compounds from both living and decomposing aquatic plants. Among these the most prominent are the long chain fatty acids (for example, chlorellin from *Chlorella*), which comprise a large portion of allelochemicals from aquatic plants. Flavonoids are also found to occur largely in several members of Lemnaceac, and most of the Nymphaeaceae members contain alkaloids with greater allelopathic potential. The allelopathic interactions in an aquatic ecosystem can be exploited for biological control of some of the noxious plants by introducing a plant with greater allelopathic potential or artificially introducing the allelochemical into a body of water infested with weeds.

6. CONCEPT OF COMMUNITY

An **ecological community** may be defined as an interactive assemblage of species (producers, consumers, and decomposers) occurring together within a particular area; a set of species whose ecological function and dynamics are in some way interdependent (Putman 1994). These interactions include overt and critically determining competitive interaction and feeding relationships as well as more subtle manifestations, such as reliance of plants on animals for pollination and seed dispersal or of animals on plants for meeting habitat requirements. Because of the basic function of plants as producers, and also because they usually form the structural habitat within which animals live, plant groupings are treated primarily here, even though they form only one part of the living component of the ecosystem.

The term **vegetation** denotes all the plant species in a region (the **flora**) and it is concerned with the pattern of how all these species are spatially or temporally distributed. A vegetation type which extends over a large region is called a **formation**. For example, a tropical rain forest is a formation dominated by broadleaf evergreen trees, and it is characteristic of thousands of square kilometers in humid tropical regions on several continents. Part of the formation classification scheme adopted by UNESCO (1973) is given in Table 6.1.

Formations may be subdivided into **associations**. For example, an extensive area of vegetation varies from one part to another because of differences in topography such as north and south facing slopes, soil conditions, or stages of succession. Such a large area presents too much complexity in vegetation and environmental conditions to study as a whole, so it must be separated into subdivisions that are usually recognized by dissimilarities in physiognomy and species composition. Each basic unit is an association - the collection of all plant populations coexisting in a given habitat. An association has the following attributes: (i) it has a relatively fixed floristic composition, (ii) it exhibits a relatively uniform physiognomy, and (iii) it occurs in a relatively consistent type of habitat. The same species tend to recur together wherever a particular habitat repeats itself. Associations are usually named by their dominant or most characteristic taxa; thus the *Shorea robusta* forest is an association in the lowland (Terai) forest formation of Nepal. Typically, several to many associations may belong to the same formation, all sharing a similar physiognomy, but each differing qualitatively or quantitatively in species composition. The term **community** is a very general one which can be applied to vegetation types of any size or durability. It can, for example, be applied to one stratum of plants in a very local area, such as the herbs, woody seedlings, and mosses on the floor of a streambank forest; or to a very widespread, regional vegetation type; or to a transitory plot of vegetation undergoing rapid change in the species which compose it; or to very stable vegetation which has exhibited no significant change for hundreds of years.

The zone of vegetation separating two different types of communities is called **ecotone** or **tension zone**. These are the marginal zones sometimes easily recognizable. The development of a community is intimately related to the prevailing environment, hence difference in community structure and physiognomy reflects a difference in their environments such as the moisture condition, biotic influence or any other factors. Therefore, an ecotone is a region where the influence of two different patterns of environment work together and hence the vegetation of ecotones is somewhat specialized. The width of an ecotone may be narrow or wide. For instance, the ecotones between adjacent plots-one fenced and protected from grazing and other openly exposed to grazing, or between a pond and an adjacent land, are quite sharp and narrow whereas among many other types of communities ecotones are very wide and community differentiation is vague and continuous. Usually in ecotones the variety of species is larger than in any of the adjacent communities. The phenomenon of increased variety and density of plants at the community junctions is called the **edge effect** and is essentially due to a wider range of suitable environmental conditions.

TABLE 6.1
PART OF THE FORMATION CLASSIFICATION SCHEME ADOPTED BY UNESCO (1973).

Classification number	*Formation*	*Description*
I	Closed forest formation class	Dominants 5+ m tall, crowns interlocking
IA	Mainly evergreen forest	Individual trees may shed leaves, but canopy as a whole remains green
IA1	Tropical ombrophilous forest (=tropical rain forest)	Dominants mainly broadleaved, evergreen, with drip tips; neither cold nor drought resistant
IA2	Tropical and subtropical evergreen seasonal forest	Number of drought-deciduous trees intermediate between IA1 and IA3
IA3	Tropical and subtropical semi-deciduous forest	Most upper canopy trees drought-deciduous; many understory trees evergreen and sclerophyllous; leaves without drip tips
IA4	Subtropical ombrophilous forest	Local variant, grading into IA1
IA5	Mangrove forest	Intertidal location of tropics and subtropics; dominated by evergreen sclerophyllous broadleaved trees with stilt roots or pneumatophores; vascular epiphytes rare
IA6	Temperate and subpolar evergreen ombrophilous forest	Occurs in extremely oceanic, frostfree climates of Southern Hemisphere, as in *Nothofagus* or *Podocarpus* forests of New Zealand

Classification number	*Formation*	*Description*
IA7	Temperate evergreen seasonal broadleaved forest	Dominated by hemisclerophyllous evergreen trees, rich in herbaceous undergrowth, but few epiphytes and lianas; grades into IA4, IA6, or IA8
IA8	Winter-rain evergreen broadleaved sclerophyllous forest	Dominated by sclerophyllous, evergreen trees with little understory but some lianas
IA9	Tropical and subtropical evergreen needleleaved forest	Dominated by needleleaved or scaleleaved evergreen trees; vascular epiphytes and lianas absent
IA10	Temperate and subpolar evergreen needleleaved forest	As IA9, but to the north
IB	Mainly deciduous forest	Majority of trees shed foliage simultaneously in connection with an unfavourable growing season
IB1	Tropical and subtropical drought-deciduous forest	Foliage shed during dry season (usually winter)
IB2	Cold-deciduous forests with evergreen trees	Foliage shed during frost season; deciduous trees dominant but evergreens present as in hemlock-hardwood forest
IB3	Cold-deciduous forests without evergreen trees	Deciduous trees absolutely dominant, vascular epiphytes absent

IC	Extremely xeromorphic forest	Dense stands of xeromorphic trees, with succulents and xeromorphic shrubs, often grading into woodland (II)
II	Woodland formation class	Dominants 5+ m tall, crowns not usually touching, but canopy 40+ %, herbaceous layer may be present
IIA	Mainly evergreen woodland	Dominants evergreen as defined in IA
IIB	Mainly deciduous woodland	Dominants variously deciduous, as in IB
IIC	Extremely xeromorphic woodland	Similar to IC, but trees less dense
III	Scrub formation class	Dominants shrubs or dwarf trees 0.5-5 m tall
IIIA	Mainly evergreen scrub	Evergreen as defined in IA, includes chaparral
IIIB	Mainly deciduous scrub	Deciduous as defined in IB, includes riparian thickets
IIIC	Extremely xeromorphic (subdesert) shrubland	Very open stands of shrubs with xerophytic adaptations, some plants with thorns
IV	Dwarf-scrub and related communities	Dominants less than 0.5 m tall, includes arctic-alpine tundra, bogs, heaths
V	Herbaceous vegetation	Dominated by graminoids or forbs, more or less continuous cover, woody sinusia less than 40% cover
VA	Tall graminoid vegetation	Dominant graminoids 2+ m tall when in flower, forb cover less than 50%

Classification number	*Formation*	*Description*
VA1	Tall grassland with tree sinusia 10-40% cover	An open woodland with graminoid cover greater than 50%
VA2	Tall grassland with tree sinusia less than 10% cover	Savannas, sometimes with shrubs
VB	Medium tall grassland	Dominant graminoids 0.5-0.2 m tall when in flower, forb cover less than 50%
VC	Short grassland	Dominant graminoids less than 0.5 m tall when in flower, forb cover less than 50%, includes meadows, some types of tundra
VD	Forb vegetation	Forb cover greater than 50%, graminoid cover less than 50%

6.1 Nature of The Plant Associations

The study of community ecology is pervaded by an important controversy over the nature of the plant associations: is the association an organized system of recurrent species (**discrete view**) or a haphazard collection of populations with minimal integrations (**continuum view**)?

The famous plant ecologists Clements, Tansley and Braun-Blanquet endorse the view that plant associations can be identified as discrete (individually distinct) units. As mentioned above, an association has a relatively consistent floristic composition, physiognomy, and habitat range; whenever a particular habitat repeats itself in a given region, the same cluster of associated taxa is found. Certain species (called **differential species**) show large degree of association (presence) with each other and with a particular habitat. If clusters of species do repeatedly associate together, that is indirect evidence for strong interactions between them. Some of the interactions such as mycorrhizal unions, pollinators and certain plants, parasitic plants and their hosts etc. are also the evidence that communities are units. The discrete view assumes that associations are *closed systems*, with interdependent species which synchronously peak in abundance. The ecotones between adjacent associations are narrow, with very little overlap of species ranges, except for a few accidental and ubiquitous taxa found in many associations. In brief, discrete view considers "association" as a **superorganism** which exists in nature and which can be discovered and classified like a species.

In contrast to discrete view, the "continuum"view" (also known as "the individualistic concept of the plant association") proposed by Henry Gleason (1926) considers an association as an arbitrary unit, unlike a species. According to this view changes in species abundance and presence along an environmental gradient occur so gradually that it is not practical to divide the vegetation in associations. The vegetation of an area is merely the resultant of two factors, the fluctuating and fortuitous immigrations of plants and an equally fluctuating and variable environment. As a result, there is no inherent reason why any two areas of the earth's surface should bear precisely the same vegetation, nor any reason for adhering to ideas of definiteness and distinctiveness of plant associations. The continuum/

individualistic view assumes that associations are *open systems*, with independent species that happen to associate together wherever their range limits and chance arrival of propagules overlap; consequently, associations are at most arbitrary units along a continuum.

It is now apparent that the method used to sample vegetation will determine whether associations appear as distinct units or an arbitrary segments along a continuum. If sampling is subjective, such that **stands** (localized vegetation) are searched for which show the presence of differential species, then the discrete view will be supported, for only similar stands will be sampled, and intermediate stands will be ignored. It is possible that if only climax stands are sampled, many intermediate stands will also be ignored. If sampling is done less subjectively, ecotonal stands will be included and all the stands will be seen to lie along a continuum. Exceptions to the continuum view do exist, of course, where there are sudden discontinuities in the environment, such as a change in soil parent material, a change in elevation or slope aspect, the advance of fire, or the presence of landslide rubble.

6.2 Community Attributes

Whether associations are real or not, however, stands (communities) of vegetation certainly do exist. The stands exhibit certain emergent attributes beyond those of the individual populations which make them up. These emergent attributes include physiognomy, species importance, spatial and niche patterns, species richness, species evenness, species diversity, the rate of nutrient cycling, the pattern of nutrient allocation to above- and belowground parts, and the community change over time.

6.2.1 Physiognomy, species composition, and spatial patterns

Physiognomy is a combination of the external appearance of vegetation, its vertical structure (**architecture**: height and canopy coverage of each layer within the community, or biomass structure) and the **life forms** (life span, location of the perennating buds, phenology) of its dominant taxa.

The species composition of a community is also extremely important, because communities are partly defined on a floristic basis. Several communities may belong to the same vegetation type (that is, have similar physiognomies), yet differ in the identity of dominants or other species. The abundance, importance, or dominance of each species can be expressed numerically, so that different communities can be compared on the basis of species similarities and differences.

The relative spatial arrangement of species within a community is another community trait. Individuals within a species or individuals of different species may be distributed at *random* with respect to each other, *clumped* (positive interactions), or *overdispersed* (negative interactions).

One theory of community stability suggests that stable, long-lasting communities exhibit more species interactions and more species than transient seral communities. If this is so, then the niches of species in stable communities must be narrower, with less overlap in niche boundaries, than those of less stable communities.

Community descriptions based on physiognomy, life form, niche overlap, and other functional traits (as opposed to taxonomic traits such as the identity of species in the communities) are useful because they permit comparison of widely disjunct stands which have little or no floristic similarity. These comparisons often show a convergence of vegetation types, given a similar macroenvironment. For example, **chaparral** (temperate shrubland) stands in southern California and Chile have little floristic similarity even at the family level, but they share a similar Mediterranean-type climate and they exhibit similar total numbers of species, growth forms, leaf phenology and size, and percentage canopy cover by succulent and spinescent species.

6.2.2 Species richness, evenness, diversity and dominance

Species richness is simply the number of species in a community. Each species is not likely, however, to have the same number of individuals. One species may be represented by 1000 plants, another by 200, and a third by only a single plant. The distribution of individuals among the species is called **species evenness**, or **species equitability**. Evenness is maximum when all species have the same number of individuals. **Species diversity** is a product of richness and eveness. For example, community A, with five species but uneven numbers of individuals in each species, has a lower diversity than community B, with four species which have a very similar number of individuals in each. Community A has higher species richness, however.

The inverse of evenness is dominance or concentration of: (i) individuals, (ii) cover, or (iii) biomass of a few species. It is commonly found that during succession there is a progression in variety of species or diversity increases to a certain extent. Finally towards maturity or relatively stable state some species assume dominance. It is often found that with increase in dominance the diversity decreases.

Several indices have been proposed to indicate the similarity, dissimilarity and species diversity of plant communities.

6.2.2.1 Similarity and dissimilarity indices

Sorensen (1948) gave a simple formula to establish the **index of similarity** between two stands of vegetation as:

$$S = \frac{2C}{A+B}$$

where, S= similarity index, A and B are number of species on stand A and B, and C is the number of species common to both the stands. For example, if there are 10 species on site A and 10 on site B, and 7 species are common for both the stands, then:

$$S = \frac{2x7}{10+10}$$

$$= 0.7$$

On the basis of S, dissimilarity index (D) is calculated as:

$$D = 1\text{-}S = 1\text{-}0.7 = 0.3$$

6.2.2.2 Indices of species diversity

(a) Species richness or variety indices (d): It can be calculated from any one of the following formulae:

$$d = \frac{S-1}{\log 10}, \text{ or, } d = \frac{S}{\sqrt{N}}, \text{ or, } d = \frac{S}{1000 \text{ individuals}}$$

where, S = number of species,

N = number of individuals

(b) **Evenness index (e):**

$$e = \frac{\overline{H}}{\text{Log } S}$$

Where, $\overline{H}$ = Shannon index,

S = number of species

(c) Shannon-Wiener's index: The Shannon-Wiener index of general diversity $(\overline{H})$ is:

$$\overline{H} = -\Sigma(\text{ni/N}) \log_2 (\text{ni/N})$$

or,

$$\overline{H} = -\Sigma \text{ Pi} \log_2 \text{Pi}$$

Or,

$$\overline{H} = -\Sigma \text{ Pi ln Pi}$$

where, ni = importance value of each species,

N = total of importance values,

Pi = importance probability for each species = ni/N.

(d) **Simpson's index:**

It is calculated as: $D = 1 - \Sigma (P_i)^2$

In Table 6.2, $\overline{H}$ and D have been calculated based on the data of density (that is, the importance value) per sq m of forage species occurring in a seral grassland at Biratnagar (lat. N 26°20', long. E 87°16', altitude 72 m, msl) during the summer season. $\overline{H}$ varies from zero (0), for a community of one species only, to values of 7 or more in species rich forest. D saturates very close to one (1) as the number of species in a community increases above 50 or so.

6.2.3 Nutrient cycles and allocation patterns

Communities differ in their utilization of certain essential nutrients; they also differ in the rate at which nutrients are returned to the soil in litter fall, and the efficiency of the plant - soil - plant cycle. Early successional communities, for example, may require little soil nitrogen, accumulate very little of any nutrient in their tissues, and return nutrients rapidly in the soil, but erosion removes a large fraction of the returned nutrients because of their low cover or seasonal absence. Climax

TABLE 6.2

CALCULATION OF SHANNON-WIENER'S INDEX ($\overline{H}$) AND SIMPSON'S INDEX (D) OF SPECIES DIVERSITY.

Plant species	*density*	*Pi*	*(Pi)²*	*Pi ln Pi*
Chrysopogon aciculatus	311.08	0.3050	0.0931	0.3622
Brachiaria distachya	305.53	0.2996	0.0897	0.3611
Axonopus compressus	126.65	0.1242	0.0154	0.25906
Evolvulus nummularius	126.65	0.1242	0.0154	0.25906
Imperata cylindrical	118.87	0.1165	0.0136	0.2505
Desmodium triflorum	21.11	0.0207	0.0004	0.0803
Cyperus rotundus	6.66	0.0065	0.00004	0.0327
Saccharum spontaneum	2.22	0.0022	0.000005	0.0135
Alysicarpus vaginlis	1.11	0.0011	0.000002	0.0075
$\overline{H}$ =1.625 D = 0.772	N= 1019.88		Σ (Pi)²= 0.227647	Σ=1.625

communities may require greater quantities of some nutrients, store great quantities of nutrients in wood, and return only a small fraction to the soil in leaf litter, but prevent erosive losses by shielding the soil with a permanent, closed canopy. Thus, climax communities have fewer leaks in nutrient cycles and more efficiently hold the nutrients in the plant - soil - plant cycle.

If vegetation types are ranked according to the amount of standing biomass per hectare, they range from desert communities of only 100 kg/ha to tropical rain forests of 500,000 kg/ha. Generally, greater biomass indicates greater leaf area, which means that more radiant energy can be trapped each year, and a faster growth rate will result, with greater net productivity. Net productivity ranges from 100 kg/ha/yr for desert communities to 40,000 kg/ha/yr for tidal zone, mangrove, marsh, and swamp communities. **Efficiency**, the fraction of radiant energy converted into kilocalories of tissue, ranges from 0.04% in deserts to 1.5% in tropical rain forests.

The allocation of net productivity to different organs also differs from community to community. Grassland communities channel much of their energy to belowground biomass, scrub communities less so, and forests even less so. If we compare the distribution of carbon in a subalpine conifer forest, a temperate broadleaf forest, and a tropical rain forest, there are significant differences. Most carbon in the subalpine forest is in the form of humus atop or in the surface soil. This is because the acidic, cold soil is not favourable to decomposers. Litter half-life time is 10 or more years. Most carbon in tropical rain forest, in contrast, is locked up as inert wood. This means that leaf litter has a major function in the tropical rain forest, for it represents the only mobile, cycleable part of the nutrient bank. This is one reason why tropical soils cleared of forest vegetation soon become infertile; the annual litter rain has been stopped and what little soil reserves there are, are soon depleted by crops or leached from the soil. The litter decay rate is very rapid beneath the rain forest canopy; litter half-life is a fraction of a year.

6.2.4 Change over time or succession

Plant communities which exhibit no cumulative, directional change for several centuries are considered to be in equilibrium with their environment, and are called *climax communities*. Other communities may exhibit significant changes in such a time period; some species decline in abundance and may disappear from the site; invasive species may increase in abundance; and the vegetation type may change, for example, from a meadow to a forest or from a pine forest to a hardwood forest. Such transient communities are called *successional*, or *seral communities*. It may be possible to recognize and describe an entire sequence of successional communities, which replace each other on one site, finally culminating in a climax community. The sequence of communities is called a *succession*, or *sere*.

Succession is thought to be driven by biological interactions, such as competition, which occur in the microenvironment created by the plants themselves; it is not driven by macroclimate change. Climate is assumed to be constant when a succession is investigated and described.

6.2.5 Evolution

Whittaker and Woodwell (1972) suggested that evolution does occur at the community level, directed by natural selection toward optimal solutions of environmental problems. The conservative opinion is that the most complex level at which the natural selection operates is with pairs of species, such as parasite and host.

Very close evolutionary relationships between individuals of two species are known as **pairwise** or **one-on-one co-evolution** (for example, pollinators and certain plants, parasitic plants and their hosts, mycorrhizal union). On the other hand, less specific evolutionary relationships are called **diffuse co-evolution** (for example, insects and entomophilous angiosperms). In brief, the evolution of a community occurs as a process of coevolution of the associated species, making the whole community an interactive assemblage.

6.3 Methods to Study Community Structure

The structure of a community can be studied by taking into consideration a number of characters which are usually grouped under two heads, *viz.,* **analytic** and **synthetic**. Certain analytical characters, *viz.,* frequency, density, abundance and dominance can be expressed quantitatively; while others, *viz.,* sociability, vitality, periodicity and stratification find only qualitative expression. Synthetic characters include presence, constance and fidelity of components and may be computed from analytical characters of several stands of a community.

6.3.1 Qualitative characteristics

6.3.1.1 Floristic composition (kinds of species occurring in the community)

A complete list of species is the most essential characteristic of a stand, and making of such a list is the first step in its study. In order to secure a complete list, inspection and collection throughout the year are required so that all species appearing in different seasons will be included. Floristic lists are valuable for characterization of the community, for the absence of some and the presence of others may indicate present conditions and future trends.

6.3.1.2 Stratification

Stratification (architecture), or layering, is the occurrence of organisms, or their parts, at different levels in a stand. Stratification is readily seen aboveground, but is present also in the unseen root systems and rhizomes. Stratification usually occurs because life-forms such as trees, shrubs, herbs and mosses differ in their requirements and amplitudes, and therefore grow at various levels which differ in light intensity, temperature, moisture conditions, organic content of the soil, and other factors. The number of strata above ground vary according to the kind of community. In early stages of succession usually only one stratum is present, comprising such plants as lichens, mosses, or annual herbs; but as succession proceeds additional strata appear, so that in a mature grassland there may be three strata, the two highest comprising grasses and forbs, and the lowest short herbs, mosses, lichens, and occasionally algae. Five to seven strata may be found in forests: two or three of trees (often called **canopy layers**), one or two of shrubs, an herb or field layer, and a ground or moss-lichen layer. Occasional trees that project above the general canopy are called **emergent** or **supercanopy** trees. Most often, each layer of vegetation is composed of different life forms; for example, the top layer may be composed of a tree and a climber. To distinguish between morphologically distinct groups of species the term synusiae is used. Plants of similar morphological forms are put together as for instance we may speak of a synusia of trees or a **synusia** of climbers or of epiphytes. In general, stratification and synusiae both contribute to **physiognomy** (outer appearance) of vegetation, and each vegetation type has its own characteristic physiognomy.

6.3.1.3 Periodicity (phenology, aspection)

Periodicity refers to the regular seasonal occurrence of various processes such as formation of leaves, elongation of shoots/growth, flowering, pollination, ripening of fruits and dissemination of seeds. Periodicity results from inherent genetic characteristics of each species under the influence of a particular combination of environmental conditions. Periodicity means particularly the recurrence at certain times of these processes and their manifestations, while **phenology** refers more to the appearance of the manifestations at certain seasons of the year, rather than to their cyclic nature. **Aspection** is the appearance or aspect of the community as a whole at different seasons. Most communities have a definite aspect at different seasons of the year. The chief aspects are the **prevernal**, or early spring; the **vernal**, or spring; the **aestival**, or summer; the **serotinal**, or autumn; and the **hiemal** or **hibernal**, winter.

It is interesting to keep records of phenological events such as the dates of appearance of certain kinds of seedlings in various years, the inception of growth of perennials, the opening of leaf and flower buds on certain trees, the full expansion of the leaves, and the appearance of first flowers, first ripe fruits, last flowers, and the completion of the dissemination of the fruits or seeds. Phenologic behaviour influences competition and association of species. The occurrence of vegetative growth, flowering, and fruiting at different seasons, or only slight overlapping of these activities, often reduces the intensity of competition and favours association. Further, a long growing season with ample soil moisture throughout is conducive to the formation of a community rich in species because great opportunity is provided for their growth at different times. It may be concluded that the various expressions of periodicity and phenology of plants are means of adaptation to seasonal changes in the physical environment. Usually, in order to live in a given location, the plant must make full use of the favourable environmental periods and endure or tolerate, by one means or another, the unfavourable periods.

6.3.1.4 Vitality (vigour)

Vitality relates to the condition of a plant and its capacity to complete the life cycle, while **vigour** refers more specifically to the state of health or development within a certain stage; a seedling or mature plant may be vigorous, or it may be feeble and poorly developed. The vigour of species in a stand often merits special attention because they may indicate the end of a previous stage of succession, the beginning of a new stage, a change in conditions such as an increase or decrease in soil moisture, attack by parasites, or some kind of harmful grazing practice.

A number of criteria may be used in determining the vigour of plants such as the rate and total amount of growth, especially in height; rapidity of growth renewal in the spring or following mowing or grazing; quantity or area of foliage; colour and turgidity of leaves and stems; degree of damage by disease or insects; time of appearance and number and height of flower stalks; rate of growth and extent of the root system; appearance and development of new stems and leaves; and the extent of dead portions, especially in bunch- or mat-formers.

The following classification of vitality has been widely used:

Class 1: Well-developed plants which regularly complete their life cycles

Class 2: Vigorous plants which usually do not complete their life cycles or are poorly developed, and sparsely distributed plants that do spread vegetatively

Class 3: Feeble plants that never complete their life cycles but do spread vegetatively

Class 4: Plants occasionally appearing from seed but which do not increase in number, such as ephemeral plants

6.3.1.5 Life forms

Perhaps the least technical characterization of a community's structure, particularly as applied to plants, employs commonly recognized **growth forms** such as evergreens and deciduous trees, woody and herbaceous shrubs, herbs, and so on. Within each of these general categories we can distinguish further between needle-leaved, broad-leaved, and sclerophyll-evergreens, or succulents, thorn, and rosette among shrubs, and grasses and forbs among herbs. Dansereau's (1951) system recognizes six growth forms (trees, shrubs, herbs, bryoids, epiphytes, and lianas); three sizes (tall, medium, low); four functions or leaf habits (deciduous, semideciduous, evergreen, evergreen-succulent or evergreen-leafless); six leaf shapes (needle or spine, graminoid, medium or small, broad, compound, thalloid); four leaf textures (filmy, membranous, sclerophyll, succulent or fungoid); and four aspects of coverage (barren or very sparse, discontinuous, in tufts or groups, continuous).

A variation of plant-growth form, referred to as **life form**, was developed at the turn of the twentieth century by Danish

TABLE 6.3
THE SYMBOLS, CATEGORIES, AND DESCRIPTIONS OF RAUNKIAER'S LIFE FORMS

Symbol	*Life form*	*Description*
P	Phanerophyte	Shrubs, trees, and vines with buds or shoot apices located on upright shoots. The class is divided into four subdivisions
MM	(i) Megaphanerophyte	Perennating bud above 30 m high
	(ii) Mesophanerophyte	Between 8 m and 30 m
M	(iii) Microphanerophyte	Between 2 m and 8 m
N	(iv) Nanophanerophyte	Under 2 m
Ch	Chamaephyte	Herbaceous or low woody plants whose perennating buds are borne just above ground level - up to 0.3 m
H	Hemicryptophyte	Perennating buds close to ground (rather half hidden in soil)
	Cryptophyte	Perennating organs below soil surface or water. The class has three subdivisions
G	(i) Geophyte	Perennating buds underground; includes plants with bulbs, tubers, and rhizomes, that die aboveground each year
HH	(ii) Helophyte	Marsh plants with perennating buds in waterlogged mud
	(iii) Hydrophyte	Perennating buds beneath the water
Th	Therophyte	Survival in unfavourable season through seeds or spores; annuals
S	Stem succulent	
E	Epiphyte	

botanist Raunkiaer (1934) based on the relation of the ground surface to the position of the highest perennating bud or organ, by which the plant survives an unfavourable season of cold or dryness. Raunkiaer's life forms are described in Table 6.3.

The ratio of the life forms of different species in terms of numbers or percentages in any floristic community is called the **biological spectrum** (life form spectrum) or **phyto-climatic spectrum**. Since the life form is related to the environment around the plants, the biological spectrum is also regarded as indicative of the prevailing environment. For example, grasslands are usually rich in hemicryptophytes, tropical deserts (long dry seasons) in therophytes, and arctic and alpine regions (extremely cold climate) in chamaephytes and hemicryptophytes. Table 6.4 gives biological spectra found in several environments, in which Raunkiaer's normal spectrum is based on an analysis of random sample of four hundred species representing the world flora.

Another useful system of classification of life forms was designed by Kuchler (1967) to prepare maps of world vegetation. It may be used in nearly every kind of vegetation, and conveniently describes the appearance of the landscape. Abbreviations are given in Table 6.5. Each dominant type of plant in an area is given a series of symbols. Symbols for different life forms are separated from each other by placing a period between them in the finished formula. For each basic life form, the symbols are given in standard order: life-form, leaf size (if given), height class(es) in which it occurs, and coverage class. As an example, dense forests of 8 m tall *Rhododendron arboreum*, with scattered taller *Abies spectabilis* and small *Lyonia ovalifolia* occur in central Nepal, often with a fairly dense understory of a large bamboo-like grass and a sparse cover of mosses on the floor. Such a community would be symbolized: E7r. Bh5c. D5b. V4i. Llr. The symbol groups refer to the *Abies, Rhododendron*, Lyonia, bamboo, and mosses, respectively.

6.3.1.6 Sociability (gregariousness)

Sociability refers to the proximity of plants or shoots to one another. Because of their means of reproduction (vegetative means such as rhizomes or runners), or because of high specificity for a particular environment, some species tend to occur in clumps.

TABLE 6.4
COMPARATIVE RAUNKIAER'S BIOLOGICAL SPECTRA

	Th	*G*	*H*	*Ch*	*P*
Raunkiaer's normal	13	6	26	9	46
Rajasthan (desert)	40	3	16	19	10
Kumaon Himalaya					
Pinus roxburghii forest	51	2	25	6	17
Mixed forest	37	2	10	8	44
Oak forest	13	3	26	3	56
Grassland	33	14	43	10	0
Swiss Alps (alpine)	3.5	4	68	24.5	0
Mary's Peak, Oregon (Subalpine; elevation 1249 m)	0	25	54	2	19
Great Smokey Mountains (low, moist, deciduous forest)	3	26	30	4	36
Nepal's Terai (Morang district)	30	14	6	15	35

TABLE 6.5
KUCHLER'S SYSTEM OF PHYSIOGNOMIC SYMBOLS USED TO DESCRIBE VEGETATION.

(i) Life forms		10-20 m	6
Basic life forms	**Symbol**	5-10 m	5
Woody plants		2-5 m	4
Broadleaf evergreen	B	0.5-2 m	3
Broadleaf deciduous	D	0.1-0.5 m	2
Needleleaf evergreen	E	Less than 0.1 m	1
Needleleaf deciduous	N	(iii) Leaf characteristics	
Aphyllous	O	Type	
Semideciduous (B+D)	S	Hard (sclerophyll)	h
Mixed (D+E)	M	Soft	w
Herbaceous plants		Succulent	k
Graminoids	G	Large (>400 cm^2)	l
Forbs	H	Small (<4 cm^2)	s
Lichens, mosses	L	(iv) Coverage	
Special life forms		Type	
Climbers (lianas)	C	Continuous (>75%)	c
Stem succulents	K	Interrupted (50-75%)	i
Tuft plants	T	Parklike, in patches (25-50%)	P
Bamboos	V	Rare (6-25%)	r
Epiphytes	X	Barely present, sporadic	
(ii) Stratification		(1-5%)	b
Height		Almost absent, extremely	
More than 35 m	8	scarce (<1%)	s
20-35 m	7		

Other species with efficient dispersal mechanisms (for example, weedy species that invade open ground), often occur singly. Classification of these species by a sociability scale can provide insight into their systems of reproduction, dispersal, or help identify unique habitats. The Braun-Blanquet scale, given below, for rating sociability of species, has been widely used in analyzing vegetation.

Class 1: Individuals (or shoots) growing singly in a place

Class 2: Small groups of plants growing in scattered tufts

Class 3: Small groups of plants growing in scattered patches or cushions

Class 4: Large patches or broken mats

Class 5 : Very large mats or stands of nearly pure

populations that almost completely cover a large area such as a cattail marsh

6.3.1.7 Association of species

Association of species or interspecific association is the growing together of two or more species in close proximity to one another as a rather regular occurrence. It may be brought about by the similarity in ecological amplitudes of two or more species; similarity in geographical ranges; differences in life-form (such as shallow and deep root systems) so that excessive competition can be avoided; dependence of one species upon another for shade, or for food as in parasites; or dependence for protection from grazing, as in grasses growing in dense clumps of cactus. A number of methods have been used to measure the degree of interspecific association. The **association index** is secured by dividing the number (a) of random samples of a given stand in which species A occurs into the number (n) of samples in the same stand in which species A and B occur together. For example, if species A occurs in 40 sample areas (a = 40), and species B occurs together with A in 30 of these sample areas (n = 30), then the association index of species A is B/A, or 30/40, or 0.75; that is, on three-fourths of the sample areas in which species A occurred it was associated with species B. But if species B is used as the base, the association index is 1. At least 100 samples should be taken for these calculations.

6.3.2 Quantitative characteristics

6.3.2.1 Releve method

An investigator familiar with the vegetation of a region begins to develop concepts about the existence of certain community types that appear to repeat themselves in similar habitats. A number of stands that represent a given community are subjectively chosen. The investigator walks through as much of each stand as possible, compiling a list of all species encountered. Next, an area which best represents the community is located. It is then necessary to determine the **minimal area** - the smallest area within which the species of the community are adequately represented. The minimal area may be determined by a **species-area curve**. The resulting sample quadrat, based on the concept of minimal area, is called a **releve**.

A species-area curve is compiled by larger and larger quadrats on the ground in such a way that each larger quadrat encompasses all the smaller ones, an arrangement called **nested quadrats**. As each larger quadrat is located, a list is kept of additional species encountered. A point of diminishing return is eventually reached, beyond which increasing the quadrat area results in the addition of only a very few more species. The point on the curve where the slope most rapidly approaches the horizontal is called the minimal area. For example, to determine minimal area (size of the quadrat) for a pasture vegetation, an area of 10 cm x 10 cm was differentiated in the field with the help of string and nails, and the number of plant species present within that area was counted. Thereafter, area was enlarged to 20 cm x 20 cm and additional species were counted. The process of increasing the area and counting the additional species was continued until 2 m x 2 m large quadrat was formed. Thereafter, the number of species was plotted on the Y-axis against the areas on the X-axis on a graph paper. The minimal area of the quadrat for sampling that pasture was 40 cm x 40 cm, as the species-area curve flattened at that point (Fig. 6.1). Minimal area is thought by some ecologists to be an important community trait which is just as characteristic of a community type as the species which make it up. Table 6.6 shows how minimal area correlates with vegetation type. In the releve method, each species is recorded and several parameters are estimated within the releve: frequency, density, cover, importance value index (IVI), sociability, vitality, periodicity, topographic characteristics, and environmental characteristics.

6.3.2.2 Transect method

A transect is a sampling strip extending across a stand or several stands (Oosting 1958). It is useful where one is concerned with the analysis of vegetation changing in composition through an ecotone or along the gradient of some environmental factor. Depending upon the objective one may employ "line transect" (also known as "line intercept") or "belt transect".

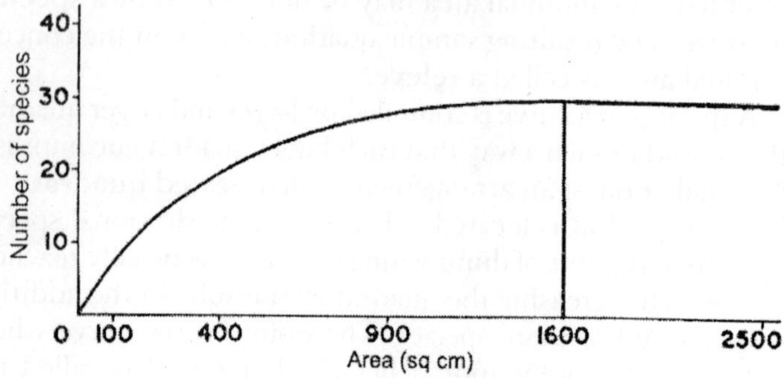

FIGURE 6.1
SIZE OF QUADRAT IN SQ CM

TABLE 6.6
MINIMAL AREA FOR VARIOUS VEGETATION OR COMMUNITY TYPES (MUELLER-DOMBOIS AND ELLENBERG 1974)

Type	*Minimal area (m^2)*
Tropical rain forest	1000-50000
Temperate forest	
overstory	200-500
understory	50-200
Dry temperate grassland	50-100
Heath	10-25
Wet meadow	5-10
Moss and lichen communities	0.1-4

Line transect: A measuring tape or calibrated string is run across the vegetation to be analyzed and the plants touching and along the tape are recorded as also the length of the tape intercepted by the individual plant cover. The total fraction of the line covered by each species, multiplied by 100, is equal to its percent cover. Just as with quadrats, total cover can be more than 100%. Disadvantages of the method are the loss of density and frequency measures, because there is no area involved (although frequency can be expressed on a cover basis if the line is broken up into segments). The method is useful especially for bunch or tussock grasses.

Belt transect: In this case instead of the line, a belt of definite width (one meter or more) is studied across the vegetation. This strip is divided into a number of small sections and plant species and numbers of individuals of each species are recorded. In simple words, it is laying of quadrats along a line. This method has been most often applied to forest vegetation. The construction of a profile or bisect diagram (**bisects** are scale drawings of the vegetation within belt transects) is also possible along such a transect. By marking the sections on graph paper (bar graphs) a useful picture of the vegetation is obtained.

6.3.2.3 Point method

If a quadrat (sampling unit) is reduced to no dimension, it becomes an infinitely small point. On this basis two main methods for the vegetation analysis have been developed: (i) point-frame method, and (ii) point-centered-quarter method.

Point-frame method: Ten movable pins are inserted in a wooden frame which rigidly limits the pin to a vertical path perpendicular to the ground (Fig. 6.2). The usual size of the pin is 50 cm and the length of the frame is also the same. The height at which this frame is fixed on two folding wooden-crosses is also 50 cm. The apparatus is placed at a number of places in the field, usually at random. As the pin is lowered, the first plant it touches is recorded, then the pin is lowered more until it touches another leaf (of the same or a different plant than the first touch), and so on until bare ground is reached. If no plant is hit, then the point is tallied as bare ground. These data permit two calculations. One is percent cover:

$$\%\text{Cover} = \frac{\text{No. of pins which hit species A at least once}}{\text{Total no. of pins}} \times 100$$

The other calculation is percent of sward, which weights each species by its canopy thickness, or cover repetition, at each point:

$$\%\text{Sward} = \frac{\text{No. of contract with species A}}{\text{Total no. of contacts}} \times 100$$

Disadvantages of the point method are that density and frequency cannot be measured (although cover frequency can), and it is limited to low vegetation, such as grasslands, for obvious reasons.

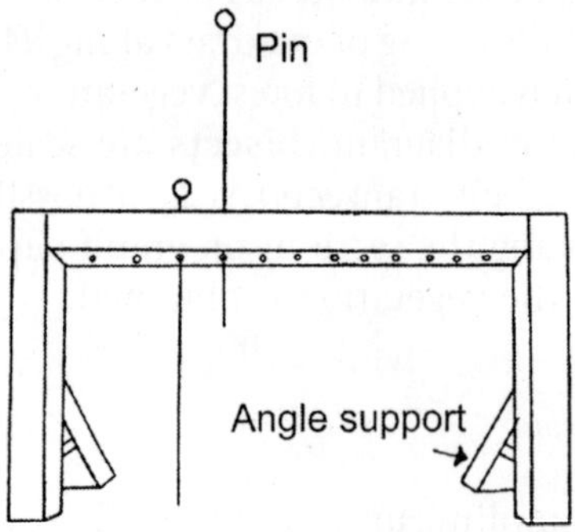

FIGURE 6.2
THE POINT FRAME

6.3.2.4 Point-centered-quarter method

In this method (also called **distance method**) the distance of the nearest plant in each quarter from the base of the pin (needle) and its diameter are recorded. A surveyor's needle (a brass rod of 3 mm thickness and 30 cm length) fitted on top with a rubber cork and compass (Fig. 6.3) is fixed by the worker at a point in the grass field. Around this point the area is visually divided into four quarters. In each quarter the distance between the needle and the nearest occurring plant and its diameter are recorded on the data sheet. In case the needle directly hits some plant, the distances in all the four quarters are recorded as zero and only the diameter of that particular plant is recorded. For adequate sampling the number of such points should be as high as to make provisions for the measurement of at least 30 individuals of a particular plant species. The points are usually distributed at random. The method is especially suited for study of bunched grasses.

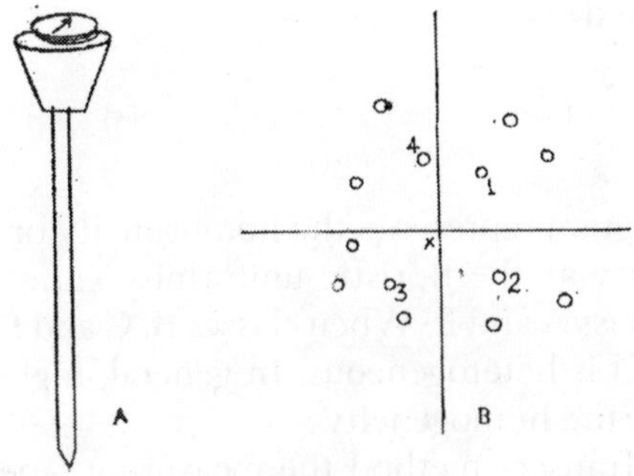

FIGURE 6.3
THE POINT-CENTERED-QUARTER

(a) Frequency: Frequency as introduced by Raunkiaer (1934) indicates the number of sampling units in which a given species occurs; and thus expresses the distribution or dispersion of various species in a community. From this, percentage frequency is calculated as follows:

$$\% \text{ frequency} = \frac{\text{No. of sampling units in which the species occurred}}{\text{Total no. of units studied}} \times 100$$

In releve method, ten quadrats of appropriate size are laid randomly in the grassland to be studied, and presence or absence of each species is noted by '+' or '-', respectively, in each of the quadrats:

1. SN	2. Plant species	3. Quadrats 1 2 3 4 5 6 7 8 9 10	4. Total no. of quadrats of occurrence	5. Total no. of quadrats studied	6.Percentage frequency	7.Frequency class

The value in column 4 of the above table is divided with that in column 5 and multiplied with 100 to obtain percentage frequency of each species. For the preparation of frequency diagram (Fig. 6.4), the species are divided into 5 frequency classes based on their percentage frequency: class A: 1-20%, class B: 21-40%, class C: 41-60%, class D: 61-80%, and class E: 81-100%.

Raunkiaer's law of frequency (on the basis of over eight thousand quadrats studied) is:

$$A>B>C \gtreqless D<E$$

Frequency diagram represents the homogeneity or heterogeneity of a community as the floristic uniformity varies directly with the value for class A and E. When classes B, C and D are relatively high, the stand is heterogeneous. In general, higher the class E, greater will be the homogeneity.

In Line Transect method the measuring tape is run across the grassland in north-south direction, and the plant species touching the tape in alternate one meter segments are recorded.

1. SN	2. Plant species	3. Segments 1 2 3 4 5 6 7 8 9 10	4. Total no. of segments of occurrence	5. Total no. of segments studied	6. Percentage frequency	7. Frequency class

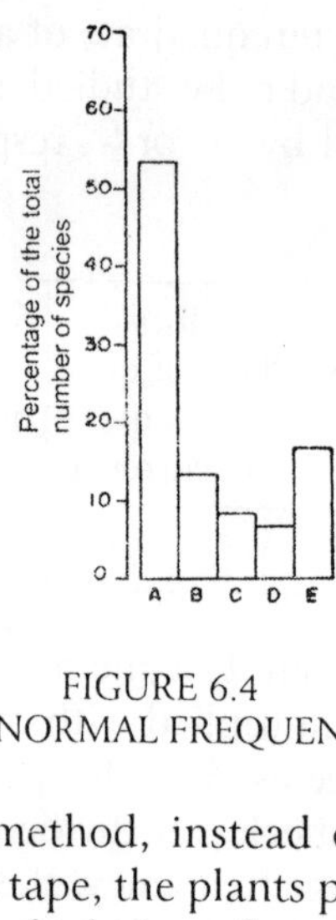

FIGURE 6.4
RAUNKIAER'S NORMAL FREQUENCY DIAGRAM

In Belt Transect method, instead of recording the plant species which touch the tape, the plants present in alternate belt segments are listed for calculation of percentage frequency.

In Point-frame method, the point-frame apparatus is placed at twenty spots (if the area is very large the number of spots may be increased), and the plant species hit by the needles each time are noted. The number of hits secured by each species is also recorded.

1. SN	2. Plant species spot numbers	3. Number of hits per spot	4. Total no. of hits secured	5. Total no. of hits	6. Percentage frequency	7. Frequency class

Percentage frequency for each species is calculated as:

$$\% \text{ frequency} = \frac{\text{Total No. of hits secured by the species}}{\text{Total no. of hits taken}} \times 100$$

(b) Density: The terms abundance and density represent the numerical strength of the species in the community. The former, if considered along with frequency, gives an idea of the distribution pattern of the species while the latter represents the number of individuals per unit area.

Ten quadrats of appropriate size are laid in the field randomly, and individuals of each species in each quadrat are counted:

1. SN	2. Plant species	3. No. of individuals (quadrat number)	4. Total no. of individuals	5. Total no. of quadrats of occurrence	6. Total no. of quadrats studied	Abundance	Density per sq m

The abundance for each species is calculated as:

$$\text{Abundance} = \frac{\text{Value in column 4}}{\text{Value in column 5}}$$

whereas, density for each species is calculated as:

$$\text{Density} = \frac{\text{Value in column 4}}{\text{Value in column 6}}$$

In distance method, random points are located in a stand. At each point the distance to the nearest tree of any species is recorded, the species is identified, and its basal area is measured. The distance between sampling point to the nearest tree in many forests is 20 m. Only one measurement is made from each random point along the transect. All distances for all species are summed and divided to yield one average distance.

Density per hectare (10000 m^2) for all trees is then:

$$\text{Density} = \frac{10000}{2(\text{average distance, in meters})^2}$$

The 2 in the denominator is a constant correction factor. Relative density of each species is:

$$\text{Relative Density of species A} = \frac{\text{No. of trees of species A encountered}}{\text{No. of trees encountered}} \times 100$$

(c) Cover: The term cover or herbage cover signifies primarily the area of ground occupied by aboveground parts of plants, such as leaves, stems, and inflorescence, as viewed from above. Cover is not measured precisely but rather is placed in one of the six categories of Braun-Blanquet's cover-class-scale by a visual estimate:

Cover class	*Range of Cover (%)*	*Mean*
1	<5	2.5
2	5-25	15
3	25-50	37.5
4	50-75	62.5
5	75-95	85
6	>95	97.5

For example, if in a series of ten plots, the cover class estimates for a species are: 1, 1, absent, 2, 1, absent, absent, 4, 1, and 5; the average cover of the species is (2.5 + 2.5 + 0 + 15 + 2.5 + 0 + 0 + 62.5 + 2.5 + 85) / 10, or 17.3%.

In the Line Intercept method, a measuring tape is run in the grass field, and the length of the tape intercepted by individual plants is measured for the estimations of cover (%) and relative cover of each plant species. For example, if two 15 m long sampling lines are placed in the field and the lengths intercepted by the species are: **Line 1:** A - 1.9 m, 1.5 m; B - 0.6 m; and C - 0.2 m, 0.5 m; **Line 2:** A - 2.4 m; B - 2.3 m; C - 0.5 m; and D - 0.1 m; then,

$$\text{Percent cover of A} = \frac{1.9 + 1.5 + 2.4}{30}$$

Likewise, cover of B = 9.7%, C = 4%, and D = 0.3%.

$$\text{Relative cover of A} = \frac{1.9 + 1.5 + 2.4}{1.9 + 1.5 + 0.6 + 0.2 + 0.5 + 2.4 + 0.5 + 0.1} 100 = 58\%$$

Likewise, relative cover of B = 29%, C = 12%, and D = 1%.

(d) Basal area: Basal area refers to the ground actually penetrated by the stems, and is readily seen when the leaves and stems are clipped at the ground surface. This is measured either at 2.5 cm above ground or on ground level. One hundred stems of each species (in case of herbaceous vegetation) are cut at the ground surface by sharp scissors and the diameters of their cut ends are measured. In the case of a tree, trunk size is expressed as its circumference (cbh) or diameter (dbh) at breast height (1.37 m above ground surface). If the tree has a base that is buttressed or otherwise swollen at breast height (bh), it is measured just above the swelling. The tree diameter is calculated as:

$$\text{dbh} = \frac{\text{cbh}}{\pi}$$

Where, π = 3.14; and the basal area (BA) is calculated as:

$$\text{BA} = \pi r^2 = \frac{\pi(\text{dbh})^2}{4} = \frac{(\text{cbh})^2}{4\pi}$$

where, r = dbh / 2. Dbh and cbh are usually expressed in cm, and BA in cm^2 / plant or m^2 / tree. Basal area of a tree population, the sum of BA of all the individual trees in the population, is given as m^2 / ha.

Importance Value Index (IVI): In order to express the dominance and ecological success of any species, with a single value, the concept of IVI has been developed. The IVI of a species is the sum of its relative dominance (relative cover or relative basal area), relative density, and relative frequency in a stand. These relative values are calculated as:

$$\text{Relative cover} = \frac{\text{total cover of the species}}{\text{total cover of all species}} \times 100$$

$$\text{Relative basal area} = \frac{\text{total basal area of the species}}{\text{total basal area of all species}} \times 100$$

$$\text{Relative density} = \frac{\text{Number of individual of the species}}{\text{Number of individual of all species}} \times 100$$

An example of calculation of IVI is given in Table 6.7. It should be noted that IVI of any species in a community ranges between 0 and 300 as IVI is the sum of three relative parameters.

TABLE 6.7
CALCULATION OF IMPORTANCE VALUE INDEX (IVI) FOR A RATHER OPEN TROPICAL RAINFOREST AT 450 M ELEVATION NEAR HONOLULU, HAWAII (MUELLER-DOMBOIS AND ELLENBERG 1974)

Species	*Relative density*	*Relative cover*	*Relative frequency*	*IVI*
Koa tree (*Acacia koa*)	30.0	78.4	30.8	139.2
Ohia lehua (*Metrosideros collina*)	20.0	13.9	23.1	57.0
Ohia (*M. tremuloides*)	5.0	5.8	7.7	18.5
Guava (*Psidium guajava*)	45.0	1.9	38.5	85.4

The IVI may also be graphically presented in the form of a **phytograph**. One such method is to draw a circle and divide it into four equal segments by drawing two lines at right angles to each other passing through the center. Each of the three radii is divided into 100 parts from the center to the circumference and the fourth into 300. On the 0-100 scales are marked the values of relative frequency on A, relative density on B, relative dominance on C, and the IVI value on 0-300 scale on D. All these points are joined as shown in Fig. 6.5.

6.3.3 *Synthetic characteristics*

Synthetic characteristics are generalizations or abstractions that are derived from data on analytic qualities of several similar stands. The synthetic traits, therefore, are traits of a community rather than of a single stand. These are: presence and constance, fidelity, dominance, and physiognomy and pattern.

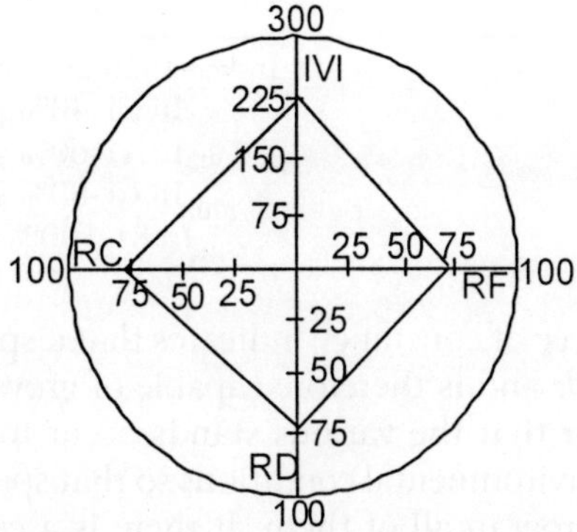

FIGURE 6.5
PHYTOGRAPHIC REPRESENTATION OF IVI OF A SPECIES HAVING 75 RELATIVE COVER, 75 RELATIVE DENSITY, AND 75 RELATIVE FREQUENCY

6.3.3.1 Presence and constancy

Presence is the percentage of all stands which contain a given species. It is calculated by dividing the number of stands in which a given species occurs by the total number of stands and multiplied by 100. For example, if species A occurs in 8 of 10 stands, the species has 80% presence. Constance, in contrast, is based on species encountered in releves. If all the stands being considered have been sampled with plots of the same size, then, when the number of stands containing a species is compared to the total number of stands, the resulting quantity is known as constancy (sometimes called *constance*). Thus, the term "constancy" is employed when equal, measured sample areas are used in each stand, "presence" is used when area of the sampling unit varies from stand to stand, and especially when it is not measured. Generally, presence is higher than constance. Species A may have been present in 8 stands, but in only 6 of the 10 releves, thus having 60% constancy.

Species may be classified into five classes of constancy according to the percentage of stands in which they occur, as follows:

Constancy class	Presence of species
1	In less than 20% of the stands
2	In 21-40% stands
3	In 41-60% stands
4	In 61-80% stands
5	In 81-100% stands

A high degree of constancy indicates that a species has wide *ecological amplitude* and is therefore capable of growing in various microhabitats, or that the various stands occur in sites that are very similar in environmental conditions so that species of narrow amplitude can grow in all of them. If there is a combination of several species with a high constancy percentage the stands belong to one association.

6.3.3.2 Fidelity

Fidelity is defined as exclusiveness or the degree to which species are confined to or do better in one or a few associations. This concept is used to determine the characteristics species of an association as, although species with low fidelity occur in a number of different communities, those with high fidelity in a few or in only one kind of community. Five grades or classes of fidelity are used in the Braun-Blanquet scheme of classification:

I. **Characteristics species**
   Fid. 5 Exclusive species: Species limited almost completely to one community
   Fid. 4 Selective species: Species occurring most frequently in one kind of community, but also, though rarely, in other communities
   Fid. 3 Preferential species: Species occurring more or less abundantly in several kinds of communities but with optimum conditions for abundance and vitality in one certain kind of community.

II. **Companion species**
   Fid. 2 Indifferent species: Species without pronounced affinities for any community

III. **Accidental species**

Fid. 1 Strange species: Species that are rare and accidental intruders from another community or relicts from an earlier stage of succession

6.3.3.3 Dominance

Generally, the dominant species of a community is that overstory species which contributes the most cover or basal area to the community, compared to other overstory species. This definition of dominance is based on physiognomy. If *Shorea robusta* has the highest relative cover in a lowland Terai forest of Nepal with *Shorea robusta*, *Adina cordifolia*, and *Terminalia alata* in the overstory, then *Shorea robusta* is said to be the *dominant* species. If all three species contribute about the same amount of cover, or if the balance shifts from one to the other depending on the stand, then the three species are *codominants*. In a savanna or a semidesert woodland, where tree canopies may contribute only 10-30% cover and understory plants such as grasses or shrubs contribute more, density enters the definition and some grass or shrub species will usually be considered the dominant species.

Another view of dominance is *sociologic dominance* (Kershaw 1973). Sociologic dominants control the reproduction and continued existence of a community, and they may be understory species. For example, the regeneration of ponderosa pine saplings in some ponderosa pine forests is inhibited by root competition for moisture by the understory grass *Festuca arizonica*, but not by another grass *Muhlenbergia montana* (Pearson 1942). The grasses are the sociologic dominants of the ponderosa pine community, even though ponderosa pine is the physiognomic dominant. For this reason, some methods of community description name communities by both overstory and understory species and separate the two species by a slash, for example, the *Pinus ponderosa / Muhlenbergia montana* community, or the *P. ponderosa / Purshia tridentata* community (Daubenmire 1952).

Foresters call any individual tree whose canopy is more than half exposed to full sun a dominant, even though it may not be a member of species which is a physiognomic or sociologic dominant of that community. In this sense, dominant is a synonym for *emergent*. In some other forest studies, dominance is equivalent to trunk basal area; the species with the most basal area per hectare is called the dominant.

The term *aspect dominance* is applied to species which are very noticeable and at first glance appear to dominate a community cover. Careful sampling would reveal, however, that other, less conspicuous species in the same canopy layer contribute more cover and are actual dominants. Aspect dominance is most common in herbaceous communities, such as grasslands or meadows, where all members of one species will flower synchronously and in this way stand out from the rest of the vegetation.

6.3.3.4 Physiognomy and pattern

Physiognomy, the appearance or "look" of a stand, may be considered a synthetic quality because the appearance of vegetation is based on a number of qualitative and quantitative characteristics such as the kinds and dominance of species, life form, population density, cover, height, sociability, stratification, association of species, and colour.

Pattern in vegetation occurs in the form of groups or clumps of individuals, or in any other nonrandom arrangement of plants. Physiognomic contrast between groups, such as shrubs in grasslands, or zones around a lake, accentuates pattern and is readily seen; but slight differences in density, cover, or frequency often require quantitative methods of determination before the pattern is detected. Causes of pattern may be grouped under three headings: (i) morphological, in which the growth of a propagative organ such as a rhizome is very important; (ii) sociological, in which competition and association of species are of great importance; and (iii) physiographic, in which topographic variations in soil moisture, concentration of nutrients, soil texture and structure, and others, are concerned. In the initial colonization of an area the distribution of species may be at random and without noticeable pattern unless some spots are more favourable for growth than others, but contagious aggregation soon appears and results in morphological and physiographic patterns; following this, competition, replacement of species, and association of species lead to sociologic patterns. However, as the climax is approached the pattern becomes less pronounced, but even in the climax the three kinds of pattern usually exists.

7. ECOLOGICAL SUCCESSION

When stripped of its original vegetation by fire, flood, or glaciation, an area of bare ground does not remain devoid of plants and animals. The area is rapidly colonized by a variety of species that subsequently modify one or more environmental factors. This modification of the environment may in turn allow additional species to become established. Succession is a directional cumulative change in the species composition of vegetation at one location over the course of about 1-500 years. If significant changes in species composition for a given area do not occur within such a period, the community is said to be **mature**, or **climax community**. If a community does exhibit some directional, cumulative, nonrandom change in a period of 1-500 years, it is said to be **successional** or **seral community**. Seral communities or species replace one another until a climax community is achieved. The entire progression of seral stages from the first one which occupies bare ground (the **pioneer community**) to the climax community, is called a **succession** or a **sere**. Many communities undergo significant, spectacular change through the seasons of one year. A grassland, for example, may be dominated by annual or perennial dicotyledonous herbs in the spring, when the grasses are just beginning their growth; then, in mid- to late-summer, the taller grasses dominate. In the hot deserts, annual herbs may form a relatively dense ground cover in spring or summer, providing sufficient rain has fallen some months earlier to trigger germination. These seasonal changes and also some long-term changes caused by climatic and evolutionary changes which occur over thousands or millions of years are usually not included in the definition of succession.

7.1 Types of Succession

(a) Primary versus secondary: New land is continuously being exposed to colonization by plants, such as when volcanic eruptions create new islands or cover a pre-existing landscape with a new surface of ash or lava, when a landslide buries a swath of vegetation beneath coarse rubble, or when an advancing sand dune smothers a coastal forest. New land is also exposed by mountain building, by the filling of ponds, by the changing of river courses, and by the expansion of tropical deltas under the influence of silt-trapping mangroves.

The establishment of plants on land not previously vegetated is called **primary succession**. If the pioneer community becomes established on a wet substrate, such as the edge of a gradually filling bog, then the invasion is a **hydrarch** primary succession. If the pioneer community becomes established on a dry substrate, such as exposed granite, then the invasion is a **xerarch** primary succession. In either case, the direction of succession is usually towards the most mesic site and the most mesophytic community possible, given the limitations of regional climate, topography, and the soil parent material.

Secondary succession is the invasion of land that has been previously vegetated, the pre-existing vegetation having been destroyed by natural or human disturbances, such as *wind-throw* (uprooting or breaking of vegetation by high winds), fire, logging, or cultivation. However, the barren surface is not as severe as the surface in primary succession because much of the soil remains (although it is sometimes depleted of nitrogen or other nutrients), and many plant disseminules (seeds, bits of rhizomes, etc.) are already present in the soil. Consequently, secondary successions often proceed five to ten times faster than primary successions (Mueller-Dombois and Ellenberg 1974). Secondary succession on abandoned cropland is called **old-field succession**.

(b) Autogenic versus allogenic: One of the driving forces behind succession is the effect plants may have on their habitat. Plants cast shade, add to the litter, dampen temperature oscillations, increase humidity, and their roots change the soil structure and chemistry. Some of these modifications put the seedlings of overstory species that are less well-adapted to shade at a competitive disadvantage with the seedlings of other species better adapted to the moderated conditions. With time, then, a different association of species will come to dominate the site. Tansley (1935) called succession that is driven in this way **autogenic (biotic) succession**. Both the environment and the community change, and this metamorphosis is due to the activities of the organisms themselves.

Allogenic succession, on the other hand, is due to major environmental changes beyond the control of the indigenous organisms. For example, a slight drying trend over the past 100 years in the southwestern United States may be partly responsible for succession from desert grassland to desert scrub (Barbour *et al.* 1980).

(c) Progressive versus retrogressive: Succession often leads to communities with greater and greater complexity and biomass and to habitats which are progressively more and more mesic (moist). This type of succession is called **progressive succession**.

Retrogressive succession leads in the opposite direction, toward simpler, more depauperate communities (communities with fewer species). Some retrogressive successions are allogenic. For example, the introduction of cattle or fire may result in degenerated rangeland. Other retrogressive successions are autogenic. Drury (1956) noted that succession in Alaskan flood plains may at first be progressive, leading from a sedge meadow to a white spruce forest with low shrubs of cranberry and blueberry. The dense shade, however, encourages the growth of a dense moss carpet and the encroachment of a shallow *permafrost* (frozen soil and water) layer. As the soil moisture rises, sphagnum moss invades, white spruce is replaced by black spruce, and ultimately retrogression to a sedge meadow can result (Fig. 7.1).

(d) Cyclic versus directional: Directional succession is characterized by an accumulation of changes that leads to community-wide changes. Even in a climax community, however, there continue to be **cyclic** successional changes on a very local scale. These changes occur because the life span of overstory plants is finite, and their disappearance from the canopy may open the site to an invasion by new species. In some climax communities, the juvenile forms of overstory plants are well adapted to life beneath the parent and when the parent dies they will replace it in overstory; in such cases, there will be no local, cyclic succession. In other communities, however, the overstory may inhibit the growth of juveniles beneath it-juveniles of its own kind or juveniles of any species-and in such cases, local, cyclic succession will occur when the overstory plant dies. Shrub-dominated communities often exhibit such cyclic succession.

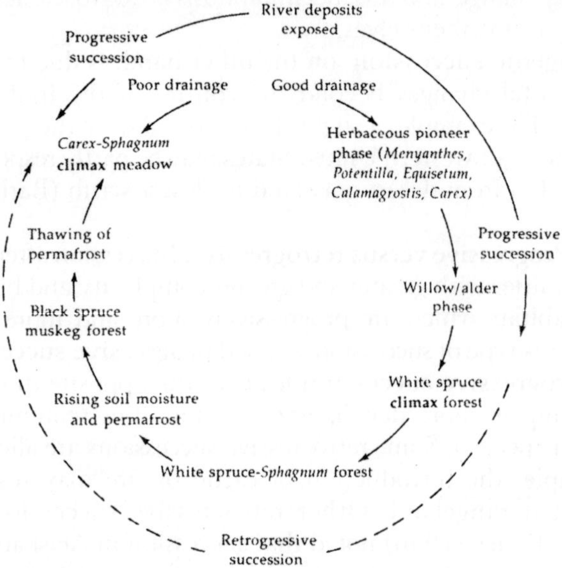

FIGURE 7.1
PROGESSIVE AND RETROGRESSIVE SUCCESSION ON ALASKAN FLOOD PLAINS (DRURY 1956).

(e) Chronosequence versus toposequence: Typically, many plant communities coexist in a complex mosaic pattern. That is, one climax community does not cover an entire region. Sometimes the mosaic reflects a periodic, local disturbance, such as fire, or it might reflect the progressive exposure of new land, as behind a retreating glacier. In such cases, different patches of vegetation represent different stages of recovery (seral stages) from fire, wind throw or other disturbances to the matrix type. The mosaic expresses a successional relationship and is called a **chronosequence**.

In other cases, the mosaic reflects topographic differences, such as south-facing versus north-facing slopes, basins with poor drainage and fine-textured soil versus upland slopes with good drainage and coarser soil, or different distances from a stress such as salt spray. In such cases, the communities within the mosaic do not bear a successional relationship to one another; they constitute a **toposequence**.

Each community in a toposequence may, in fact, be a climax community. A climax community on level mesic ground which reflects the regional climate is called a **climatic climax community**. Another community may be on atypical parent material or in a poorly drained depression so that the soil will not support the climatic climax; this is an **edaphic climax community**. Still other climax communities occur on north- or south- facing slopes where a unique microclimate will not support the regional (climatic) climax. Each region then is characterized by a mosaic of climax types; each region has a **polyclimax** landscape.

7.1.1 Hydrarch primary succession

Hydrarch primary succession or **hydrosere** begins in aquatic environments such as in lakes. The early changes in plant succession are allogenic and very slow as deep lake has to become shallow enough for rooted aquatics to grow. This may take hundreds of years depending on lake depth, erosion and drainage patterns in the catchment area.

The open water of lakes may be colonized by free floating aquatic plants like the duckweeds (*Lemna* spp.) or the fern *Azolla*. At the edges of the lake where the light penetrates to the floor submerged aquatics, with their roots in the sediment, grow. These often include genera such as *Ceratophyllum*, *Hydrilla, Potamogeton, Myriophyllum, Vallisneria*, etc. Once plants like this root in a lake, the biomass is increased and productivity increases too. More organic matter falls to the lake floor creating a fine mud.

Once submerged aquatics colonize a lake, the successional changes become more rapid and are mainly autogenic as organic matter accumulates. Inorganic sediment is still entering the lake and this is trapped more quickly by the net of plant roots and rhizomes growing on the lake floor. Areas of the lake soon become sufficiently shallow for floating-leafed aquatics to grow. These plants are rooted in the mud, but some or all of their leaves float on the water surface. These include *Nelumbo nucifera, Nymphaea* spp., *Nymphoides* spp., *Trapa quadrispinosa*, etc. The large surface leaves cast deep shade which reduces the submerged species. The plants decay to from organic muds which, as the lake becomes shallower, are invaded by emergent vegetation, such as *Phragmites* and *Typha*, to form *reed swamp*. These plants have creeping rhizomes which knit the mud together and produce large quantities of leaf litter as they die back each autumn. This litter is resistant to decomposition and reed peet rapidly builds up, accelerating the autogenic change. If the peat becomes raised above the water surface at dry times of the year, the community changes to **marsh meadow stage** (Fig. 7.2) dominated by sedges (*Carex, Cyperus, Eleocharis, Fimbristylis, Juncus, Schoenoplectus*, etc.). Sedge peat accumulates above the water level and so is no longer totally waterlogged. It can then be invaded by trees, called **wet woodland (fen carr) stage**. Some notable tree species of wet woodland include *Ficus semicordata, Salix tetrasperma, Terminalia alata*, etc. The development of wet woodland is often seen as the last stage in the hydrosere before the area dries out completely and typical regional climax community (formation) develops. Thus, the sequence of communities in a hydrosere may be summarized as: pioneer stage → submerged stage → floating-leafed stage → reed swamp stage → marsh meadow stage → woodland stage → climatic climax community.

FIGURE 7.2
MARSH MEADOW STAGE OF A HYDRARCH PRIMARY SUCCESSION IN AN AGED LAKE AT BIRATNAGAR, NEPAL

7.1.2 Xerarch primary succession

Xerarch primary succession or **xerosere** begins on any kind of dry substrate like rock (**lithosere**), sand (**psammosere**), or physiologically dry areas (**halosere**).

Bare rocky areas are often invaded first by animals such as spiders which can hide in the cracks between boulders or stones. Such animals are carnivores or scavengers and they live by feeding on insects which have been blown in or flown in by mistake. There is as yet no plant matter for herbivores to survive. The process of succession usually starts when autotrophs begin to live in the area. Both the structure and the nutrient content of exposed ground will affect which species become the first colonizers. If the ground consists of solid rock or very large boulders, there is no space for rooting plants and the first colonizers are usually lichens and mosses. These can cling directly to the rock surfaces. Slowly, the presence of moss or lichens promotes the build-up of soil in shallow depressions or cracks in the rocks as wind-blown particles are caught and added to by direct weathering of the rock surface. This creates small pockets where vascular plants can grow, but the whole process is extremely slow and may take hundreds or thousands of years before the pioneers are succeeded by another community.

If finer particles, such as silts or clays, are present in quantity in the substrate, then the initial pattern of colonization is very different from that on rock. Such areas already have more nutrients and a more balanced water regime than bare rock. The substrate can be penetrated by roots, so that vascular plants can colonize immediately. Such sites tend to be invaded by dictotyledonous herbs, or, if they are quite nutrient rich, by grasses (Grubb 1986). Even clay and silt substrates are poorly structured and nutrient poor compared with mature soils, so it is not surprising that the second community in these sites is rich in plants with nitrogen-fixing bacteria held in root nodules. Such communities include some horsetails (*Equisetum* spp.), legumes like *Astragalus*, *Lupinus* and *Trifolium*, and, in areas where trees can grow, alders (*Alnus nepalensis* and others of the birch family).

Later in the xerosere, the grass and herb mixture is invaded by scrub. Early invasion of scrub is slow, but once a few bushes have become established, they form an ideal site for roosting birds. Once birds are common, bush invasion speeds up as the birds deposit many seeds on the site in their droppings which are also rich in nutrients. Colonizing trees such as the wind-dispersed birches (*Betula* spp.) and bird-dispersed oaks (*Quercus* spp.) may also establish at this stage as in the high altitude Himalayas. Birch and oak are both species which require high light intensities for their seedlings to establish. They are pioneer trees. The birches are usually short lived and fail to regenerate once the oaks mature and cast too much shade. Once woodland is established it may be invaded by more shade-tolerant species such as limes (*Tillaea* spp.), ashes (*Fraxinus* spp.) and hornbeams (*Carpinus* spp.).

7.2 Methods of Documenting Succession

The most direct, unambiguous way to document succession is to make repeated observations of the same area over time. Permanent quadrats or exclosures can be established, and measurements of plant cover, biomass, density, diversity, demography, or the like can be taken every year, every decade, or over longer periods of time. In a successional study at Biratnagar (Nepal), a heavily grazed pasture spreaded over 3.5 ha land area was divided into three segments of nearly equal sizes. The occasionally and moderately grazed segments were subjected to grazing/mowing at fortnightly and weekly intervals, respectively; whereas there was no restriction on cattle grazing in the heavily grazed segment. After three years of commencement of the study, the species composition, importance value index (IVI), and oven dry aboveground biomass of all the three segments were determined (Table 7.1). The tufted grass *Imperata cylindrica* (IVI 107) dominated the occasionally grazed segment, whereas the prostrate grass *Chrysopogon aciculatus* dominated moderately grazed and heavily grazed segments with IVI 51 and 103, respectively. The aboveground biomass was highest in the occasionally grazed and least in heavily grazed segments.

In forested areas, a comparison of sapling density to overstory density by species may reveal successional patterns. Figure 7.3 shows that an oak-beech forest near Washington, D.C., is unlikely to remain stable, for there are no saplings of the most common overstory oak, and very few saplings of the other oaks; instead, beech and maples dominate the sapling layer. One could extrapolate from this information, collected at one moment in time, and predict that the oak-beech forest is seral to a beech-maple forest, but it is important to realize that two assumptions are being made. One assumption is that the sapling population will not significantly change from year to year or decade to decade. However, it is not always safe to assume that the pattern of reproduction for all tree species through time is same. It is possible that some years will yield a large seed crop for one species and many seedlings and saplings for several years thereafter, while another species has a complementary cycle. Another assumption is that the mortality rate of saplings is uniform from species to species. This assumption is not always appropriate. For example, the spruce-fir forests of the Applachians are dominated by spruce in the overstory but by fir in the understory. Nevertheless, the overstory balance does not shift to fir, because fir seedlings are more susceptible to damping-off disease and grow less well dense in shade than spruce (Oosting and Billiings 1951).

Most successional schemes have been intuitively determined from observations on nearby plots of different successional ages, rather than from observations on one plot over time. The objective is to find a series of plots which have been exposed to primary successon or disturbed and opened to secondary succession at different, known times. Obviously, the time since succession began varies from each plot, but one must assume or demonstrate that all other factors, such as slope, aspect, parent material, and macroclimate, are uniform. In many other cases, however, the sites of different ages are randomly scattered throughout a region, and their age sequence must be determined from land use records (Oosting 1942).

In primary hydrarch succession, different plant types produce different sediments beneath them. Thus, a record of what has grown in the lake is fixed in its infill deposits. Old lakes which have completely filled in, can be cored to give a long tube of sediment which goes right down to the bottom of the lake (Fig 7.4). Pollen analysis of such a sediment core reveals the successional changes which have occurred in the lake during its history (Walker 1970). Such a permanent record of successional changes cannot be obtained from terrestrial seres as the top surface of the soil is changing and being reworked rather than building up in a sequence.

TABLE 7.1
EFFECTS OF VARIOUS INTENSITIES OF GRAZING ON IMPORTANCE VALUE INDEX AND ABOVEGROUND BIOMASS IN A LOWLAND PASTURE OF NEPAL (JHA 2003)

	Occasionally grazed segment	*Moderately grazed segment*	*Heavily grazed segment*
A. Importance Value Index (IVI)			
Alysicarpus vaginalis	26	33	7
Axonopus compressus	35	49	39
Brachiaria distachya	7	20	17
Chrysopogan aciculatus	28	51	103
Cynodon dactylon	32	12	13
Croton bonplandianus	—	—	10
Cyperus rotundus	—	7	12
Desmodium triflorum	24	27	17
Eleusine indica	—	6	—
Evolvolus nummularius	12	18	20
Hemigraphis hirta	—	6	14
Imperata cylindrica	107	32	18
Indigofera linifolia	—	9	—
Kyllinga brevifolia	5	10	15
Mimosa pudica	12	—	—
Paspalum scrobiculatum	—	10	—
Rungia pectinata	—	10	15
Saccharum spontaneum	12	—	—
	300	300	300
B. Aboveground biomass (g/m²)	1197	801	568

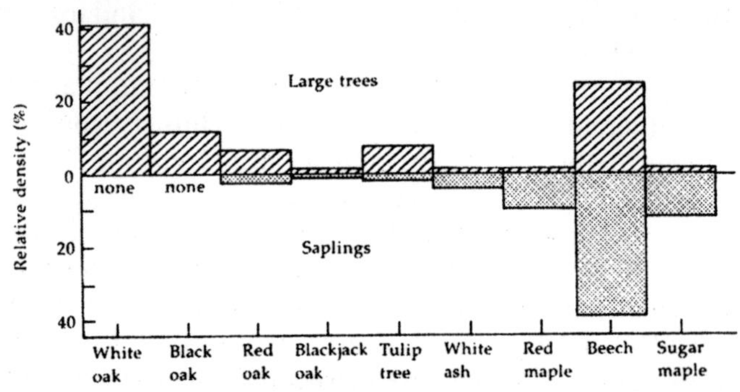

FIGURE 7.3
RELATIVE DENSITY OF OVERSTORY TREES AND SAPLINGS, BY SPECIES, FOR AN OAK FOREST NEAR WASHINGTON D.C. THE

SAPLING STRATUM IS DOMINATED BY BEECH AND MAPLE, POSSIBLY INDICATING THAT THE OAK FOREST (LARGE TREES) IS SERAL TO A BEECH-MAPLE FOREST (DIX 1957).

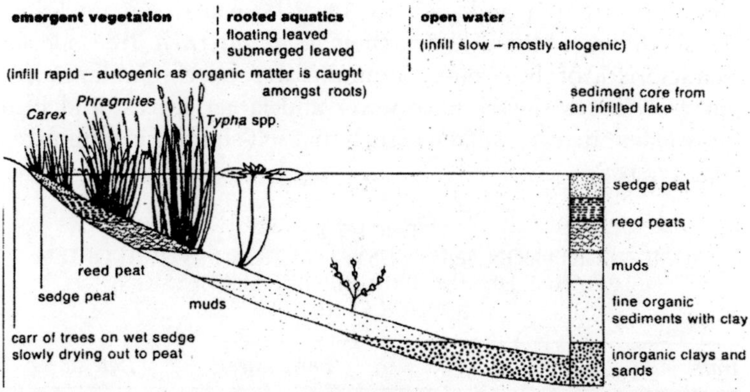

FIGURE 7.4
THE VEGETATION ZONES AND SEDIMENT TYPES USUALLY FOUND IN AN INFILLING LAKE. IF AN OLD, INFILLED LAKE IS CORED TO TAKE OUT A COLUMN OF SEDIMENT (ILLUSTRATED ON THE RIGHT), A RECORD OF THE LAKE'S HISTORY CAN BE DISCOVERED

7.3 General Trends During Succession

Changes that occur in major structural and functional characteristics of a developing community and ecosystem during progressive, directional succession are listed in Table 7.2. The status of each trait is shown for early and late stages of succession (not for pioneer and climax stages necessarily, for some trends peak at intermediate seral stages).

7.3.1 Vegetation and site quality

Biomass increases during succession. Plant cover, the density of foliage above the ground (the leaf area index), and the height of the plants also increase. If succession leads from herbs or shrubs to trees, the percentage of total biomass that is below the ground may decline, but the total biomass still increases.

Physiognomy increases in complexity because the variety of growth forms increases as succession proceeds. If successional stages involve trees, the structure and leaf orientation of those trees may change (Horn 1971, 1975). Tree species characteristic of early successional stages tend to be tall, thin, conical, and capable of rapid growth; their leaves are small, numerous, randomly oriented, and are borne in such a way that many leaves are shaded by others above them (*multilayered*). Tree species characteristic of the climax community may have a similar profile but grow more slowly, have fewer and larger leaves, and bear those leaves in a *planar fashion* such that self-shading is minimized (*monolayered*).

TABLE 7.2
SOME VEGETATION AND ECOSYSTEM TRAITS WHICH OFTEN CHANGE DURING PROGRESSIVE SUCCESSION
(BARBOUR *ET AL.* 1980)

Trait	*Early stages*	*Late stages*
Biomass	small	large
Physiognomy	simple	complex
Leaf orientation	multilayered	monolayered
Major site of nutrient storage	soil	biomass
Role of detritus	minor	important
Mineral cycles	open (leaky), rapid transfer	closed (tight), slow transfer
Net primary production	high	low
Site quality	extreme	mesic
Importance of the macroenvironment	great	moderated and dampened; less
Stability (absence or slowness of change)	low	high
Plant species diversity	low	high
Species life history character	r	K
Propagule dispersal vector	wind	animals
Propagule longevity	long	short

The major site of nutrient storage in the ecosystem shifts from soil to plant biomass (perennial roots, storage organs and tree trunks). The role of detritus (leaves, twigs, and other litter) in nutrient cycles thus becomes more important during succession, because soil nutrients have been depleted and are being stored for long periods of time in plant biomass. Detritus carries at least a fraction of the nutrient pool back to the soil where it can be trapped to satisfy new plant growth demands the next season.

The speed with which nutrients cycle from soil to plant and back again slows down during succession because many nutrients are stored in long-lived but inert parts of plants. Such storage does, however, prevent erosional losses of nutrients out of the ecosystem, so we can hypothesize that the cycling of nutrients is more efficient later in succession, even though it occurs very slowly.

Net primary production (photosynthesis minus respiration) may decline with succession for a variety of reasons: (i) there is a great deal of supporting but nonphotosynthetic tissue whose maintenance (respiration) reduces net production; (ii) nutrients may be limiting because of their storage in inert tissue; (iii) many plants in the overstory may be senescent, with lower photosynthetic rates than young plants; and (iv) the leaf orientation of climax trees may be inappropriate for high photosynthetic rates (Horn 1974).

The environment becomes more mesic during succession, and the effect of the macroclimate is dampened. As the canopy closes, diurnal fluctuations in temperature and humidity are moderated. Increased humus, increased soil depth, and a finer soil texture in later successional stages leads to greater soil moisture retention and a buffering of seasonal changes in precipitation.

7.3.2 Stability and diversity

If stability is defined as equivalent to lack of change, then stability increases with succession: the climax species are longer-lived, and any changes occurring in the undisturbed climax community are in the form of random, minor fluctuations around some long-term mean (Table 7.2). If stability is defined as resistance to minor changes in the macroenvironment, then stability also increases with succession because the plant cover dampens fluctuations and extremes in the macroenvironment. If stability is defined as the ability to return rapidly to a balance point (**homeostasis**) following a major recurring disturbance such as a fire or windstorm, then preclimax communities are the most stable while climax communities are the least stable, requiring hundreds of years to return.

Plant species diversity increases throughout early succession, but it decreases in the temperate zone in late succession as the canopy closes and a few species become major dominants. Thus, in the temperate zone, a periodic local disturbance which sets succession back to earlier seral stages is required to maintain maximum diversity (Loucks 1970). There are exceptions to this trend of increasing diversity. Old-field succession in Oklahoma, for example, shows a steady decline in species diversity from pioneer weeds through climax prairie (Perino and Risser 1972). Diversity is nearly halved, even though species richness (the total number of species) rises throughout the sere.

7.3.3 Autecology

Pioneer species are typically r-selected, with rapid growth rates, short life spans, relatively large energy allocations to sexual reproduction, and high photosynthetic rates. Climax species are typically K-selected, with slow growth rates, long life spans, small energy allocation to reproduction in any one year, lower photosynthetic rates, and heavy, animal dispersed seeds. However, r-selected species can still exist in climax communities: they may occupy sites beneath openings in the overstory and their seeds can lie dormant in the soil for long periods of time until such openings occur. Or, they may have light, abundant seeds which are imprecisely dispersed by wind.

Also, it has been hypothesized that, in general, more plant-animal, plant-plant, and plant-microbe interactions (sometimes called *information*) occur in later succesional stages than in early ones.

7.4 Driving Forces Of Succession

7.4.1 Relay floristic

Clements (1916, 1936) wrote of succession as a six step process:

(i) **Nudation,** the exposure of a new surface in primary succession or the clearing away of previous vegetation in secondary succession.

(ii) **Migration of seeds,** spores, or vegetative propagales from adjacent areas, though in secondary succession many of these are already present in the soil.

(iii) **Ecesis,** the germination, early growth, and establishment of plants.

(iv) **Competition** between the established plants.

(v) **Reaction,** the autogenic effects of plants on the habitat.

(vi) **Stabilization,** the climax.

Clements visualized succession as composed of several discrete seral communities. The associated species of each community undergo steps II-V together, changing the habitat in such a way that new species (associates of the next seral community) are at a competitive advantage and thus supplant the previous community. The process is repeated until a climax community develops which can propagate itself continuously and prevent the establishment of other taxa.

This view of succession is called **relay floristics** (Egler 1954) and it is diagrammed in Fig. 7.5. Each seral community relays the site to the next community. The driving force behind succession, then, is the reaction of the site to the plants living on it.

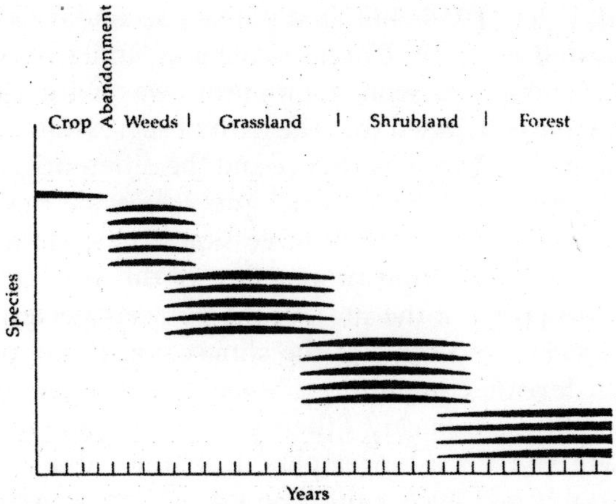

FIGURE 7.5
DIAGRAMMATIC SUMMARY OF THE RELAY FLORISTICS MODEL OF SUCCESSION. THE THICKER THE LINE, THE MORE IMPORTANT THE SPECIES AT A GIVEN TIME (EGLER 1954)

7.4.2 Chance and initial floristic composition (IFC)

If the relay floristics model of succession is entirely correct, then we should expect to find groups of species appearing and disappearing together during the course of succession. We should also have to agree with Clements that the course of succession is predictable for each seral stage relies upon the previous one. Some recent work on primary and secondary succession shows that these assumptions are not always correct.

In England, Walker (1970) took corings of sediment from the edges of 66 lakes which had been undergoing hydrarch succession for hundreds of years. Sediment had been laid down in the order of the succession that had actually occurred, with the deepest sediments harbouring the oldest plants. Because of the slow rate of decomposition, plant remains could be identified throughout the cores. Walker examined the sequence of sediments in each core to see if the sediments corresponded to the traditional sequence of succession. His results showed that there was considerable variability in the progression of seral communities. In fact, the traditional sere was never followed.

Frank Egler (1954) initiated plots of secondary succession on his forested estate in Connecticut, and similarly concluded that the progression of seral communities was neither fixed nor predictable. He concluded that the path of succession (at least in the early stages) is driven by chance and the differential longevity of plants. Chance determines which propagules are in the soil or soon reach the site at the time succession begins; he referred to this collection of starting propagules as the **initial floristic composition (IFC)** of the site. All the pioneer species, many of the seral species, and some of the climax species are present in this initial floristic composition. Some of them germinate and become established quickly, others germinate quickly but grow more slowly and for a longer period of time, and others may become established even later. The larger, longer-lived, slower-growing species ultimately outcompete the smaller pioneer species, and dominance shifts. The site is not relayed from one collection of species to another in this view of secondary succession; rather, there is a sorting out of species one at a time, based mainly on their longevity (Fig. 7.6).

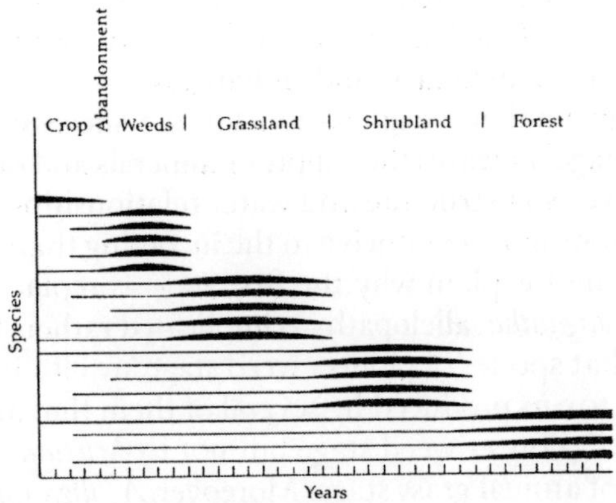

FIGURE 7.6
DIAGRAMMATIC SUMMARY OF THE INITIAL FLORISTIC COMPOSITION MODEL OF SUCCESSION. THE THICKER THE LINE, THE MORE IMPORTANT THE SPECIES AT A GIVEN TIME (EGLER 1954)

It is likely that there is some value in both the relay and the initial floristic composition models of succession, but more research is needed to formulate a realistic, comprehensive model of succession. Two other driving forces that have been proposed are species interactions (especially allelopathy) and natural selection for maximum efficiency in the cycling of energy and nutrients.

7.4.3 Allelopathy

There is evidence that allelopathy plays a role in some seres, but that evidence is so scattered and sometimes contradictory that it is not possible to reach a general conclusion.

Booth (1941) found that succession in abandoned fields in centeral Oklahoma and southeast Kansas included four stages: (i) pioneer weed, (ii) annual grass, (iii) perennial bunch grass, and (iv) true prairie. The weed stage lasted only 2 to 3 years. The annual grass stage lasted from 9 to 13 years and was dominated by prairie threeawn, *Aristida oligantha*. The perennial bunch grass stage was dominated by little bluestem, and this stage was still present 30 years after abandonment. This was the oldest abandoned field studied by Booth so he was not able to ascertain how long a period is required for the return of the true prairie, which in central Oklahoma was dominated by little bluestem, big bluestem, switchgrass, and Indiangrass.

The generalized explanation for the successional changes that each stage increases the supply of minerals and organic matter and improves soil structure and water relationships thus making the conditions more conducive to the incoming than the outgoing species, cannot explain why the first stage is replaced so rapidly by *Aristida oligantha*; allelopathy is implicated rather. The evidence indicates that species of pioneer weed stage are eliminated rapidly because of toxins produced by several of them that are inhibitory to species of pioneer weed stage but not to *Aristida oligantha,* the dominant of annual grass stage. Moreover, *A. oligantha* grows well and reproduces in soil that is still so infertile that it will not support species that come in later in succession. Rice (1984) hypothesized that low nitrogen-requiring early plant invaders of abandoned fields might inhibit growth of the nitrogen-fixing bacteria and blue-green algae. This would give such plants a selective advantage in competition over plants with higher nitrogen requirements and could conceivably slow down plant succession.

7.4.4 Natural selection

Clearly, succession occurs much too quickly for evolution to play a direct role. What some ecologists believe, however, is that climax communities have been subjected to natural selection in much the same way that populations have. They believe that community-level processes such as nutrient cycling and energy flow may act as selective forces at the community level, and that populations replace one another through time, thereby optimizing community functions. In other words, during evolutionary time, species and niches have shifted in response to selective pressure so that solutions to problems of time and energy allocation have been maximized or optimized (Cody 1974). Succession might also be viewed as a sequence of communities each one within the chain being a more efficient system than the previous one.

It is ironic that this recent view of succession and the nature of climax communities- a view which treats communities as intergrated entitles which can be affected by natural selection, much like populations or organisms- brings us full circle back to the organismic view of Clements; although there is no direct evidence that natural selection operates at the community or ecosystem level.

The simplest concept of succession is that succession is a population phenomenon, involving the gradual, inevitable replacement of opportunist species (r-selected species) with equilibrium species (K-selected species). The rate at which the transformation occurs, and the exact pathway it takes, simply reflects the rate at which propagules arrive and the nature of the macroenvironment. In the absence of any disturbance, equilibrium species are always at a competitive advantage, as dominants, over opportunists. This advantage is due to, or enhanced by, such features as longevity, large biomass, moderated microenvironment, and symbiotic or antagonistic interactions with other organisms (Drury and Nisbet 1973). The temporal and spatial frequency of disturbances, however, has always been great enough to maintain opportunist species as well as equilibrium species.

8. ENERGY FLOW IN THE ECOSYSTEM

All life on earth is supported by the stream of energy radiated by sun and flowing into the biosphere. Even the relatively small amount of radiant energy bound in the form of latent chemical energy by the photosynthesis of plants suffices to maintain the biomass and the vital processes of all members of the food chain. By far the larger fraction of radiation absorbed is transformed immediately into heat; part of this fraction is used in the evaporation of water and the rest produces an increase in the temperature on the earth's surface. Radiation is thus the source of energy underlying the distribution of heat, water and organic substances. It creates the prerequisites for an environment adequate to sustain life.

One important concept of energy flow in the ecosystem (**ecological** or **ecosystem energetics**) is that energy always flows in one direction in the ecosystem because the law of thermodynamics govern the behaviour of energy. The **first law of thermodynamics** states that energy may be transformed from one type into another but is never created or destroyed. Light, for example, is a form of energy, for it can be transformed into work, heat, or potential energy of food, depending on the situation, but none of it is destroyed. The **second law of thermodynamics** states that no process involving an energy transformation will spontaneously occur unless there is a degradation of energy from a concentrated form into a dispersed form. Hence, no spontaneous transformation of energy (for example, light) into potential energy (for example, protoplasm) is 100% efficient because some energy is always dispersed into unavailable heat energy. Organisms, ecosystems and the entire biosphere possess the essential thermodynamic characteristic of being able to create and maintain a high state of internal order, or a condition of low **entropy** (a measure of disorder or the amount of unavailable energy in a system).

The three main aspects of ecosystem energetics are: (a) quantity of energy reaching an ecosystem per unit area per unit time, (b) quantity of energy trapped by green plants and converted to a chemical form, and (c) the quantity and path of energy from green plants to organisms of different trophic levels over a period of time in a known area.

8.1 Radiation Within the Atmosphere

The biosphere receives solar radiation in the range of wavelengths between 290 to 3000 nm. Radiation at shorter wavelengths are absorbed by the ozone layer in the upper atmosphere at about 25 km above the earth's surface and atmospheric oxygen, and the long-wavelength cutoff is caused by the air's content of water vapour and carbon dioxide.

About 40-45% of the energy incident from the sun consists of wavelengths between 380-720 nm. This is the region of the spectrum that we perceive as **visible light**. The chloroplast pigments absorb radiation between 380 and 740 nm. Ultraviolet radiation (UV) bounds the photosynthetically active radiation (PhAR; usually considered as lying between 400 and 700 nm) on the short-wavelength side, and infrared (IR), on the long-wavelength side.

Extraterrestrial sunlight reaches the outer limits of the earth's atmosphere (biosphere) at a rate of about 2 cal/cm^2/min (more precisely, the "solar constant" is 1.94 cal/cm^2/min). Of this, however, only an average of 47% reaches the earth's surface. Of the incident sunlight, 25% is reflected by clouds and 9% is scattered by atmospheric particles and returned to outer space. A further 10% is absorbed by clouds and 9% by water vapour. The radiation eventually reaching the ground or the plant cover is composed of direct sunlight (averaging 24% of the incident radiation), diffuse radiation from clouds (cloudlight, 17%), and diffuse radiation from the sky (skylight, 6%). These values are long-term average values for the northern hemisphere. At sea level a horizontal surface at intermediate latitudes at noon receives total (direct plus diffuse) radiation amounting to as much as 1.3

cal/cm²/min. Depending on the latitude of a site, its altitude above sea level, the nature of the terrain and the frequency of clouds, there are large regional and local differences in the supply of radiation (Fig. 8.1). There is a gradual decrease in global radiation over the land areas from latitudes of about 30°N and S toward the poles. This decline is due primarily to the seasonality experienced in temperate regions and poleward. From the equator to 30°N and S, radiation is generally high with limited areas of very high radiation or very low radiation. Radiation over the Amazon Basin is low because of frequent heavy cloud cover and heavy rainfall. On the other hand, the high-pressure regions in the tropics, where clouds are few, receive a greater than average quantity of solar radiation; rather than 47%, an average of 70% of the incident radiation penetrates the clear envelope of air above the dry regions. Moreover, at greater altitudes, owing to the shorter optical path of the rays and the lesser degree of air turbidity, more radiation (the larger fraction being UV light) reaches the ground than in lower-lying places. Also, in hilly and mountainous country, south-facing slopes receive more and north-facing slopes receive much less solar radiation than do horizontal surfaces.

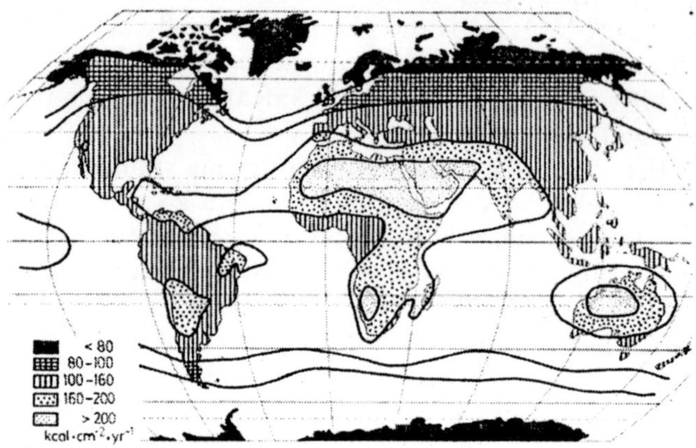

FIGURE 8.1
ANNUAL AVERAGE OF SOLAR RADIATION (300-2200NM) REACHING THE SURFACE OF THE EARTH (GEIGER 1965).

Some of the radiation striking a leaf is reflected at the surface, some is absorbed so as to be physiologically effective, and the remainder is transmitted through the leaf. The degree of reflection, absorption and transmission in leaves depends on the wavelength of radiation (Fig. 8.2).

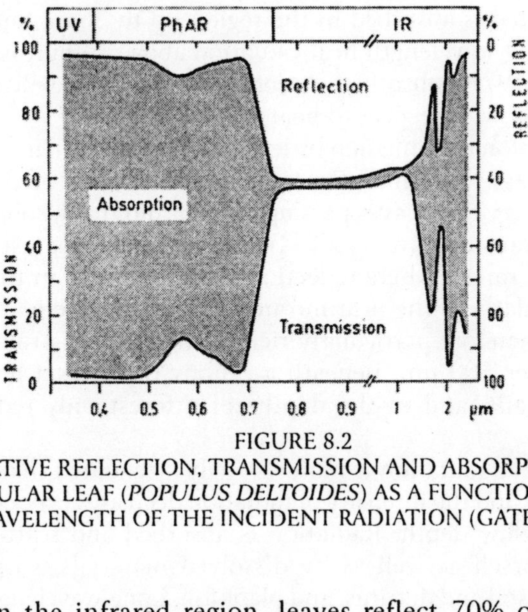

FIGURE 8.2
RELATIVE REFLECTION, TRANSMISSION AND ABSORPTION OF A POPULAR LEAF (*POPULUS DELTOIDES*) AS A FUNCTION OF THE WAVELENGTH OF THE INCIDENT RADIATION (GATES 1965).

In the infrared region, leaves reflect 70% of radiation incident perpendicularly, whereas in the PhAR range an average of only 6-12% is reflected. Green light is more strongly reflected (10-20%), orange and red light less so (3-10%). Little ultraviolet radiation is reflected; as a rule, in the UV range, leaves reflect no more than 3%. Certain flowers, however, display marked UV reflection that can be detected by insects and serves as an attractive target. Also, the capacity to reflect light depends upon the nature of the leaf surface; for example, a dense covering of hairs can increase reflection by a factor of two to three.

Radiation penetrating a leaf is to a great extent absorbed. Most of the UV is retained by the epidermis; only 2-5% of the UV radiation enters the deeper levels of the leaf. Thus the epidermis is an effective UV filter, protecting the parenchyma in which photosynthesis occurs. PhAR absorption is determined by the chloroplast pigments. Correspondingly, the spectral absorption curves of leaves show maxima wherever the absorption maxima of chlorophylls and carotenoids occur. About 70% of the photosynthetically utilizable radiation entering the mesophyll is absorbed by the chloroplasts. In its passage through the leaf, radiation is progressively attenuated, so that the amount captured by successive cell layers falls off approximately exponentially. Not much infrared is absorbed in the region up to 2 μm, but in the range of long-wavelength heat radiation above 7 μm it is almost completely (97%) absorbed. Accordingly, the plant behaves like a black body with respect to heat radiation.

Radiation transmission by leaves depends on their structure and thickness. Mesomorphic leaves pass 10-20% of the sun's radiation, very thin leaves pass up to 40% and thick, solid leaves may not transmit any at all. Transmission is greatest at the wavelength ranges where reflection is great-that is, in the green and particularly in the near infrared. Radiation filtered through foliage is therefore particularly rich in wavelengths around 500 nm and over 800 nm. Beneath a canopy of leaves a red-green shade prevails, and in the depths of a forest only red shade remains.

In water, radiation is more strongly attenuated than in the atmosphere and the intensity of radiation decreases exponentially with increasing depth. Radiation is absorbed and scattered by the water itself as well as by dissolved materials, suspended particles of soil and detritus, and plankton. Long-wavelength heat radiation is absorbed in the upper few millimeters, and infrared radiation in the uppermost centimeters. UV penetrates the top decimeters or meters. Photosynthetically active radiation (particularly that at wavelengths in the region of 500 nm) reaches greater depths, where blue-green twilight predominates in the ocean and yellow-green, in lakes. In turbid, flowing waters the light can decrease to 7% - a value comparable to that beneath the crowns of the trees in a spruce forest - at depths of little more than 50 cm. In clear lakes 1% of the incident PhAR reaches depths between 5 and 10 m, so that vascular plants can exist at 5 m and sessile algae, as deep as 20-30 m. The layer of water above the limit of existence of autotrophic plants is called the **euphotic zone.** In the open ocean the illuminated euphotic zone is deeper than in lakes; in the Mediterranean near the coast 1% of the radiation penetrates to 60 m, and in the clear water of the oceans the corresponding depth is as great as 140 m.

8.2 Concept of Productivity

Energy enters the ecosystem in the form of light and is fixed into chemical energy by photosynthesis. The total amount of energy fixed by photosynthesis per unit time is referred to as **gross primary productivity (GPP)**. Not all energy fixed by photosynthesis is converted to biomass; a significant part is released by cellular respiration to supply energy for plant metabolic activities. Gross photosynthesis minus respiration is equal to **net primary productivity (NPP)**, which is the energy that is stored in plant tissues. The dry weight of plant material present at any point in time is referred to as **standing crop biomass (phytomass)**. **Net community productivity (NCP)** is the rate of storage of organic matter not used by heterotrophs (that is, net primary production minus heterotropic consumption) during the period under consideration, usually the growing season or a year. Finally, the rates of energy storage at consumer levels are referred to as **secondary productivities**.

It is common practice to compare the efficiency by which a species or a stand transmits energy from one form or physical state to another. Efficiency values are the ratio of relative energy flow between various pathways of transfer within a community (Odum 1971). **Exploitation efficiency** is related to the ability of plants to intercept light (Table 8.1). Characteristics important in this regard include latitude, topographic location, leaf area index (LAI), and leaf orientation. **Assimilation efficiency** refers to the ability of plants to convert absorbed radiation into photosynthate. Factors modifying assimilation efficiency are those governing the photosynthetic process, such as resistance to CO_2 assimilation, water and light availability, evaporative demands of the atmosphere, and temperature. **Net production efficiency** is a measure of the capacity to convert photosynthate into growth and reproduction biomass rather than utilizing it for maintenance respiration. The amount of energy used for maintenance depends on such factors as temperature (because of the direct effect on rates of chemical reactions) and on the amount of nonphotosynthetic biomass that must be supported.

TABLE 8.1
FORMULAE DEFINING COMMON EFFICIENCY VALUES AS APPLIED TO PRIMARY PRODUCERS (BARBOUR *ET AL.* 1980)

$$\text{Exploitation efficiency (\%)} = \frac{\text{Gross primary productivity}}{\text{solar radiation}} \times 100$$

$$\text{Assimilation efficiency (\%)} = \frac{\text{Gross primary productivity}}{\text{Absorbed radiation}} \times 100$$

$$\text{Net production efficiency (\%)} = \frac{\text{Net primary productivity}}{\text{Gross primary productivity}} \times 100$$

8.3 Methods of Measuring Productivity

Some of the methods used to measure productivity include the harvest method, dimension analysis, bomb calorimetry, oxygen measurement, carbon dioxide method, disappearance of raw materials, leaf area index (LAI), radioisotope method, and chlorophyll estimation method.

8.3.1 The harvest method

Net primary productivity (NPP) is frequently measured by calculating the change in biomass through time as equal to:

$$NPP= (W_{t+1} - W_t) +D+H$$

where, $W_{t+1} - W_t$ is the difference in standing crop biomass between two harvest times, D is the biomass lost to decomposition, and H is the biomass consumed by herbivores during the period between harvests. The productivity is expressed as $g/m^2/yr$.

By the harvest method aboveground biomass can be measured with little error in herbaceous vegetation by replicate samples harvested randomly from a grid. This technique is most effective with annual vegetation where little biomass is lost to decomposition during the growing season. If herbivore activity is significant, comparisons between replicate samples taken inside and outside herbivore exclosures are often employed.

Based on Singh and Yadava's (1974) method of the summation of positive increments in biomass on successive sampling dates the NPP of a seral grassland is worked out in Table 8.2. The **turnover rate** or the rate of rebuilding of organic matter with respect to total biomass may also be estimated through the harvest method. In Table 8.2 turnover rate has been calculated following the method of Dahlman and Kucera (1965) as:

$$T = \frac{A}{B} \times 100$$

where, T= turnover rate, A= annual increment or NPP, B= maximum biomass; and turnover time as: $\frac{1}{T} \times 12$

TABLE 8.2
MONTHLY VARIATIONS IN BIOMASS, PRODUCTIVITY, TURNOVER RATE AND TURNOVER TIME IN A ROUND THE YEAR GRAZED SERAL GRASSLAND AT BIRATNAGAR (JHA 2003)

	Above-ground	*Under-ground*	*Litter*
1. Biomass (g/m²)			
December	555.5	111.1	217.5
January	491.7	103.2	173.2
February	470.0	103.4	144.3
March	308.6	109.4	123.9
April	237.3	99.3	144.0
May	149.6	93.2	67.3
June	277.8	87.8	88.6
July	498.4	124.6	85.2
August	491.4	103.2	78.0
September	575.6	128.2	146.2
October	341.9	96.9	140.0
November	491.6	98.2	265.6
2. Productivity (Annual increment)	June (128.2)+ Jul (220.6) + Sep (84.2) + Nov (149.7) = 582.7 g/m²/yr	Feb (0.2) + Mar (6.0) + Jul (36.8) + Sep (25) + Nov (1.3) = 69.3 g/m²/yr	Apr (20.1) + Jun (21.3) + Sep (68.2) + Nov (125.6) = 235.2 g/m²/yr
3. Maximum biomass	575.6 g/m²	128.2 g/m²	265.2 g/m²
4. Turnover rate	101%	54%	88%
5. Turnover time	12 months	22 months	14 months

8.3.2 Dimension analysis

It is an alternate way of estimating productivity in those situations where the volume of individual plants is very large or regrowth is so slow that extensive damage may be caused by complete harvest of sample plots. The technique is based upon the assumption that some easily measured parameter such as plant height, diameter at breast height (dbh), or plant volume can be correlated with standing crop. A few individuals need to be harvested to determine the slope of a regression line which may then be used to predict biomass from the easily measured parameter. For example, satisfactory predictability of standing crop has been obtained from basal diameter, dbh, or diameter^2 x height using the regression model proposed by Kittredge (1944):

$$\log e\ Y = a + b \log e\ X$$

where, Y is the estimate of standing crop, X is the measured parameter, a is a point at which the regression line crosses the Y axis, and b is the slope of the regression line. The values of regression constants are determined in a preliminary harvest.

8.3.3 Bomb calorimetry

A bomb calorimeter (Fig. 8.3) consists of a thick walled stainless steel cylindrical pot known as the "bomb". It has an air tight cap fitted with the help of screws and has four main connections. On the two sides of the cap are outlet and inlet valve devices for filling oxygen gas and its release. Through the other two points electricity passes. Inside the bomb a small crucible is placed in which dried plant material is kept. An ignition wire passes through the material and its ends are connected to the two electric ignition terminals. This bomb is kept in a thin-walled pot called kettle which is removable. It is filled with water sufficient to cover the bomb. This again is put in a bigger double walled plastic container. The container cap has holes through which a Beckmann thermometer passes below in the kettle water. An electric operated stirrer is also provided to stir the water.

In the bomb calorimeter a known weight of dry biological material is burnt within stainless bomb in an atmosphere of pure oxygen at about 30 atmospheric pressure. The bomb is kept immersed in bath of water all the time. The bath is kept in a

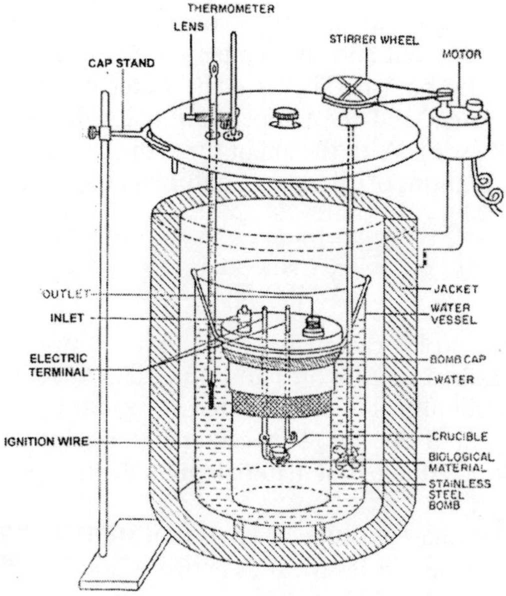

FIGURE 8.3
BOMB CALORIMETER AND ITS VARIOUS COMPONENTS

container adequately insulated against loss of heat from the water.

The biological material is first dried in an oven at 70°C for 24 to 48 h or more. The dried material is powdered in a grinder. A small quantity (1g) of the powder is then taken and pressed in the form of small pellet or tablet in a special pressing instrument. While preparing the tablet an ignition wire is inserted in the powder. The tablet is placed in an ignition crucible in the bomb, the ignition wires are connected to the terminals. About 5 ml water is poured in the bomb and the bomb is closed tight. Through an inlet device the bomb is filled with oxygen gas from a gas cylinder, driving out the air. The outlet is closed and further oxygen is fed in until the desired pressure is attained. Known quantity of water is placed in the calorimeter in which an electric stirrer is fitted. The temperature of water is carefully recorded. Through a press button electric current is passed through the tablet and it is completely ignited. The temperature of bomb and, therefore of water surrounding it, rises. This is recorded with the aid of a Beckmann thermometer. The calorific value (V) of the biological material is then calculated. In this connection, besides the dry weight (g) of material and the temperature rise (Δt) two more things should be known. They are the water equivalent value (W) of the system and the temperature rise value (Σc) due to ignition by the wire. With the aid of these the calorific value can be calculated in terms of calories per g through the formula (Lieth 1968):

$$V = \frac{W.\Delta^t - \Sigma c}{g}$$

In determining ecological efficiencies the rate of input of radiant energy is measured first for area and time. The calorific value of the plant community is then calculated at regular intervals from identical areas and the ratio between energy receipt and energy converted into the biological materials gives the *ecological efficiency* value.

Calorific values of different types of materials differ and in most cases of plant tissues it varies between 3000 to 5000 cal/g of dry matter. Leith (1975) has given approximate energy values of different biological compounds. Starch and cellulose have about 6300 cal/g, and fat 9300 cal/g. In biological materials after complete burning some ash is left. The energy values are sometimes expressed on ash free dry weight basis also by subtracting the weight of ash from the weight of biological material taken for ignition.

8.3.4 Oxygen measurement (light and dark bottle) method

The oxygen measurement method is usually used in both marine and freshwater habitats. The basic assumption of this method is that the amount of O_2 produced is proportional to gross production, because one molecule of O_2 is produced for each atom of carbon fixed. The sum of the oxygen produced in a bottle exposed to light and the oxygen used in a bottle with light excluded gives the total O_2 production, thus providing an estimate of primary productivity of the aquatic system.

Two bottles of known volume, one covered with aluminum foil and black tape (the *dark bottle*), and another uncovered (the *light bottle*) are taken. A measured equal volume of water with phytoplankton is transferred from the pond into each bottle and these bottles are placed at same level below the pond surface from which the samples were obtained. Both the bottles are left in the pond for 6-8 h and then removed. A third bottle of water is also taken at the beginning of the experiment, and its oxygen concentration is determined immediately (the initial sample).

The **Winkler** method is the standard procedure for oxygen measurement. The basic concept of the method is to bind the oxygen of water with iodine, which is released as a result of chemical reactions. This chemically bound dissolved oxygen is quantitatively measured by titration with sodium thiosulphate. The fixation and titration reactions of this method are as follows:

$$MnSO_4 + 2\,NaOH = Mn\,(OH)_2 + Na_2SO_4$$
(white ppt.; no O_2)

$$2\,Mn\,(OH)_2 + O_2 = 2\,MnO\,(OH)_2$$

$$MnO\,(OH)_2 + 2\,H_2SO_4 = Mn\,(SO_4)_2 + 3H_2O$$
(O_2 present)

$$Mn\,(SO_4)_2 + 2KI = MnSO_4 + K_2SO_4 + I_2$$

$$2\,Na_2S_2O_3 + I_2 = Na_2\,S_4O_6 + 2\,NaI$$

Manganous sulphate ($MnSO_4$) solution is prepared by dissolving 480 g $MnSO_4$. $4H_2O$ (or 364 g $MnSO_4.H_2O$) in 1 litre distilled water. Alkaline iodine (NaI or KI) solution is prepared by litre dissolving 500 g of NaOH (or 700 g KOH) and 150 g KI (or 135 g NaI) in 1 litre distilled water. Further, 10 g sodium azide (NaN_3) dissolved in 40 ml distilled water is added to the alkaline iodide solution. 0.025N sodium thiosulphate ($Na_2S_2O_3$) solution is prepared by dissolving 6.205 g $Na_2S_2O_3.5H_2O$ in 1 litre freshly boiled distilled water to which 5 ml chloroform is added as a preservative. The sodium thiosulphate solution thus prepared is stored in a brown bottle. Starch indicator is prepared by grinding 5-6 g potato or soluble starch in a mortar and dissolving it in 1 litre boiling distilled water, to which 1.25 g salicylic acid is added as a preservative.

Sample water (200 ml) is taken in a conical flask to which 1 ml $MnSO_4$ solution (by pipette, dipping the end below the surface) and 1ml alkaline iodide solution are added. Thereafter, the flask is shaken up to mix the solution thoroughly. The precipitate is allowed to settle, and 2 ml concentrated sulphuric acid (specific gravity 1.83-1.84) is added to dissolve the precipitate by shaking. This solution is titrated with sodium thiosulphate solution, using starch as indicator; and volume of titrant used in getting end point is noted.

The calculation of dissolved oxygen in the pond water (mg/l) is based on the rule that 1 ml of titrant ($Na_2S_2O_3$) contains 1 ppm O_2, hence Q ml titrant has 1 x Q ppm O_2.

In the light bottle oxygen is liberated due to photosynthesis and some of it is used in respiration whereas in the dark bottle only oxygen is used in respiration. Thus if the oxygen value of the light bottle is added to the quantity of oxygen used in respiration, the total oxygen evolved can be calculated.

8.3.5 Leaf area index (LAI)

Since LAI is a measure of how many leaves a typical vertical ray of sunlight passes through before hitting the soil (Fig. 8.4), determination of leaf area of a population or community provides a good index of primary productivity. LAI is expressed as total area of leaf per unit area of ground:

$$\text{LAI} = \frac{\text{Total leaf area, one surface only}}{\text{Unit ground area}}$$

For determination of LAI, all the leaves in the quadrat of a suitable size (25 cm x 25 cm, for example) are clipped. Many quadrats are taken to reach on a statistically significant value. The areas of the leaves are measured using any one of the following methods: (a) planimetry, (b) graph paper method, (c) weight method.

The rate of primary production usually increases with increase in LAI, up to a certain stage. Maximum net production is obtained when the LAI is about 4 (that is, the leaf surface exposed to light is 4 times the ground surface), but maximum gross production is reached when LAI reaches 8 to 10, which is the level found in mature forests. Net production is reduced at the higher levels because of the respiratory losses required to support the large volume of leaf and supporting tissues.

8.3.6 Chlorophyll estimation method

Chlorophyll content has certainly a relationship with primary productivity. In an ecosystem like a desert or an oligotrophic lake the quantity of chlorophyll is low per sq m or per ha and hence

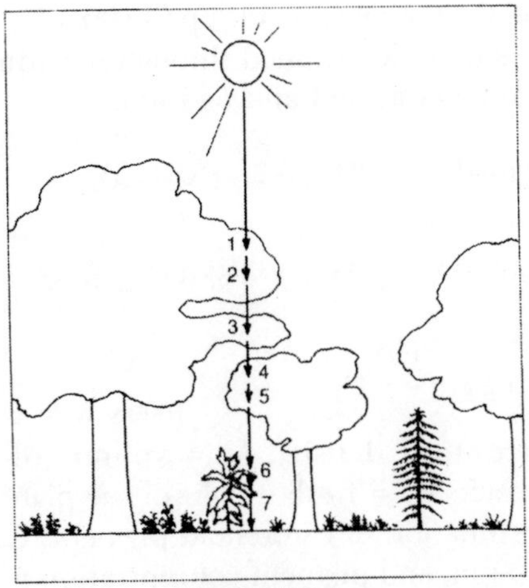

FIGURE 8.4
LEAF AREA INDEX IS A MEASURE OF THE NUMBER OF LEAVES A TYPICAL VERTICAL RAY OF SUNLIGHT HAS TO PASS THROUGH BEFORE HITTING THE SOIL. THE SINGLE RAY OF SUNLIGHT SHOWN HERE HAS TO PASS THROUGH SEVEN LEAVES BEFORE STRIKING THE SOIL

the primary productivity is also low. Similarly in multistoreyed forests in the tropics the quantity of green tissues or the quantity of chlorophyll is very high and the rate of production is also high. The relationship of primary production and chlorophyll content is shown by assimilation ratio:

$$\text{Assimilation ratio} = \frac{g / cm^2 / d}{\text{Average of chlorophyll a} / m^2}$$

For the estimation of chlorophyll content on area basis (g/m^2), leaf discs of known size are cut by using a cork borer. One to 0.5 g of plant material is weighed separately in two sets. One such set is dried in oven at 105°C to a constant weight and another set is used for chlorophyll estimation. The plant material is ground in a mortar and some quantity (5 ml) of 90% acetone (made basic by adding a few ml of magnesium carbonate) is added. The material is thoroughly crushed and then centrifuged at 250 rpm for 10-15 min. All the operations are carried out in shade without delay. The extract is decanted to a volumetric flask and the volume is made up to 25 or 50 ml using 90% acetone. If the solution is very dark green, the volume is made up to 100 ml. The chlorophyll concentration is then determined through recording absorbance in spectrophotometer at 663 and 645 nm:

$$\text{Chl a (mg/g)} = [12.7\,(OD_{663}) - 2.69\,(OD_{645})] \times \frac{V}{1000 \times W}$$

$$\text{Chl b (mg/g)} = [22.9\,(OD_{645}) - 4.68\,(OD_{663})] \times \frac{V}{1000 \times W}$$

$$\text{Total Chl (mg/g)} = \frac{OD_{645} \times 1000}{34.5} \times \frac{V}{1000 \times W}$$

where, OD = optical density, V = volume of the acetone-chlorophyll extract, W = fresh weight of the plant material.

For the estimation of carotenoid pigment, OD is recorded at 510 and 480 nm, and pigment concentration is calculated as:

$$\text{Carotenoid (mg/g)} = [7.6\,(OD_{480}) - 1.49\,(OD_{510})] \times \frac{V}{1000 \times W}$$

For calculation of chlorophyll content on area basis, following equation can be used:

$$\text{Chlorophyll content (mg/m}^2\text{)} = \frac{\text{Chlorophyll content mg/g of dry weight}}{\text{Area covered by one gramme of leaves}} \times 10000$$

In general, the amounts of chlorophyll range between 0.4-3.0 g/m^2 in forests, stratified grasslands and croplands; between 0.001-0.5 g/m^2 in winter, underwater or cave communities, lab cultures under low light intensity; between 0.02-1.0 g/m^2 in phytoplankton in lakes and oceans; and, between 0.01-0.60 g/m^2 in thin vegetation, algae mats on rocks, young crops, and lab cultures under intense light.

8.4 World Distribution of Productivity And Biomass

Different community types differ greatly in their NPP (Table 8.3). Differences result partly from variation in light, temperature, nutrients, and the availability of water, but individual species show remarkable differences under specific conditions optimum for each. Species possessing the C-4 pathway of photosynthesis generally have the highest productivity, while slow growing desert succulents exhibiting crassulacean acid metabolism (CAM) have among the slowest rates. The successional stage of the community also affects productivity; productivity increases during the pioneer and early tree stages and decreases as the community reaches maturity.

Terrestrial vegetation occupies approximately 30% of the globe's surface, and provides 65% of the total world primary productivity. However, productivity estimates are lowest in deserts and highest in tropical rainforests, spanning the entire range between 0 and approximately 3000 g/m^2/yr. Biomass estimates range from 1 kg/m^2 in desert and tundra ecosystems to 200 kg/m^2 in some temperate rainforests. In a terrestrial plant community which is well supplied with water and sunlight, plants respond by producing many leaves. These leaves trap sunlight and carbon dioxide. In a dried environment, plants cannot afford to produce as many leaves as they would risk transpiring so much water that they would wilt. In a shady environment, there is no point in plants producing great numbers of leaves as there will be not enough light for the leaves to reach their **compensation point** which refers to the amount of light necessary for a plant to have a rate of photosynthesis equal to its rate of respiration. If the compensation point is not reached, the leaf respires more than it photosynthesizes and so loses weight.

The relationship of biomass to productivity is often expressed as the **biomass accumulation ratio (BAR)**. BAR is the ratio of dry weight biomass to annual net primary productivity. BAR values are a measure of the accumulation of primary woody material, a characteristic related to environmental harshness and the potential age of the dominant species. There is a broad overlap in BAR values between ecosystems. Representative values are 2-10 for deserts, 1.3-5 for grasslands, 3-12 for shrublands, 10-30 in woodlands, and 20-50 for mature forests. In general, trends in world productivity are loosely related to biomass, in that K-selected plants in less harsh or more predictable environments are able to accumulate extensive root and branch systems. These form the bases for more efficient light interception, nutrient uptake, and in turn, productivity.

There is some evidence suggesting that temperate evergreen forests are more productive than their deciduous counterparts. Although evergreen leaves have 30-60% lower maximum photosynthetic rates per unit surface area than deciduous leaves, there is high leaf area index (LAI) for evergreen forests. Kira (1975)

TABLE 8.3
ESTIMATES OF NET PRIMARY PRODUCTIVITY (NPP) AND PLANT BIOMASS (DRY MATTER) IN VARIOUS ECOSYSTEMS (WHITTAKER 1975)

Ecosystem type	Area ($10^6 km^2$)	Net primary productivity ($g/m^2/yr$)		World net primary production ($10^9 t/yr$)	Biomass (kg/m^2)		World biomass ($10^9 t$)
		Normal range	Mean		Normal range	Mean	
1. Tropical rainforest	17.0	1000-3500	2200	37.4	6-80	45	765
2. Tropical seasonal forest	7.5	1000-2500	1600	12.0	6-60	35	260
3. Temperate evergreen forest	5.0	600-2500	1300	6.5	6-200	35	175
4. Temperate deciduous forest	7.0	600-2500	1200	8.4	6-60	30	210
5. Boreal forest	12.0	400-2000	800	9.6	6-40	20	240
6. Woodland and shrubland	8.5	250-1200	700	6.0	2-20	6	50
7. Savanna	15.0	200-2000	900	13.5	0.2-15	4	60
8. Temperate grassland	9.0	200-1500	600	5.4	0.2-5	1.6	14
9. Tundra and alpine	8.0	10-400	140	1.1	0.1-3	0.6	5
10. Desert and semidesert scrub	18.0	10-250	90	1.6	0.1-4	0.7	13
11. Extreme desert, rock, sand, and ice	24.0	0-10	3	0.70	0-0.2	0.02	0.5
12. Cultivated land	14.0	100-3500	650	9.1	0.4-12	1	14
13. Swamp and marsh	2.0	800-3500	2000	4.0	3-50	15	30
14. Lake and stream	2.0	100-1500	250	0.5	0-0.1	0.02	0.05
Total continental	**149**		**773**	**115**		**12.3**	**1873**
15. Open Ocean	332.0	2-400	125	41.5	0-0.005	0.003	1.0
16. Upwelling zones	0.4	400-1000	500	0.2	0.005-0.1	0.02	0.008

17. Continental shelf	26.6	200-600	360	9.6	0.001-0.04	0.01	0.27
18. Algal beds and reefs	0.6	500-4000	2500	1.6	0.04-4	2	1.2
19. Estuaries	1.4	200-3500	1500	2.1	0.01-6	1	1.4
Total marine	**361**		**152**	**55**		**0.01**	**3.9**
Grand total	**510**		**333**	**170**		**3.6**	**1841**

has explained that deciduous plants have very low LAI during the periods of leaf development and may miss periods of favourable moisture and temperature following photoperiodically induced leaf drop. He also suggested that the high net productivity of tropical evergreen forests is related to the increased height and greater solar radiation which allow increased LAI, and therefore higher exploitation efficiency. Needleleafed evergreens frequently have nearly twice the LAI of deciduous trees of similar biomass and height, thus allowing plants with a lower photosynthetic capacity per unit leaf area to have a higher potential stand production value. A high LAI is made possible by large size, necessitating larger amounts of nonproductive structural tissue, thereby reducing the net production efficiency of forest trees.

Grassland productivity is typically lower than in communities with trees mainly because of the absence of woody tissues resulting in low BAR values (1.3-5) and short-lived aboveground parts, and therefore a lower LAI.

Net primary productivity in arctic and alpine communities is typically 100-150 g/m^2/yr with biomass as low as or lower than desert communities. The main restrictions on production are low LAI, low temperatures (optimum temperatures for productivity coincide with 15-25°C optimum range of photosynthesis), and/ or low angle of incident radiation.

Primary production in arid lands varies from 10 to over 200 g/m^2/yr depending upon the amount and pattern of rainfall. LAI is severely limited by stomatal closure during dry periods and variations occur seasonally in response to water availability.

Nearly 70% of the world is covered in water. Yet, the aquatic communities is only responsible for approximately 35% of the world primary productivity, although some algal beds, coral reefs and estuaries are more productive than the most productive tropical rainforests or cultivated land (Table 8.3). High rates of production in such communities occur due to **energy subsidies** (continuous input of organic detritus, for example) from outside the systems that reduce the cost of maintenance, and favourable physical factors. The community type most responsible for the low overall productivity of the aquatic environment is open ocean. What limits productivity in open oceans? Obviously water is available, and temperatures in many regions are not too cold; sunlight and CO_2 are available, especially in surface waters.

Mineral nutrients are limiting factor. Many organisms settle to bottom when they die, taking their minerals with them. Thus the surface waters where light and CO_2 are most abundant become impoverished in phosphates, nitrates, and other essential nutrients. When special conditions bring upwellings with nutrients to the surface, a profuse bloom of phytoplankton often results. But most oceans are "nutrient deserts" with much lower productivities than thought several years ago (Salisbury and Ross 1986).

8.5 Litter Production and Decomposition

Biomass may remain living, serve a support function, be consumed directly by herbivores, or become *detritus* (dead organic matter/ litter). The latter possibility is more likely; more than half of annual net production is deposited as litter. Litter is the food source for decomposers and detrivores and it is the means by which nutrients are returned to the cycling pool. *Decomposition* of detritus may partly occur through leaching and fragmentation but the principal agents of decomposition are microbes that act through the process of metabolism. *Leaching* results from water percolating through an organic structure, such as a fallen tree, and dissolving some of its material content. Not only is there loss of mass form the structure, the minerals released are then available for transport as well as recycling. Leaching is an important process in the loss of mass and materials from decomposing leaf litter. *Fragmentation* is the process of pulverizing or reducing material into small particles. Physical fragmentation may occur when a dead tree falls to the ground and shatters into smaller pieces, or through abrasive actions of streams and rivers, or due to drying actions of sunrays; whereas biological fragmentation is caused by invertebrate animals which chew, ingest, excavate and tunnel the woody products. Metabolism of organic material is primarily a microbial activity, although, to a lesser degree, invertebrates play a role. In terrestrial ecosystems, basidiomycetes (fungi) are the primary agents of decomposition, whereas in aquatic ecosystem, bacteria, including actinomycetes, are most important.

Olson (1963) calculated the ratio of litter production to litter accumulation as an expression of the rate of decomposition.

The ratio is high in tropical environments, reflecting the high rate of leaf production and rapid decomposition, and decreases on a gradient toward the poles. Litter fall is generally related to LAI causing the ratio to decrease with latitude in a fashion similar to that of NPP. Within forest, litter fall decreases with latitude and/or altitude.

The rate of litter decomposition varies with the vegetation and environment. Broadleaf temperate forest requires about one year for complete litter decomposition whereas coniferous forests typically require 3-5 years. In tropical forests complete breakdown may occur in less than two months. These turnover rates depend primarily on the chemical make up of the litter, the temperature, and moisture conditions of the habitat. Species specific differences in decomposition rates may be related to the relative amount of leachable material contained in the litter (Gosz *et al.* 1973). Species with high leachability show rapid loss of weight in the first few weeks, followed by slower loss after leachable material has been removed. Organic compounds leached from newly deposited litter provide usable source of energy for decomposers. The increased activity of decomposers would also stimulate initial degradation. Nitrogen is an essential nutrient for soil organisms and its availability is therefore related to rates of decomposition. In general, the hard wood litter has higher leachable nitrogen content than conifer forest, and tropical forest litter has higher nitrogen content than temperate forest litter.

8.6 Energy Flow Beyond The Producers

A "food chain" is defined as the sequence of stages through which stored energy in the form of food is passed from the primary producers to a number of organisms. As the organisms at each stage are eaten by those of the next, a flow of energy through the ecosystem is maintained, beginning with the capture and conversion of radiant energy from the sun and ending with the complete decomposition of the organic compounds. At each step in the food chain when food energy is transferred from one trophic level to the next higher level, a large proportion (80-90%) of the potential (food) energy is lost as heat because of the inefficiencies in energy transfer from one trophic level to the next. Consequently, the number of steps or "links" in a food chain is limited, usually to four or five. The available energy is greater to organisms situated nearer to the beginning of the chain.

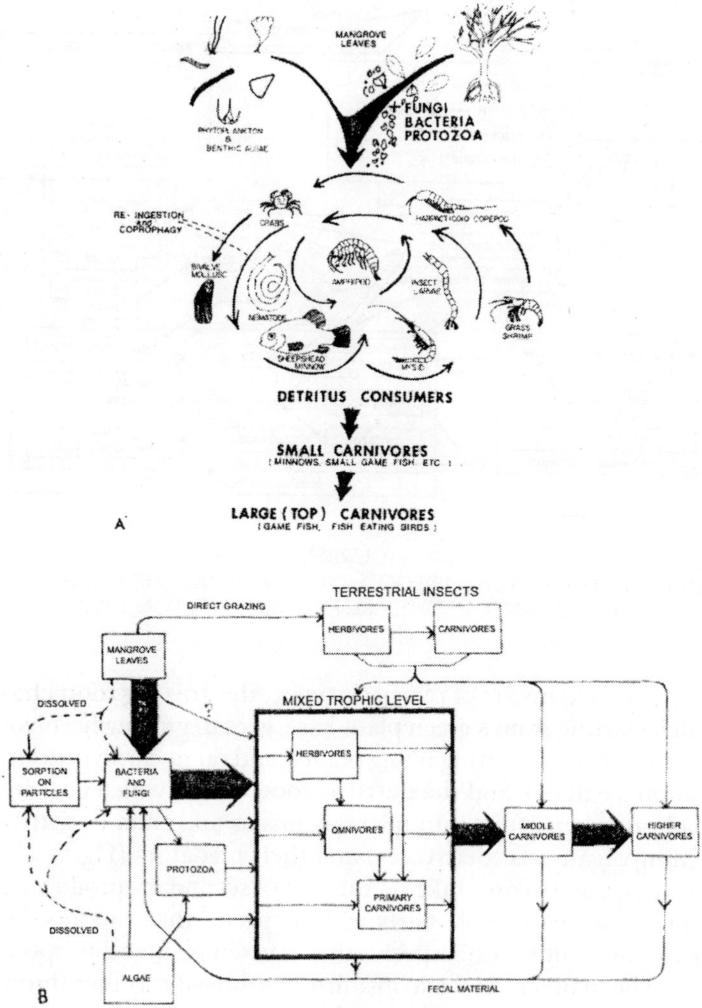

FIGURE 8.5

A DETRITUS FOOD CHAIN BASED ON MANGROVE LEAVES WHICH FALL INTO SHALLOW ESTUARINE WATERS OF SOUTH FLORIDA. LEAF FRAGMENTS ACTED ON BY SAPROTROHS AND COLONIZED BY ALGAE ARE EATEN AND RE-EATEN (COPROPHAGY) BY A KEY GROUP OF SMALL DETRITUS CONSUMERS WHICH, IN TURN, PROVIDE THE MAIN FOOD FOR GAME FISH, HERONS, STORKS AN IBIS: A. PICTORIAL MODEL (UPPER DIAGRAM); B. COMPARTMENT MODEL (LOWER DIAGRAM) (ODUM 1971)

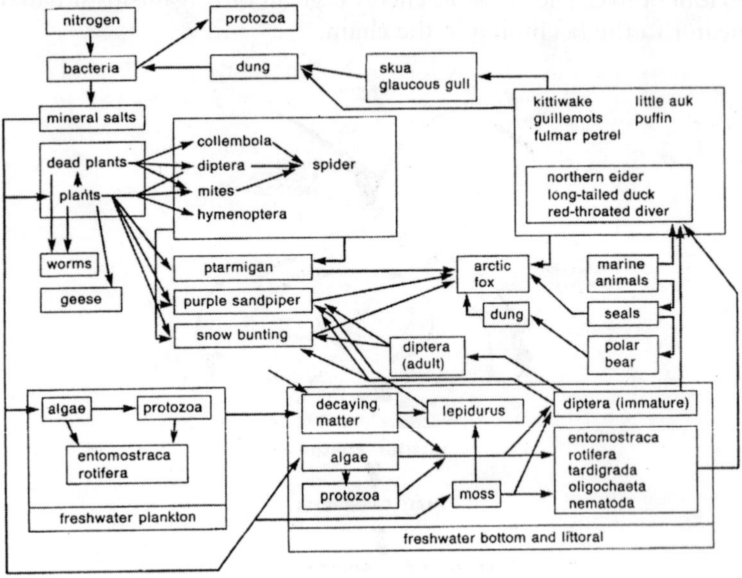

FIGURE 8.6
AN EARLY FOOD WEB (OR FOOD CYCLE) AMONG THE ANIMALS OF BEAR ISLAND, IN THE ARCTIC CIRCLE (ELTON 1927).

Food chains are of two basic types : the **grazing food chain**, which starting from a green plant base, goes to grazing herbivores (that is, organism eating living plants) and on to carnivores (that is animal eaters); and the **detritus food chain**, which goes from dead organic matters into microorganisms and then to detritus-feeding organisms (detrivores) and their predators (Fig. 8.5). In such cases, detrivores take on the role of secondary producers. A food chain, in general, starts with a species that eats no other species and ends with a species that is eaten by no other species.

The manner in which organisms are linked together through food tends to be relatively simple and linear in the very earlier stages of succession, and the heterotrophic utilization of net production tends to occur predominantly by way of grazing food chain (plant -herbivore - carnivore sequences). In contrast, food chains become complex webs (Fig. 8.6) in mature stages, with the bulk of biological energy flow following detritus pathways.

In complex natural communities, organisms whose food is

obtained from plants by the same number of steps are said to belong to the same **trophic level.** Thus, green plants (the producer level) occupy the first trophic level, plant-eaters the second level (the primary consumer level), carnivores, which eat the herbivores, the third level (the secondary consumer level), and the secondary carnivores the fourth level (the tertiary consumer level). There is always a one-way street along which energy moves in the various trophic levels. The energy that is captured by the autotrophs does not revert back to solar input; what passes to the herbivore does not pass back to the autotrophs, and so on. As it moves progressively through the various trophic levels, it is no longer available to the previous level. The immediate implication of this unidirectional energy flow in an ecosystem is that the system would collapse if the primary energy source, the sun, were cut off.

Figure 8.7 explains how the principles of the food chains are governed by the two laws of thermodynamics. In these diagrams the "boxes "represent trophic levels and the "pipes" depict the energy flow in and out of each level. Energy inflows balance outflows as required by the first law of thermodynamic (energy may change form, for example from radiant to chemical, but not amount), and each energy transfer (change in the form of energy) is accompanied by dispersion of energy into unavailable heat (that is, respiration) as required by the second law. The ecosystem tends to run down, dissipating its energy and losing its organized structure as a result of energy transformations at each trophic level. This is a tendency toward **entropy** or maximum disorganization of structure and maximum dissipation of usable energy. This trend is countered by the continual input of energy from the outside.

Figure 8.7 (a) is a very simplified energy flow model of three trophic levels, illustrating how the energy is greatly reduced at each successive level. The total energy that impinges on the green plants (L) is 3000 kcal/m^2/day. Out of this the energy absorbed (LA) by green plants is 1500 kcal/m^2/day (50% of incident light). The net amount of energy (Pn) fixed at the first trophic level is only 1% of the absorbed energy, the other 99% energy escape as heat in the process of transformation of light energy into the potential energy (Pn). This heat energy, however, is not wasted; it performs the work of driving hydrological and mineral cycles and maintaining environmental temperatures within the range that can be tolerated by organisms. Secondary productivity (P2 and P3 in the diagram) is 10 and 20%, respectively at herbivore and carnivore levels. In brief, only a very small percentage of the available sun energy is fixed by plants. Consequently, the number of consumers that can be supported by a given primary production output depends very much on the length of the food chain. As each link in the chain decreases the available energy by about one order of magnitude (order of 10), consumer populations become smaller and smaller at the successive trophic levels.

In Figure 8.7(b) the grazing and detritus food chains are sharply separated in a Y- shaped or 2-channel energy flow diagram. In the forest community, 90% or more of the NPP is normally utilized in the detritus food chain, whereas energy flow in the marine community occurs mainly through the grazing food chain. In a heavily grazed pasture or grassland 50% or more of the net production may pass down the grazing path, whereas there are many aquatic systems, especially shallow water ones, which, like mature forests, operate largely as detritus systems. Also, since not all food eaten by grazers is actually assimilated, undigested material in faeces is diverted to the detritus route.

8.7 Ecological Pyramids

There are some sort of relationship between the numbers, biomass, and energy content of the primary producers and consumers of different orders in any ecosystem. These relationships may be represented graphically by means of ecological pyramids in which the first or producer level forms the base and successive levels the tiers which make up the apex. Ecological pyramids may be of three general types :(1) the pyramid of numbers in which the number of individual organisms is depicted, (2) the pyramid of biomass based on the total dry weight, caloric value, or other measure of the total amount of living material, and (3) the pyramid of energy in which the rate of energy flow and /or "productivity" at successive trophic levels is shown. The number and biomass pyramids may be inverted (base may be smaller than one or more of the upper tiers) if producer organisms average smaller than consumers in individual size. On the other hand, the energy pyramid is always upright or triangle shaped.

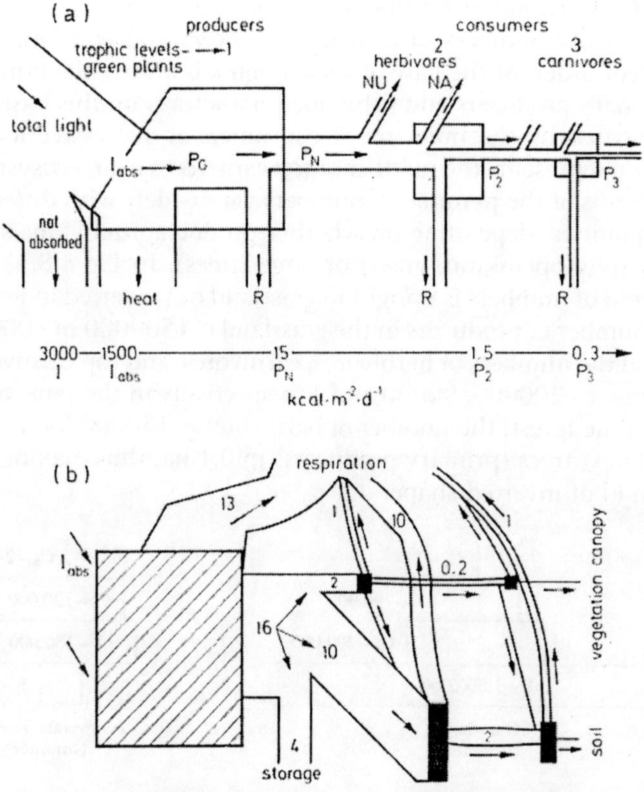

FIGURE 8.6 (A)

SIMPLIFIED ENERGY-FLOW DIAGRAM DEPICTING THREE TROPHIC LEVELS (BOXES NUMBERED 1,2,3) IN A LINEAR FOOD CHAIN. STANDARD NOTATIONS FOR SUCCESSIVE ENERGY FLOWS ARE AS FOLLOWS: I = TOTAL ENERGY INPUT, IABS = LIGHT ABSORBED BY PLANT COVER; PG = GROSS PRIMARY PRODUCTI-VITY; PN= NET PRIMARY PRODUCTIVITY; P_2, P_3 = SECONDARY (CONSUMER) PRODUCTIVITY; NU = ENERGY NOT USED (STORED OR EXPORTED); NA= ENERGY NOT ASSIMILATED BY CONSUMERS (EGESTED); R= RESPIRATION. BOTTOM LINE IN THE DIAGRAM SHOWS THE ORDER OF MAGNITUDE OF THE ENERGY LOSSES EXPECTED AT MAJOR TRANSFER POINTS, STARTING WITH A SOLAR INPUT OF 3000 KCAL PER SQUARE METER PER DAY. (B) TWO-CHANNEL ENERGY FLOW DIAGRAM THAT SEPARATES A GRAZING FOOD CHAIN (VEGETATION CANOPY) FROM A DETRITUS FOOD CHAIN IN A HYPOTHETICAL FOREST. ESTIMATES IN KCAL/M²/DAY (ODUM 1971)

(a) Pyramid of numbers: This deals with the relationships between the numbers of primary producers and consumers of different order. At the base of such a figure is always the number of primary producers and subsequent structures on this base are represented by the numbers of consumers at successive levels. The top represents the number of top carnivores in an ecosystem. The forms of the pyramid of numbers vary widely with different communities, depending on whether producing individuals are small (phytoplankton, grass) or large (trees). In Fig 8.8(a) the pyramid of numbers is upright in grassland but inverted in forest. The number of producers in the grassland is 1500000 in 1000 sq m, and the numbers of herbivores, carnivores, and top carnivores decrease to 200000, 90000 and 1, respectively in the same area. But in the forest, the number of herbivores is 150000 for a mere 200 forest trees (primary producers) in 0.1 ha, thus making the pyramid of inverted shape.

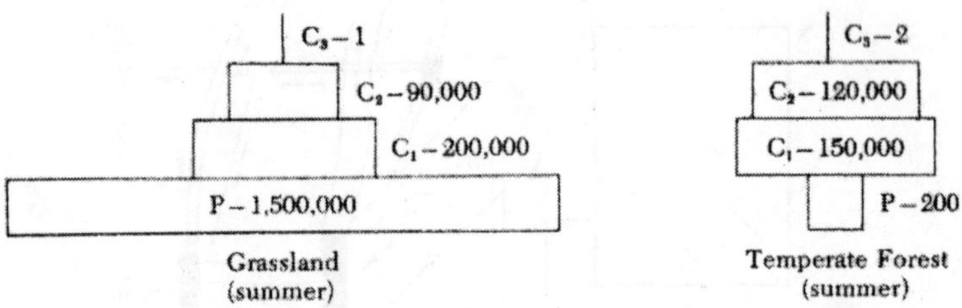

(a) Pyramid of numbers. Individuals (exclusive of microorganisms and soil animals) per 0.1 hectare

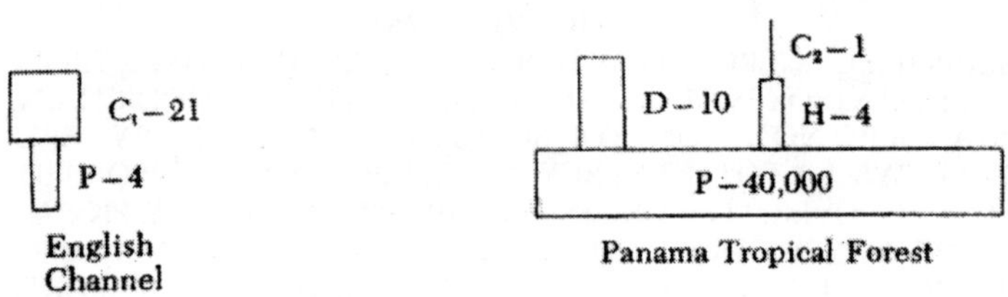

(b) Pyramid of biomass (g dry weight per sq m).

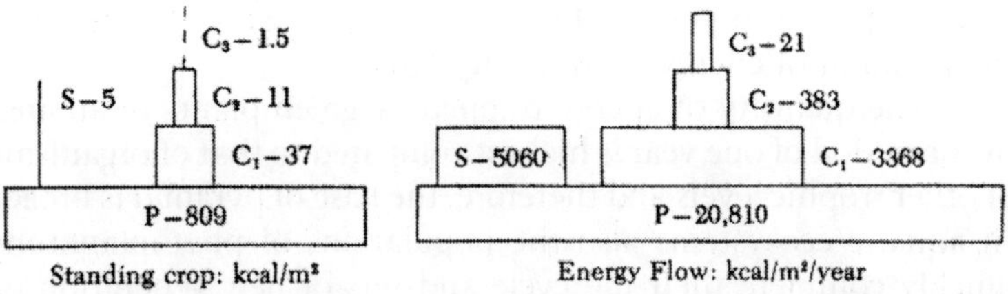

(c) Comparison of standing crop and energy-flow pyramids for Silver Springs, Florida.

FIGURE 8.8
ECOLOGICAL PYRAMIDS OF (A) NUMBERS: GRASSLAND (LEFT), TEMPERATE FOREST (RIGHT); (B) BIOMASS: ENGLISH CHANNEL (LEFT), PANAMA TROPICAL FOREST (RIGHT); (C) COMPARISON OF STANDING CROP (LEFT) AND ENERGY FLOW (RIGHT) PYRAMIDS. P=PRODUCERS, C_1= PRIMARY CONSUMERS, C_2= SECONDARY CONSUMERS, C_3= TERTIARY CONSUMERS (TOP CARNIVORES), S= SAPROTROPHS (BACTERIA AND FUNGI), D= DECOMPOSERS (BACTERIA, FUNGI AND DETRIVORES)

(b) **Pyramid of biomass**: The weight or biomass of primary produçers forms the base of pyramid of biomass. If the size or biomass of producers is large (as in forest trees) and that of consumers small, the pyramid of biomass becomes upright; on the other hand, if the size of producers is very small and that of consumers large, the pyramid of biomass becomes inverted in shape.

In terrestrial and shallow water ecosystems, where producers are large and relatively long-lived, a broad-based, upright pyramid is to be expected; and pioneer communities have similar trends. For example in a Panama forest for 15 g of heterotrophs in a square meter area, 40000 g of primary producers are reported to be available (Fig. 8.8b). In contrast, in open water or deep water situations the biomass of produces (phytoplanktons) is quite negligible as compared to herbivore fish that feed on them, and therefore the shape of the pyramid of biomass is inverted. For example, in the English Channel, the biomass of primary producers is only 4 g/m^2 in contrast to 21g/ m^2 biomass of consumers (Harvey 1950).

(c) **Pyramid of energy**: The pyramid of energy represents the total quantity of energy utilized by different trophic level organisms of an ecosystem per unit area over a set period of time (usually per sq m per yr). Due to this time factor, the shape of the pyramid of energy is always upright.

The quantity of energy trapped by green plants in an area over a period of one year is highest compared to that of organisms of other trophic levels and therefore, the base of pyramid is broad. In aquatic ecosystems also the populations of phytoplanktons quickly complete their life cycle and sets of new generation or crops of phytoplanktons are formed every few hours or days. The cumulative energy content that these generations of phytoplanktons trap in course of a year is certainly much more than of only a few generations of herbivore fishes in the corresponding time and space. The energy content of top carnivores (utilized in one year) is the least. Therefore, the pyramid of energy can never be of any other shape except upright pyramidal. In the event of destruction of primary producers or organism of any trophic level, the organism of next trophic level will automatically die for want of food (the source of energy) and ultimately the upright pyramidal shape will be maintained. In Silver Springs (USA), the energy contents of different trophic level organism over a period of one year make an upright pyramid (Fig 8.8c). The energy contents from 20810 kcal/m^2/yr in primary producers decrease to 3368 to 383 to 21 kcal/m^2/yr in consumers of the first, second and third orders (Odum 1957). Thus a top carnivore for the organic production of 21 kcal/m^2/yr needs a broad base of green plant equivalent to 20810 kcal/m^2/yr of energy.

9. BIOGEOCHEMICAL CYCLES

The **biosphere**, the portion of the earth which is inhabited by living organisms, includes the hydrosphere, the atmosphere, and the lithosphere. The *hydrosphere* consists of the earth's liquid water (both surface and underground), ice (polar ice, icebergs, and ice in frozen soil layers or *permafrost*), and water vapour in the atmosphere. The part of the *atmosphere* with which living organisms come into contact is the troposphere, the weather zone of the earth's envelope of air. It mediates the balance of water through the processes of rain, condensation and evapotransipiration, and provides living organisms with carbon dioxide and oxygen. The *lithosphere* is the earth's crust which is the inexhaustible storehouse for the variety of chemical elements of which organisms are composed (Fig. 9.1), and the basic material for the formation of the soil. In general, biosphere is the narrow band, about 100 m thick, above the earth and below the surface of the ocean. It is true that birds can fly as high as 2000 m, and that there are bacteria on the floor of 1000-m-deep marine trenches, but abundant life is limited to a much more restricted region near the surface of the earth. Trees stand no more than 70-100 m tall and sink their roots but a few tens of meters into the earth; in water the layer penetrated by light (and hence densely populated) ordinarily extends to a depth of 30 m, and at most to 100 m.

The chemical elements, including all the essential elements of protoplasm, tend to circulate in the biosphere in characteristic paths from environment to organism and back to the environment. These more or less circular paths are known as **biogeochemical cycles** (literally, life-earth-chemical cycles; "bio" refers to living systems and "geo" to the rocks, water and air of the earth). **Nutrient cycling** is the movement of those elements and

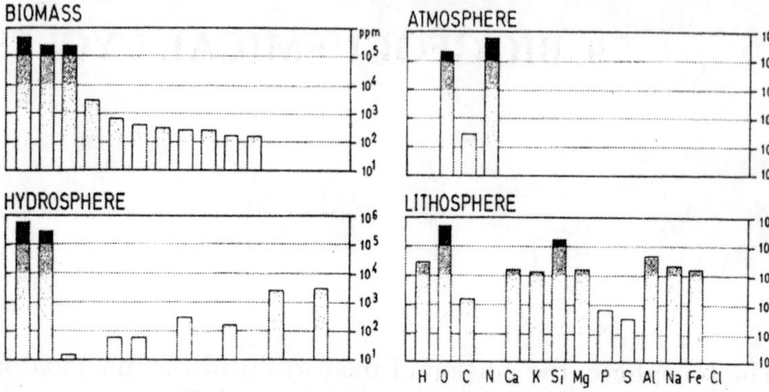

Figure 9.1 Compositon of the biomass, atmosphere, hydrosphere and lithosphere, in terms of the relative numbers of atoms (atoms per million atoms, not the proportion by weight) of the various chemical elements. The composition of living organisms is clearly distinct from that of the three components of their environment; they select from the available elements, according to their needs. The scale of the ordinate is logarithmic. For example, in the biomass H, O, C and N are present in the greatest proportions: $4.98 \cdot 10^5$ atoms per million (that is, about 50% of all atoms) are hydrogen atoms; oxygen and carbon atoms each comprise $24.9 \cdot 10^5$ atoms per million (about 25%), and $2.7.10^3$ (about 0.3%) are nitrogen atoms (Deevey 1970).

inorganic compounds that are essential to organisms to live, grow, or reproduce. In Fig. 9.2, a biogeochemical cycle is superimposed on a simplified energy-flow diagram to show how the one-way flow of energy drives the cycle of matter.

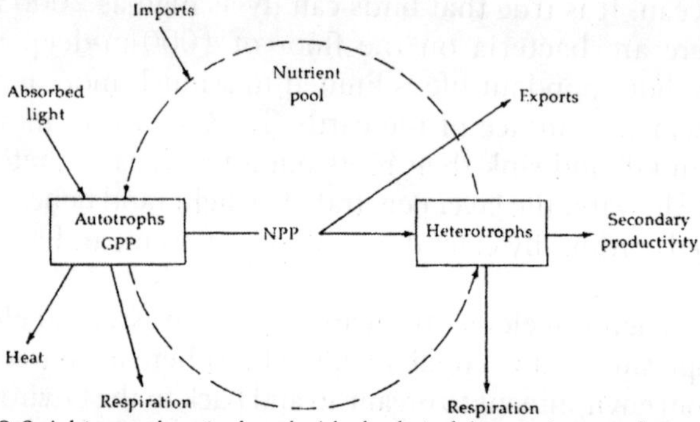

Figure 9.2 A biogeochemical cycle (dashed circle) superimposed on a simplified energy flow diagram contrasting the cycling of material and the one-way flow of energy. GPP = gross primary producitivity and NPP = net primary productivity, which may be consumed within the system by heterotrophs or exported from the system (Odum 1963).

9.1 Plant Nutrients

There are two principal criteria by which an element can be judged essential or nonessential to any plant (Epstein 1972): *first*, an element is essential if the plant cannot complete its life cycle (that is, form viable seeds) in the absence of that element. *Second*, an element is essential if it forms part of any molecule or constituent of the plant that is itself essential in the plant (for example, N in proteins and Mg in chlorophyll; Table 9.1). Either criterion is sufficient to demonstrate essentiality, and most elements listed in Table 9.2 have met both. The first seven elements listed in Table 9.2 are often called **trace elements** or **micronutrients** (needed in tissue concentrations equal to or less than 100 μg/g of dry matter), and the last nine are the **macronutrients** (needed in concentrations of 1000 μg/g of dry matter, or more). In addition to the 16 elements essential for plants, higher animals require sodium, iodine, cobalt, selenium, and apparently also nickel, silicon, chromium, tin, vanadium, and fluorine, but not boron (Miller and Neathery 1997; Mertz 1981).

TABLE 9.1
ELEMENTAL COMPOSITION OF THE SIX MAJOR CLASSES OF CONSTITUENTS OF LIVING MATTER

Plant constituent	*Elements present*
Carbohydrates	Carbon, hydrogen, oxygen. (Nitrogen and/or phosphorus are also present in certain derivatives.)
Amino acids	Carbon, hydrogen, oxygen, nitrogen. (Sulphur is also present in some amino acids.)
Nucleotides	Carbon, hydrogen, oxygen, nitrogen, phosphorus
Porphyrins	Carbon, hydrogen, oxygen, nitrogen. (Magnesium is present in porphyrins known as *chlorophylls*; and iron is present in porphyrins known as *cytochromes*.)
Lipids	Carbon, hydrogen, oxygen. (Phosphorus and/or sulphur are also present in some lipids.)
Enzymes	Carbon, hydrogen, oxygen, nitrogen, phosphorus, sulphur. (A heavy metal-copper, iron, molybdenum or zinc-is also an integral component of the reactive site in about 150 metalloenzymes.)

TABLE 9.2
ESSENTIAL ELEMENTS FOR MOST HIGHER PLANTS AND INTERNAL CONCENTRATIONS CONSIDERED ADEQUATE*.

Element	*Chemical symbol*	*Forms available to plants*	*Concentration in dry tissue* (*ppm*)	(*%*)
Molybdenum	Mo	$MoO4^{2-}$	0.1	0.00001
Copper	Cu	Cu^{+}, Cu^{2+}	6	0.0006
Zinc	Zn	Zn^{2+}	20	0.0020
Manganese	Mn	Mn^{2+}	50	0.0050
Boron	B	H_3BO_3	20	0.002
Iron	Fe	Fe^{3+},Fe^{2+}	100	0.010
Chlorine	Cl	Cl^{-}	100	0.010
Sulphur	S	SO_4^{2-}	1000	0.1
Phosphorus	P	H_2PO^{4-}, HPO_4^{2-}	2000	0.2
Magnesium	Mg	Mg^{2+}	2000	0.2
Calcium	Ca	Ca^{2+}	5000	0.5
Potassium	K	K^{+}	10000	1.0
Nitrogen	N	NO_{3-}, NH_4^{+}	15000	1.5
Oxygen	O	O_2, H_2O	450000	45
Carbon	C	CO_2	450000	45
Hydrogen	H	H_2O	60000	6

* If one of two forms is more common than the other, this is indicated by boldface type.

Whether we speak of major or minor nutrients, the nutritional state of a plant in regard to elements present in the growth medium may be viewed as: (i) deficient, (ii) adequate, or (iii) injuriously excessive (Fig. 9.3). In the range of low concentrations called the *deficient zone*, growth increase dramatically as more of the element is provided and its concentration in the plant increases (Moorby and Besford 1983). Above the *critical concentration* (minimal tissue concentration giving almost maximal growth), increases in concentration do not appreciably affect growth (*adequate zone*). The adequate zone represents *luxury consumption* of the element, during which storage in vacuoles occurs. This zone is fairly wide for macronutrients but is narrower for micronutrients such as boron, zinc, iron, copper, manganese, and molybdenum. Continued increases of any element lead to toxicities and reduced growth (*toxic zone*).

Plants respond to an inadequate supply of an essential element by forming characteristic *deficiency symptoms*. Such visually observed symptoms include stunted growth of roots, stems, or leaves and chlorosis or necrosis of various organs (Table 9.3).

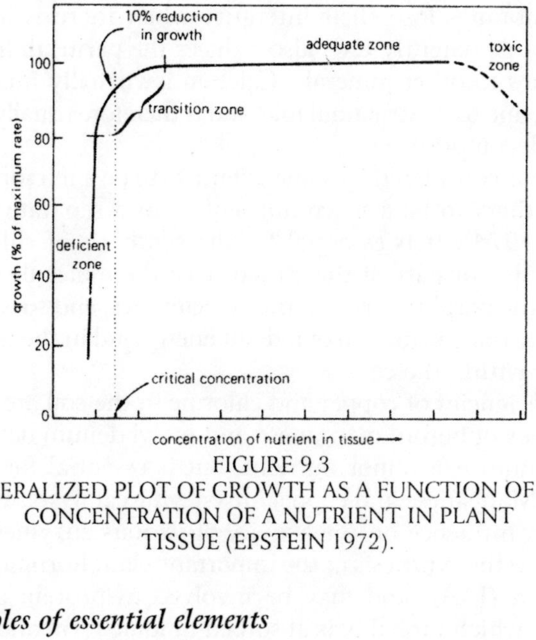

FIGURE 9.3
GENERALIZED PLOT OF GROWTH AS A FUNCTION OF THE CONCENTRATION OF A NUTRIENT IN PLANT TISSUE (EPSTEIN 1972).

9.1.1 Roles of essential elements

Carbon and oxygen in CO_2 and hydrogen in water are assimilated in photosynthesis and subsequently metabolized into a variety of plant constituents. Nitrogen is a constituent of proteins, nucleic acids, and is structurally involved in most catalytic molecules. It accumulates in young tissues, seeds and storage organs. Phosphorus is involved in the structure of many vital molecules, such as nucleic acids, coenzymes, and phospholipids. The energy that is used in the cell is released largely by hydrolysis of various phosphate bonds. Sulphur forms part of some amino acids, and is crucial to the stability of the tertiary structure of enzymes and other proteins. Potassium is very soluble and is easily leached from light or sandy soils. Although it has no apparent structural role, potassium plays a catalytic role: it is used primarily in meristematic zones in young tissues, and is important in regulating guard cell turgor. Magnesium is a structural part of the chlorophyll molecule, and serves to activate many enzymatic reactions which transfer phosphates. It tends to accumulate in young leaves, and will be translocated out before leaf abscission. Calcium is essential for the formation and metabolism of mitochondria and nucleus. Cell membranes lose their intergrity when there is a calcium deficiency. Apparently Ca^{2+} also reduces the permeability of cell membranes to other minerals. Calcium is virtually immobile in the plant, due to its structural role, and is therefore usually released only by decomposition.

Iron is considered by some scientists to be a micronutrients, and by others to be a macronutrient or in a separate category (Bidwell 1974). It is essential for the synthesis of chlorophyll, though it is not part of the structure of the molecule. It is the center of the porphyrin ring of the cytochromes, and so is involved both in the transformation of radiant energy and in the utilization of energy within the cell.

Deficiencies of copper and chlorine in the soil are rare, but deficiencies of boron, manganese and molybdenum occur often. Molybdenum is a minor nutrient, but is essential for nitrogen fixation. Manganese and copper are enzyme catalysts, and boron may exert influence on the activity of various enzymes. Zinc is essential to the synthesis of the important plant hormone, indole acetic acid (IAA), and may be involved in protein systhesis. Chloride, which probably is absorbed in ionic form and remains so, plays a vital role in photosynthesis.

Certain elements behave as analogs to nutrients, and will replace them to some extent. The uptake of strontium, which mimics calcium and will relieve deficiency symptoms for a time, has been of great interest to nuclear scientists and nutritionists in recent years. Radioactive strontium was released by atmospheric testing of nuclear devices and tended to be concentrated in the Northern Hemisphere. Lichens in the tundra took up the strontium as if it were calcium. Reindeer fed upon the lichens, and Laplanders, who were nourished by caribou (reindeer), showed higher concentrations of the harmful isotope than peoples of other lands.

9.2 Gaseous and Sedimentary Cycles

There are two major types of biogeochemical cycles: *gaseous* and

TABLE 9.3
SITES OF ACCUMULATION, TRANSPORT ABILITY, AND DEFICIENCY SYMPTOMS OF ESSENTIAL NUTRIENTS

Element	*Sites of accumulation*	*Transport ability*	*Deficiency symptoms*
N	Young shoots, leaves, buds, seeds, storage organs	Good, primarily in organically bound form	Stunting or dwarfism; spindly appearance; shoot: root ratio shifted toward roots; premature yellowing of old leaves
P	More in reproductive organs than in vegetative	Good, in organically bound form	Disturbance of reproductive processes (delayed flowering), stunting, bronze-violet discoloration of leaves and stems
S	Leaves, seeds	Good, in organic form, poor as ion	Similar to N-deficiency
K	Division zones, young tissue, bark parenchyma, sites of intense metabolism	Good	Disturbed water balance (tip drying), curling of leaf edges
Mg	Leaves	Good in part	Stunted growth, interveinal chloroses of old leaves
Ca	Leaves, tree bark	Very poor	Disturbance in growth by division (small cells)
Fe	Leaves	Poor	Interveinal chloroses, in extreme case white coloration of young leaves (veins green)
Mn	Leaves	Poor in part	Inhibition of growth, necroses
Zn	Roots, shoots	Poor	Stunted growth, discoloration of leaves, disturbances in fructification
Cu	Woody axes of shoots	Poor	Tip drying, chloroses of young leaves
Mo		Poor	Disturbance of growth and deformation of shoots
B	Leaves, tips of shoots	Poor	Disturbance of growth (meristem necroses), diminished branching of roots, phloem necroses, disturbances of fructification
Cl	Leaves	Good	

sedimentary. In a gaseous cycle, the atmosphere constitutes a major reservoir of the element that exists there in a gaseous phase. The gaseous nutrients are part of what Odum (1971) termed *perfect cycles*, in that local perturbation is quickly compensated for, and equilibrium is rapidly reestablished. Carbon and nitrogen are prime representatives of biogeochemical cycles with a prominent gaseous phase. Those elements which do not have a prominent gaseous phase, such as calcium, phosphorus and iron, share the earth's crust as their reservoir. Such elements cycle quite slowly, moving mostly from the land to sediments in the seas, in some ponds, or in a lake, and only returned to the nutrient pool by geological activity-*orogeny* (mountain building) or some other tectonic phenomenon. Exceptions to this limitation include the short-term recycling which occurs when predatory birds return to land after feeding on ocean fishes. Their rookeries become covered with gleaming white deposits of *guano*, which is rich in phosphorus and other materials. *Anadromous fishes* (which have matured in the sea) return to spawn and die inland. This migration also recycles nutrients quickly but the contribution of such transfer is minimal.

Elements in nature are never, or almost never, homogeneously distributed nor are they present in the same chemical form throughout the ecosystem. They reside in compartments, which represent a defined space in nature. *Compartments* can be defined very broadly or very specifically. For example, all of the plants in an ecosystem can be included in one compartment, or each plant species may constitute separate compartment or even the leaves and the stem of a single plant can be considered as separate compartments. A compartment contains a certain quantity, or *pool*, of nutrients. Compartments exchange nutrients, and the rate of movement (uptake and outflow) of nutrients between two compartments is called the *flux rate*. It is measured as the quantity of nutrient passing from one pool to another per unit time. The flux rates and pool sizes together define the nutrient cycling within any particular ecosystem. Ecosystems are not isolated from one another, and nutrients come into an ecosystem through meteorological, geological, or biological transport mechanisms and leave an ecosystem *via* the same routes. Meteorological inputs include dissolved matter in rain and snow, atmospheric gases, and dust blown by the wind; geological inputs include elements transported by the surface and subsurface drainage; and biological inputs include movements of animals between ecosystems. Figure 9.4 gives a diagrammatical view of nutrient cycling.

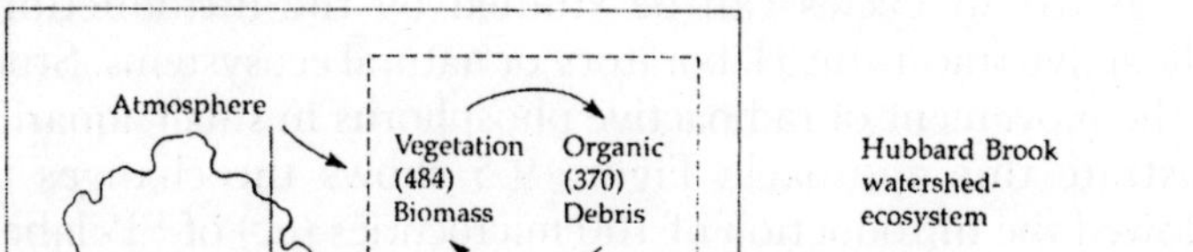

Figure 9.4 A nutrient cycling model. Within the ecosystem a nutrient element, such as calcium, may be found in the organic compartment, where it is a part of living and dead biomass; in the available nutrients compartment, where it occurs as a cation on exchange surfaces or is dissolved in the soil solution; or in the soil and rock minerals compartment, where it occurs as a component of primary or secondary minerals. Calcium does not reside in the atmospheric compartment. Nutrients flow between compartments along pathways such as nutrient uptake, decomposition of organic matter, weathering of minerals, or formation of new minerals. The ecosystem is connected to the larger biogeochemical cycles by meteorologic, geologic, and biologic vectors that move nutrients across ecosystem boundries. Data are from Hubbard Brook watershed ecosystem, where input of nutrients with sedimentary cycles is almost wholly meteorologic, while output is almost wholly geologic. All the data are in kg ha^{-1} and kg ha^{-1} yr^{-1} (Likens et al. 1977).

The time frame for nutrient cycling depends on the compartment. Metabolic processes may turn over elements in minutes. For example, in photorespiration carbon is moved from the atmosphere, into the plant, and back in a matter of minutes. Nutrients in leaf litter in a forest may turn over within an ecosystem annually, or require several years or decades to turn around. Some biochemical process (deposition, tectonic movement which exposes sediments, and subsequent retrieval by some root systems) may require hundreds of million of years.

Nutrient cycles can be studied by the introduction of radioactive tracers into laboratory or natural ecosystems. Studies on the movement of radioactive phosphorus in small aquariums illustrate this approach. Figure 9.5 shows the changes that followed the introduction of 100 microcuries (μc) of ^{32}P- labelled phosphoric acid in a 200-liter aquarium. There is a very rapid initial uptake of ^{32}P by phytoplankton. One half of the ^{32}P had been taken up by the phytoplankton within 2 h, and within 12 h there was an equilibrium of uptake and excretion of ^{32}P between the phytoplankton and the water. Filamentous algae attached to the sides and bottom of the aquarium slowly picked up ^{32}P, and crustaceans grazing on phytoplankton began to accumulate ^{32}P even more slowly. As the experiment progressed an increasing fraction of the radioactive tracer began to accumulate in the bottom mud and was tied up in less-active or bound form in the sediments. Some ^{32}P does move from the sediment back into the water column, but more moves down into the sediment and thus accumulates.

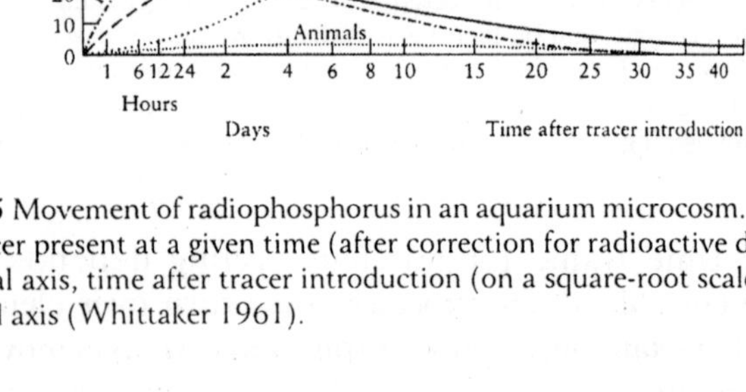

Figure 9.5 Movement of radiophosphorus in an aquarium microcosm. Percentage of the tracer present at a given time (after correction for radioactive decay) is on the vertical axis, time after tracer introduction (on a square-root scale) is on the horizontal axis (Whittaker 1961).

9.2.1 Carbon cycle

Next to water, carbon is the most significant element in organisms, constituting 49% of their dry weight; and, the fixation of carbon by plants in photosynthesis through geological time accounts for the oxygen in the earth's atmosphere. The gaseous reservoir of carbon is the atmosphere, which has an average concentration of 0.03% carbon dioxide; whereas carbon is found in water as hydrogencarbonate (HCO_3^-) and carbonate (CO_3^{2-}) ions. Sedimentaion gives rise to carbonate rocks such as limestones and dolomites.

The amount of CO_2 in the air has increased slowly since about 1850 from about 280 ppm to slightly over 350 ppm at present. The main reason for this increase is the burning of fossil fuels, but removal of forests and grasslands that use more CO_2 in photosynthesis than they release in respiration has also contributed. The small increases show, overall, that return of the gas to the air is nearly balanced by its use in photosynthesis or that its concentration is otherwise maintained fairly constant. Important return agencies are respiration of plants, microorganisms, and animals, and activities of volcanoes, factories, and automobiles (Fig. 9.6), but oceans and seas are even more important in preventing rapid rises in atmospheric CO_2 levels. In these, solid and dissolved carbonates equilibrate with CO_2, a change in one eventually affecting the other, so that the concentration of CO_2 on a worldwide scale is buffered by the carbonates in water. Within plant canopies, as in a maize field for example, the CO_2 content can fall to 260 ppm or less during daylight hours, whereas the content there can reach 400 ppm during darkness because of respiration by plants and soil microbes.

Worldwide increases in atmospheric CO_2 are of concern because they could lead to warming of the earth's surface (Bazzaz 1990). Such warmings could, within hundreds of years, cause melting of so much ice in polar regions that the ocean would rise and flood many coastal cities. Other accompanying climatic changes, especially in rainfall patterns, would greatly alter both agriculture and natural vegetation. The situation remains uncertain, however, because atmospheric particulate pollutants (the ever-present *haze*) reflect the sun's rays, and increases in these pollutants could lead to global cooling.

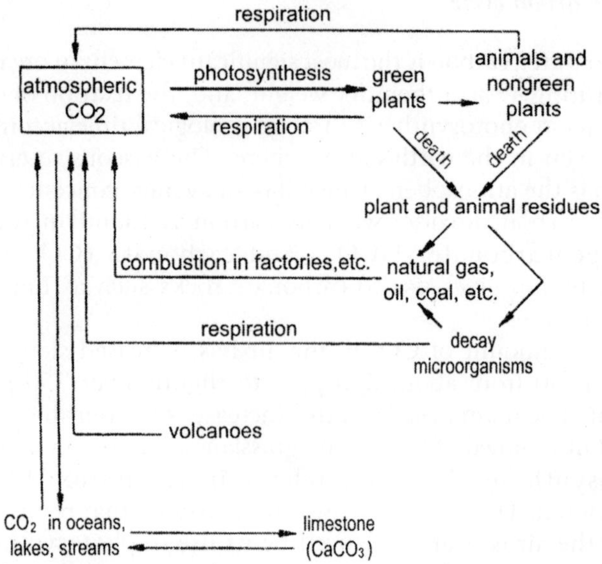

FIGURE 9.6 THE CARBON CYCLE IN NATURE

9.2.2 Nitrogen cycle

The cycle of fixation of gaseous N_2, assimilation of fixed N_2 by plants and then by animals, and return of gaseous N_2 to the atmosphere by denitrifying bacteria is called the *nitrogen cycle* (Fig. 9.7).

Vast amounts of N_2 occur in the atmosphere (78% by volume), yet it is energetically difficult for living organisms to obtain the N atoms of N_2 in a useful form. Although N_2 moves into leaf cells along with CO_2 through stomates, enzymes are available to reduce only the CO_2, so N_2 moves out as fast as in. Most of the N_2 in living organisms arrives there only after **fixation** (reduction of atmospheric N_2 to NH_4^+) by prokaryotic microorganisms, or by industrial fixation to form fertilizers.

Principal N_2-fixer prokaryotes include certain free-living soil bacteria, free-living cyanobacteria (blue-green algae) on soil surfaces or in water, cyanobacteria in symbiotic associations with fungi in lichens or with ferns, mosses, and liverworts (Peters 1978), and by bacteria or other microbes associated symbiotically with

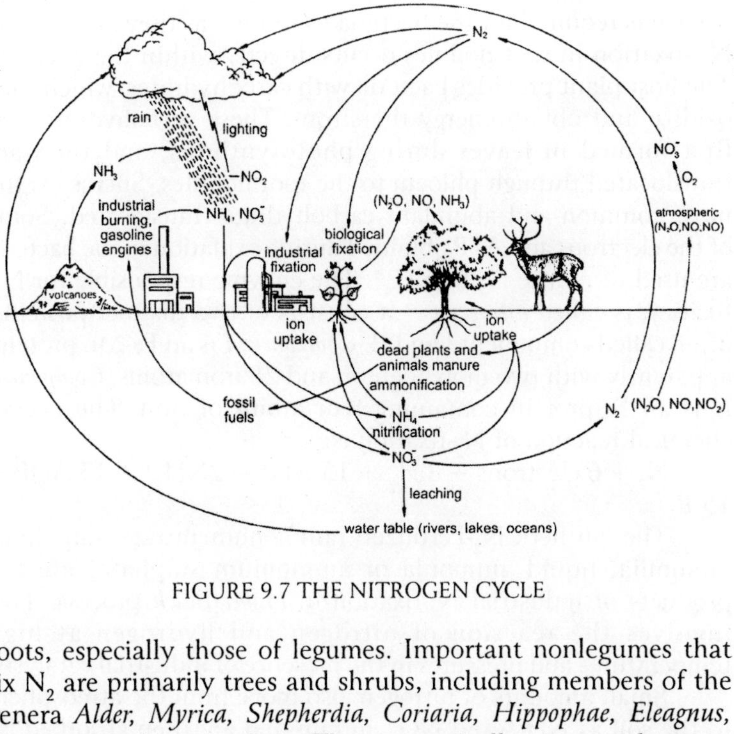

FIGURE 9.7 THE NITROGEN CYCLE

roots, especially those of legumes. Important nonlegumes that fix N_2 are primarily trees and shrubs, including members of the genera *Alder, Myrica, Shepherdia, Coriaria, Hippophae, Eleagnus, Casuarina* and some others. They are typically pioneer plants on nitrogen-deficient soils.

The microorganisms responsible for N_2 fixation in roots of many species are actinomycete-like organisms (filamentous bacteria), although cyanobacteria also fix N_2 in some tropical trees. In the legumes, bacterial species of the genus *Rhizobium* are responsible. A particular *Rhizobium* species is generally effective with only one legume species. *Rhizobia* are aerobic bacteria that persist saprophytically in the soil unit they infect a root hair or damaged epidermal cell. The plant responds to this infection by producing *root nodules* made largely of tetraploid cells containing bacteria and some diploid cells without bacteria. The root nodules contain *leghaemoglobin*, a red molecule that gives legume nodules a pink colour. Leghaemoglobin is thought to transport O_2 into the bacteria at carefully controlled rates. Too much O_2 inactivates the enzyme that catayzes N_2 fixation, possibly because nitrogen and oxygen resemble each other in size and shape. However, some oxygen is required for the bacteria to respire, as they are aerobic. N_2 -fixation in root nodules occurs directly within the bacteria. The host plant provides bacteria with carbohydrates, which they oxidize and obtain energy therefrom. These carbohydrates are first formed in leaves during photosynthesis, and then are translocated through phloem to the root nodules. Sucrose is the most common and abundant carbohydrate translocated. Some of the electrons and ATP obtained during oxidation in the bacteria are used to reduce N_2 to NH_4^+. The enzyme responsible for N_2-fixation is called *nitrogenase*. It consists of two distinct proteins, often called component I and II. *Component I* is an Fe-Mo protein, apparently with two molybdenum and 28 iron atoms; *Component II* is an Fe-protein containing four atoms of iron. The overall chemical reaction of N_2-fixation is:

$$N_2 + 6 \text{ electrons} + 8 H^+ + 15 \text{ ATP} \rightarrow 2NH_4^+ + 15 \text{ ADP} + 15 \text{ Pi}$$

The synthetic N_2-Fertilizers (ammonium nitrate, anhydrous ammonia, liquid ammonia or ammonium sulphate) are the products of industrial N_2-fixation in *Haber-Bosch* process. This involves the reaction of nitrogen and hydrogen at high temperatures and pressures in the presence of industrial catalysts.

Small amounts of nitrogen also move from the atmosphere to the soil as NH_4^+ and NO_3^- in rain and are then absorbed by roots. This NH_4^+ arises from industrial burning, volcanic activity, and forest fires, while NO_3^- arises from oxidation of N_2 by O_2 or ozone in the presence of lightning or ultraviolet radiation. Another source of NO_3^- is the oceans. Wind-whipped white caps produce minute droplets of water called *aerosols*, from which the water evaporates, leaving ocean salts suspended in the atmosphere near coastlines. These salts can be brought to the land in rainwater. They are called *cyclic salts*, because they eventually cycle through streams back to the oceans.

The amount of N_2 removed from the atmosphere by N_2-fixation is closely balanced by the amount returned to the atmosphere by soil microorganisms that convert organic nitrogen to gaseous N_2. Several steps are involved. First, ammonia produced by N_2-fixation is assimilated by plants and microorganisms into amino acids, proteins, and other nitrogenous products. Some plants may be consumed by animals. When animal wastes and dead plants, animals, and microorganisms are subjected to decay, their nitrogen is converted to ammonium ions by the action of saprotrophic fungi and bacteria. This process of conversion of organic nitrogen to NH_4^+ by soil microbes is called **ammonification**. The saprotroph microbes use the ammonia to synthesize their own proteins and other nitrogen-containing compounds; or the ammonia formed by ammonification may react with other chemicals present in the soil such as CO_2 and water, to form ammonium salts (ammonium carbonate, for example). The next step in the sequence is **nitrification**, a process that can only be carried out by certain nitrifying bacteria. In warm, moist soils with pH near 7, ammonium ions (NH_4^+) are oxidized within a few days of their formation or their addition as fertilizer by nitrifying bacteria. This oxidation benefits the bacteria performing the reactions by releasing energy which the bacteria can use for the synthesis of ATP. Nitrification takes place in two stages. In the first stage of nitrification, NH_4^+ is oxidized to nitrite (NO_2^-) by bacteria of the genera *Nitrosomonas, Nitrosospira, Nitrosococcus* and *Nitrosolobus* (Hamilton 1988). The first reaction involves the addition of oxygen to ammonia, giving rise to hydroxylamine (NH_2OH). This is then further oxidized to nitrite. In the second stage of nitrification, nitrite is oxidized to nitrate by bacteria of the genera *Nitrobacter*, *Nitrospira* and *Nitrococcus*. The reaction proceeds by the addition of water followed by the removal of hydrogen. The association between *Nitrosomonas* and *Nitrobacter* has been described as one of commensalism (Gooday 1988). The two species exist in the same soil environment without much mutual influence except that *Nitrobacter* depends on *Nitrosomonas* for its nitrite. Most plants absorb the majority of their nitrogen as nitrate. However, many plants also absorb ammonium. In acidic soils, conversion of ammonium to nitrate is slow and in forests on acidic soils, many trees absorb most of their nitrogen as ammonium. Finally, nitrates in the soil are reduced by certain anaerobic soil bacteria to nitrogen dioxide (NO_2), dinitrogen oxide (N_2O), nitrogen monoxide (NO) or nitrogen (N_2), which escape into the atmosphere. This process is called **denitrification**. Denitrifying bacteria are especially active under anaerobic conditions in wet soils with high organic matter content. Not only denitrification of nitrate may occur in soils, but soil nitrate may also be lost to rivers and oceans. Oxidized forms of nitrogen in the atmosphere are important ecologically, because when converted to NO_3- they contribute HNO_3 to acidic rainfall ("acid rain").

9.2.3 Sulphur cycle

The sulphur cycle involves the uptake of divalent sulphate anions (SO_4^{2-}) by producers, largely through their roots, and the release and transformation of the sulphur in a number of different steps and variety of forms, including sulfhydryl (-SH), hydrogen sulphide (H_2S), sulphites (SO_3^{2-}), and molecular sulphur (S) (Fig. 9.8).

There are three major natural sources from which plants can be supplied with available sulphur: (a) organic matter, (b) soil minerals, and (c) sulphur gases. In humid region surface soils, most of the sulphur (90% and more) is present in organic form (proteins and specific amino acids such as cysteine, cystine, and methionine). These materials are bound with the humus and clay fractions, which protects them from microbial attack. Over time, however, the microorganisms can simplify the bound sulphur into soluble inorganic forms. In arid and semiarid regions, the proportion of organic sulphur is not so high because of the lower organic matter level of these soils. The inorganic forms of sulphur are not as plentiful as the organic forms, but they include the soluble and available compounds such as calcium sulphate, sodium sulphate, etc. on which plants and microbes depend. In addition, the combustion of fuels, especially high-sulphur coal, releases sulphur dioxide and other sulphur gases and solids into the atmosphere. Some of these materials are oxidized in the atmosphere to sulphates, forming sulphuric acid by reacting with water and sulphate salts such as calcium and magnesium sulphate. Some of the solids and gases return to the earth as dry particles, *dry deposition*, whereas some are brought down with precipitation, *wet deposition*. Atmospheric sulphur becomes part of the soil-plant system in three ways. The wet deposition materials, which are usually high in H_2SO_4, are absorbed by soils. Part of the dry deposition is also absorbed directly by soils, while some is absorbed directly by plants.

The organic forms of sulphur must be mineralized (the release of organically bound forms of an element into inorganic

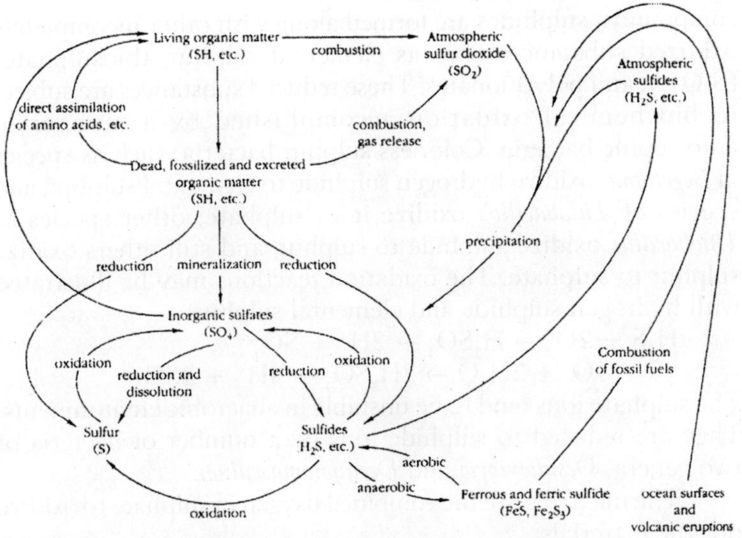

FIGURE 9.8 THE SULPHUR CYCLE

mineral forms is termed **mineralization**) before it is absorbed by plants and microorganisms and moves through the sulphur cycle. When conditions are proper (good soil drainage and aeration, adequate soil moisture, 25-35°C temperature range, and nearly neutral pH) for general microbial activity, sulphur mineralization occurs:

$$\underset{\text{(protein and other organic combinations)}}{\text{Organic sulphur}} \rightarrow \underset{\text{(}H_2S\text{ and other sulphides)}}{\text{decay products}} \xrightarrow{O_2} \underset{\text{Sulphates}}{SO_4^{2-} + H^+}$$

Immobilization (conversion of inorganic forms into organic forms) of inorganic forms of sulphur occurs when low-sulphur, energy-rich organic materials are added to soils not plentifully supplied with inorganic sulphur. The energy-rich material stimulates microbial growth, and the inorganic sulphate is synthesized into microbial tissue. Only when the microbial activity subsides does the inorganic sulphate form again appear in the soil solution.

During the microbial decomposition of organic sulphur

compounds, sulphides are formed along with other incompletely oxidized substances such as elemental sulphur, thiosulphates ($S_2O_3^{2-}$), and polythionates. These reduced substances are subject to biochemical oxidation accomplished by a number of autotrophic bacteria. Colorless sulphur bacteria, such as species of *Beggiatoa*, oxidize hydrogen sulphide to elemental sulphur, and species of *Thiobacillus* oxidize it to sulphate; other species of *Thiobacillus* oxidize sulphide to sulphur, and still others oxidize sulphur to sulphate. The oxidation reactions may be illustrated with hydrogen sulphide and elemental sulphur:

$$H_2S + 2O_2 \rightarrow H_2SO_4 \rightarrow 2H^+ + SO_4^{2-}$$

$$2S + 3O_2 + 2H_2O \rightarrow 2H_2SO_4 \rightarrow 4H^+ + 2SO_4^{2-}$$

The sulphate ions tend to be unstable in anaerobic environments. They are reduced to sulphide ions by a number of bacteria of two genera, *Desulfovibrio* and *Desulfotomasculum*.

The microbes use the combined oxygen in sulphate to oxidize organic materials:

$$\underset{\text{Organic alcohol}}{2R\text{-}CH_2OH} + SO_4^{2-} \rightarrow \underset{\text{Organic acid}}{2R\text{-}COOH} + 2H_2O + \underset{\text{Sulphide}}{S^{2-}}$$

In poorly drained soils, the sulphide ion will react immediately with iron or manganese, which in anaerobic conditions would be present in the reduced forms. This reaction might be expressed as:

$$Fe^{2+} + S^{2-} \rightarrow \underset{\text{Ferrous sulphide}}{FeS}$$

$$Mn^{2+} + S^{2-} \rightarrow \underset{\text{Manganese sulphide}}{MnS}$$

Sulphide ions will also undergo hydrolysis to form gaseous hydrogen sulphide, which is the cause of the rotten egg smell of swampy or marshy areas. Sulphur reduction may take place with sulphur-containing ions other than sulphates. For example, sulphites (SO_3^{2-}), thiosulphates ($S_2O_3^{2-}$), and elemental sulphur (S) are rather easily reduced to the sulphide form by bacteria and other organisms. The highly insoluble ferrous sulphide (FeS) and ferric sulphide (Fe_2S_3, pyrite) or relatively insoluble calcium sulphate ($CaSO_4$) represent the sedimentary aspect of the sulphur cycle. Of considerable ecological significance is ferrous sulphide, which is formed under anaerobic conditions. It is insoluble in netural or alkaline water; consequently, sulphur has the potential for being bound up under these conditions to the limits of the amount of iron present. Once buried, pyrite appears to be geologically stable and is a primary reservoir of both iron and sulphur in salt-marsh as well as other sediments (Luther *et al.* 1986).

Sulphur is absorbed from soils as divalent sulphate cations (SO_4^{2-}). It is metabolized by roots only to the extent that they require it, and most sulphate is translocated unchanged to the shoots in the xylem. Because enough sulphate is present in most soils, sulphur-deficient plants are fairly uncommon. Nevertheless, they have been observed in several parts of Australia, some regions of Scandinavia, south-western grain-producing regions of Canada, and in scattered parts of the northwestern United States. Deficiency symptoms consist of a general chlorosis throughout the entire leaf, including vascular bundles (veins). Sulphur is not easily redistributed from mature tissues in some species, so deficiencies are usually noted first in younger leaves. In other species, however, most leaves become chlorotic at about the same time, or even first in older leaves. Many crop plants, including the roots contain about one-fifteenth as much total sulphur as nitrogen, and this appears to be a useful guide to evaluate nutritional needs. Most of the sulphur in plants occurs in proteins, specifically in the amino acids cysteine and methionine that are building blocks for proteins. Other essential compounds that contain sulphur are the vitamins *thiamine* and *biotin* and *coenzyme A*, a compound essential for respiration and for synthesis and breakdown of fatty acids.

9.2.4 Phosphorus cycle

Phosphorus circulates through water, the earth's crust, and living organisms in the phosphorus cycle (Fig. 9.9). In this sedimentary cycle, crystalline rocks are the major natural source of phosphorus which mainly occurs as phosphates. There is no atmospheric phase in the phosphorus cycle. By slow weathering of rocks the phosphates are released. Second to nitrogen, phosphorus is most often the limiting element is soils and waters.

Plants obtain their phosphorus from the soil primarily as the monovalent dihydrogenphosphate anion ($H_2PO_4^-$) and less rapidly as the divalent hydrogenphosphate anion (HPO_4^-). The soil pH controls the relative abundance of these two forms, H_2PO^{4-} being favoured below pH 7 and HPO_4^{2-} above pH 7. The absorption of phosphate, along with ammonium, nitrate and potassium ion (K^+), is greatly aided by the presence of mycorrhizae. Much phosphate is converted into organic forms upon entry into the root or after transport through xylem into the shoot. In contrast to nitrogen and sulphur, phosphorus never undergoes reduction in plants and remains as phosphate, either free or bound to organic forms of esters. Phosphorus-deficient plants are stunted and, in contrast to those lacking nitrogen, are often dark in colour. Anthocyanin pigments sometimes accumulate. Oldest leaves become dark brown as they die. Maturity is often delayed compared to plants containing abundant phosphate. In many species there is a close interaction between the phosphorus and nitrogen regarding maturity, excess nitrogen delaying and abundant phosphorus speeding maturity. If excess phosphorus is provided, root growth is often increased relative to shoot growth. This, in contrast to effects with excess nitrogen, causes low shoot-to-root ratios. Phosphate is easily redistributed in most plants from one organ to another and is lost from the older leaves, accumulating in younger leaves and in developing flowers and seeds. As a result, deficiency symptoms occur first in more mature leaves. Phosphorus is an essential part of many sugar phosphates involved in photosynthesis, respiration, and other metabolic processes, and it is also part of nucleotides, as in RNA and DNA, and the phospholipids present in membranes.

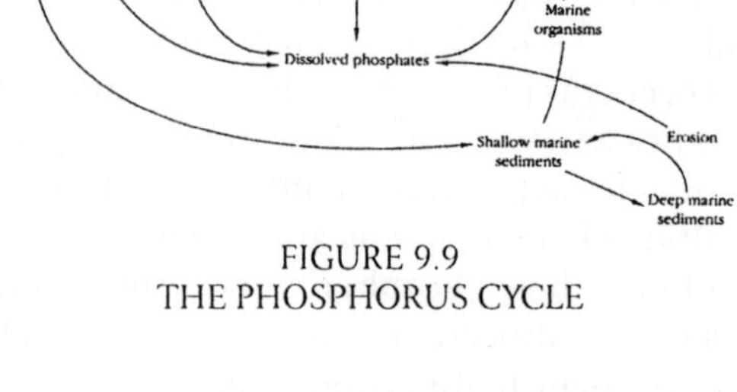

FIGURE 9.9
THE PHOSPHORUS CYCLE

Animals obtain their phosphorus from plants, if they are herbivores, or from other animals, if they are carnivores. Decomposers return phosphorus to the soil as the phosphate ion. Phosphorus held in organic from can be mineralized and immobilized by the same general processes pertinent for nitrogen and sulphur:

$$\xleftarrow{\text{Immobilization}}$$

$$\text{Organic P forms} \underset{\text{microbes}}{\overset{\text{microbes}}{\rightleftharpoons}} H_2PO_4^- \overset{Fe^{3+}, Al^{3+}, Ca^{2+}}{\rightleftharpoons} \text{Fe, Al, Ca phosphates (Fixed phosphorus)}$$

$$\xrightarrow[\text{Mineralization}]{}$$

Soluble phosphorus compounds are released as organic residues and humus are decomposed. The resulting soluble inorganic phosphate ion ($H_2PO_4^-$) is subject to uptake by plants or to fixation into insoluble forms. Should organic residues low in phosphorus but high in other nutrients be added to a soil, rapid microbial activity would take place and available $H_2PO_4^-$ in the soil solution would temporarily disappear, just as is the case for soluble NH_4^+, NO_3^- and SO_4^{2-} ions.

When phosphates form compounds with metals like aluminum, iron or calcium, phosphorus becomes unavailable to plants and is thus lost to the cycle. Under very acid conditions, enough soluble aluminum and iron is present to precipitate phosphate as iron and aluminum phosphates, a form of phosphorus that is unavailable to the plants:

$$Al\,(OH)_3 + H_2PO_4^- \rightarrow Al\,(OH)_2.H_2PO_4$$
$$Fe\,(OH)_3 + H_2PO_4^- \rightarrow Fe(OH)_2.\,H_2PO_4$$

If an $H_2PO_4^-$ containing fertilizer such as superphosphate is added to a dry alkaline soil (pH 8) with excessive amounts of calcite ($CaCO_3$), the $H_2PO_4^-$ ion quickly reacts to from almost insoluble tricalcium phosphate:

$$Ca\,(H_2PO_4)_2.2H_2O + 2CaCO_3 \rightarrow Ca_3\,(PO_4)_2 + 2CO_2 + 2H_2O$$

Biological processes, such as the formation of teeth and bone - both of which are more or less resistant to weather - and excretion, also account for considerable losses from the biotic portion of the phosphorus cycle. The tremendous deposits of *guano* (seabird excreta) that have accumulated over hundreds of years along the western coast of South America and are "mined" for fertilizer are a case in point in such removal.

9.2.5 Other minerals

Calcium, magnesium and other minerals are also cycled like other materials. These elements are slowly released from the rocks by water and wind actions. These are either blown into the air as dust or absorbed by plants through their roots. Recycling occurs between the organisms and their surroundings. Some organisms such as molluscs and corals, deposit a substantial quantity of calcium in their shells and skeleton that are not available for quick cycling. All sedimentary elements return very slowly to the cycle *via* the sea food chain or through coastal sea water sprays, in general.

9.3 Mineral Cycling In Land Ecosystems

9.3.1 Circulation between organisms and soils

Under natural conditions, the cycling of minerals between the organisms in the ecosystem and the soil is, to a great extent, a closed system (Fig. 9.10). The mineral nutrients taken up by the plants are returned to the soil either directly or through the food chain, packaged in organic material; there they are mineralized by the decomposers and taken up by the soil colloids. The mechanisms for this recycling of minerals is a crucial part of the circulation; if the organic wastes (of which there are more as the productivity of the plant canopy increases) were removed, the store of nutrients represented by the incorporated minerals would be lost to the ecosystem.

Mineralization and the production of organic matter must keep pace with one another. Where the decomposition of litter and humus proceeds too slowly, the growth rate of the plants usually slows until it is in equilibrium with the rate of mineralization. If the minerals bound in organic matter are liberated rapidly and returned in abundance, the primary producers are better supplied with nutrient ions and can build up a greater mass. The remarkable productivity of tropical rain forests is thus determined by the rapid mineral turnover there. The constant favourable temperature and the high humidity promote decomposition by microorganisms at an extraordinary rate, so that the nutrient elements in the soil are only briefly tied up in organic compounds and soon are made available to the plants again in inorganic form. Excessive liberation of the inorganic elements or a reduction in withdrawal by the vegetation (for example, after the system has been disturbed by fire or the actions of man) can disturb the mineral balance of the ecosystem. Nutrient elements in the mineralized state are not only more accessible to the plants; they are also more mobile in the soil and thus can be more easily leached out. Therefore the organically bound minerals in the biomass and in the soil are important in that they constitute relatively secure stores; all members of the ecosystem benefit.

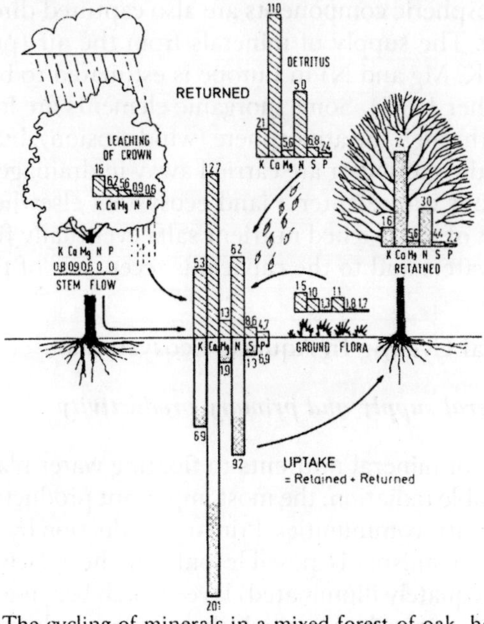

Figure 9.100 The cycling of minerals in a mixed forest of oak, hornbeam and beech in Belgium. All quantities are given in units of kg ha^{-1} yr^{-1}. The yearly uptake of minerals by the stand was calculated by summation of the quantities retained in the phytomass (dotted areas) and those returned to the soil (cross-hatched) by way of washing and leaching of the canopy, stem flow, and the litter from trees and ground flora (Larcher 1975).

9.3.2 Mineral transport to and from the ecosystem

The system of functional interactions represented by plants, microorganisms and soil is regularly supplied with inorganic elements from external sources, and there is a continuous loss of these elements from the ecosystem, so that the rate of microbial mineralization in the soil is not the sole determinant of mineral availability. The *input* of inorganic elements to the ecosystem comes from the underlying rock, the water and the air; to these are added fertilization in the case of agricultural areas. Weathering brings to the soil and the plants quantities of mineral nutrients that are hard to estimate, but no doubt significant. In raw soils it is primarily by weathering that the nutrient requirements of plants are met. Minerals are moved into the root zone by groundwater and by water rising by capillarity. Precipitated water carries inorganic materials contained in the atmosphere as gases (SO_2, NH_3, and gases containing nitrogen oxides), dust, fog or aerosols. These atmospheric components are also captured directly by the plant cover. The supply of minerals from the air (primarily Cl, Na, Ca, S, K, Mg and N) in Europe is estimated to be 25-75 kg/ha/yr (Larcher 1975). Some inorganic elements are lost from the ecosystem through the atmosphere (wind erosion), but most seep down into the ground or are carried away in drainage water. The dissolved ions can re-enter a land ecosystem elsewhere, but the greater part of the leached nutrient salts eventually finds its way to inland waters and to the capacious receptacle of the ocean.

9.4 Mineral Cycling In Aquatic Ecosystems

9.4.1 Mineral supply and primary productivity

The supply of mineral nutrients to floating water plants is, next to the available radiation, the most important production-limiting factor in aquatic communities. Primary production by autotrophic planktonic organisms is possible only in the relatively narrow *euphotic* (adequately illuminated) layer, which becomes noticeably depleted of minerals (particularly nitrogen, phosphorus and silicon) when the phytoplankton multiplies rapidly. Even before such depletion, nitrogen and phosphorus are present only in trace amounts in fresh and sea water. The planktonic algae manage initially with this limited supply due to their special ability to concentrate minerals. Algae can accumulate stores of phosphate so large that they suffice for synthesis and metabolism throughout successive generations.

The onset of greatest productivity in waters with seasonal layering occurs in the spring, when the higher elevation of the sun and the increase in day length permit a greater photosynthetic yield, and when the concentration of nutrients is maximal as a result of the preceding mixing of the water (Fig. 9.11). At this time the spring blooms of plankton occur. The growing population soon uses up the supply of nutrients; this supply is no longer adequately replenished because in the meantime the thermocline has formed and prevents the movement of mineral nutrients from the deep water into the euphotic zone. In summer there is a reduction in the population density of phytoplankton, brought about primarily by the lowered rate of reproduction and the intensive use of phytoplankton as food by consumers, which themselves have now increased in number. Not until autumn does productivity increase, as mixing of the water is renewed; then the population density again rises somewhat before it falls back to the winter low because of lack of light.

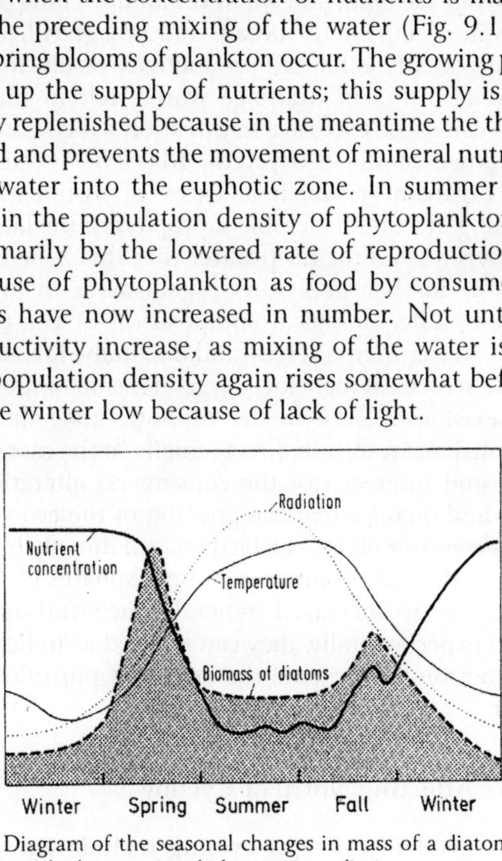

Figure 9.11 Diagram of the seasonal changes in mass of a diatom population (dotted area), with the associated changes in radiation, temperature, and the concentration of mineral nutrients in the euphotic zone of the ocean. Because of the increased irradiation and warming in the spring, there is at first an adequate supply of nutrients, and production of phytoplankton mass proceeds at a high rate. The nutrients are, however, rapidly used up, which in turn soon leads to a decline in productivity. Meanwhile, the zooplankton population has greatly increased; the resulting increased consumption, together with sinking of the floating

algae, in summer, reduces the biomass of the diatom population. In late fall and winter, mineral nutrients are carried up by water circulation from the deeper layers, but then it is too cold and dark for high-yield production by the phytoplankton.

9.4.2 Eutrophication and its consequences

Wherever mineral salts appear in the surface waters, having been washed in from the land, and wherever an upwelling of nutrient-rich deep water occurs in the oceans, *eutrophic* (fertilized) regions of water with greatly enhanced productivity result. In rivers and lakes, a great influx of drainage water rich in nitrates and phosphates can cause excessive fertilization which endangers the overall balance of the ecosystem. Massive development of planktonic organisms results in an excess of organic wastes and a corresponding increase in aerobic saprophytes; the latter remove so much oxygen from the deep water that this zone is rendered uninhabitable for fish and other organisms with high oxygen requirements. The biological composition of the deep-water regions of lakes is then altered fundamentally, favouring forms that can live with extremely low oxygen tension or are capable of anaerobic existence. Even in the euphotic zone the increased supply of nutrients results in direct (chiefly in the case of primary producers) and indirect (for the consumers) alterations in the qualitative and quantitative composition of the ecosystem.

In rivers, water plants can be damaged directly by the influx of drainage water rich in nitrates and phosphates (Table 9.4). If their sensitivity to increased mineral concentration has been determined experimentally, they can be used as indicators, from their distribution, to estimate the degree of purity or pollution of the river.

9.5 Factors Affecting Nutrient Cycling

Factors involved in nutrient cycling include fire and earth movement and the various aspects of the hydrologic cycle: rainfall, leaching, evapotransipiration, runoff, infiltration, and salt spray. Biotic factors include grazing, decomposition, root exudation, and retrieval of deep-lying nutrients.

Nutrient budgets are strongly correlated with the hydrologic

TABLE 9.4
DISTRIBUTION OF FLOWERING PLANTS IN A RIVER (THE MOOSACH IN BAVARIA) THROUGHOUT ITS COURSE, AS A FUNCTION OF THE AMMONIA AND PHOSPHATE CONCENTRATIONS (LARCHER 1975)

	Upper reaches	*Middle zone*	*Lower reaches*
Indicator plants	*Potamogeton coloratus** *Juncus subnodulosus*	*Potamogeton densus* *Hippuris vulgaris*	*Potamogeton pectinatus* *Potamogeton crispus* *Elodea canadensis* *Ranunculus fluitans* *Callitriche obtusangula*
Average content of NH_4^+ in the water (mg/l)	0.03	0.09	0.28
Average content of PO_4^{3-} in the water (mg/l)	0.04	0.07	0.45

* In experiments, *Potamogeton coloratus* is damaged at NH_4^+ concentrations above 10 mg/l.

cycle: precipitation carries nutrients in solution, run off and infiltration remove nutrients from a system or move them down the soil column, and evapotransipiration of water concentrates and conserves nutrients. Groundwater and water rising by capillary action transport nutrients into the root zone.

Leaching refers to the removal of soluble constituents from the soil or litter by percolating water. The primary cause of nutrient loss from a system is leaching, and the most easily leached nutrients are K and Ca (Tappeiner and Alm 1975). Hunt (1977) calculated that 45-65% of the *labile component* (the component which readily undergoes chemical change) may be leached from grassland soils and litter in as short a time as 4 hours. This process is positively correlated with increasing temperature. For example, up to 36% of the labile component will be transferred by leaching at 25°C. Leaching from attached leaves, on the other hand, adds nutrients to the soil. As rain runs over the foliage, it passively absorbs nutrients. The nutrients are then carried to the ground in **throughfall** (rainwater which falls through canopy) and

stemflow (the fraction of precipitation that flows down the stem or trunk of a plant and then enters the soil).

Rather than promoting nutrient loss, evapotransipiration actually conserves mineral elements because excess water, which would promote loss through leaching, is removed without harm to the system, leaving the nutrients behind.

In general, the successional status of community can be deduced from *nutritional budgets*, a measurement of nutrient input and output. If elements are being stored, as indicated by a positive input/output balance, new net biomass is being accumulated, and the system is aggrading (that is, it is a seral stage). A small negative balance may represent weathering of parent material, and suggests that the system is at equilibrium (a mature, or climax community). Rich, fertile systems show a greater loss of nutrients than do impoverished soils as precipitation increases, because nutrient loss is proportional to the volume of runoff. The acidity of rainwater also influences nutrient loss: cations such as K and Ca may be displaced by the hydrogen ions of the acid, and will be leached away by seepage and surface runoff. Nutrient-unsaturated soils tend to lose a lower proportion of nutrients even with increased runoff (Jordan and Kline 1972).

In chaparral regions, fire may serve to: (i) break down sclerophyllous litter, which is resistant to biotic decomposition; (ii) remove inhibitors of microbial decomposition; and (iii) alter wettability of the soil. Nutrients are relased in the form of soluble mineral ash by slash and burn agriculture. Thus, in a tropical rain forest, slash and burn temporarily enhances the nutrient regime. Crop yields are good at first, but diminish quickly. Reasons for this include: (i) the fire destroys such recycling mechanisms as mycorrhizae within the soil ; (ii) the nutrients are released in a single large pulse, and their availability probably exceeds the exchange capacity of the soil, thus, nutrients are quickly leached out of the root zone (Jordan and Kline 1972).

Earth movement such as mountain building (*orogeny*), movement along a lateral fault, volcanism, and other types of tectonic activity exert an enormous influence on mineral cycling. The impact may be sudden or require aeons of time. For example, volcanic ash dissolve easily in water, and in tropical regions it may enhance the nutrient regime within a short time. On the other hand, an atom of phosphorus may be lifted from the sea by tectonic movement over aeons of time, and then exposed to weathering, ultimately to enter the biosphere and resume cycling, at least for a time.

Two factors determine the amount of litter in an ecosystem: the total litter produced in a unit of time, and the rate at which it is decomposed. In boreal forests, such as the jack pine (*Pinus banksiana*) system in northern Ontario, Canada, the annual litter production is about 4000 kg/ha/yr. This organic material decomposes very slowly, requiring as much as 16 years for some elements to recycle (Foster and Morrison 1976). The acid soils that are produced contain a well-defined surface layer of organic material in which nutrients are taken up and immobilized in the biomass of microbes. In contrast to the slowly turning mineral cycle of the boreal forest, litter in the tropical rain forest decomposes quickly, and there is nearly complete turnover each year. Litter production in a tropical rain forest can be staggeringly high; leaf fall alone is 45-126 kg/ha every day (Walter 1973). The deciduous forest of the temperate zone stands midway between these two extremes. It exhibits moderate turnover times, such that a year's litter production is decomposed in 2-3 years (Larcher 1975). As Fig. 9.12 shows, the bulk of the carbon in the tropical rain forest is tied up primarily in the plant mass; the litter is rapidly converted and there is only a little organic matter in the soil. In the subalpine spruce forest, on the other hand, more organic matter is found in the litter and humus than in the plant canopy, chiefly because under the prevailing cool, damp conditions decay in the soil takes place much more slowly than in the tropics.

There are three stages of microbial decomposition: (i) formation of particulate detritus; (ii) production by saprophytes of immobile humus and the concomitant release of immobile, though soluble, organic compounds (in a relatively short time); and (iii) mineralization (mobilization) of humus (a slower process). Although the products of stages one and two are not available as nutrients, humic substances and other organics play a vital role in nutrient storage, forming complexes with minerals that enhance uptake by plants. Immobilization through decomposition serves to retain nutrients beyond the period of dormancy to the time of growth.

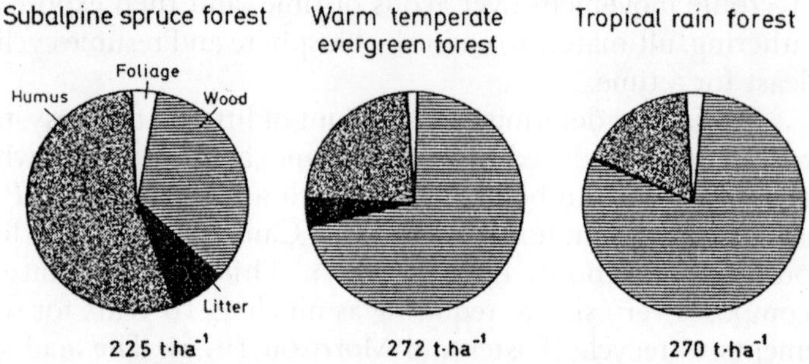

Figure 9.12 Content of organic carbon in the foliage, the wood, and the ground litter and humus in woodlands of regions differing in climate. In the tropical rain forest the carbon is tied up primarily in the plant mass; the litter is rapidly converted and there is only a little organic matter in the soil. In the subalpine spruce forest, on the other hand, more organic matter is found in the litter and humus than in the plant canopy, chiefly because under the prevailing cool, damp conditions decay in the soil takes much more slowly than in the tropics.

Stability is a measure of system's response to some outside perturbation. A stable system returns to steady-state conditions in a short-time following some irregularity in either the physical environment or in the flow and cycle of energy and materials. The shorter the recovery time, the higher the stability. Over ecological time, populations of organisms accumulate, concentrate and recycle nutrients. Up to a certain point, this accumulation enables succeeding populations to occupy the site. The rate of accumulation of even one element may limit succession. The accumulation process is governed by internal feed back and is, to a certain extent, self-reinforcing. For instance, climax forest is supported by a very large reserve of nutrients. The forest shields the soil against erosive nutrient loss and enhances evapotransipiration, which reduces streamflow and thus reduces mineral loss due to leaching. In general, evergreenness, size of storage compartments, erosion control, and immobilization of nutrients are properties that are involved in the stability of ecosystem nutrient cycles. Evergreenness, with progressive leaf fall, may be correlated with site characteristics (Monk 1966). Progressive leaf fall may be a significant conservation mechanism in dry, sterile sites. In a sense, forests on mesic, fertile sites can afford the "luxury" of deciduousness (that is, synchronous leaf fall). If a big pulse of nutrients released in a very short time period cannot be accommodated in the soil storage compartment, the nutrients may be lost through leaching. Furthermore, nutrients are more easily leached from broadleaf litter than from needle litter. Progressive needle fall results in a slowly decomposing litter. In fact, coupling of metabolism and mineral cycling is a stabilizing force within an ecosystem. The bridge between metabolism and mineral cycling is the availability of large storage compartments within the system. Large storage capability tends to buffer nutrient cycles against perturbation. A large standing crop (trees in tropical rain forest, for example) provides a spacious storage compartment; so do ample exchange sites for cations (such as young volcanic ash). The organically bound component in soil provides secure storage for nutrients also. In brief, there are two basic ecosystem strategies for ameliorating the effects of perturbation: maintenance of resistance to erosion (proliferation of understory shrubs, ferns and herbs following a fire in a forest ecosystem insulates the soil against erosion), and enhanced primary productivity.

10. BIOMES: THE MAJOR ECOSYSTEMS

In the broadest sense, there are two major types of ecosystems-aquatic and terrestrial. In turn, subdivisions can be recognized in each type, thus we can distinguish freshwater, mangrove, and marine aquatic ecosystems and several major types of terrestrial ecosystems, such as grassland, forest, and tundra. The former are distinguished on the basis of a major chemical difference (salt content), the latter generally on the basis of the predominant type of vegetation (grass, trees, etc.).

10.1 Terrestrial Biomes

The major terrestrial ecosystems occur on a major regional or subcontinental level and are often referred to as *biomes*, although terms like *provinces, biochores, regions*, and *formations* are used by some ecologists. The biome includes not only the climatic climax vegetation, which is the key to recognition, but the edaphic climaxes and the developmental stages as well, which in many cases are dominated by other life forms. Thus, grassland communities are temporary developmental stages in the deciduous forest biome where the broad-leaved deciduous tree is the climax life form. Many organisms require both the developmental and the climax stages in succession or the ecotones between them; therefore, all of the communities in a given climatic region, whether climax or not, are natural parts of the biome.

The nature and extent of each biome is determined by latitude and altitude (the principal determinants of climate-temperature range, major wind patterns and rainfall), nature of soil, and barriers. The terrestrial biomes of the world are shown in Fig. 10.1, whereas world climatic types and the associated biomes are summarized in Table 10.1. Fig. 10.2 is an attempt to map the biomes on scales of temperature and precipitation, the two master limiting factors of vegetation.

The series of the biomes occurring parallel to the lines of latitude can also be noted along the slopes of high mountain ranges like the Himalaya (Asia), Andes (South America) and the Rockies. In these mountain ranges the gradual change in the biome types follows the altitude rather than the latitude. The horizontal sequence of biomes extending over thousands of kilometres latitudinally is telescoped altitudinally into a few thousand metres. Temperature and precipitation rates (rainfall or snowfall) are most important determinants of these biomes. Temperature decreases with increasing altitudes from the earth's surface at a normal lapse rate of about 10°C per km.

In the tropical mountains the sequence of communities and conditions from the base of the mountain to the snow-line are: tropical forest (Terai), deciduous forest, coniferous forest and tundra. The higher the latitude in which the mountain is situated, the more northern (latitudinally) type of community and conditions are found at its base. Therefore, the number of habitats covering its slopes is reduced (Table 10.2).

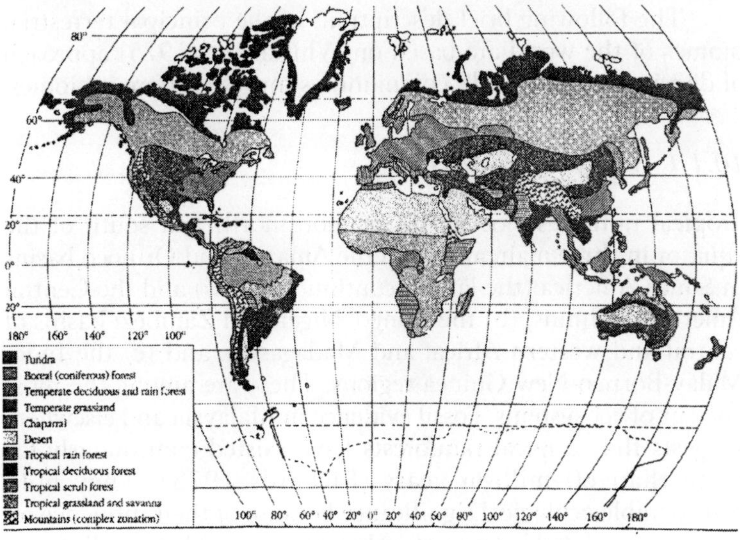

FIGURE 10.1
THE MAJOR BIOME TYPES OF THE WORLD (ODUM 1959)

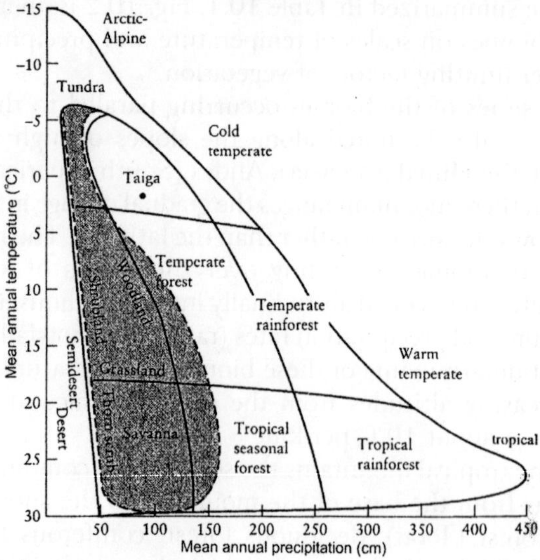

Figure 10.2 A pattern of world formations in relation to precipitation and temperature. Boundaries between formations are approximate. Soil and fire effects can shift the balance between woodland and grassland formations in the shaded region (Whittaker 1975).

The following brief descriptions of the principal terrestrial biomes of the world are based on Whittakar's (1975) approach of division of the world's communities into 36 different biomes.

10.1.1 Tropical rainforests

Tropical rainforests occur 10° or more north and south of the equator in three main areas: (a) the Amazon and Orinoco basins in South America (the largest continuous mass) and the Central American isthmus, (b) the Congo, Niger, and Zambezi basins of central and western Africa, and Madagascar, and (c) the Indo-Malay-Borneo-New Guinea regions. These are among the most ancient of ecosystems. Fossil evidence in Malaysia and elsewhere suggests that tropical rainforests have existed continuously for more than 60 million years (Richards 1973). They hold considerable ecological significance because of their influence on climate, the global balance of carbon and atmospheric pollutants, and their highly diverse pools of species with considerable

TABLE 10.1
WORLD CLIMATIC TYPES AND ASSOCIATED BIOMES.

Climatic zone	*Latitude (approximate)*	*Climatic type*	*Rainfall regime (approximate total)*	*Biome*
1. Equatorial zone	0-10° N and S	Hot, wet equatorial	Rainfall all year round: 2000 mm	Tropical rainforests
2. Hot Zone	10-30° N and S	(i) Tropical monsoon	Heavy summer rain: 1500 mm	Monsoon forests
		(ii) Tropical marine	Much summer rain: 1800 mm	
		(iii) Sudan type	Rain mainly in summer: 750 mm	Savanna
		(iv) Desert	Little rain: 125 mm	Desert vegetation and scrub
		(a) Saharan type (b) Mid-latitude type		
3. Warm temperate zone	30-45° N and S	(i) Western margin (Mediterranean type)	Winter rain: 900 mm	Mediterranean forests and shrubs
		(ii) Central continental (Steppe type)	Light summer rain: 500 mm	Steppe or temperate grassland
		(iii) Eastern margin	Heavier summer rain : 1100 mm	Warm, wet forests and bamboo
		(a) China type (b) Gulf type (c) Natal type		

Climatic zone	*Latitude (approximate)*	*Climatic type*	*Rainfall regime (approximate total)*	*Biome*
4. Cool temperate zone	45-65° N and S	(i) Western margin (British type)	More rain in autumn and winter: 750 mm	Deciduous forests
		(ii) Central continental (Siberian type)	Light summer rain: 600 mm	Evergreen conIe-rous forests
		(iii) Eastern margin (Laurentian type)	Moderate summer rain: 1000 mm	Mixed forests (coniferous and deciduous)
5. Cold zone	65-90° N and S	Arctic or Polar	Very light summer rain: 250 mm	Tundra

TABLE 10.2
CHANGE IN VEGETATION WITH ALTITUDE AT DIFFERENT LATITUDES (GOOD 1953)

Latitudinal zonation	*Altitudinal zonation*			
	0-1000 m	*1000-2000 m*	*2000-4000 m*	*4000-6000 m*
Tropical (0°-20°)	tropical	sub-tropical	temperate	arctic-alpine
Subtropical (20°-40°)	subtropical	temperate	arctic-alpine	—
Temperate (40°-60°)	temperate	arctic-alpine	—	—
Arctic and Antarctic (60°-80°)	arctic-alpine	—	—	—

potential for new sources of food, fiber, and medicinal and industrial products (Jordon 1985).

In tropical rainforests, the annual rainfall, which exceeds 2000 to 2250 mm, is generally evenly distributed throughout the year. Temperature and humidity are also relatively high throughout the year.

The flora is highly diversified: 2.5 sq km may contain 300 different species of trees, a diversity unparalleled in any other biome. Tropical rainforests are also typically well stratified into a continuous canopy of trees 25 to 35 m tall, with an interrupted emergent layer of shallow rooted very tall trees with buttressed (swollen) bases. Trees in the canopy are profusely covered by epiphytic orchids and woody lianas which often hide the outline of the trees. The understory stratum of shrubs and herbs tends to be less massive because of the dense shade, but responds quickly to any opening in the overstory layer. Many trees produce aerial roots, and most plants are evergreen with large, dark green, leathery leaves.

As is true of the flora, the fauna of the rainforest is incredibly rich in species, and a much larger proportion of animals live in the upper layers of the vegetation. Fruits are found throughout the year and specialized fruit-eaters have evolved among the insects, the birds and the primates. Other animals concentrate on leaves.

The **elfinwood** (synonyms: *mountain rainforest, cloud forest*) is a variant of the lowland rainforest occurring in the mountainous areas of the tropics (Africa, South America and New Guinea) where the climate is cold but nonseasonal. Elfinwoods get their name from **elves**, the mythical little people recorded in many European fairy stories. The cold leads the trees to be small and stunted, with contorted branches and a low canopy of broad evergreen leaves. The branches are covered with curtains of lichens, mosses and filmy ferns.

Still another variant of the rainforest occurs along banks and floodplains of rivers and is called the **gallery forest**, or sometimes the **riverine** forest.

Decomposition in the tropical rainforest is rapid and the soils, **oxisols** throughout much of the biome, are subject to heavy leaching and so tend to be acidic and nutrient poor (Lucas *et al.* 1993). The potential nutrient loss is offset by extremely rapid uptake by the plants (that is, high productivity), with most of the nutrients being tied up in the biomass. It is the nature of the soil, both its potential for high leaching as well as its chemical composition, that promotes a rocklike quality (*laterites*) when exposed to air that largely has prevented western-style agriculture from being applied to the tropical forests.

The tropical rainforest is perhaps the only major vegetation type in which fire is not a factor; in fact, the rainforest might be defined as the "forest that never burns" in its natural or intact condition.

When the rainforest is removed, a secondary forest often develops that includes soft wood trees, such as *Musanga* (Africa), *Cecropia* (America), and *Macoranga* (Malaysia). The secondary forest looks lush but is quite different from the virgin forest, both ecologically and floristically. The "climax" is very slow to return mainly due to rapid loss of soil nutrients.

10.1.2 Tropical seasonal forest

Tropical seasonal forests (also called **monsoon forests**) occur in humid tropical climates with a clear dry season during which trees may lose their leaves. They are found in India, South-east Asia, West and East Africa, South and Central America, the West Indies and Northern Australia. It is their seasonality that probably accounts for their being less diverse than the rainforests. Tropical seasonal forests are often found where monsoons occur as these provide seasonal rain.

10.1.3 Tropical broad-leaved woodland

Tropical broad-leaved woodlands contain small trees and replace tropical seasonal forests when the climate gets drier and the soils poorer. The canopy is typically only 3-10 m high and consists of trees or shrubs with twisted branches and thick, fire-adapted bark. Tropical broad-leaved woodlands occur in northern South America, the West Indies, southern Africa and Myanmar.

10.1.4 Thornwood

Thornwoods (synonyms: *thorn forests, tropical scrub*, "the bush" in

Africa or Australia, "Caatinga" in Brazil) occur in the same parts of the world as tropical broad-leaved woodlands and in other parts of Asia and in Madagascar. The climate, though, is more arid. The name of the biome comes from the presence of spiny species of trees in the genus *Acacia* and in other genera of the pea family (Leguminosae). The small hardwood trees are often grotesquely twisted and thorny; leaves are small and are lost during dry seasons.

10.1.5 Temperate rainforest

Temperate rainforests occur along the Pacific Coast of North America and New Zealand, Australia and Chile. Their climate is cool and maritime, lacking great variation in temperature and with abundant summer rain and much cloudiness and fog. In common with tropical rainforests, they have rain throughout the year, though at some times of the year the "rain" is condensed fog. This fog comes from moisture brought in by winds from the sea.

The trees in temperate rainforest are the tallest in the world. In Australia, the dominant tree of these forests is the mountain ash (*Eucalyptus regnas*) which can grow to over 90 m in height (Bergamini 1965). In North America, the dominant tree is the redwood (*Sequoia sempervirens*) which may reach 100 m in height. The New Zealand forests include trees in the genus *Podocarpus*.

10.1.6 Temperate deciduous forest

Temperate deciduous forests grow in continental climates with summer rainfall and severe winters. They are dominated by broad-leaved deciduous trees and are found in Europe, Asia and the Americas. They represent one of the most important biotic regions of the world because "white man's civilization" has achieved its greatest development in these areas. This biome is, therefore, greatly modified by man and much of it is replaced by cultivated and forest edge communities.

Notable genera of trees in temperate deciduous forests are alder (*Alnus*), beech (*Fagus*), cedar (*Cedrus*), champ (*Magnolia*), cottonwood (*Populus*), elm (*Ulmus*), fir (*Abies*), hazel-nut (*Corylus*), hemlock (*Tsuga*), hickory (*Castanea*), juniper (*Juniperus*), laurel (*Rhododendron*), maple (*Acer*), oak (*Quercus*), pine (*Pinus*), spruce (*Picea*), sycamore (*Platanus*), walnut (*Juglans*), willow (*Salix*) and yew (*Taxus*). The understory of shrubs and herbs in a mature deciduous forest is typically well developed and richly diversified. Here a considerable portion of the photosynthesis and flowering is attuned to the short days of the spring season, prior to the leafing out of and consequent shading by the tree canopy. **Inceptisols** (brown forest soils) characterize the biome.

Where the moisture remains high but temperature differences between winter and summer become less pronounced, the temperate deciduous forest gives way to **broad-leaved evergreen subtropical forest** biomes. This community is well developed in the warm-temperate marine climate of central and southern Japan, in the "hammocks" (an area of mature soils which has had some protection, at least from fires, thus allowing a climatic climax community to develop) of Florida (USA), and in the live oak forests along the Gulf and South Atlantic coasts. Plant dominants of these communities range from the more northerly live oaks (*Quercus virginiana*), magnolias, bays, and hollies to more tropical species such as strangler-fig (*Ficus aurea*), wild tamarind (*Lysiloma*), and the gumbo limbo (*Bursera*). Palms such as the sabal or cabbage palm (*Sabal palmetto*) are also often prominent. Vines and epiphytes are characteristic. The latter group includes many ferns, the bromeliads (pineapple family), and orchids which support characteristic fauna of small animals.

A variant of temperate deciduous forest is the **temperate woodland** which occurs when the climate is too dry for true forest, that is, on a continuum between forest and shrubland. It is dominated by small trees, but these often provide only an incomplete and open canopy, so that the appearance may be of grassland with occasional trees. An example of temperate woodland is **pygmy conifer woodland** in the western United States (Whittaker 1975). Here, open woodland dominated by pines and junipers is found with shrubs and grasses beneath.

10.1.7 Temperate shrubland

Around the shores of the Mediterranean Sea, the annual rainfall is between about 300 and 800 mm, and during the summer there is usually no rain for about four months. In these months the mean temperature is 20-25°C. In the coldest months the mean temperature is about 10°C and frosts are only sporadic. Very similar climates occur in Southern California, parts of Chile, the Cape of Good Hope in southern Africa and South-western Australia. The vegetation is very similar in these separated regions, even though the individual plant species differ greatly between the regions.

The shrubs are between 1 and 5 m high, and have small thick drought-resistant leaves. **Mediterranean maquis** and **Californian chaparral** are among the best known examples of this biome. **Heathland** is also included in this classification, although it can only just be classified as shrubland. All of these types of shrubland are greatly affected by fire which burns off the above-ground stems. Regrowth then occurs from root systems that survive the fire and from buried seeds (Gimingham 1972).

Californian chaparral contains shrubs of *Quercus, Ceanothus, Arctostaphylus* and many others forming a low **scrub** (dwarf or stunted vegetation) which is highly inflammable. Fires are often started by lightning. If a fire occurs every 12 years or so, the chaparral can perpetuate itself. However, if fires occur too often, those species which rely on germination from seed, rather than on regrowth from the roots, are eliminated and the composition of vegetation changes. If fires occur much less often than once every 12 years, fire-intolerant species such as *Prunus ilicifolia* and *Rhamnus crocea* invade (Walter 1979).

Heathland occurs in Ireland, Scotland, England, Denmark and parts of northern Germany and southern Sweden. The vegetation is dominated by low shrubs in the heather family (Ericaceae). These heaths generally occur on very poor and shallow sandy podzols. The origins of heathland have been widely debated. One theory is that they are derived from the forest by grazing and burning. Another theory is that they have never been forest; rather, they represent a "sub-climax" community. Evidence from archaeological sites and from pollen analysis now favours the former theory.

10.1.8 Boreal forest

Boreal forest or **taiga** (a Russian word for coniferous forest) extends from north-eastern Europe across Russia to the Pacific Ocean, and right across North America from Alaska to Newfoundland (that is, largely between 45 and 57°N latitudes). To the north it merges into tundra; to the south it grades into deciduous forest or grassland. The climate is characterized by a bitterly cold winter of long duration, and a cool brief summer (that is, sub-Arctic type). **Permafrost** is widespread, so that the deeper layers of the soil remain frozen all year round, and for much of the year snow lies on the ground. The soils are acidic and mineral deficient *podzols*, characterized by deep litter layer and slow decomposition.

The vegetation of the boreal forest is dominated by *coniferous* trees. The important genera are pines (white pine, red pine, Scots pine, Jack pine, lodgepole pine), firs (Douglas fir and balsam fir), spruces and larches. With the exception of larches, these are evergreen. Some broad-leaved species also occur, notably poplars and birches. The coniferous forest belts of Eurasia and North America are the richest sources of softwood for use in building construction, furniture, matches, paper and pulp, rayon and others. The understory is relatively limited as a result of continual low light penetration. However, among common understory associates are orchids and ericaceous shrubs.

10.1.9 Savanna

Savannas are tropical grasslands with scattered, drought-resistant trees that generally do not exceed about 10 m in height. As savanna is subject to fire, these trees have thick bark which insulates the living cambium from the heat of the fire. Savannas constitute extensive areas in eastern Africa-areas that support the richest diversity of grazing mammals in the world, and also occur in Australia, South America, and Asia. The climate is generally characterized by a rainy season (May-October) accounting for more than 90% of the annual rainfall, and a dry season (November-April).

Since both trees and grasses must be resistant to drought and fire, the number of species in the vegetation is not large, in sharp contrast to adjacent equatorial forests. Grasses belonging to such genera as *Panicum, Pennisetum, Andropogon*, and *Imperata* provide the dominant cover, while the scattered trees are of entirely different species from those of the rainforest. In Africa, the picturesque acacias, baobab trees (*Adansonia*), arborescent euphorbias, and palms dot the landscape. Often single species of both grass and trees may be dominant over large areas.

East African savanna is unexcelled anywhere in the world for its abundance of wildlife. Despite the activities of hunters, killing elephants for their ivory, rhinoceroses for their horns and cats for their skins, enormous numbers of large herbivores and carnivores still roam East Africa. Among these the notable large grazers are wildebeest (*Connochaetes taurinus*), zebra (*Equus burchelli*), buffalo (*Syncerus caffer*), and Thomson's gazelle (*Gazella thomsoni*); browsers include impala (*Aepyceros melampus*), giraffe (*Giraffa camelopardalis*), eland (*Tautotragus oryx*), black rhinoceros (*Diceros bicornis*) and gerenuk (*Litocranius walleri*); and the mammalian carnivores include spotted hyaena (*Crocuta crocuta*), lion (*Panthera leo*), leopard (*P. pardus*), cheetah (*Acinonyx jubatus*), African wild dog (*Lycaon pictus*) and three species of jackal (*Canis* spp.).

10.1.10 Temperate grasslands

Temperate grasslands (synonyms: **steppes** in Eurasia, **prairies** in North America, **Pampas** in Argentina and Uruguay, **velds** in South Africa, **downs** in Australia) are found across large areas of eastern Europe, Asia, central North America and Argentina. The climate is moderately dry (between 250 and 750 mm of rain each year) and continental, so the summers are hot and winters cold. The flora of these temperate grasslands is dominated by perennial graminoids (grasses, sedges, rushes). Associated with graminoids are broad-leaved perennials. The smaller broad-leaved plants flower early in the growing season, before the graminoids reach their maximum height; the larger broad-leaved plants flower towards the end of the growing season, after the grasses have begun to die down.

A large portion of mammals of the grassland are either running or burrowing types. Aggregation into colonies or herds is characteristic; this life habit provides some measure of protection in the open type of habitat. The important large grazing animals native to grasslands in various biogeographic regions are as follows:

North America	Eurasia	Africa	Australia
Bison, pronghorn antelope	Saga antelope, wild horses, wild asses	Numerous species of antelope, zebra, etc.	The large Kangaroos

In grasslands, a precipitation (P)/ evaporation (E) ratio of 1 permits weak podzolization to occur as in tall grass prairie. Soil pH is close to neutral, however, and decay of fibrous roots gives the soil a dark brown colour. Where precipitation is less than evaporation, *mollisol* soils become prominent. A layer of calcium carbonate accumulates in the subsoil, deposited at shallower and shallower levels as the P/E ratio declines. The most mesic mollisol is *chernozem* with up to 16% organic matter in the nearly black A horizon, a clay loam texture, high cation exchange capacity, very high percent base saturation, pH slightly basic, and carbonate accumulation only at 100 cm depth. Chernozem is a Russian word, meaning black earth. The colour is a result of richness in organic matter.

Until the end of the nineteenth century, the temperate grasslands were dominated by nomadic and semi-nomadic peoples like the *Kirghiz* of the Asiatic steppes. They roamed far and wide with their herds and earned a precarious living from whatever pastures they could find. The **Red Indians** of North America were mostly hunters who moved around after the bison and other animals. Cultivation was unknown and the region was one of the most sparsely populated parts of the world. In recent years great changes have taken place in the grasslands and only in few areas they retain their original landscape. The grasslands have been ploughed up for extensive, mechanized wheat cultivation and are now the "granaries of the world". In addition, the tufted grasses particularly those of southern hemisphere have been replaced by more nutritious species (Lucerne or alfalfa, for example) for cattle and sheep rearing. These constitute the leading ranching regions of the globe exporting large quantities of beef, mutton, wool and hides.

10.1.11 Alpine shrubland and grassland

Alpine shrublands as well as alpine grasslands are sometimes found above the *treeline* on mountains. They occur in South America, Africa, the Himalayas and New Zealand.

In the Himalayas, the lower limit of the treeline is between altitudes 3300 to 3600 m above sea level. In some localities where the glaciers descend to lower levels or where the winter snow remains in deep pockets, the treeline occurs even at altitudes of 3000 m or less as in Kashmir. The upper limit of the alpine zone is up to 5600 m.

In the alpine zone, a fringe of shrubs may be seen in some places and in others, continuous or scattered dwarf or gnarled trees may be present immediately above the treeline. In the absence of tree cover, the vegetation is exposed to the fury of cold winds, blizzards and snow storms and as such the stunted trees often show a characteristic wind-swept or snow-swept appearance. Birches, rhododendrons and willows, among others, may be seen in such situations.

The shrubs found above the treeline are usually species of *Berberis, Myricaria, Lonicera, Rhododendron, Salix* and a dwarf conifer, *Juniperus*. In the more favourable areas where adequate water is available, the rhododendrons are frequently seen along with *Berberis, Lonicera* and others. In the drier inner valleys where the vegetation is of a scrub type, the chief components are species of *Artemisia, Astragalus, Caragana* and *Ephedra*. The junipers are generally seen on limestone. In extreme conditions where precipitation is very scanty, the vegetation is closely similar to that of the Tibetan highlands. In the alpine shrubby zone, it is common to see the very dwarf recumbent shrubs with large number of long prostrate branches spreading over a considerable area; often they are closely packed together. Topography and disposition of the slopes are important factors in determining the composition and extent of distribution of the shrubby species. In northern Kumaon (India), for example, *Rhododendron anthopogon*, an aromatic shrub, is particularly common on the northern slopes whereas the vegetation on the southern slopes is more xeric in character with the presence of cotoneasters, *Ephedra*, junipers and others (Rau 1974). A similar topographic distribution has also been recorded by Polunin (1950) in west Nepal where the north facing slopes are forested with *Abies* and *Betula* and the dry and sunny southern slopes support herbs, grasses, and scattered bushes of junipers.

Above the treeline is usually present a grassy belt with lush meadows, grassy slopes and flower laden pastures. One often finds many prostrate shrubby species like willows, heathers and, here and there, grassy tussocks with bare soil between them. These pastures, meadows and grassy slopes form an important habitat for alpine species, and may occur higher than the shrubby zone or alongside it. The herbs are dominant here and include many species well known for their attractive flowers. These herbs are mostly perennial and show well marked preference for high altitudes. They exhibit also a very interesting cycle of growth; with the melting of snow late in May or early in June, a profusion of early summer flowering herbs appear including several species of crucifers, primulas, anemones and saxifragas, and later in summer and towards autumn appear the gentians, swertias and others with the saussureas attaining their best development late in autumn.

Another aspect of interest is the strange association of some plants. For example, Rau (1974) reported an association of the beautiful, delicate blue corydalis (*Corydalis cashmeriana*) growing in the midst of tussocks of a grass (*Danthonia* sp.) which is not grazed by cattle.

The alpine grassy meadows form a lush carpet in many localities. The meadows are extensive and sloppy and often possess plants of bushy habit besides a rich source of grasses and sedges in association with beautiful herbs. These pastures are usually found up to the snowline. The over-grazed alpine pastures often possess toxic plants unpalatable to grazers.

At the higher limits of vegetation, the habitats are varied. Conditions are severe and exposure to cold and blizzards prevent many herbs from occupying exposed situations. The herbs are generally restricted to sheltered slopes, rock crevices and beneath rock ledges. The commonest adaptation seen in this zone is the *cushion habit*. Cushion-like plants also predominate in this cold scree desert, stony waste and rubble of the exposed drier localities. Species of *Androsace, Draba, Paraquilegia, Saxifraga* and *Sedum* form soft cushions. Rigid mat forming shrubs, often spinescent, are represented by the species of *Acantholimon, Arenaria, Astragalus,*

Caragana, Thylacospermum and others.

The most curious of all flowering plants are, however, the woolly Compositae: *Saussurea graminifolia, S. gossypiphora, S. simpsoniana*, and *Soroseris glomerata*. Some species of *Anaphalis, Leontopodium* and *Tussilago* have also soft white hairs but they do not attain the extensive development of wool that one sees in the species named above.

The high alpine grasslands generally show mixed association of grasses, sedges and colourful herbs belonging to various families but in some localities, pure formation of grasses may occur. *Helictotrichon virescens, Danthonia* spp., and *Stipa orientalis* are among the grasses which occur in such pure communities.

The faunal diversity of the Himalayan alpine zone includes wild yak (*Bos grunniens*), takin (*Budorcas taxicolor*), wolf (*Canis lupus*), markhor (*Capra falconeria*), ibex (*Capra ibex*), roe deer (*Capreolus capreolus bedford*), white-lipped deer (*Cerus albirostris*), MacNeil's deer (*Cervus elephas macneilli*), Tibetan wild ass (*Equus kiang*), musk deer (*Moschus moschiferus*), argali (*Ovis ammon*), urial (*Ovis orientalis*), Tibetan antelope (*Pantholops hodgsonii*), Tibetan gazelle (*Procarpa picticaudata*), blue sheep (*Pseudois nayaur*), snow leopard (*Uncia uncia*), fox (*Vulpes ferrilata*) and others.

The Himalayan alpine zone, in general, is too cold and arid to support cultivated agriculture. However, it has been used for livestock rearing since about 4000 years. In the dry western Tibet where good grazing land is sparse, sheep and goats predominate in nomad's/ pasturalist's herds. In the lush grasslands on the north-eastern edge of the Tibetan plateau yaks are the primary livestock raised, and in much of the alpine Himalaya where herds must be moved to low-elevation grazing lands in winter, yak-cattle hybrids are common. The yaks and yak-hybrid crosses provide milk and milk products, meat, fiber, and hides. Yaks are also used as pack and draft animals and for riding.

10.1.12 Tundra

The polar type of climate and vegetation is found mainly north of the Arctic circle (north of latitude 60°N) in the northern hemisphere. The *ice-caps* are confined to Greenland and to the highlands of these high-latitude regions, where the ground is permanently snow-covered. The lowland, with a few months ice-free, have tundra vegetation. They include the coastal strip of Greenland, the barren grounds of northern Canada and Alaska and the Arctic seaboard of Eurasia. In the southern hemisphere, the virtually uninhabited continent of Antarctica is the greatest single-stretch of ice-cap where the permanent ice are as thick as 3000 m.

Although there are considerable variations in climate, ice cover, soils, size of the flora, and composition of plant communities, arctic tundra can be characterized by the absence of trees, the predominance of dwarfed plants (5-20 cm high), and an upper ground surface that is spongy and uneven, or hummocky, as a result of freezing and thawing of this poorly drained land. *Spodosols, inceptisols*, and *histosols* are the characteristic soils of arctic tundra.

Also characteristic of the tundra is the presence of permanently frozen soil (permafrost) at a depth of a few centimeters to several meters. The permafrost may be relatively dry in the case of course gravels or largely filled with ice in less porous soils. According to Bliss (1988), thickness of the permafrost is about 400 m at Barrow, Alaska, 500-600 m in the Mackenzie River Delta region, and 400-650 m throughout the High Arctic. The permafrost line is the ultimate limit of plant root growth, but the immediate control is the depth to which soil is thawed in summer.

The vegetation consists of relatively few species: a few grasses but mostly sedges are characteristic of the numerous marshes and poorly drained areas; large areas consist of ericaceous heath plants, low flowering herbs, and lichens. Perhaps the most characteristic arctic tundra plant is lichen known as "reindeer moss" (*Cladonia*), a symbiotic mutualism of a fungus and an alga. The principal herbivorous mammals are reindeer (caribou), musk ox, arctic hare, voles, and lemmings. Carnivores include the arctic fox and wolves. Characteristic birds are longspurs, plovers, snow bunting, snow owls, and horned larks. Reptiles and amphibians are few or absent, and insects are rare with the exception of blackflies and mosquitoes during the summer when they are extremely abundant.

In Greenland, northern Canada and Alaska live the **Eskimos** who lead a nomadic way of life due to seasonal changes in climate. During winter they live in compact **igloos** and in summer when they move out to hunt they pitch portable tents of skins by the side of streams. Their food is derived from fish, seals, walruses, and polar bears. Other Eskimos hunt reindeer and other animals to secure a steady supply of their meat, milk, fat, skins and bones.

Tundralike areas, usually called **alpine tundra** or **alpine meadow**, occur on high mountains in the temperate zone.

10.1.13 Warm, cool, and arctic-alpine semi-deserts

Warm semi-desert scrubs occur throughout the world in dry warm-temperate and subtropical climates. The vegetation is characterized by scattered bushes, while the fauna is often rich in small mammals, birds, lizards and snakes.

Cool semi-deserts are found in parts of North America, central Asia and Iran, Australia, Patagonia and the Andes. The vegetation often contains scrub and is found in places too dry for grassland.

Arctic-alpine semi-deserts are found in arid regions either in the Arctic or on mountains above the treeline. The vegetation usually consists of low cushion plants.

10.1.14 Deserts

Deserts are regions of scanty rainfall (less than 250 mm annually) which may be hot like the hot deserts of the Saharan type; or temperate as are the mid-latitude deserts like the Gobi. The aridity of hot deserts is mainly due to the effects of off-shore trade winds, hence they are also called **trade wind deserts**. The temperate deserts are rainless because of their interior location in the temperate latitudes, well away from the rain-bearing winds.

The major hot deserts of the world are located on the western coasts of continents between latitudes 15° and 30°N and S. They include the Sahara Desert, the largest single stretch of desert, which is 5150 km long from east to west and at least 1600 km wide. Its total area of 5.6 million sq km is larger than all the 50 states of USA put together. The next biggest desert is the Great Australian Desert which covers almost half of the continent. The other hot deserts are the Arabian Desert, Iranian Desert, Thar Desert, Kalahari and Namib deserts. In North America, the desert extends from Mexico into USA and is called by different names at different places-Mojave, Sonoran, Californian and Mexican deserts. In South America, the Atacama or Peruvian Desert is the driest of all deserts with less than 15 mm of rainfall annually.

Among the mid-latitude deserts, many are found on plateau and are at a considerable distance from the sea. These are the Gobi, Turkestan and Patagonian deserts. The Patagonian desert is more due to its rain-shadow position on the leeward side of the lofty Andes than to continentality.

The deserts are some of the hottest spots on earth and have high temperatures throughout the year. Days are unbearably hot, and in the open barren sands, 35°C is often recorded. The diurnal range of temperature in the deserts is very great. Intense insolation by day in a region of dry air and no clouds causes the temperature to rise with the sun. But as soon as the sun sets, the land loses heat very quickly by radiation and temperature drops to well below 10°C. Intense evaporation increases the salinity of the soil so that the dissolved salts tend to accumulate on the surface forming **hardpans**. Absence of moisture retards the rate of decomposition and desert soils are very deficient in humus.

There are three life-forms of plants that are adapted to deserts: (i) the annuals, which avoid drought by growing only when there is adequate moisture; (ii) the succulents, such as the cacti which store water; and, (iii) the desert shrubs which have numerous branches ramifying from a short basal trunk bearing small, thick leaves that may be shed during the prolonged dry periods. All desert vegetation has a highly characteristic "spaced" distribution in which individual plants are scattered thinly with large bare areas in between. In some cases, allelochemics play a part in keeping plants spaced apart. In any event, spacing reduces competition for a scarce resource; otherwise intense competition for water might result in the death or stunting of the plants.

Ants, locusts, wasps, scorpions, spiders, lizards, rattle snakes, several insect-eating birds, seed-eating quails, doves, desert rats, rabbits, foxes, jackals and various cats are the common desert animals. They are adapted to low water levels through drought avoidance (possess impervious integuments), water conservation (excrete uric acid and guanine instead of urea), and reduction in direct water intake (using and conserving water produced in metabolism-that is, from breakdown of carbohydrates) along with behavioural adaptations. Most animals tend to be **nocturnal**, many burrowing during the heat of the day, or are **crepuscular**-that is, active only at sunrise and sunset.

10.2 Wetland And Freshwater Biomes

10.2.1 Cool temperate bog

Bogs are wetlands that receive water only from rainfall. They tend to be nutrient poor as rainwater has very little nutrient content, and as it drains through the soil profile it tends to leach out any remaining nutrients. The dominant species in bogs are mosses in the genus *Sphagnum*. The mosses grow upwards and their lowermost leaves decay and join the **peat** (partially decomposed plant material) building up beneath. The main areas where bogs occur are in the temperate and boreal regions.

Wherever rainfall is high and temperatures cool, so that water loss by evapotranspiration is less than water gain from rain, *Sphagnum* bog tends to develop. The bog may slowly grow up to form a huge dome of peat called a *raised bog*. Raised bogs are usually confined by some feature of topography such as hollow formed by an infilled lake. If rainfall is high (>1000 mm a year), *blanket bog* can develop even on shallow slopes. The peat in blanket bog can build up to several meters thick. The peat is vulnerable to erosion as it is easily washed away once the surface vegetation has been damaged by trampling, grazing or pollution, leaving deep gullies. The bog pH is very low, often about 3-4, caused by humic acids and sulphuric acid formed when organic sulphur is oxidized. The whole bog community is slow growing and short, with low primary productivity.

10.2.2 Floodland ecosystems

This type of wetland ecosystems obtains water from rivers. These are often extremely seasonal. They flood deeply with water when the river overflows its banks. This may happen during a rainy season, or when winter snows melt in the catchment area. Floodlands tend to occur in lowland, flat-bottomed valleys through which a large river meanders. Such areas can be very large on the flood plains of great rivers. These large lowland rivers are associated with a number of features illustrated in Fig. 10.3. Normally, the mature river will meander across a wide, flat plain which was created by sediment deposited in the past by the river. The channel in which the river flows builds up at the side into sandy banks called *levees*. It is only when the water reaches the top of these banks and overflows that the valley is flooded. The flood plain often has small permanent wetland areas which form when the river takes a new path on the valley floor, leaving behind an *oxbow lake*. These lakes slowly fill in with silt. Often a flood plain will have several oxbow lakes in various stages of infill. When the river floods the valley, it carries with it clays, silt and mineral nutrients which are deposited on the flood plain.

In many parts of the world floodlands have been cleared of their natural vegetation. The frequent flooding constantly enriches the soils and the proximity of the river makes it ideal for irrigation. These areas are very valuable for cultivation and many such areas have been farmed for thousands of years. In temperate zones, flood plains make excellent wet meadows of rich grazing land, although the wet environment can harbour diseases and parasites which may infect grazing animals.

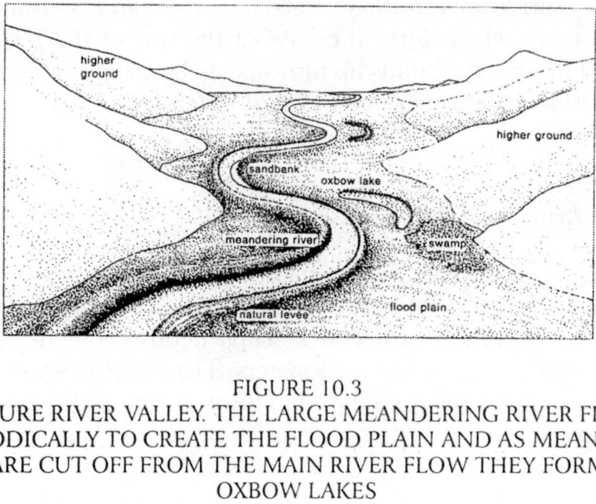

FIGURE 10.3
A MATURE RIVER VALLEY. THE LARGE MEANDERING RIVER FLOODS PERIODICALLY TO CREATE THE FLOOD PLAIN AND AS MEANDERS ARE CUT OFF FROM THE MAIN RIVER FLOW THEY FORM OXBOW LAKES

10.2.3 Swamp and marsh ecosystems

The ecosystems in this category are found in areas of impeded drainage, where water runs off the surrounding land and collects, or where groundwater lies close to the surface. In some cases rivers and streams may also feed into the area. The land is flooded all the year round, except in very dry years. These ecosystems are very variable in size and form, depth of soil and plant community structure. In general they can be divided into two major types: **swamps**, in which trees are the dominant vegetation; and **marshes**, which have large open areas of grasses (Poaceae), sedges (Cyperaceae) and reeds.

The best developed tropical freshwater swamp forests are those of the Amazon basin. Minerals in the waters and accompanying soils tend to be scarce, so that primary productivity is low. Consequently, there are few animals. In the black-water tributaries of the Amazon, the water is dark brown due to the presence of humic acids which also lower the pH. On the other hand, temperate freshwater swamp forests such as those of Everglades, Florida (USA) are rich in faunal diversity.

The movement of groundwater, runoff and the addition of stream water make marshes nutrient rich with a fairly high pH. Most leaf litter falls to the wet marsh surface and is attacked by decomposers which form the base of the rest of the food web. The leaf litter often builds up into a peaty layer. This is especially so in **fenland**, although the thickness of fen peat is never as great as seen in bogs.

10.2.4 Lentic ecosystems

Lakes and ponds are lentic (stagnant) freshwater bodies and occur practically in every biome. They vary in size from less than a hectare to hundreds of hectares. Their depth varies from a few centimeters to over a hundred meters. The water in small ponds may dry up after one season only. However, large lakes may exist for many centuries. Lakes also differ in their origin. Many are direct or indirect result of glaciation. Others are natural or man-made depressions filled with water. Ox-bow lakes are parts of meandering rivers cut off from the mainstream.

The relatively shallow lakes, called **eutrophic lakes**, have a rich accumulation of organic products. Generally deep lakes, often with steep and rocky sides, are poor in circulating nutrients such as phosphates, and are called **oligotrophic lakes**.

The physical factors of the lakes are dependent on location, latitude, altitude and the surrounding biome. Although classified as freshwater, saline lakes with sodium as the dominant cation are found throughout much of the world and particularly in arid climates (Sambar Lake in Rajasthan, India, for example).

The freshwater pond autotrophs include microscopic phytoplankton. Other free-floating plants occupying the surface that move with the wind are *Azolla, Pistia*, and *Lemna*. Macrophytes such as *Ceratophyllum* and *Utricularia* are the submerged hydrophytes whereas *Nelumbo* and various species of *Nymphaea* are floating-leafed species.

10.2.5 Lotic ecosystems

Freshwater streams (springs, rivulets, creeks, brooks, etc.) and rivers change over their course from being narrow, shallow, and relatively rapid to become increasingly broad, deep, and slow moving. As water winds its way along a stream or river from its source to its mouth, certain generalizations usually hold: (a) the speed at which the water moves (the current) decreases; (b) the volume of water passing down increases; (c) oxygen levels fall; (d) the bed becomes composed of smaller particles (gravel gives way to silt); (e) the bed becomes less steep; (f) the summer temperature of the water often increases; (g) eutrophication often occurs; (h) invertebrate diversity often decreases; (i) human influences increase. Most streams are characterized by a repeating sequence of **rapids** (shallow water where velocity of current is great enough to keep the bottom clear of silt and other loose materials, thus providing a firm substrate) and **pools** (deep water where velocity of current is reduced and silt and other loose materials tend to settle to the bottom, thus providing a soft substrate) that decrease in frequency downstream (Fig. 10.4).

As would be expected, only organisms capable of adhering to an exposed surface such as rocks are found in the upper reaches of a stream. Such adhering organisms are termed **periphyton** or **aufwuchs**. The periphyton includes attached clumped and filamentous green algae, cyanobacteria and various sessile

FIGURE 10.4
A TYPICAL LOTIC ECOSYSTEM OF RAPIDS AND POOLS

invertebrates, including larvae of insects such as blackflies and other midges, flattened mayflies and stonefly nymphs, and planarians. Farther downstream floating and emergent vegetation may be found, along with sessile invertebrates and those that burrow in the softer substrate, such as calms and burrowing mayfly nymphs.

10.3 Coastal and Marine Biomes

Covering more than 70% of the earth's surface, with an average depth of 3750 m and a salinity averaging 3.5%, the major oceans of the world (the Pacific, Atlantic, Indian, and Arctic) account for more than 90% of the total water surface of the earth.

Unlike land and freshwater ecosystems, the sea is continuous and is in continual circulation via the major surface currents (Fig. 10.5), upwelling seasonal turnover in the upper waters, waves, and tidal action. In waves, water oscillates back and forth without actually changing position. Tides consist of a rise and fall of water level caused by the gravitational effects of the moon and the sun. When earth, moon, and sun are, as at the time of a new and a full moon, aligned, the gravitational pull is strongest, resulting in the highest, or **spring tides**. When the moon and sun are at right angles to each other, as at the time of the first and last quarters, their gravitational fields counteract, resulting in the smallest, or **neap tides**. Each of these tides occurs monthly. Tidal range varies from less than one-third of a meter to 15 m in some enclosed bays.

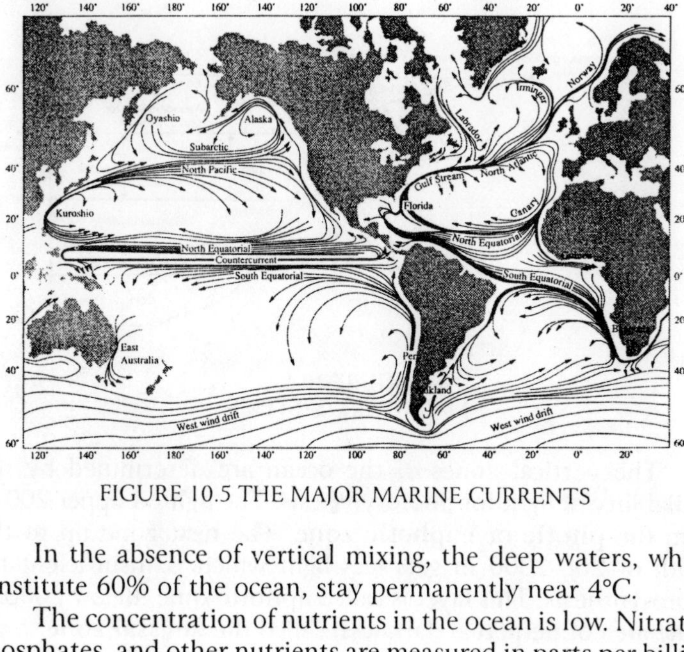

FIGURE 10.5 THE MAJOR MARINE CURRENTS

In the absence of vertical mixing, the deep waters, which constitute 60% of the ocean, stay permanently near 4°C.

The concentration of nutrients in the ocean is low. Nitrates, phosphates, and other nutrients are measured in parts per billion in contrast to salts, such as sodium chloride, which are measured in parts per thousand. After hydrogen and oxygen, which constitute about 96% of the ocean's elements, the major elements (those constituting 0.3 to 1.9%) in descending order are chlorine, sodium, magnesium, sulphur, calcium, and potassium (Klemm *et al.*, 1990). Minor elements (.00013 to .0067%), also in descending order, are bromine, carbon, nitrogen, strontium, boron, silicon, and fluorine. There is a wide spectrum of trace elements (less than .000063%) from argon to zirconium, some 45 in all.

All the ocean basins are roughly of the shape of a wash-basin or an inverted hat (Fig. 10.6). From the coastline a gradually sloping region extends to an average of 160 km into the sea. This zone with a depth of 8-200 meters is called the **continental shelf**. The angle of the slopes then abruptly steepens to form the

continental slope. The continental slopes often have ridges, basins, and trenches formed of mud and sand, deposited by slow-flowing rivers. The continental slope levels off into a more or less horizontal expanse forming the **ocean floor**. The depth of the ocean floor is in the range of several thousand meters.

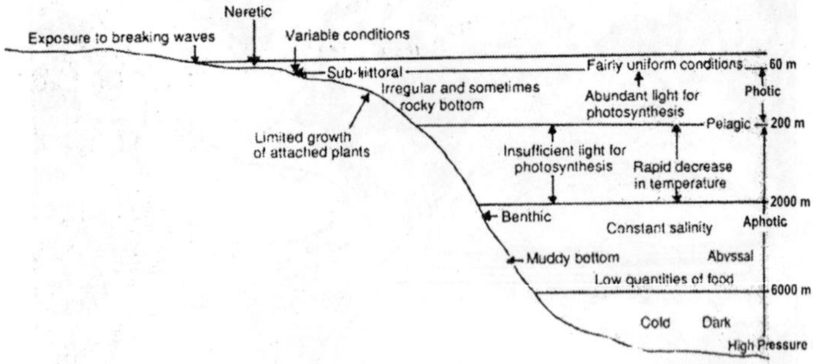

FIGURE 10.6
MARINE ECOSYSTEMS

The vertical zones of the ocean are determined by the availability of light for photosynthesis. The lighted upper 200 m form the **photic** or **euphotic zone**. The next zone up to the depth of 200-2000 m gets less light which is insufficient for photosynthesis. This layer is called **aphotic** zone. Below 2000 m is the area of perpetual darkness, called the **abyssal zone**.

Three major environments may be recognized in the ocean basin: (i) the **littoral zone**, comprising the sea floor from the shore to the edge of the continental shelf; (ii) the **benthonic zone**, formed of the sea floor along the continental slope and the aphotic and abyssal zones; (iii) the **pelagic zone**, constituting the water of the ocean basin.

Marine life can be classified into three main categories: plankton, nektons and benthonic organism. **Plankton** consists of passively drifting or floating organisms. Most of these minute organisms, with very weak or no locomotory organs, are confined to the euphotic zone. The plankton includes photosynthesizing organisms such as diatoms (phytoplankton) and heterotrophic organisms such as small crustaceans (zooplankton). **Nektons** consist of actively moving organisms with well-developed locomotory organs. They include actively swimming organisms found in the surface or deep waters such as fishes, squids, whales, etc. They feed mainly on planktons or smaller nektons. **Benthonic** organisms are found along the floor of the sea-bed and include creeping, crawling or sessile organisms. Many of them are detrivores.

10.3.1 Marine rock pools

Marine rock pools can occur on any rocky coastline and are found in both temperate and tropical climates. The shoreline is an extremely variable place as it is the area between extreme high and low tides. The tides advance and retreat in a cycle lasting 12 h 26 min, so high and low tide slowly work themselves round the clock. This means that not only is the degree of exposure changing on the shore as the tides move in and out, but the time at which the exposure occurs also changes. So sometimes areas will be exposed at night, when it is cold and dark, and sometimes during the heat of a midday sun.

When the tide goes out, the sea level falls and the shore becomes exposed. Seawater is trapped and held in the rock pools until the next incoming tide covers the shore. If the pools are exposed in the daytime, the water will be warmed up by the sun. Water will evaporate making the pool more saline (salty). If it is raining heavily, the pool may become less saline, although usually the less dense rainwater floats in a layer on top of the salty pool water, so that only the top of the pool is affected.

Vascular plants (pteridophytes and spermatophytes) and mammals are totally absent. This is because the energy of the waves crashing on the rocks would probably break up a large rigid organism and wash it away. The primary fixers of solar energy are photosynthetic multicellular algae or seaweeds (brown algae: *Fucus*, etc.; red algae: *Corallina, Chondrus*, etc.; green algae: *Enteromorpha*, etc.). Most of the animals are invertebrates, with the exception of a few fish, and the occasional visit from a predatory bird such as the Oystercatcher (*Haemantopus ostralegus*) or purple sandpiper (*Calidris maritima*).

10.3.2 Marine sandy beach

Marine sandy beaches occur around the world in the littoral belt.

To us they may look like an inviting environment. However, they pose considerable problems for the organisms living there. For a start, sand is abrasive and constantly on the move. That is sufficient to ensure that in the intertidal region no rooted plants can establish themselves. Further, sand is nutrient poor, being composed almost entirely of silica (SiO_2), though there may be considerable amounts of calcareous matter from seashells. Diatoms and other algae occur in the surface layers of the sand where they can photosynthesize, but in the intertidal region the animals depend largely on organic matter brought in by the waves.

Tiny creatures up to 2 mm long like the mollusc *Caecum glabrum* and the sea cucumber *Leptosynapta minuta* live between the sand grains. Larger organisms are either flattened such as bivalve molluscs including rounded cockles (*Cardium* spp.), flattened tellinids (*Tellina* spp.) and elongated razor shells (*Ensis* spp.), or worm-like such as lugworms (*Arenicola* spp.) which produce characteristic casts (Barrett and Yonge 1958). The rich invertebrate life often supports large numbers of wading birds. To some extent, birds with bills of different lengths specialize on different prey, as different invertebrates tend to occur at different depths in the sand.

10.3.3 Marine mud flat

Marine mud flats are composed of fine silts. These tend to retain fine organic matter and the relative abundance of organic matter together with the fine particle size of the muds combine to produce anaerobic conditions only a few mm below the surface of the mud. This can be seen by scraping at the surface of the mud: the grey surface lies immediately above a thick black anoxic mud. Few species are adapted to survive in these conditions, but those that are may reach large densities because of the high productivity of the mud flats. Depending on where in the world the mud flat is situated, the fauna may include clams, crabs, gastropod molluscs, annelid worms and amphipod crustaceans. Marine mud flats often support even larger number of waders than are found on sandy beaches.

10.3.4 Mangrove swamps

The coastline in the tropical and subtropical regions is fringed with a strip of swampland which is inundated every high tide with marine or *brackish waters*. Wherever the wave action is not too strong to prevent regeneration, these coastal wetlands are densely vegetated with thickets of mangrove trees. There are about 70 species of mangrove around the world, the most important genera being *Rhizophora* and *Avicennia*. Mangroves are capable of growing in freshwater, but it is thought that the saline conditions of the intertidal zone give these poor competitors an advantage over other species. They are well adapted to the salty conditions as they can prevent high concentrations of salt entering the roots and can secrete excess salt from their leaves. They may also jettison extra salt when they drop their leaves. Just like the ecosystems on the rocky shore, mangrove swamps are influenced strongly by the tides. Incoming tides import nutrients to the system, and tides are also responsible for dispersing seeds. Mangrove seeds often germinate while still held on the tree **(viviparous)**. They produce a thick spear-like hypocoty so when the seed falls off the tree, the hypocotyls is embedded in the mud like spear and then grows anchoring roots (Mitsch and Grosselink 1986).

A variety of animals live within the mangrove including fiddler crabs (*Uca* spp.) and mudskippers (*Periophthalmus* spp.). The food web is based on the presence of many detrivores, but ascends to major carnivores like alligators, crocodiles and big cats such as tigers (*Panthera tigris*).

The mangrove trees have a whole series of aerial roots arising quite high up on the trunk and plunging into mud beneath. These **stilt roots** help to support the trees and lessen the wave action of incoming tides. Sometimes considerable amounts of organic mud build up under the mangrove forest. In such cases mangrove trees produce many stick-like structures (called **pneumatophores**) protruding from the mud. Both these and supporting aerial roots have little pores on them that can take in oxygen from air.

10.3.5 Temperate salt marsh

In higher latitudes the mangrove ecosystems disappear because the trees cannot tolerate even minor frosts. They are replaced on the coastal strip by salt marshes. Like the mangrove swamps, the salt marshes are within the tidal influence of the sea and are inundated with salt water every high tide. They consist of a patchwork of low vegetation, and they tend to develop in sheltered intertidal regions where wave action is not too strong. Salt marsh vegetation is dominated by grasses such as *Spartina* and rushes (*Juncus* spp.).

Salt marsh soils are high in phosphates but low in nitrogen, which may limit plant growth. Sulphur from sea water often collects in marsh soils as iron pyrites and hydrogen sulphide. When the soils dry out, sulphuric acid may form which lowers soil pH. In these marshes up to 70% of the net primary productivity is due to chemosynthetic bacteria which utilize the sulphur (Howarth *et al.*, 1983). Mats of cyanobacteria on the mud are important as fixers of light energy. Salt marshes have high primary productivity, but much of the organic matter produced gets washed out of the marsh with the retreating tides.

10.3.6 Coral reef

In terms of their high productivity, high species diversity, complexities of co-evolution and sheer beauty, coral reefs are the tropical rainforests of the oceans. Corals are cnidarians (coelenterates), relatives of the sea anemones. Most corals are colonial and secrete a protective limestone skeleton from which they extend. At times of danger, the individual polyps can pull themselves down into this skeleton out of harm's way.

Corals are both autotrophic and heterotrophic. The polyps have tiny tentacles which catch food and stuff into a central mouth. However, the polyps also contain symbiotic unicellular dinoflagellates *Symbiodinium microadriaticum* in their endoderm. These exist in huge numbers and fix carbon dioxide in photosynthesis. It is the presence of the dinoflagellates which accounts for the high productivity of coral reefs.

The reef building corals survive best under the following conditions: (a) the water temperature must not fall below 20°C. This virtually limits the distribution of corals to the tropical and subtropical zones; (b) the depth of water should not exceed 50 m, because beyond this depth sunlight is too faint for photosynthesis of the endozoic algae on which survival of the coral polyps depend; and (c) the water should be free from sediment. Corals therefore survive best in the moving ocean water well away from the silty coasts or muddy mouths of streams. The corals are best developed on the seaward side of the reef, where constantly moving waves, tides and currents maintain an abundant supply of clear, oxygenated water. They also bring an adequate supply of food in the form of planktons.

There are three main categories of coral reef: fringing reef, barrier reef and atoll (Fig.10.7). A **fringing reef** is a coralline platform around islands lying close to the shore. It is sometimes separated from the shore by a shallow lagoon. The outer edge facing the sea grows rapidly because of splashing waves that continuously renew the supply of fresh food. The reefs may be 1-2 km wide, lying just above the low water and sloping steeply downwards on the seaward side to a depth of about 30 m. A **barrier reef** is separated from the continental coast by a much wider and deeper channel or lagoon. The reef is partially submerged. Where it lies above the water level and sand can accumulate on it, a little vegetation is possible. The barrier reefs have narrow gaps at several places to allow the water from the enclosed lagoon to return to the open ocean. Such gaps are very useful for shipping and provide the only entrances for ships to enter or leave the lagoon. The best known barrier reef is the Great Barrier Reef off the coast of Queensland, Australia. It is 1900 km long, separated from the coast by a channel 1600 km wide in places and over 60 m deep. **Atolls** are similar to barrier reefs except that they are horsehoe-shaped, enclosing a shallow lagoon without any land in the centre. The encircling ring is usually broken in a few places to allow the free flow of water. On the inside of the reefs, sand and limestone debris collect and palm trees like coconuts may grow. Such palm trees thrive well in the brackish water of the lagoon. The nuts fall into the water and are distributed widely by floating from one coral island to another. The calm waters are useful for fishing and canoeing. Some of the large atolls, for example, Suvadiva in the Maldives, have a lagoon over 60 km across. A number of atolls provide essential air bases for trans-Pacific aircraft.

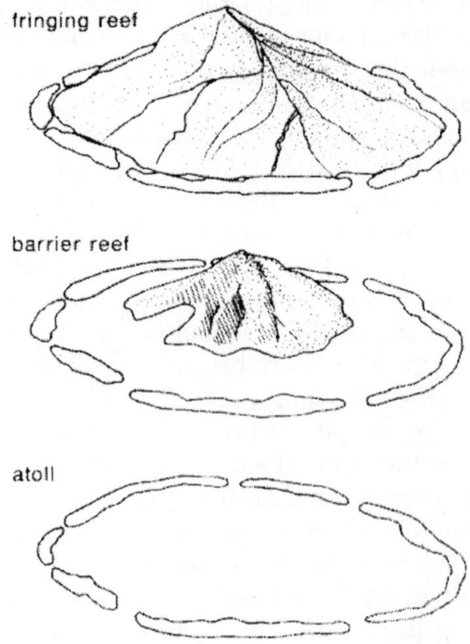

FIGURE 10.7
BARRIER REEFS AND ATOLLS ARE FORMED FROM FRINGING REEFS EITHER BY A RISE IN SEA LEVEL OR BY THE SUBSIDENCE OF THE LAND THEY LIE NEXT TO (BARNES AND HUGHES 1988)

Charles Darwin's **subsidence theory** explains the probable origin of coral reefs. Darwin assumed that all coral reefs began as fringing reefs around an island or the topmost portions of extinct volcanoes that stood above the ocean bed. Due to a general downwarping of the earth's crust, the islands gradually subsided. The corals continued to grow upwards to keep pace with the subsidence. The growth was more vigorous at the outward edge than the landward edge because of the more favourable living conditions for corals, so the encircling reef widened. It then formed a barrier reef, with a lagoon between the island and the reef. Eventually, when the land completely submerged, only the outer rims of the reefs were seen, forming an atoll. The submerged island was covered by a layer of sediment so that the characteristic circular lagoon is generally shallow. Thus atolls mark the position of the former islands.

Many species of corals live together within a single reef. In turn these corals support many other animals. For a start, a number of species simply eat the corals. Many fish, including parrot fish, have teeth and jaws which enable them to attack the coral. Polyps are also eaten by a number of invertebrates including the crown-of-thorns starfish (*Acanthaster planci*).

10.3.7 Marine surface pelagic

Pelagic communities inhabit the ocean waters, not the sea floor. The marine surface pelagic biome includes ocean waters only down to depths penetrated by light. The maximum depth at which photosynthesis occurs is approximately 250 m, and this is only in full sunlight in the most clear and tranquil open ocean waters (Barnes and Hughes 1988). Figures of less than 150 m are more typical for open ocean waters, and ones of 30 m for coastal waters. Phytoplanktons (diameter ranging between 2 and 200 μm) in the photic zone of the marine surface pelagic biome include cyanobacteria, dinoflagellates, diatoms and green algae. They support a variety of nektons including many species of fish, turtles, penguins and seals, though many of these animals also forage in the marine deep pelagic biome.

10.3.8 Marine deep pelagic

Marine deep pelagic communities are those found beneath the photic zone but still in open water: bottom-dwelling organisms are excluded. As no photosynthesis is possible, the creatures found here chiefly rely on the gravitational shower of food from the communities above. Carnivorous copepods and other crustaceans are frequent, along with Foraminifera and Radiolaria. Because so little light penetrates from above, the organisms here are either blind or produce their own light. These light-emitting organs serve a variety of functions. They can help in the detection of prey, confuse predators and help individuals to find each other for mating purposes.

The further down one goes in the oceans, the less food there is. Organisms have responded to this lack of food in a variety of ways. Many deep-sea pelagic fish have hugely distensible stomachs and enormous mouth, so that fish twice as long as themselves may be taken as prey. Most deep-sea pelagic fish minimize their metabolic rate: they move slowly and have poorly developed muscles.

10.3.9 Continental shelf benthos

Marine benthic organisms live on the bottom of the ocean. The continental shelves that surround many of the continents lie, on average, 130 m below sea level, so that in their shallower regions benthic algae and plants can photosynthesize. Here are found the impressive kelps- the large brown algae such as *Laminaria*. They attach themselves to the substratum by means of a holdfast and have one or more flat blades or fronds which can grow to more than 50 m in length. Kelps are often found in areas where ocean currents and the actions of waves ensure a plentiful supply of nutrients. Their productivity may be high, with individual fronds growing 25 cm a day. Surprisingly few animals feed on them. Most of their production enters the food web in the form of detritus.

The continental shelf benthos supports large numbers of animals from a variety of phyla. Polychaete worms, nemertine worms, molluscs, sea squirts, bryozoans, sponges, sea spiders, crustaceans and echinoderms are all found along with a number of fish.

10.3.10 Deep ocean benthos

The deep ocean benthos extends from the edge of the continental shelf to the deepest ocean trenches. Until recently these communities were assumed to be wholly heterotrophic, but it is now known that whole communities depend on the sulphur spewed from deep volcanic vents. The other deep ocean benthic organisms depend on the slow rain of food from above that has escaped the attentions of the deep pelagic organisms. A surprising variety of organisms are found on the deep floor of the ocean despite the perpetual cold, the continual darkness and the enormous pressures caused by the tremendous weight of water above.

Inevitably, our knowledge of the deep ocean benthos is rather sparse (Edwards 1988). The limited data available suggest that growth rates may be 50 times less than those typical in shallow waters, so that small molluscs are thought to take 50 years to reach adulthood (Friday and Ingram 1985). Currently, the deep ocean benthos must be one of the world's few biomes still relatively unaffected by human activities. Whether that continues to be the case remains to be seen. Every few years proposals are put forward for the deep sea mining of the nodules of manganese and other elements found on the ocean floor. Thankfully, such proposals remain uneconomic (Chapman and Reiss 1992).

11. BIODIVERSITY IN NEPAL

Nepal extends along the great Himalayan range between the latitudes 26°20' to 30°27' N, and longitudes 80°04' to 88°12' E, covering a distance of 885 km (from east to west), average width of 193 km (from south to north), and estimated area of 1,47,181 sq km. The elevation of the country rises from 60 m, mean sea level in the southern Terai to the crest of the Himalayas reaching up to 8848 m at Mt. Sagarmatha (Everest) in the north.

Geographically, the country can be divided into three zones each characterized by their dynamism: the **High Mountains** (above 2100 m altitude) with glaciers and deep-cutting streams; the **Hills** (350 to 2100 m altitude) with their cultivated steep slopes and frequent landslides; and the **Terai** (60 to 350 m altitude) with its floodplains and shifting rivers. These three zones comprise 35%, 44% and 21% of the total land area, and have 7.3%, 44.3% and 48.4%, respectively of the country's total population of 2,31,51,423 people (CBS 2001). Only 2% of the land in the Mountains is cultivable. The respective figures of arable land are 10% in the hills and 40% in the Terai.

There is a wet season, from June to October, during which the intense monsoon can cause considerable erosion and flooding. The dry season often results in drought, and can be severely limiting to agriculture. Within this general pattern there are great many variations, arising from the diverse combinations of altitude and topography. For example, many valleys remain exceptionally dry because a continuous wind keeps them clear of moist air. The time of change between wet and dry seasons has recently become less predictable, and this often causes chaos to crops. There are serious suggestions that this may be due to recent environmental disturbances. Surface water is channeled through three main watersheds: the Karnali, Kali-Gandaki, and Koshi.

All three start in the Tibetan plateaus and end in the Ganges plain.

Nepal is an agricultural country; nearly 80% of the economically active population is involved in agriculture and oil and grains mills. Tourism is the major source of foreign exchange, but studies must be made on tourist management and environmental protection before it is further developed. Extreme poverty is the major constraint to both conservation (of nature and natural resources) and development. There is a negative trade balance and most expenditure on development comes from foreign aid.

11.1 Physiography

In general, the mountain chains subdivide Nepal from south to north into five geographical zones: (a) Terai plain, (b) Churia (Siwalik) hills, (c) Mahabharat ranges, (d) Midland or central hills, and (e) higher or Great Himalayas.

Terai is the part of the Indo-Gangetic plain and is about 30-40 km wide lowland. Its average elevation is 100 to 300 m above mean sea level and it occupies 17% of the total land of the country. It is composed of old and new alluvium deposits. The northern limit to Terai adjacent to Churia/Mahabharat ranges is called **Bhabar** zone. It is about 12 km wide and consists of deposits of gravels and boulders.

The **Churia hills** which stand out in front of the Bhabar zone have elevation range between 700 to 1500 m above mean sea level. The hills are narrow in the east but wider in the west and far-west parts of the country, enclosing dun-type lensoidal valleys (**Bhitri Madesh**). The hills are composed of sedimentary rocks of the Oligocene to Pleistocene age which are indurated at some places and are loose friable at other places (Sharma 1999).

The **Mahabharat range** which lies between Churia and Midland hills, has altitudinal variations between 1500 to 2700 m. It is well-developed in eastern and central Nepal but poorly developed in western Nepal. It is cut at many places by antecedent rivers, *viz.* Koshi, Gandaki (Narayani), Karnali and Mahakali. It is composed of hard rocks like granite and limestone.

Situated between the snow-covered peaks in the north and Mahabharat range in the south, the **Midland hills** have elevation variations between 600 to 3000 m. This zone is a wide belt of low-lying valleys and tectonic basins arranged in a linear fashion from the east to west of Nepal.

The zone of **Higher (Greater) Himalayas** lies in the northern part of the country stretching from east to west. Its highest peaks are perpetually clothed in ice and snow and reach between 6755 to 8848 m altitudes. There are many glacial lakes in this zone and the topography is smooth, especially where a glacial lake culminates, otherwise most of the areas have canyons, steep cliffs and hanging valleys. This zone has several inner valleys with desert conditions, *viz.* Mustang and Mugu valleys with altitude more than 3600 m. Table 11.1 depicts the physiographic locations of all the administrative districts of Nepal.

TABLE 11.1
DISTRICTS INCLUDED IN THE FIVE PHYSIOGRAPHIC REGIONS

Physiographic regions	*Districts*
1. Terai (Tropical zone)	Banke, Bara, Bardiya, Dhanusha, Jhapa, Kailali, Kanchanpur, Kapilvastu, Mahotari, Morang, Parsa, Rauthat, Rupendehi, Saptari, Sarlahi, Siraha, Sunsari (17 Districts)
2. Siwaliks (Tropical to Subtropical zones)	Chitwan, Dang, Makwanpur, Nawalparasi, Sindhuli, Surkhet, Udaypur (7 Districts)
3. Middle Mountain (Subtropical to Temperate zones)	Achham, Arghakhanchi, Baglung, Baitadi, Bhaktapur, Bhojpur, Dadheldhura, Dailekh, Dhading, Dhankuta, Doti, Gorkha, Gulmi, Ilam, Kaski, Kathmandu, Kavrepalanchok, Khotang, Lalitpur, Nuwakot, Okhaldhunga, Palpa, Panchthar, Parbat, Pyuthan, Ramechhap, Rolpa, Salyan, Syangja, Tanahum,Terhathum (31 Districts)
4. High Mountain (Temperate to Sub-alpine zone)	Bajhang, Bajura, Dolkha, Jajarkot, Kalikot, Lamjung, Myagdi, Rukum, Sankhuwa-sabha, Sindupalchok, Solukhombu, Taplejung (12 Districts)
5. High Himal (Sub-alpine to Alpine zones)	Darchula, Dolpa, Humla, Jumla, Manang, Mugu, Mustang, Rasuwa (8 Districts)

11.2 Vegetation and Faunas

Altitude is the most important of the various factors which combine to create contrasts in climate, habitat, flora and fauna in Nepal. Accordingly, Dobremez (1972) divides Nepal into six bio-climatic zones as follows: (a) Tropical (up to 1000 m altitude), (b) Sub-tropical (from 1000 to 2000 m altitude), (c) Temperate (from 2000 to 3000 m altitude), (d) Sub-alpine (from 3000 to 4000 m altitude), (e) Alpine (from 4000 to 5000 m altitude), and (f) Nival (above 5000 m altitude).

11.2.1 Tropical zone

Phytogeographically and floristically this zone is related to the Indian and south-east Asia-Malaysian region; and, out of the total species of flowering plants (5833) recorded so far from Nepal, 1499 species of which 29 are endemic (total endemic = 246), are represented here (Shakya *et al.*, 1997). It is a part of the *cis*-Gangetic or Indian faunal zone (Bhatt 1977).

Climax vegetation of this zone is dominated mostly by *Shorea robusta* (Sakuwa/Sal), in association with *Terminalia alata* (Asana/Saaj), *T. bellirica* (Barro/belliric myrobalan), *T. chebula* (Harro/chebulic myrobalan), *Phyllanthus emblica* (Amala/Indian gooseberry), *Lagerstroemia parviflora* (Bot-Dhamgero), *Adina cordifolia* (Karma), *Syzygium* spp. (Jamun, Kyamun) etc., and big lianas (woody vines) like *Bauhinia vahlii* (Bhorla/camel's foot-climber), *Millettia* spp. (Kurkus/Gonjo) etc. The forest in its virgin state is highly stratified and trees generally form there layers: (a) top scattered emergent layer of very tall buttressed trees, 50 m or more in height; (b) a continuous dense canopy of the second layer of medium-sized trees, 25 to 30 m tall; and (c) an understory stratum of the third layer of vegetation that becomes dense only where there is a break in the canopy (Fig. 11.1). Because species richness is high and a single species may generally not account for more than one-fourth of the total importance value of the forest stand, it is not possible to recognize distinct dominant types in any real sense.

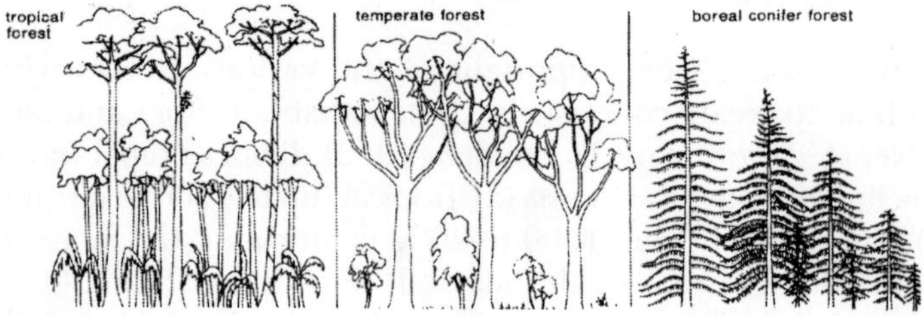

FIGURE 11.1
GRAPHIC PRESENTATION OF STRATIFICATION IN THE MAJOR TERRESTRIAL VEGETATION ZONES IN NEPAL

Tiger, leopard, elephant, rhino, rhesus monkey, long-tailed entellus monkey, bison, blue-bull, wild bear, wild boar, sambar, barking-deer, spotted-deer, hog-deer, black-buck, jackals, rodents, etc. among the mammals ; and a large number of arboreal snakes, ants, moths, butterflies, and birds feeding on fruits, seeds, buds, nectar or insects constitute the faunal diversity of the tropical forest. Tall termite mounds are also common on the forest floor. In general, a much larger proportion of animal lives in the upper layers of the vegetation than the temperate forests where most of the life is near the ground level.

There is a shortage of certain nutrients essential for development of life, for example minerals such as potassium, calcium, phosphorus and magnesium in the tropical zone. That is why the vegetation present, in spite of its abundance, contains a comparatively small amount of these nutrients. Consequently it requires considerable effort from larger animals to collect enough salts to keep their bodies in condition. This explains why only a relatively small density of higher animals is found in the tropical forest. Another reason why these animals need a large area in order to find enough food is that, as a result of there being a larger number of different plant species, only a few individuals of any single species of tree are found, per hectare. Many animals depend on certain trees for their existence, and so have to travel great distances. Thus, large tracts of the forest need to be maintained in order to cater for these species.

In tropical forest nearly all the nutrients and organic matter are in the standing crop, that is, in the vegetation. Any leaves, twigs, or fruit cases which fall to the forest floor quickly decompose in the hot, humid conditions and are immediately cycled back into the vegetation *via* the surface roots of the trees. Hence, if a tropical forest is cut down and burned, much of the nutrient and organic content of the ecosystem is lost as gases, wind-blown ash and leached out of the soils by the heavy rains. The soil beneath may be quickly exhausted and become susceptible to rapid erosion so that crops will only flourish for a few years.

11.2.2 Sub-tropical zone

Phytogeographically the lower part (1000 to 1500 m altitude) of this belt is related to the Indian and south-east Asia-Malaysian region like the tropical one while the upper part (1500 to 2000 m altitude) along with the temperate Himalayan belt closely to the Sino-Japanese region. The angiospermic flora is represented by 2028 species including 50 endemics.

This zone is represented by *Schima wallichii* (Chilaune) - *Castanopsis indica* (Dhale Katus) - *C. tribuloides* (Musure Katus) forest in the wet eastern half of the country and *Pinus roxburghii* (Khote Salla/Rani Salla/Chir-Pine) in the western half. The major associates in the former are *Engelhardia spicata* (Mauva), *Acer oblongum* (Yali/Maple), *Pyrus pashia* (Mayal/Wild Pear), *Stranvaesia nussia, Eurya acuminata* (Tegaar), *Myrica esculenta* (Kafal), *Cornus oblonga, Persea duthiei* (Mahilo Kaulo) etc. In the latter the associates are less in number and kinds and are usually composed of *Myrica esculenta, Lyonia ovalifolia* (Angeri), *Quercus lanata* (Baanjh/Wooly Oak), *Q. leucotrichophora, Rhododendron arboreum* (Lali Gurans/Rosy Bay), etc. The oaks *Quercus lanata* and *Q. leucotrichophora* on dry slopes particularly in far western part of the country form dominant forest in association with *Rhododendron arboreum, Lyonia ovalifolia*, etc. In the far humid east *Castanopsis hystrix* (Patle Katus) along with *C. tribuloides* takes the similar position. This belt has a moderate climate, congenial to human settlement. Hence, the impact of human activities on natural vegetation is distinctly visible.

11.2.3 Temperate zone

This belt is characterized by evergreen oaks, rhododendrons, conifers, etc. in association with deciduous magnolias, maples, etc. Broad-leaved forests are predominant over coniferous one in the eastern and central parts of the country. Among the broad-leaved species like *Quercus lamellosa* (Bajrat/Bull Oak), *Q. glauca* (Falaat/Blue Japanese Oak), *Daphniphyllum himalense*, various laurels, *Magnolia campbellii* (champ), various maples, etc. are conspicuous in the east and central Nepal. Board-leaved forest of *Lithocarpus pachyphylla* (Thulo Katus/Oak) is localized in the extreme east bordering with Sikkim Himalaya. West Himalayan oak like *Quercus floribunda* (Tilonj Oak), and others like *Aesculus indica* (Lekh Pangro), *Juglans regia* (Hade Okhar/Walnut) etc. are the species that dominate the broad-leaved forest in western Nepal. *Pinus wallichiana* (Gobre-Salla/Blue Pine), *Picea simithiana* (Jhule-Salla/ West-Himalayan Spruce), *Cedrus deodara* (Deodar/ Himalayan Cedar), and *Abies pindrow* (Thigo/Fir) are the main conifers that dominate the needle-leaved forest. *Tsuga dumosa* (Thinge Salla/Hemlock), a pan Himalayan pine, forms patches of forest is moist valley slopes. Similarly, *Quercus semecarpifolia* (Kharsu oak/Brown Oak of Himalaya), a pan Himalayan oak, forms dominant forest on rather drier slopes. The associated trees of the forest in this belt are various laurels, *viz. Rhododendron arboreum, Lyonia ovalifolia, Corylus* spp. (Dante Okhar/Hazel-Nut etc.), *Carpinus viminea, Taxus baccata* (Loth Salla/Talispatra/ Himalayan Yew), *Betula alnoides* (Saur), etc.

Alnus nepalensis (Uttis/Alder) forest is common in mid-mountainous Himalayan ecosystem extending from 1300 m to almost 2700 m altitude. This species can quickly colonize in barren slopes newly exposed by landslides or erosion.

From the temperate zone 1989 species of flowering plants including 113 endemics are recorded.

Species richness in temperate forest is fairly high but less than that of the tropical forest; and *architecture* of the forest is generally two layered (Fig. 11.2): (i) a closed interlocking canopy of the overstory layer of 20 to 35 m tall trees, and (ii) a second canopy of vegetation consisting of genetically smaller trees, shrubs and saplings.

In the Himalayas, seasonality increases with increasing altitude. The most obvious change is in the minimum temperatures, which drop from above 10°C in lowland tropics down to below freezing point (0°C) in temperate regions. Low winter temperatures coincide with short days and low light intensities. Below about 4°C, plants find it difficult to take in water, so plants in temperate communities may experience water shortages even though the soil is quite wet. Trees with broad-leaves in temperate forests lose a lot of water by transpiration. This means that keeping the leaves during winter would result in too much transpiration leading to damaging water loss. So the leaves are dropped; trees with broad-leaves in temperate forests are *deciduous*. Because these leaves are dropped when the cold winter approaches, they never suffer from heavy frosts, so the leaves do not have to survive freezing temperatures. This means that expensive biochemical mechanisms to prevent frost damage are not needed in the leaves.

The *aspect* of the temperate forest changes dramatically with the season as most species in each stratum are deciduous. The warmer spring season setting out in mid-March releases spring-flowering herbs from dormancy. At this time, the overstory tree canopy is leafless and more than half full sunlight reaches the ground. By late March to early April, the ground is green from renewed vegetative activity by herbs that have overwintered as bulbs and rhizomes. By mid-April, the herbs flower and the overstory trees resume meristem activity and begin to leaf out. As the canopy closes through May, light intensity on the forest floor is reduced to a minimum 2-5% full sun. By July, many of the spring-flowering herbs die back and become dormant. Leaves of shrubs and trees turn colour in early October and are shed in November.

In the brown soils of deciduous forests in Temperate zone, the surface litter is thin due to rapid decomposition, earthworms being active in this process as well as in mixing the soil. The upper soil layer is mildly acidic (pH 5-6), but not as acidic as Sub-alpine belt, partly because the pH of deciduous leaves is 5-7 rather than 4, the pH of needle-leafed forests in Sub-alpine zone. Other differences are the higher levels of nitrates and other soil nutrients that are held by clays and organic particles, making these nutrients accessible to plant roots in the Temperate deciduous forests than the soils of the Sub-alpine zone.

Temperate forests provide habitats for the wildlife of both the Indo-Malayan Realm (represented in lowlands) and the Palearctic Realm (Himalayan mountains) (Upreti 1991).

11.2.4 Sub-alpine zone

This zone, together with Alpine and Nival zones is related to the Central Asiatic region in phytogeography, and houses 1645 species of flowering plants including 177 endemics.

This zone is characterized by extensive silver fir (*Abies spectabilis*) forest in combination with the Himalayan birch/ bhojptra (*Betula utilis*). The birch usually concentrates along water courses and forms pure forests, otherwise it forms the second layer of canopy along with rhododendrons under the fir. Deciduous Himalayan larch/Bogre Salla/Lekh Salla (*Larix griffithiana*) forms forest in pockets in central Nepal. Pure juniper forest composed of *Juniperus indica* (Dhupi/Himalayan cedar) is a feature of inner Himalayan valleys and dry southern aspect. The usual associates in the forest are *Acer caudatum, A. pectinatum* (Kapasi/Putaliphul/Firfire), *Sorbus cuspidata, S. microphylla, S. foliolosa* (Najo/Beam), *Prunus* spp. (Arupate, Jungali Painyu, etc.), etc. The juniper extends well beyond tree line in the alpine belt, and the plant height goes on decreasing with the increase of altitude-it becomes bushy and forms mats/cushions above before it vanishes.

The trees are closely spaced, stunted and low in height (below 15 m), profusely branched and with dense canopies. Branching may start from less than one meter above the ground level. However, birch is able to maintain its clean bole and may grow taller. But the boles generally are bent due to heavy snowfall and strong winds (Singh and Singh 1992). Shrubs and subdominant trees are not common, but a ground layer of dwarf-shrubs, ferns, club mosses, true mosses and lichens may reach 100% cover. The forest, then, is essentially a two-layered community. Between the forest stands pastures (**'Kharkas'**) of varying sizes may exist.

Soils are acidic (pH 3-5) whose upper layers have been leached of clay, organic matter, and many nutrients. Conifers (needle-leafed plants) are characterized by shallow root systems which lie above zones of nutrient and clay accumulation in the soil. However, nutrient uptake is improved by *ectomycorrhizae*. All Sub-alpine conifers possess ectomycorrhizae and there is experimental evidence which shows that tree growth and survival would be minimal without them. Soils are acidic in part because of leaching, and in part because conifer foliage is tough and acidic.

The conifers are evergreen trees; they keep their leaves through the winter and usually for several years, and they drop only a portion of their leaves at a time. The leaves of evergreen have to be frost tolerant and need to reduce their water loss in winter. To cut down evapotranspiration the leaves are needle-like and have thick cuticle; usually the stomates are in grooves too. Because of the antifreeze mechanisms and thick cuticles, they are expensive to make. So the plant does not throw them away the following spring, but keeps the needle as long as possible. They therefore have to be tough so as to last many years without disintegrating under heavy snowfall or in high winds. This also reduces damage by insect or fungi. Having such resistant leaves has problems: when eventually they are dropped, the needles do not break down easily. The Sub-alpine soils are very cold during the winter which also reduces the rate at which leaf litter is broken down. So the needle leaves tend to lie around on the surface of the soil where they lock up nutrients for years, making the soil underneath nutrient poor.

Some notable wild animals of the Sub-alpine zone are black bear, blue sheep, goral, Himalayan tahr (jharal), marten, musk deer, pika mouse, red panda, serow, snow leopard, weasel, wild sheep, wolf, etc. With its harsh winters, the Sub-alpine zone poses severe problems for the animals found in it. Some of the birds and mammals migrate, some remain active during the long winter and some of the mammals hibernate. Most of the invertebrates survive as resting stages, typically eggs or pupae (Friday and Ingram 1985).

Between the northern tundra and southern Sub-alpine zone, there is a **timberline/tree-line** where trees become dwarfed, twisted, shrub-like and restricted in patches. These tree patches are surrounded by continuous matrix of tundra.

11.2.5 Alpine and Nival zones

Hardy perennials with low height (usually not more than 2 m), a high density of branches and their contorted form which can withstand harsh and severe cold and the annuals which complete their life-cycles within the short snowless period of the year are the characteristic physiognomical features of the Alpine (tundra) vegetation. They endure excessive windy conditions with substantial amounts of radiation of shorter wavelengths. Precipitation is almost entirely in the form of snow, and melting snow is the principal source of water to plants. From this area 1071 species of flowering plants including 189 endemics have been recorded.

The shrubby vegetation is usually dominated by rhododendron bushes. Far more number of *Rhododendron* species is found in eastern Nepal. Other common shrubs include *Potentilla fructicosa* (Bajaradanti/silver leaf), *Hippophae tibetana* (Ashuk/seabuck thorn), *Berberis* (Chutro/Berberry) spp. etc. *Cotoneaster microphyllus* (Dhalke Phul) and *Salix calyculata* are the common low shrubs trailing over rocks. Fields of bright coloured flowers resulting from overgrazing are quite common. The common ones are primeroses, gentians, poppies, marsh marigolds, buttercups, etc. Loose scree vegetation is usually composed of *Arenaria, Androsace, Waldheimia*, etc.

In Alpine zone, only the top 0.5 m of soil thaws in summer - the short growing season of 60 to 90 days. The soil is coarse textured, brown from undecomposed organic matter, and contains all the root mass of the vegetation above. The permanently frozen subsoil impedes drainage, root growth, and decomposer activity. The dwarf vegetation of the Alpine zone remains covered with snow in winter which protects the plants like a blanket.

Above the Alpine zone in north is a barren region of permanent ice and snow called the **Nival zone**.

In general, species diversity declines with increasing altitude. One of the reasons for this is the change in community structure. The major community type changes from tall tropical forest, with two or more layers of trees, through shorter temperate forest, to sub-alpine forest, which has a poorly developed understory and ground flora because of deep shade, to alpine zone (tundra) with no trees and few bushes. This results in the decline in the total number of niches available for species from lowland to alpine tundra.

The apparent importance of community structure might explain why there are more climbers, birds and insects in tropics, but it does not explain why, among the tree themselves, there are many more species in tropical forests than in temperate or sub-alpine ones. There may, however, be an additional factor involved in tree diversity: **pollination biology**.

All the conifers generally distributed at higher altitudes are *wind pollinated*. They rely on breezes to pick up pollen grains from male cones and transport them to little droplets on female cones which catch pollen and draw it inside the cone to the eggs where fertilization occurs. The action of the wind is random; it cannot take pollen precisely to one tree. Instead the pollen blows about and chance dictates where it lands. This means that conifers are limited to living in groups of one or a few species. If many species were mixed up together, then almost all the pollen a female cone received would be from another species. The dominant broad-leaved trees such as oaks of the temperate forests are also wind pollinated.

The vegetation of tropical forests is dominated by broad-leaved angiosperms. They are pollinated mostly by insects, birds, bats or other animals. Some species flower all year round and have very specific pollination relationship with a single species of pollen vector. Others use more general pollinators, but the whole population flowers all at once, so that animals which are attracted to the flowers are bound to visit another tree of the same species in flower and pass on the pollen. Because animal pollen vectors can be relied on to visit flowers directly, the randomness of using wind is eliminated. Trees do not have to grow in clumps of one species to ensure pollination; they can be distributed more sparsely in the forest.

Biodiversity have a crucial role in man's existence and development. They furnish him food. They provide his livestock with forage. They supply the air he breathes with oxygen, preserve soil fertility and regulate water cycles through a gradual release of rainwater for streams and rivers, irrigation, power stations and inland navigation. From the biodiversity man obtains fibres for clothing, wood for shelter and furnishings, and medicines which he uses to alleviate his ailments, and through hunting and trapping man obtains protein, horn, skins, wool and other products of wild animals. Biodiversity supply genetic stocks of breeds in arable farming, livestock breeding, commercial timber production and horticulture, or the improvement of existing breeds through cross-breeding. They are the source of fuel-energy and raw material for structures, paper and countless other purposes. They trap toxic substances from air and water. They influence on the climate both globally and locally and prevent natural disasters. They are the source of inspiration for art and culture, and wonderful sights to visit for tourist traffic. Besides these, biodiversity are the guarantee of potential benefits which are still unknown, and they are the only source of gradual or abrupt creation of new organisms (*organic evolution*).

11.3 Status of Biodiversity

Biodiversity, in general; are classified into three categories: genetic, species, and ecosystem diversity.

11.3.1 Genetic diversity

The heritable variations existing among the varieties/cultivars or different populations of a species due to gene and /or genome variations constitute the genetic diversity. In livestock breeding and arable farming it is most important that the breeds used possess certain lasting characteristics, for example the ability to produce many offspring, to grow rapidly, or to show resistance to diseases and pests. These characteristics can be lost over a long period and attempts must be made to retrieve them by cross-breeding with other stocks for which the climax formations are the only sources. The food of millions of people depends on whether we can continually improve our most important agricultural crops or livestocks in this manner. Scientists working on the cultivation of crops, and organizations concerned with food supply, such as the FAO and UNESCO are much worried at the loss of much useful material with the depletion of wilderness areas. This concerns not only the ancestors of our present domesticated species but, perhaps more particularly, those species which will be eradicated even before their properties are known.

Over 400 species of agro-horticulture crops are estimated to be available in Nepal. There are over 100 varieties of 15 major fruit species, 200 varieties of 50 vegetable species and about 10 varieties of potato under commercial cultivation (HMG/N 2002). There are many horticulture species with ecotypes in Nepal; many of these species are yet to be cultivated for their economic values. Seasonal fruits harvested from forests belong to over 37 genera and 45 species. Many of the unexploited group of plants can be directly brought to cultivation and/or can be used as genetic resources for particular desirable characters. Some of the wild genotypes have been identified and domesticated for their economic values by the local people through their own experiences and wisdom.

11.3.2 Species diversity

The total number of species in the world has been estimated to range from 5 to 30 million, and approximately 1.7 million living species of all kinds of organisms have been described (WCMC 1992). Nepal has an amazing wealth of floras and faunas and, in the six phytogeographical provinces of the country, scientists have identified 118 ecosystems, 75 vegetation types and 35 forest types (Dobremez 1972; Stainton 1972) harbouring about 6000 species of flowering plants and hundreds of lower taxons; and, the butterflies and moths, freshwater fishes, amphibians, reptiles, birds and mammals of the country account for 2.6, 1, 1, 1.6, 9.4 and 4.5 %, respectively for the total world species (Table 11.2). The mountainous region of the country has provided an abode and shelter to the Himalayan faunas and floras including the elusive *yeti* (the abominable snowman) while the Terai forests are rich in tropical biodiversity.

A native species confined to an isolated geographical region is known as **endemic** and this situation is referred to as **endemism**. In total, 342 species of plants and 160 species of animals are endemic to Nepal, the number being maximum in the central region, decreasing gradually towards eastern and western regions of the country.

An environmentally threatened region rich in the endemic species is referred to as **hot-spot** (Fig. 11.2). The lower slopes of the Himalayas (including parts of eastern Nepal, Sikkim, Bhutan, and western provinçe of China), Western Ghats of India, and Sri Lanka are the globally important hot-spot regions in South Asia. The Himalayan hot-spot contains a mixture of tropical and temperate plant species, about 9000 in number, and 39% of these species are endemic. It is the centre of origin for Bermuda grass (*Cynodon dactylon*), black pepper, cardamom, dwarf wheat, mango, paddy, tree cotton, etc. among the crop species and chicken among the domestic fowls. Thick forest of the region which was spreaded over 34000 sq km in past is rapidly disappearing under increasing human pressures (Table 11.3). Similarly, tropical forest covering about 1700 sq km on the seaward mountain slopes of the Western Ghats of India contains about 4000 plant species, 40% of which are endemic; and the Sinharaj forest spreaded over 56 sq km in south-western Sri Lanka has about 1000 plant species, 50% of which are endemic.

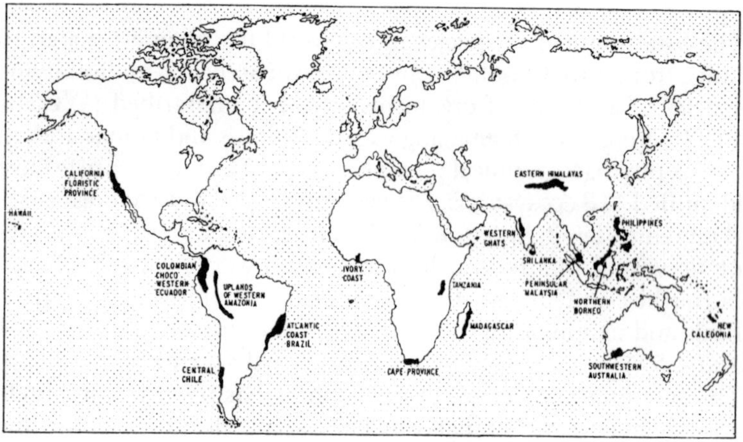

FIGURE 11.2
HOT-SPOT AREAS OF THE WORLD.

The **Survival Service Commission** of the IUCN keeps records and publishes the **Red Data Book** on species at risk including vascular plants, mammals, birds, fish, amphibians and reptiles. It recognizes the following four categories of risk status:

(a) Extinct (ex): Species not found in areas they once inhabited or in other likely habitats (after repeated searches for the past 50 years).

(b) Endangered (E): species in danger of extinction and whose survival is unlikely if the causal factors continue operating; included are species whose numbers have been reduced to a critical level or whose habitats have been so drastically reduced that they are deemed to be in immediate danger of extinction.

(c) Vulnerable (V) : Species likely to enter into the endangered category in the near future if the causal factors continue operating; included are:(i) species of which most or all the populations are decreasing because of over-exploitation,

TABLE 11.2
SPECIES RICHNESS IN NEPAL. FIGURES IN PARENTHESIS INDICATE PERCENTAGE OF THE WORLD TOTAL (CHAUDHARY 1990; MAJUPURIA AND MAJUPURIA 1998; SCHLEICH AND KASTLE 2002, SHRESTHA 2001, SHRESTHA 2001)

Taxons	*Species numbers*	
	Global	*Nepal*
1. Bacteria	3000-4000	—
2. Lichens	20000	465 (2.3)
3. Fungi	69000	1822 (2.6)
4. Algae	26000	687 (2.6)
5. Bryophytes	16600	853 (5.1)
6. Pteridophytes	11300	380 (3.4)
7. Gymnosperms	529	28 (5.3)
8. Angiosperms	220000	5856 (2.7)
9. Platyhelminthes	12200	168 (1.4)
10. Spiders	73400	144 (0.2)
11. Insects	751000	5052 (0.7)
12. Butterflies and moths	112000	640 (2.6)
13. Fishes	18150	182 (1.0)
14. Amphibians	4184	43 (1.0)
15. Reptiles	6300	100 (1.6)
16. Birds	9040	852 (9.4)
17. Mammals	4000	185 (4.5)

extensive destruction of habitat or other environmental disturbances, (ii) species with populations that have been seriously depleted and whose ultimate security is not yet assured, (iii) species with populations that are still abundant but are under threat from serious adverse factors throughout their range, and (iv) species whose populations are beginning to recover as a result of remedial action, but whose recovery is insufficient to justify their transfer to another category.

(d) Rare (R) : Species with small world populations that are not at present endangered or vulnerable, but are at risk; these species are usually localized within restricted geographical areas or habitats (endemics) or are thinly scattered over a more extensive range.

The general term **threatened** is used to refer to a species belonging to any one of the above categories.

TABLE 11.3
HUMAN THREATS TO BIODIVERSITY

1. Clearing of vegetation for agriculture and settlements
2. Logging and exploitation of forests for timber and firewood
3. Browsing and overgrazing
4. Fire
5. Dam construction, hydro-electric schemes and/or associated flooding of large areas
6. Road building
7. Mining and quarrying
8. Industrialization and urbanization
9. Water pollution
10. Drainage
11. Disturbance, especially from motor vehicles, and trampling
12. Coastal pressures and development, especially for tourism
13. Traditional rural uses, often greatly increased as a result of tourism, population increase etc.
14. Changes in agricultural methods
15. Species dependent on prevailing agricultural systems and under threat from factors such as regeneration of scrub and lack of grazing
16. Forestry (changes in practice, monocultures of exotic species)
17. Pressures from introduced species
18. Selective removal as unwanted/poisonous species
19. Collecting for botanical / zoological specimens
20. Picking of the flowers
21. Collecting (usually uprooting) for horticulture
22. Critically low population with subsequent danger of breeding collapse

The main objective of the **Convention on International Trade in Endangered Species of Wild Fauna and Flora (CITES)** to which Nepal is a Party is to protect certain species of wild flora and fauna against over-exploitation through international trade. For this purpose the IUCN provides scientific inputs to CITES through over 5000 scientists who function under its **Species Survival Commission** and other relevant commissions such as the **Law Commission** and the **World Conservation Monitoring Centre (WCMC).**

CITES functions under the following three appendices:

Appendix I contains species that the Parties have deemed to be threatened with extinction and which are, or may be, affected by trade. For these species, trade is strictly regulated and only permitted in special circumstances. Appendix I includes some highly threatened species, such as the tiger, the rhino, the giant panda, the Madagascar lemur, some crocodiles, some orchids, and some cacti.

Appendix II contains species that, although not necessarily threatened, could become so if their trade is not properly controlled. Some of these species are parrots, wild cats, some butterflies, and stony corals. Trade in appendix II species is only permitted if the exporting country issues an export permit, while in the case of appendix I species, where extra safeguards are necessary, an import permit is also required.

Appendix III includes species that are identified by any Party as being subject to regulation in that country and which require international cooperation to control trade.

Nepal still can boast of her rich and varied biodiversity, however, many commercially important species of wildlife have been brought to the very brink of extinction through a greater harvest than is normally possible. The increase in human population which requires conversion of forests into agricultural lands and irresponsible felling of the forest trees has badly affected the wildlife in Nepal. Further, excessively large herds of domestic animals have also played a significant role in the reduction of wildlife population through overgrazing, converting fertile areas into wastelands. Poaching for the purposes of meat, skin, ivory, tiger bones and rhino horns has considerably reduced wildlife, even to the extent of annihilation. Feeling the urgency to save this natural wealth, Government of Nepal has laid stress to a great extent on conservation and management of biodiversity. Laws have been formulated against exploitation of forests and legislation has been introduced for the protection of 27 species of mammals, 9 species of birds, and 3 species of reptiles under the National Parks and Wildlife Conservation Act 1973 (Table 11.4). Legal protection has been granted to 19 plant species under section 70 (a) of the Forest Act 1993 (amended 2001), under three categories; (a) species banned for collection, use, sale, distribution, transportation and export; (b) species banned for export; and (c) timber trees banned for felling, transportation and export. The plant species included under this Act are: *Acacia catechu* (Khair), *Bombax ceiba* (Simal), *Cinnamomum glaucescens* (Sugandh-Kokila), *Cordyceps sinensis* (Yarsa-Gumba), *Dactylorhiza hatagirea* (Paanch-ounle), *Dalbergia latifolia* (Satisaal), lichens (Jhyau), *Michelia* sp. (Chaamp), *Nardostachys grandiflora* (Jataamasi), *Rauvolfia serpentina* (Sarpgandhaa), *Shorea robusta* (Sakuwaa/Saal), *Taxus wallichiana* (Talis-Patra/Loth-Salla), *Valeriana wallichii* (Sugandhbaal), and others.

11.3.3 Ecosystem diversity

Major ecosystem diversity in Nepal include protected areas, forest, rangeland, agriculture, wetland and mountain ecosystems.

11.3.3.1 Protected areas (PAs)

There are two basic types of PAs: (a) totally protected (b) partially protected.

Totally protected areas are maintained in a natural state and are closed to extractive uses. They encompass the following three management categories:

Category I. **Scientific reserves** and **Strict Nature Reserves** possess outstanding representative ecosystem. A reserve's size is determined by the area required to ensure the integrity of the site. In many reserves, natural perturbations (for example, insect epidemics, forest fires) are allowed. Generally public access is limited, with only scientific research and educational use permitted.

Category II. **National Parks** and **Provincial Parks** are relatively large areas of national or international significance not materially altered by humans. These PAs include a minimally disturbed *core area* for conservation and research and may be surrounded by *buffer zones* where traditional land uses, experimental ecosystem research, and ecosystem rehabilitation may be permitted. Visitors may use these national parks for recreation and study.

Category III. **Natural monuments** and **Natural landmarks** contain unique geological formations, special animals or plants, or unusual habitats.

Partially protected areas are areas that may be managed for specific uses, such as recreation or tourism, or that provide optimum conditions for certain species or communities of wildlife.

TABLE 11.4 PROTECTED FAUNAS OF NEPAL

Scientific name	*Local name*	*Common name*	*IUCN status*	*CITES Appendix*
Mammals				
1. *Ailurus fulgens*	Habrey	Red panda	V	I
2. *Antilope cervicapra*	Krishnasagar	Black buck	V	III
3. *Bos gaurus*	Gaurigai	Gaur	V	I
4. Bos mutus	Chaurigai	Wild yak	E	I
5. *Bubalus bulalis arnee*	Arna	Wild water buffalo	E	III
6. *Canis lupus*	Bwanso	Tibetan wolf	V	I
7. *Caprolagus hispidus*	Hispid kharayo	Hispid hare	E	I
8. *Cervus duvauceli*	Barasingha	Swamp deer	E	I
9. *Elephas maximus*	Hatti	Asiatic elephant	E	I
10. *Felis lynx*		Lynx	E	II
11. *Hyaena hyaena*	Hundar	Striped hyaena		
12. *Macaca assamensis*	Assame rato bandar	Assamese monkey		
13. *Manis crassicaudata*	Salak	Indian pangolin		II
14. *Manis pentadactyla*	Salak	Chinese pangolin		II
15. *Moschus chrysogaster*	Kasturi mriga	Musk deer	E	I
16. *Ovis ammon*	Nayan	Great Tibetan sheep		I
17. *Panthera tigris*	Bagh	Bengal tiger	E	I
18. *Panthera uncia*	Hiun chituwa	Snow leopard	E	I
19. *Pantholops hodgsoni*	Chiru	Tibetan antelope		I
20. *Pardofelis nebulosa*	Dwanshe chituwa	Clouded leopard	V	I
21. *Platanista gangetica*	Suns	Gangetic dolphin	V	I

Scientific name	*Local name*	*Common name*	*IUCN status*	*CITES Appendix*
22. *Prionailurus bengalensis*	Chari bagh	Leopard cat		I
23. *Prionodon pardicolor*	Silu	Spotted linsang		I
24. *Rhinoceros unicornis*	Gainda	Asian one-horned rhino	E	I
25. *Sus salvanius*	Pudke bandel	Pygmy hog	Ex (?)	I
26. *Tetracerus quadricornis*	Chauka	Four-horned antelope	V	III
27. *Ursos arctus*	Himali rato bhalu	Brown bear		I
Birds				
1. *Buceros bicornis*	Raj dhanesh	Giant hornbill		I
2. *Catreus wallichii*	Cheer	Cheer pheasant	E	I
3. *Ciconia ciconia*	Seto sarus	White stork		
4. *Ciconia nigra*	Kalo sarus	Black stork		II
5. *Eupadotis bengalensis*	Khar majur	Bengal florican	E	I
6. *Grus grus (G. antigone)*	Sarus	Common crane		II
7. *Lophophorus impejanus*	Danfe	Impeyan pheasant		I
8. *Sypheotides indica*	Sano khar majur	Lesser florican	E	II
9. *Tragopan satyra*	Munal	Crimson-horned pheasant		III
Reptiles				
1. *Gavialis gangeticus*	Ghadial gohi	Gharial	E	I
2. *Python molurus*	Ajingar	Asiatic rock python	V	I
3. *Varanus flavescens*	Sun gohoro	Golden monitor lizard	I	I

Ex = Extinct; E = Endangered; V = Vulnerable

Some extractive use within these areas is allowed. They encompass two management categories:

Category IV. **Managed nature reserves** and **Wildlife sanctuaries** are protected for specific purposes, such as the conservation of a significant plant or animal species. Manipulative management techniques are permitted to this end, such as predator control to protect an endangered species or livestock grazing to maintain grassland communities. Management for harvestable renewable resources is permitted.

Category V. **Proctected landscapes** and **Seascapes** may be entirely natural or may include cultural landscapes (for example, scenic attractive agricultural areas). Examples would include coastlines, lakeshores, and hilly or mountainous terrain along scenic highways.

Besides these five IUCN management categories of PAs, others include Biosphere reserves, wetlands of international importance, and World Heritage Sites.

Biosphere reserves are representative of terrestrial and coastal environments that have been recognized internationally under the **Man and Biosphere (MAB)** programme of the UNESCO. They have been selected for their value to conservation and are intended to foster the scientific knowledge, skills, and human values necessary to support sustainable development. Each reserve essentially contains a diverse natural ecosystems of a specific biogeographic province large enough to be an effective conservation unit.

Any Party to the **Convention on Wetlands of International Importance especially as Waterfowl Habitat** (Ramsar, Iran, 1971) who agrees to respect the site's integrity and to establish wetland reserves can designate wetlands of international importance.

World heritage sites are areas of "outstanding universal value" inscribed either for their natural features, for their cultural value, or for both natural and cultural values. Any Party to the World Heritage Convention may nominate natural sites that contain examples of a major stage in the earth's evolutionary history; a significant ongoing geological process; a unique or superlative natural phenomenon, formation, or feature; or a habitat for threatened species.

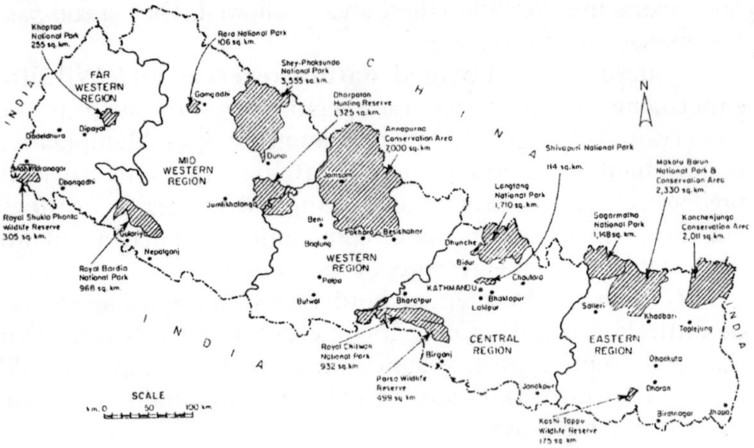

FIGURE 11.3 PROTECTED AREAS OF NEPAL

The main objectives of PAs in Nepal is the preservation of natural, historic, scenic, cultural and wildlife value. They cover more than 18% (about 2.7 million hectares) of country's land (Fig. 11.3, Table 11.5). The PAs designated as World heritage site are Sagarmatha National Park for the outstanding natural landscapes that includes Mt. Everest (8848 m altitude), and the Royal Chitwan National Park for the outstanding and beautiful Siwalik valley ecosystem and for being the habitats of tiger, rhinoceros, gaur, ghariyaal and the gangetic dolphin. Shey Phoksunda National Park is being considered for designation as natural and cultural heritage. The **Ramsar Sites** (wetlands of international importance especially as waterfowl habitat) of Nepal include Koshi Tappu Wildlife Reserve, Ghodaghodi Lake Area, Beeshazaar Lake, and Jagdishpur reservoir.

11.3.3.2 Forest ecosystems

The forest formations of Nepal which have evolved to present state through the long process of natural succession in the last many millions of years are seriously threatened under human pressures. There is a widening gap between demand and supply of wood for diverse uses. The strategy to meet the shortages has been chalked out by the Department of Forest, Government of

TABLE 11.2
PROTECTED AREAS OF NEPAL

Category and the year of establishment	*Area (sq km)*	*Altitude (m)*
1. National Park		
a) Royal Chitwan National Park (1973)	932	150-815
b) Royal Bardia National Park (1976/1988)	968	152-1494
c) Shivpuri National Park (1984/2001)	144	1366-2732
d) Khaptad National Park (1984)	225	1000-3276
e) Makalu Barun National Park (1991)	1500	435-8463
f) Sagarmatha National Park (1976)	1148	2800-8848
g) Langtang National Park (1976)	1710	792-7245
h) Shey-Phoksundo National Park (1984)	3555	2000-6885
i) Rara National Park (1976)	106	1800-4048
2. Buffer Zone of National Park		
a) Royal Chitwan National Park	750	
b) Royal Bardia National Park	328	
c) Makalu Barun National Park	830	
d) Langtang National Park	420	
e) Shey-Phoksundo National Park	449	
f) Sagarmatha National Park	275	
3. Wildlife Reserve		
a) Royal Suklaphanta Wildlife Reserve (1976)	305	90-270
b) Koshi Tappu Wildlife Reserve (1976)	175	90
c) Parsa Wildlife Reserve (1984)	499	150-815

Category and the year of establishment	*Area (sq km)*	*Altitude (m)*
4. Hunting Reserve		
a) Dhorpatan Hunting Reserve (1987)	1325	2850-7000
5. Conservation Area		
a) Kanchenjunga Conservation Area (1997)	2035	1200-8598
b) Manaslu Conservation Area (1998)	1663	1360-8163
c) Annapurna Conservation Area (1986/1992)	7629	1000-8092

Nepal against two major objectives of forestry: (a) affording long-term ecological security, and (b) supply of goods and services to the people and industry through a well-thought-out plan of production. To achieve these objectives Government of Nepal classifies forests into five management categories; special forests, private forests, leasehold forests, community forests, and state forests.

Special forests are managed for the priority objective of wildlife and nature conservation, and include all categories of the protected areas, some sacred religious groves as well as some army barracks. The bottlenecks for the effective management of the protected areas seem to be the lack of other land for surrounding people to derive their forestry goods and services, necessity of posting army units within these forests for preventing poaching, lack of a holistic ecosystem approach in their use and management, and so on. In addition, major income and employment opportunities generated from these forests are grabbed by outsiders including hotel and safari entrepreneurs than to the local people.

Farm trees and other patches and plots of forests under private ownership constitute the **private forests**. They obviously contribute significantly towards the economic and environmental benefits to the respective owners and the local communities. These contributions can be further enhanced if these owners are provided with technical and market supports in increasing the productivity and values of their products. In some cases significant innovations including planting of shelterbelts, windbreaks, contour rows etc, may be integrated into their farming and other landuse systems. The main thrust of these supports to the owners should be oriented in enhancing income and employment to themselves in commensurate with the environmental amelioration in the locality.

The National Forestry Plan of 1976 has made a provision for **leasing** the government forests to individuals and industrial firms for more intensive management. Due to the lack of provision to lease out large chunks of forests, industrial firms did not or could not participate into this programme. Nevertheless, many landless farmers or small groups, particularly in the mid-hills, have been able to take advantage of the leasehold forest management. An integrated approach of bringing together agricultural and forestry line agencies with the rural financing bank has been adopted with an aim to improve the ecological conditions as well as economy of the people below the poverty line.

Community forest is a part of the government forest handed over to the "Forest Users Groups (FUGs)" for collective conservation and utilization. Since the introduction of community forest management in 1970s the programme has become quite successful and beneficial, especially in the hills and mountains of Nepal. The formal programme has strengthened the traditional system of forest use in these regions where extensive subsistence agriculture has already converted the remaining forests into small galleries and patches. However, out of the estimated 2 million hectares of potential community forests in the country, less than 10% have been handed over to the respective FUGs. The FUGs are able to protect their forests effectively through cooperative efforts but are not able to enhance the economic and environmental productivity due to the lack of adequate technical and market supports. The two major deficiencies are: (a) the Community Forestry Division (CFD) of the Department of Forest (DOF) is not able to muster adequate financial and human resources to reach many thousands of the already formed FUGs and hundreds of thousands of other potential ones, and (b) the DOF has not adequately emphasized the productive, protective and regulative roles of community forests beyond meeting the subsistence requirement of the forest users. Under these circumstances, the most effective policy guideline for Government of Nepal would be to mobilize capable NGOs to provide the required extension supports in collaboration with the CFD/DOF with or without financial supports from the international donor agencies. In the case of lack or inadequacy of external financial support it will be necessary for the government to charge a small proportion of income made by FUGs on commercial scale. It will make the community forestry much more beneficial and successful, sustainable and continuous.

The remaining forests of Nepal which comprise most of the productive forests in the Terai and the Siwalik valleys as well as other significant areas of watershed forests in the High Mountains and the Himalayas are recognized as **State forests**. These are under the direct managerial jurisdiction of DOF of the Ministry of Forests. DOF has adequate number of forestry graduates and technicians even then the performance of DOF is less than satisfactory mainly due to the insufficiency of budgetary allocation; paucity of survey and research information for scientific management ; apathetic and lethargic attitudes of concerned officials and technicians; frequent hacking, poaching and smuggling of forest products from the more accessible and productive forests, and occasional flares of corruption and irregularity induced from various levels of the government and society.

11.3.3.3 Grassland ecosystem

Grasslands occur where rainfall is too low to support the forest life form but is higher than that which results in desert life forms. Generally this means between 250 to 750 mm, depending on temperature and seasonal distribution. However, grasslands also occur in regions of forest climate where edaphic factors such as high water table or fire favour grass in competition with woody plants. Grasslands cover about 11.5% of Nepal's total land area and about 70% of the country's grasslands are located in alpine and sub-alpine regions. The high altitude grasslands of Nepal contain 131 endemic plant species (Shrestha and Joshi 1996) and provide shelters to several endangered species of wildlife including snow leopard, Tibetan wolf, Tibetan argali, lynx, brown bear, Tibetan wild ass, wild yak, and many threatened birds.

11.3.3.4 Wetland ecosystems

Known as "Simsar" in Nepali, wetlands are those areas which lie between the land and deepwater and remain waterlogged or submerged under water, seasonally or throughout the year. Generally the land is so muddy that one cannot easily walk over it, and water is so deep that one can neither swim nor get drowned. River floodplains, shallow margins of lakes and reservoirs, shallow and seasonal ponds, islands in rivers, deepwater paddy fields, and sea-coasts are typical examples of wetlands.

Wetlands have a special historical and cultural importance in Nepal. The name of the country "Nepal" is perhaps derived from the Maithili word "Nipal" meaning "a sacred piece of land, cleansed by the holy rivers Mahakali, Koshi, Gandaki, Bagmati, etc". The river floodplains in Nepal are important sites of culture and history. The river banks are the sites for cremating the dead bodies as per the Hindu tradition. People bathe in holy streams/ ponds and the famous **Chhathi** festival of eastern Nepal is celebrated in the wetlands (rivers and ponds). These are also important as bird's paradise, and out of 39 nationally protected species of fauna under the National Parks and Wildlife Conservation Act 1973, the wild water buffalo, Gangetic dolphin, tiger, elephant, and greater one-horned rhinoceros are wetland dependent mammals; white stork, black stork and sarus crane are wetland dependent birds; and the ghariyaal is a wetland dependent reptile. The most important value of wetlands for human use in Nepal include water retention, replenishment of underground aquifers, and water quality improvement by removing or retaining nutrients, processing organic wastes, and reducing sediment before it reaches open water. Wetlands are crucial for food production: rice-paddy, fish and several types of vegetables; for the occurrence of medicinal plants, wild food, etc; and for the wild genetic sources of agrobiodiversity. The fisherfolk and other ethnic communities depend on wetlands for fishing, river transport, and traditional crafts produced from wetland plants-Korhilaa (*Aeschynomene asper*), Pater (*Typha angustifolia*), Mothi (*Cyperus procerus*), Baint (*Calamus* spp.), and Siki (*Vetiveria zizanioides*). These wetland dependent communities are some of the poorest and most marginalized people in Nepal. Reduced access to wetlands, loss and degradation of wetlands, increasing population and lack of alternative livelihood options are some of the key challenges facing these communities.

11.3.3.5 Agro-ecosystems

The root cause of decline in agricultural production in recent years is the large scale deforestation in the country. When a temperate forest is removed, the soil retains nutrients and structure and can be farmed for many years in the conventional manner, which involves ploughing one or more times a year, planting annual species, and applying inorganic fertilizers. During the winter, freezing temperatures help hold in nutrients and control pests and parasites. In the humid tropics, on the other hand, forest removal takes away the land's ability to hold and recycle nutrients (as well as combat pests) in the face of high year-around temperatures and periods of leaching rainfall. Too often crop productivity declines rapidly and the land is abandoned, creating the pattern of shifting agriculture. Considering the magnitude of problem, immediate conservation measures need to be undertaken for conserving soil through various management practices including reforestation/ afforestation, composting, manuring, use of legumes, azolla, crop rotations, and maintenance/development of irrigation facilities to sustain the crops throughout the dry season.

In spite of the adverse environmental effects of high-input high-output agriculture technology, for the foreseeable future, the country will have to depend on this model, which cannot be suddenly changed in favour of environmentally sustainable but low input and relatively low output models like organic farming, no-till need to minimum-till agriculture. Further, there is little scope for extending land under crops as the remaining uncultivated lands are prone to erosion due to the steepness of slopes and thin soils, and they are the only refuge for much of Nepal's wildlife. Agroforestry, beekeeping and sericulture yield good returns and yet protect the soil on slopes by avoiding the need for tillage. Aquaculture (fishponds) can make use of resources which might otherwise go to waste. Horticultural crops can take advantage of many of the different site types in the hills, but are at present little used; low-weight, high-value crops such as spices and medicinal herbs take little land, do not exhaust the soil, and yield high prices.

To meet the present and future needs of agriculture Nepal has 500 species/varieties of food plants including more than 95 aromatic and fine rice-paddy landraces. However, they are being replaced by monoculture of high-yielding-varieties (HYVs) of crops and habitat destruction. For example, *Samundraphinj*-a suitable rice-paddy landrace grown in swampy lands around the lakes in Pokhara valley is threatened as swamps are being converted into ordinary agriculture land for cultivation of HYVs of rice-paddies. Similarly, preference for more productive exotic breeds of livestock is threatening the indigenous breeds: *Lulu* and *Achhame* cattle are on the verge of extinction, *Yak* population is declining, *Lime* buffalo is endangered, and *Lampuchhre* and *Kage* sheep are at risk.

11.3.3.6 Mountain ecosystems

About 83% of the total land area of Nepal composed of hills and mountains is a part the Hindu Kush-Himalayas (HKH). This landscape has attracted the attention of man since ancient times. Sages, philosophers, poets and artists have sought refuge in its recesses, and have found stimulus from the loftiness of its peaks and divine grandeur. To a large extent, it has shaped the culture of the Indian subcontinent, and has contributed towards making the climate benign. It is the cradle of important mighty river systems which have been responsible for the fertility of the vast expanses of the plains. The Himalayan ranges support a great variety of vegetation ranging from forests to climax grasslands which vary in species composition from east to west and from low to high elevations. These vegetation types have traditionally played a key role in safeguarding the environment and have provided the much needed long-term ecological security to the subcontinent such as protection to wildlife and soil, stability to climate, perennial yield of fresh water, and sustenance to man. However, human onslaughts such as expansive as well as shifting agriculture (**Khoriya**), overgrazing, felling of trees for fuelwood and timber, construction of roads and other developmental infrastructures, and tourism have led to indiscriminate, unscientific and overexploitation of the forests and other types of vegetation to such an extent that their rate of renewal has been exceeded by rate of exploitation. The consequent depletion of the vegetation cover has accelerated landslides in the hills, and floods in the plains. Many of the erstwhile abundant species have become rare and several of them are on their way to extinction. The delicate relationship between man and vegetation has been shattered. All this calls for a sound policy of management and conservation.

Based on natural ecological features such as soil, slope, and natural biotic communities, the mountainous landscape is classified into eight land (use) types (Fig. 11.4). Land types I and II comprise level areas with good agricultural soils that can be continuously cultivated with only simple precautions such as crop rotations and strip cropping, whereas land types III and IV (steeper slopes) require greater restriction if cultivated, for example, periodic fallowing, perennial crops, or rotated pastures.

Types V and VII are not suitable for cultivation and should be used for permanent pasture, tree crops, or retained in their natural state. Type VIII (steepest slopes, thinnest soil) is productive only in its natural state as habitat for game, furbearers, non-timber forest products, or is valued for recreational, scenic, watershed protection. In the Himalayan mountains and hills, there is no class I and II land, little class III land and some class IV land. As a consequence large areas of class V to VII land are used for cultivation. Cultivating slopes greater than 30 degrees, ploughing up and down the slopes, constructing terraces without proper consolidation and drainage, and leaving land bare of vegetation makes the thin hill soils very prone to erosion. Heavy monsoon rains can easily break apart the soil and carry it downstream.

FIGURE 11.4
LAND USE PATTERN

Marginal lands of the hills are used for **shifting cultivation** with a fallow period between five to seven years. This time span is insufficient for natural regeneration of vegetation, hence shifting cultivation is a source of watershed and environment degradation.

Animals grazing in the Himalayan landscapes are many times more the level that the land can sustain. The effects of increased cattle and goat grazing and growing demand for forest trees for fuelwood, construction, furniture and other industries can only lead to greater deforestation.

In recent times the increasing pace of developmental activities such as mining, quarrying, road construction and power transmission towers constructed unscientifically without consideration for conservation and environment, have contributed to the degradation of vegetation and watersheds. The adverse effect of tourism is that fuelwood is cut for use in lodges and tea shops, food prices in the region rise and cultural values are weakened and often degraded.

Water resources present perhaps the greatest hope for Nepal's development. At present, our knowledge of how to handle them in the unique Nepalese context is lacking. We must keep options for their use open, avoiding soil erosion and pollution of watercourses which could lessen their potential and create further difficulties. When water resources are developed, care must be taken to protect other land-uses. Prime farmland should be avoided when choosing a valley to dam up and flood for a reservoir for hydro-electric power projects.

Chapter 13 of Agenda 21 ratified during the 1992 earth summit of the United Nations Conference on Environment and Development (UNCED) at Rio de Janeiro, Brazil, has given focus to sustainable mountain development. Government of Nepal intends to promote this aspect of development through integrated watershed management in collaboration with national and international agencies including the National Trust for Nature Conservation (NTNC) and the International Center for Integrated Mountain Development (ICIMOD).

11.4 Conservation of Biodiversity

To preserve the biosphere as a stable habitable system, *conservation* is the management of human use of the biosphere so that it may yield the greatest sustainable benefits to present generation, while maintaining its potential to meet the needs and aspirations of future generations. The biodiversity need to be maintained as a resource of immeasurable value. Many undomesticated species have proved useful to humans in the past for food or as sources of medicinal compounds and building materials. Wild plants may provide useful genes for drought tolerance and disease resistance which can be bred into our crops. The Himalayan forests are thought to contain many potential useful species from which we could develop insecticides, anti-cancer drugs, and so on. All these protective functions cease once a species has become extinct.

In geological time the fossil record shows that new species are continually evolving and others become extinct. Occasionally a major extinction event occurred when very large numbers of species disappeared suddenly from the fossil record. These catastrophic events are thought to have been caused by such factors as changes in world climate, loss of habitat (for example when continental drift destroyed large areas of marine shelf), or the effects of giant meteors or volcanic eruption. The greatest known extinction event happened at the end of the Permian about 250 million years ago when over 90% of fossilisable marine species went extinct.

Since the evolution of human society man has impoverished the biodiversity and nature both directly and indirectly. Flightless birds suffered greatly from human predation as they were easy to catch and often fat and tasty. Hence the dodo (*Raphus cucullatus*) of Mauritius was killed by sailors and New Zealand moas (*Diornis maximus*) probably met the same fate at the hands of invading Polynesians. Some animals are hunted not for food but luxury items such as elephants for ivory, rhinoceroses for horns, parrots and coral reef fish for pet trade, and so on. The most serious threat to species is the loss of available habitat as this affects the whole community of organisms. Some species, such as weeds, a few birds and other adaptable animals, have managed to thrive in the new environments produced by cultivation and urban life. Most species, however, have suffered a loss of available habitat. For some this has reduced numbers to the point of extinction. Apart from individual threats to species or ecosystems, the whole world is under threat from more extensive changes which transcend national boundaries. Many of these global threats are the result of human activities. The ozone layer already has seasonal 'holes' over the poles and is in serious danger of breaking down further because of the activity of chlorofluorocarbons (CFCs) produced and released from aerosols and refrigeration units as well as industrial processes. The forests of Europe are threatened by pollutants in the atmosphere. Many rivers are also polluted and the oceans, especially shallow seas such as the North Sea and the Mediterranean, are increasingly affected. The excess release of carbon dioxide from the burning of fossil fuels and forest destruction is producing a reflecting blanket which is causing *global warming*. Most of the extensive amounts of remaining forest are under threat from logging companies even with our present knowledge of the importance of the forests for the future of our planet.

In every organism the decline in the numbers of individuals and the local extinctions of populations may finally lead to a situation where there is only one population left. A single population is susceptible to accidents: freak weather, a virulent disease, fire, flood, and so on. If a species is widespread then the destruction of a population produces a temporary fluctuation in numbers: emigrants from other populations will probably found a new population to replace the old one. If a species is rare, the loss of a whole population may be a disaster from which the species never recovers. This is because of the genetic phenomenon called **inbreeding depression.**

Every organism has a collection of genes consisting of lengths of DNA (deoxyribonucleic acid). This DNA can be read like a book and is used by the organism to construct proteins including enzymes. Each cell has a copy of this genetic material stored in the nucleus. The genes are arranged in a set order on several chromosomes. Most cells have two sets of chromosomes and, usually, one set comes from each of the two parents of the individual. The genes are responsible for the growth, development, maturation and reproduction of the individual.

A single gene can have many different forms called **alleles.** For example, a gene controlling the characteristic 'height' in peas may have two alleles, one for 'tall' and another for 'short'. In general, a gene with more than one allele is said to be **polymorphic.** Because many genes are polymorphic considerable variation is possible between different organisms of the same species. The set of alleles an individual possesses on its chromosomes is called its genotype, and it is the genotype which causes the organism to interact with its environment. Together the **genotype** and the environment are responsible for the physical appearance of the individual. The outward expression of the genotype is called its **phenotype**.

An organism has no say in the genotype it inherits from its parents. It can, however, affect the genes its offspring inherit by the method it uses to pass on its genes.

Organisms pass genes on to their offspring by different breeding systems: by outcrossing (cross-fertilization), self-fertilization, seed apomixis, parthenogenesis, and vegetative reproduction. Outcrossing results in individuals obtaining half their genes from one parent, half from the other. Obviously, such individuals are **heterozygous** having two different alleles for the same gene. On the other hand, offspring produced by self-fertilization or, asexual reproduction (seed apomixis, parthenogenesis, vegetative reproduction) are **homozygous** having the same two alleles for a given character because they inherit all their genes from a single parent.

In general, organisms with a high degree of heterozygosity are usually very vigorous and healthy. They show what is called **hybrid vigour** or **heterosis**. Another advantage of being heterozygous is that if one of the inherited alleles is disadvantageous (*deleterious*) or fatal (*lethal*), then the other allele may be expressed instead.

If we self-pollinate a normally vigorous, outcrossing plant so that it is self-fertilized and try to grow the seed produced, we find some quite serious problems. Many seeds will be empty with no embryo present. Of those with embryo present, some may fail to germinate, or when the seedlings emerge some will grow poorly and die before maturity. The surviving plants will show a decrease in heterosis, the vigour of variability lost due to decreased heterozygosity and conversely increased homozygosity. This inbreeding depression (high mortality) is the result of the previously hidden deleterious and lethal alleles being expressed because they have been brought together in the homozygous state. In nature, sometimes an existing population suffers a crash in numbers due to a disaster of some kind. Disease epidemics are probably the most frequent cause of large population crashes, but drought, fire, flood, predation or lack of food may also be involved. Such times when a population is unusually small are known as **bottlenecks**. During such a population crash, some alleles will decrease in frequency, and may be lost altogether due to the excessive mortality of individuals. If only a few individuals survive in the population, then inbreeding will occur after the event producing increased homozygosity in the resulting genotypes. An extreme example of the results of a genetic bottleneck is provided by the cheetah (*Acinonyx jubatus*). The bottleneck in the cheetah population perhaps occurred as long as 1000000 years at the beginning of the last ice age when four other species of cheetah became extinct. Since then the cheetah has lost its widespread distribution. Only 20000 years ago it lived in Asia, Europe and North America. Now it is restricted to parts of Africa and may have a population size of less than 5000. The cheetah is obviously now a vulnerable species. There is evidence of inbreeding depression as male fertility is ten times lower than in other cat species. Also, infant mortality is as high as 70% in the wild; this is high for a K-selected mammal. Increased levels of homozygosity mean that if a disease attacks one animal because it has no resistance, then it is likely to be able to attack all the population. This vulnerability was shown in 1983 among captive animals in an American wildlife park. Within the colony 90% developed an infectious disease which attacks cats; despite intensive medical treatment, 18 animals died (O'Brien *et al.*, 1983, 1985). The inbreeding depression is the sole reason why there are cultural restraints on marriage between close relatives in human societies.

Nepal has an impressive array of protected areas covering more than 18% of country's land for *in-situ* conservation of biodiversity. However, unless these protected areas are linked together through biodiversity-friendly corridors their roles in maintaining ecosystem processes will be limited as the forest cover which existed outside the protected areas until a decade ago has been virtually wiped out under ever-increasing human pressures. Animals establish themselves in habitats which provide them with the basic needs for their normal life, and different animals have different requirements for food, water, resting place, escape cover, etc. If the animals are forced to live in a habitat that does not have all the requirements for their life, then natural selection will weed them out.

In view of the present scenarios of habitat loss and species extinctions, *ex-situ* conservation of endangered species is urgently needed. This can be achieved through the establishment of botanic gardens, zoos, seed banks, *in-vitro* ("in glass") storage of germplasm, and DNA libraries.

The botanic gardens and zoos are suitable for *ex-situ* conservation at the level of population and individuals. In Nepal, Royal Botanical Garden established in 1962 at Godavari has a notable live collection of indigenous ferns, orchids, and rhododendrons. Similarly, the National Zoo of Nepal located at Jawalakhel possesses some endangered wild animals including rhinoceros, tiger, panther, crocodile and others. However, breeding programmes of these endangered species have not yet been planned in this zoo.

A population perpetuates, multiplies and disperses through propagules, which are seeds in most of the plant species. After attaining physiological maturity seeds either dehydrate or desiccate (**orthodox seeds**), germinate on the plant (**vivipary**), or bypass complete desiccation (**recalcitrant seeds**). The orthodox seeds are either medium-lived or long-lived. The medium-lived seeds remain viable for periods of 2-15 years, provided that seeds are stored at low humidity and at low temperature. Seeds of most conifers, fruit trees, and commercially grown vegetables, flowers, and grains fall into this category. The long-lived seeds generally have hard seed coats that are impermeable to water. Plant families that produce seeds with hard seed coats include the legumes, geranium, and morning glory families. For such seeds storage in hermetically sealed, moisture-resistant containers is advantageous for long storage, but the moisture content of the seeds must be low (4 - 6%) at the time of sealing.

The recalcitrant seeds normally are short-lived and retain viability for as little as a few days, months, or at most a year after harvest. Species producing such seeds are : (a) certain spring-ripening, temperate zone trees such as poplar (**Populus**), maple (**Acer**), willow (**Salix**), and elm (**Ulmus**); their seeds drop to the ground and normally germinate immediately; (b) many tropical plants grown under conditions of high temperature and humidity; these include such plants as sugarcane, rubber, jackfruit, citrus, many palms, litchi, mango, tea, coffee, etc; (c) many aquatic plants of temperate zones, such as wild rice (**Zizania aquatica**), pondweeds, arrowheads, and rushes; (d) many tree-nut and similar species with large fleshy cotyledons, such as birch (**Betula**), beech (**Fagus**), oak (**Quercus**), walnut (**Juglans**), etc. Germination in recalcitrant seeds must proceed soon after maturity or the seeds must be stored under conditions that prevent drying.

Reduced temperature invariably lengthens the storage life of seeds, and in general, can offset the adverse effects of high moisture content. Sub-freezing temperatures, at least down to -18°C, increase storage life of most kinds of seeds but moisture content should be in equilibrium with 70% relative humidity or lower, or the free water in the seeds may freeze and cause injury. Survival of seeds exposed to ultra-low temperature has been known since 1987 (Callan *et al.*, 1990). Seeds are **cryopreserved** by immersion and storage in liquid nitrogen at -196 °C.

Modern technologies to preserve plant organs (embryos, shoots, roots, flowers), and plant tissues (cells, callus, and protoplasts) in aseptic culture to regenerate new plant from them are collectively called **tissue and organ culture**, *in vitro* culture, **micropropagation**, and **biotechnology**. Storage of leafy shoots (DNA rich material) or extracted DNA in liquid nitrogen could become sources of DNA libraries from where a relevant gene or combination of genes can be recovered. In Nepal conservation of valuable genotypes is neglected and little seed storage (National Agricultural Research Council of Nepal has seed banks of paddy, wheat, buckwheat, and maize) of numerous varieties of farm and forestry crops occurs. Nepal is thus discarding genetic material which could be of high value to future development.

Today's problems of biodiversity degradation, soil erosion, droughts and floods, pollution and environmental instability, terrorism and many other socio-economic conflicts reflect serious waning of the harmonious concept of life based on ecological principles: *"Ahimsa (non-violence)"* and *"live and let live"* philosophies. If the benefits of nature and natural resources are used only by the privileged class of society, inequalities in wealth and opportunity will continue to grow between the haves and the havenots, majority of people will lose confidence in the system becoming frustrated and indifferent to life, and ultimately there will be no scope for the conservation of biodiversity and therefore, human existence. Mahatma Gandhi said, "An economy that inculcates mammon worship and enables the strong to amass wealth at the cost of the weak, is a false and dismal science..., it spells death. True economy stands for social justice".

Raising self-confidence of deprived people through village reconstruction, development of cottage industries, promotion of traditional skills, and an egalitarian access to natural resources is essential to let them lead a meaningful life, and in return, seek their assured participation in conserving the biodiversity. This is indispensable for the fulfillment of country's commitment to **Convention on Biological Diversity** whose basic objectives are: protection of biodiversity, sustainable use of its components, and the fair and equitable sharing of the ensuing benefits.

Part Two
ENVIRONMENTAL FACTORS

12. ENVIRONMENTAL FACTORS

Dictionaries define **environment** as the circumstances, objects, or conditions by which one is surrounded .The definition includes everything by which an organism is surrounded although several of these factors may not affect the organism. G.G. Spomer (1973) defined the **operational** environment applying the concepts of thermodynamics. He pointed out that an environmental factor interacts with an organism only when the factor heats or does work upon the organism or the organism heats or does work on the factor. Such an interaction involves a direct transfer of mass (matter) or energy across the boundary between an organism and its environment. If the radio waves pass through a plant unchanged, for example, then the plant is also unchanged, and there could be no interaction; the radio wave would not be part of the plant's operational environment.

Factors of the environment that fit Spomer's definition of operational factors include light, heat, water, various gases, mineral elements, and organic substances. These factors may be directly transferred across the boundary between the organism and its environment. Other factors are not operational-temperature, pH, partial pressure of gases, concentrations, water potential-because they are not themselves transferred across boundaries. Instead, they indicate a **potential** for transfer. If the temperature inside an organism is lower than the temperature outside, the difference indicates the potential for transfer of heat across the boundary. Different pH values indicate a potential for transfer of hydrogen ions, different partial pressures of a gas indicate a potential for transfer of the gas, concentrations indicate potentials for transfer of dissolved substances, water potentials indicate potentials for transfer of water, and so on.

12.1 Some Principles of Plant Response to Environment

12.1.1 Saturation, the cardinal points, and limiting factors

Perhaps the most fundamental principle of response of an organism to environment is that of **saturation**. Organisms respond to virtually any environmental parameter according to a common pattern: as a parameter increases, it reaches a threshold above which it begins to have an effect, after which the response increases until the system becomes saturated by the parameter. Then, as the parameter level or concentration continues to increase, response remains constant or begins to decrease if the parameter at its high levels becomes toxic or inhibitory (Fig. 12.1). The organism simply utilizes the factor being considered until its capacity for this utilization is used up or saturated.

In 1840, Justus von Liebig in Germany formulated the **law of minimum** stating that the yield of any crop depended on the soil nutrients most limited in amount. The principle was discussed and the term **limiting factor** was proposed by F.F. Blackman in 1905. Figure 12.2 illustrates the principle with a simple experiment in mineral nutrition involving two levels of phosphorus given to plants over a wide range of nitrogen concentrations. The threshold for nitrogen is extremely low, but below that level plants do not grow at all. As nitrogen increases, plants respond the same at both phosphorus levels until phosphorus becomes limiting at the nitrogen saturation level. A higher concentration of phosphorus leads to a higher saturation level for nitrogen.

The validity of the law of minimum has been shown in many parts of the world. The poor growth of many clover pastures in parts of Australia, for example, was shown to be a result of deficiencies in the micronutrients copper, zinc, or molybdenum. Addition of only 6.8 kg/ha of copper or zinc sulphate every 4 to 10 years increased plant growth 300% and increased the wool harvest from sheep grazing the vegetation even more. As little as 140 g/ha sodium molybdate, applied every 5 to 10 years, increased pasture yield six to sevenfold (Moore 1970). In the light of many such findings the law of minimum may be loosely defined: *the growth and/or distribution of a species are dependent on the one environmental factor most critically in demand.*

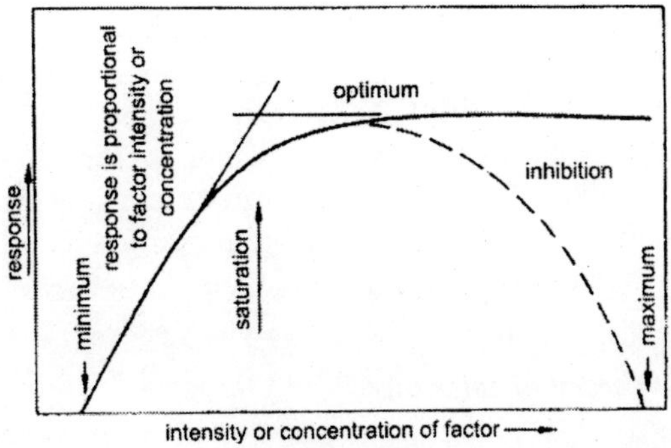

FIGURE 12.1
A GENERALIZED CURVE SHOWING ORGANISM RESPONSE TO AN ENVIRONMENT PARAMETER. MINIMUM, OPTIMUM, AND MAXIMUM ARE CALLED CARDINAL POINTS

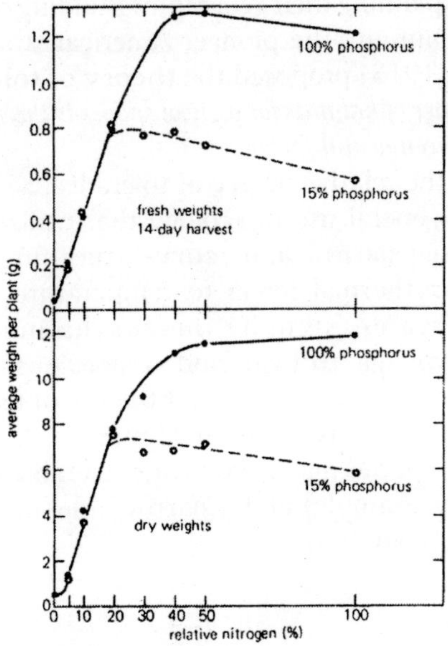

FIGURE 12.2
RESULTS OF AN EXPERIMENT IN MINERAL NUTRITION. TOMATO PLANTS WERE GROWN IN VERMICULITE AND WATERED WITH NUTRIENT SOLUTIONS CONTAINING VARIOUS CONCENTRATIONS OF NITRATE (NO_3^-) AND ONE OF TWO CONCENTRATIONS OF DIHYDROGEN PHOSPHATE ($H_2PO_4^-$) AS SHOWN. CURVES REPRESENT FRESH AND DRY WEIGHTS OF 14-DAY-OLD PLANTS. DATA ARE AVERAGES OF SIX PLANTS (SALISBURY, 1975)

Two limitations to the law of minimum have become evident, however. First, organisms have an upper tolerance limit to every factor as well as lower limit. Second, most factors act in concert rather than in isolation; a low level of one factor can sometimes be partially compensated for by appropriate levels of other factors, or the influence of one factor may be magnified as other factors reach their maximum or minimum levels.

12.1.2 The theory of tolerance

The law of minimum was developed from the perspective of control exerted by the environment. It is the organism, however, which responds to minimum and maximum factors, and thus the ecological and physiological attributes of organisms assume significance in explaining their geography. Building on the concept of the law of minimum, the pioneer American animal ecologist Victor Shelford (1913) proposed the **theory of tolerance** which states that *the range of habitats for a given species is the sum of tolerance limits for each environmental factor*.

To express the relative degree of tolerance, a series of terms have come into general use in ecology that utilize the prefixes "**steno-**", meaning narrow and "**eury-**", meaning wide. Thus, stenothermal-eurythermal refers to temperature, stenohydric-euryhydric to water, stenohaline-euryhaline to salinity, stenophagic-euryphagic to food, and stenoecious-euryecious to habitat selection. These concepts, and the use of terms in regard to temperature, are illustrated in Figure 12.3. As might be expected, a given species may have a wide tolerance for one factor (eurythermal, for example) and a narrow tolerance for another (stenohaline, for example).

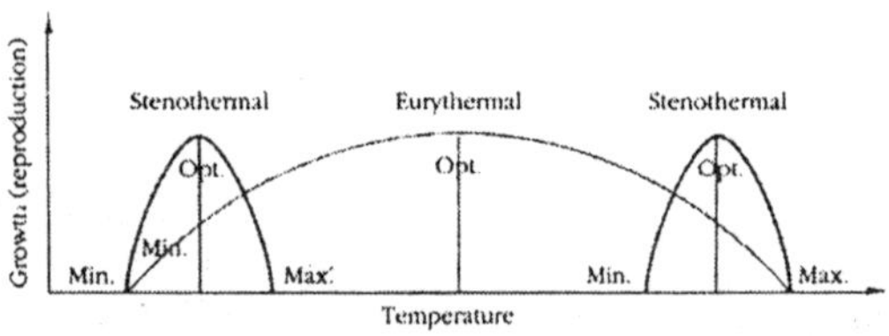

FIGURE 12.3
THE RANGES OF TEMPERATURE TOLERANCE IN STENOTHERMAL AND EURYTHERMAL ANIMALS (RUTTNER 1963)

The conclusion that organisms with a wide range of tolerance to many factors are likely to be very widely distributed should not be surprising. What may not be immediately self-evident is that when one factor becomes less than optimum the limits of tolerance for other factors may become limiting. For example, more water is required for healthy plant growth at low levels of nitrogen than at high nitrogen levels.

The conditions under which a species can exist and grow best in isolation are its **potential range** and **potential optimum,** respectively. These may differ from its observed ecological range and observed ecological optimum in nature, where the species grows in competition with other species. There have been some experiments which dramatically show how the tolerance range or optimum for a physical factor is modified by competition (Ellenberg 1958, Harper 1964). For example, when the common weedy annuals wild radish *(Raphanus raphanistrum)* and spurrey *(Spergula arvensis)* are grown in separate pots in controlled conditions, their growth curves exhibit similar pH tolerance ranges and optima. Radish exhibits optimum growth at pH 5, and spurrey shows an optimum at pH 6 (Fig. 12.4). When grown together, however, the optimum of spurrey shifts to pH 4 and its range for good growth becomes narrow, while the optimum for radish shifts slightly towards pH 6 and its tolerance range remains much as when grown alone. According to Walter (1973), the natural limit of distribution of a particular species is reached when, as a result of changing physical environmental factors, its ability to compete is so much reduced that it can be ousted by other species.

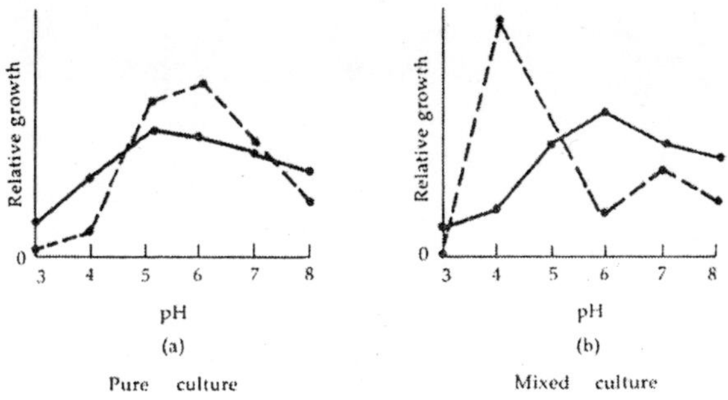

Figure 12.4 The relative growth of wild radish (*Raphanus raphanistrum*) (solid line) and spurry (*Spergula arvensis*) (dashed line) as a function of substrate pH, (a) when grown alone and (b) when grown in competition with each other (Ellenberg 1958).

12.1.3 The holocoenotic concept of the environment

The holocoenotic concept (*holocoen* is a synonym for ecosystem) states that it is impossible to isolate the importance of single environmental factors to the distribution or abundance of a species, because the factors are interdependent and synergistic (Billings 1952). Therefore, single factor ecology (the search for one all-important environmental factor which best determines the distribution of organisms) is short-sighted and naïve according to this concept. However, this concept does not mean that all factors are necessarily equal, only that they are interactive. Certain factors in any ecosystem are of overriding importance, such as moisture in desert scrub, fire in chaparral scrub, or moving sand in coastal dune scrub. Billings (1970) named these important factors **trigger factors**.

Major environmental factors and their effect on physiological and community processes are discussed separately in the following chapters although separate discussions of these factors is artificial; the interaction of all environmental factors determines the response of the organism.

13. SOLAR RADIATION

The sun is essentially a thermonuclear reactor with a temperature and composition such that hydrogen is transmuted to helium with a release of considerable radiant energy in the form of electromagnetic waves. This radiation extends from high frequency X-rays, gamma rays and cosmic rays to low-frequency, longwave radio waves (Fig. 13.1). The earth, however, intercepts only a small fraction of the energy poured into space by the sun. Region of photophysiology lies between the wavelengths 280 and 1000 nm whereas visible portion of spectrum (called **light**) lies between 380 and 780 nm. In biological systems the molecule absorbing radiant energy in the visible range is referred to as a **pigment**. Pigments are complex organic molecules composed of many atoms of hydrogen, oxygen, carbon, nitrogen, etc.; examples are chlorophylls (green), carotenoids (red or yellow) and phytochrome (blue). In general, a pigment will be black to the human eye if it absorbs all the visible spectrum; if all but the wavelengths in the green part of the spectrum is absorbed, the pigment appears green.

Light is an important environmental factor controlling plant growth and development. A principal reason for this, of course, is that light causes photosynthesis. Furthermore, light influences development by causing phototropism. Numerous other effects of light that are quite independent of photosynthesis also occur; most of these effects control the appearance of the plant, that is, its development or **morphogenesis** (origin of form). The control of morphogenesis by light is called **photomorphogenesis**. The phytochrome pigment absorbs light (red and far-red) effective in causing photomorphogenesis. One of the major differences between photomorphogenesis and photosynthesis is that the former is initiated by low levels of radiant energy. In many plants, for example, photosynthesis may be driven by solar flux densities

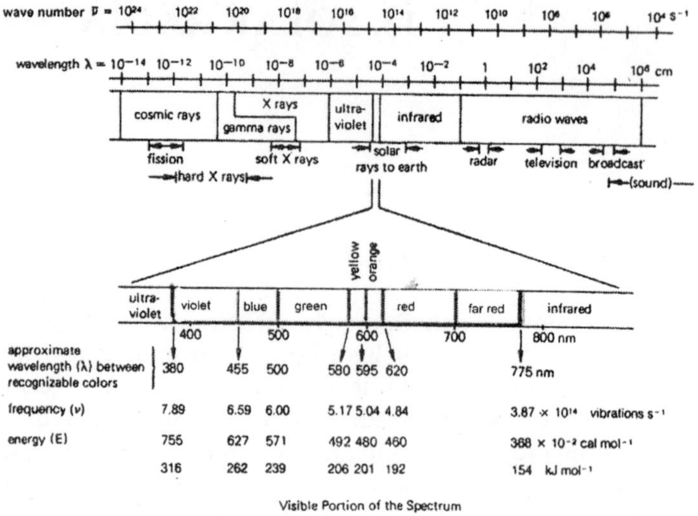

FIGURE 13.1
THE ELECTROMAGNETIC SPECTRUM, USING BOTH WAVE NUMBER ($\bar{\nu}$) AND WAVELENGTH (λ) IN CM. VARIOUS PORTIONS OF THE SPECTRUM ARE SHOWN, AND THE VISIBLE PORTION IS EXPANDED TO INDICATE THE REGION THAT APPEARS TO THE HUMAN EYE TO HAVE VARIOUS COLOURS

in the range of 1000 to 1200 W/m^2 whereas a photomorphogenetic reaction, such as seed germination, is triggered at a flux densities of around 0.01 to 0.1 W/m^2. Further, in photosynthesis radiant energy must be supplied continuously whereas in photomorphogenesis a brief exposure to an appropriate radiation source may suffice to set the process in motion.

13.1 Photosynthesis

The photosynthetic process is divided into light reactions and dark reactions. Light reactions convert radiant energy into chemical energy and are common to all higher land plants. Dark reactions, which do not require light, convert CO_2 into sugars and starches. In the **light reaction** (Fig. 13.2) radiant energy is absorbed by pigment systems within the chloroplasts where it is converted into chemical energy in the form of ATP and NADPH (reduced NADP), which are the sources of the energy for the dark reactions. Briefly, the process is as follows: water is oxidized and acts as the source of hydrogen to reduce NADP and the source of oxygen released during photosynthesis. ATP is also formed in this process from ADP and inorganic phosphate:

$$H_2O + NADP^+ + ADP + P \xrightarrow[\text{Chloroplast}]{\text{Light}} NADPH + O_2 + H^+ + ATP$$

Plants fixing CO_2 only by the *pentose phosphate pathway* (the **Calvin cycle**) are often referred to as C_3 photosynthesizers because the first stable product formed is phosphoglyceric acid (PGA) with a skeleton containing three carbon atoms. A series of reactions follow (Fig.13.3) in which sugars and starch are formed as the products of photosynthesis. The energy necessary for the incorporation of atmospheric CO_2 into photosynthate is obtained from ATP and reduced NADP formed in the light reactions. Additional ATP from the light reactions provides energy and phosphate to regenerate ribulose bisphosphate (RuBP, a 5-carbon sugar), which then can combine with CO_2 to form PGA. This process takes place during the day within the mesophyll cells and cortical cells of stems which contain chloroplasts. Anatomically the leaves of C_3 plants are typically divided into palisade and spongy layers with the chloroplasts widely distributed within the mesophyll.

Coupled with the Calvin cycle is a series of reactions which leads to the light stimulated release of CO_2 known as **photorespiration**. Photorespiration is not the same as true respiration, and both processes may go on independently of each other in the same cell at the same time. True respiration oxidizes carbohydrates, such as glucose, and generates energy in the form of ATP and NADH (Fig. 13.4 b,c). Photorespiration oxidizes organic acids, using up oxygen and releasing CO_2, but it uses more energy that it produces. It is wasteful of the plant's resources; it may reduce the rate of net photosynthesis by 30-50% (Goldsworthy 1976). In the presence of atmospheric levels of oxygen (21%), the enzyme responsible for fixation of CO_2 *(RuBP carboxylase)* may fix oxygen instead of CO_2 to the receptor molecule of RuBP. Thus, O_2 and CO_2 compete for the same enzyme. For convenience, the enzyme is called RuBP carboxylase when it fixes CO_2, and *RuBP oxygenase* when it fixes oxygen but it is the same enzyme. When O_2 is fixed, only half as much phosphoglyceric acid in initially produced; the remaining carbohydrate is converted into phosphoglyconic acid. A series of energy-requiring steps converts this to the amino acid glycine and then two molecules of glycine are combined into one molecule of serine, with the loss of CO_2 and the consumption of O_2. Serine can be chemically altered back to PGA with additional consumption of energy. Thus, glucose sugar is produced during photorespiration, as with photosynthesis, but photorespiration requires the expenditure of a great deal more energy.

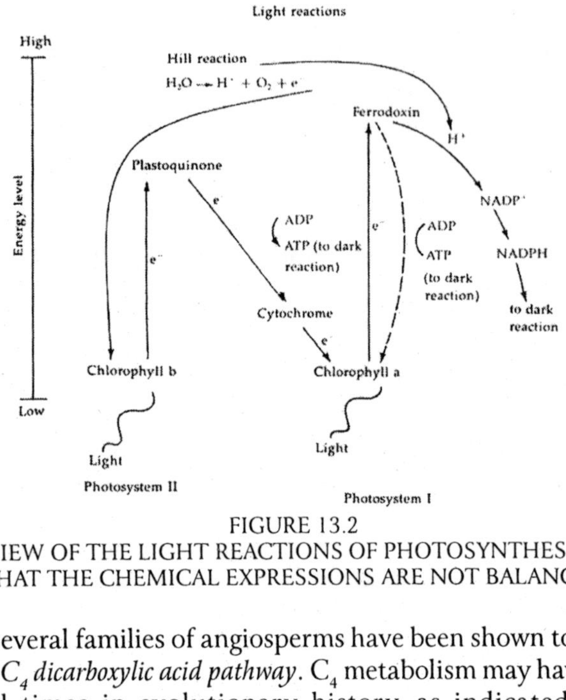

FIGURE 13.2
OVERVIEW OF THE LIGHT REACTIONS OF PHOTOSYNTHESIS. NOTE THAT THE CHEMICAL EXPRESSIONS ARE NOT BALANCED

Several families of angiosperms have been shown to fix CO_2 by the *C_4 dicarboxylic acid pathway*. C_4 metabolism may have arisen several times in evolutionary history, as indicated by the conspicuous lack of evolutionary relationship between the C_4 families.

In the C_4 pathway, CO_2 is fixed by the enzyme *PEP carboxylase* to form 4-carbon acids such as malate or aspartate in the mesophyll cells. These acids are then transported from the mesophyll cell into a bundle sheath cell where CO_2 is released to enter the Calvin cycle. The 3-carbon carrier molecule returns to

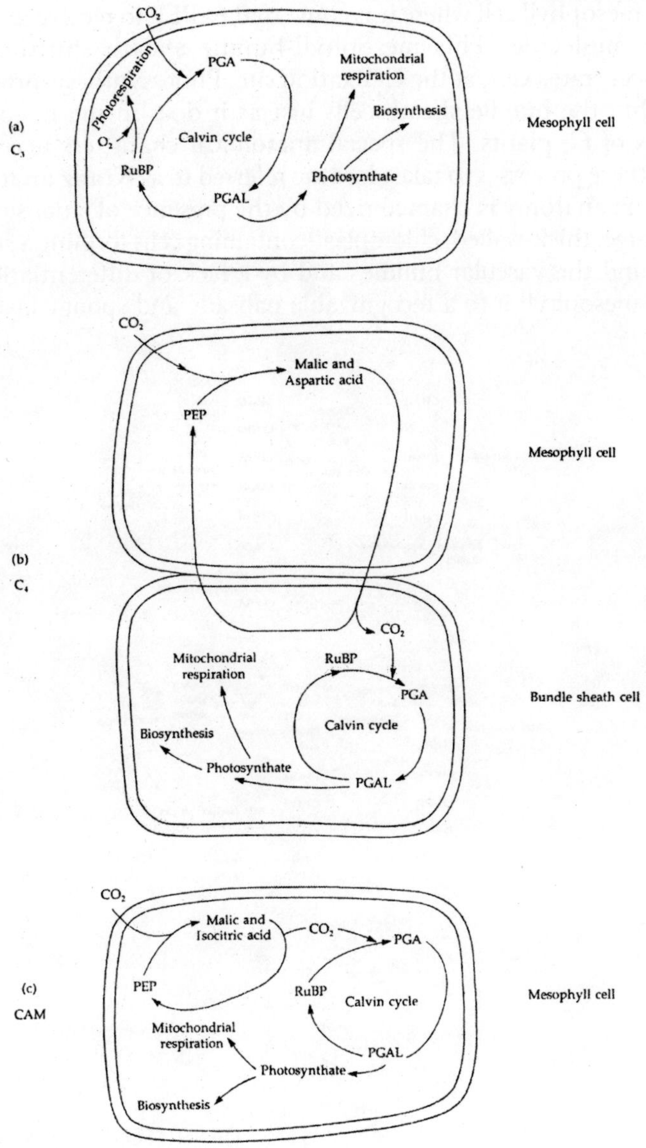

FIGURE 13.3
BASIC DARK REACTIONS AND PRODUCT FATE OF (A) C_3, (B) C_4, AND (C) CAM PHOTOSYNTHESIS

the mesophyll cell where it is converted to PEP to receive another CO_2 molecule. This mesophyll-bundle sheath shuttle thus concentrates CO_2 at the C_3 fixation site. Photosynthesis proceeds within the bundle sheath cells just as it does in the mesophyll cells of C_3 plants. The special anatomical characters necessary for these processes to take place are referred to as **Kranz anatomy**. Kranz anatomy is characterized by the presence of a dense layer of large, thick-walled, chloroplast-containing cells forming a sheath around the vascular bundles and by a lack of differentiation of the mesophyll into a recognizable palisade and spongy layer.

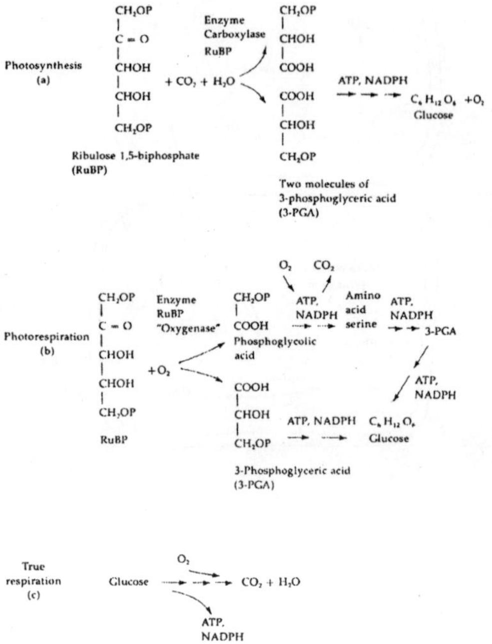

FIGURE 13.4
COMPARISON OF CHEMICAL STEPS OF PHOTOSYNTHESIS, PHOTORESPIRATION, AND TRUE RESPIRATION.

The oxygen inhibition noted in C_3 plants is not a problem in plants with C_4 photosynthesis. The lack of oxygen inhibition is apparently due to the high concentration of CO_2 at the site of RuBP carboxylase, which prevents RuBP oxygenase activity and allows C_4 plants to reduce the concentration of CO_2 to nearly zero. The CO_2 *compensation point* (the point at which the concentration of CO_2 is sufficient to support net photosynthesis) has been used to distinguish between C_3 and C_4 plants.

Many succulent plants belonging to families Agavaceae, Orchidaceae, Crassulaceae, Cactaceae, etc., exhibit a carbon fixation scheme referred to as **crassulacean acid metabolism (CAM)**. All CAM plants have succulent photosynthetic tissues but do not have specialized bundle sheath cells as do C_4 species. The stomates of CAM plants are open at night and CO_2 is fixed at night by PEP carboxylase, forming malic or isocitric acid. In contrast, the stomates of C_3 and C_4 species are open during the day and CO_2 is also fixed during the day. In CAM photosynthesis, acids are accumulated within the vacuoles of cells at night, which results in a pattern of low nighttime pH followed by increasing pH during the day when photosynthesis proceeds by the Calvin cycle. During the day when stomates are closed, CO_2 from the organic acids is fixed by the C_3 cycle. The elevated CO_2 concentrations within the cell counteract the inhibitory effect of oxygen by photorespiration (Osmond 1976). Table 13.1 summarizes and compares some traits of C_3, C_4 and CAM photosynthetic pathways.

13.1.1 Factors affecting the rate of photosynthesis

13.1.1.1 Gas exchange

Since water also leaves the plant when the stomates are open, the balance between water loss and CO_2 uptake is important in determining the relative success of terrestrial plants. The steps in the movement of CO_2 can be visualized as a series of resistances and potential gradients defined by a restatement of **Fick's law** for gas diffusion:

$$\Phi = \frac{\Delta c}{\Sigma r}$$

where CO_2 flux (or the flux of any diffusing substance), ô, depends on the change in concentration from the source to reaction site

(Δc) and the sum of the resistances (Σr) to diffusion. Figure 13.5 depicts the pathway along which CO_2 diffuses from external air to reaction site within a chloroplast-containing cell. The concentrations and resistances dictate the rate of CO_2 assimilation. Therefore, the consideration of relative values is important in understanding photosynthetic adaptations. **Boundary layer resistance** (ra) is encountered near the leaf surface where a zone of air may become depleted of CO_2 and through which CO_2 must move by diffusion rather than by turbulent mass transfer as it does further from the surface. **Stomatal** and **cuticular resistance** (rl) restricts CO_2 diffusion at the leaf surface. Cuticular resistance has been shown to be so great in most mature leaves that we may assume that essentially all of the CO_2 which enters a leaf diffuses through the stomates. Resistance at the point where CO_2 leaves the gas phase in the air and is dissolved in the cytoplasm of the photosynthesizing cell is referred to as **liquid phase resistance** (rw). Limitations placed on CO_2 fixation by the photosynthetic process itself represent the final resistance to CO_2 transport and are called the **biochemical resistance** (rk). The sum of rw and rk is called **mesophyll resistance**. It can be equal to or greater than stomatal resistance, depending on the plant and environmental conditions.

The concentration of CO_2 in the atmosphere (ct), in the intercellular spaces (ci), and within mesophyll cells (cw), in conjunction with the various resistances, determine the rate of CO_2 uptake. In a rapidly photosynthesizing cell intercellular CO_2 concentrations (cw) are quite low relative to ct, and so a gradient exists. This gradient becomes greater toward the outside of the leaf where normal CO_2 concentrations are approximately 320 ppm. Carbon dioxide released by photorespiration and mitochondrial respiration adds to the CO_2 concentration within the leaf. The concentration becomes high enough at night to reverse the gradient (except in CAM plants), and CO_2 passes into the atmosphere.

Stomatal aperture (and therefore stomatal resistance) appears to be responsive to light, intercellular CO_2 concentrations and leaf water status. Stomatal aperture responds to these factors by active transport of ions into and out of the guard cells, thereby modifying turgidity by changing osmotic concentrations within the cell. At high (about 700 ppm) intercellular concentrations of

TABLE 13.1
SOME PHOTOSYNTHETIC CHARACTERISTICS OF THREE MAJOR PLANT GROUPS (BLACK 1973)

Characteristic	*C-3*	*C-4*	*CAM*
1 Leaf anatomy	No distinct bundle sheath of photosyn-thetic cells	Well-organized bundle sheath, rich in organelles	Usually no palisad e cells, large vacuoles in mesophyll cells
2. Carboxylating enzyme	Ribulose bisphosphate carboxylase	PEP carboxylase, then ribulose bishosphate carboxylase	Darkness: PEP carboxylase; Light: mainly ribulose bisphos-phate carboxylase
3. Theoretical energy requirement (CO_2: ATP:NADPH)	1:3:2	1:5:2	1:6.5:2
4. Transpiration ratio (gH_2O/g dry weight increase)	450-950	250-350	18-125
5. Leaf chlorophyll a to b ratio	2.8	3.9	2.5
6. Requirement for Na^+ as micronutrient	No	Yes	Yes
7. CO_2 compensation point (ppm CO_2)	30-70	0-10	0-5 in dark
8. Photosynthesis inhibited by 21% O_2 ?	Yes	No	Yes
9. Photorespiration dectable?	Yes	Only in bundle sheath	Detectable late afternoon
10. Optimum temperature for photosynthesis	15-25°C	30-47°C	35°C
11. Dry matter production (t/ha/yr)	22	39	Low and highly variable

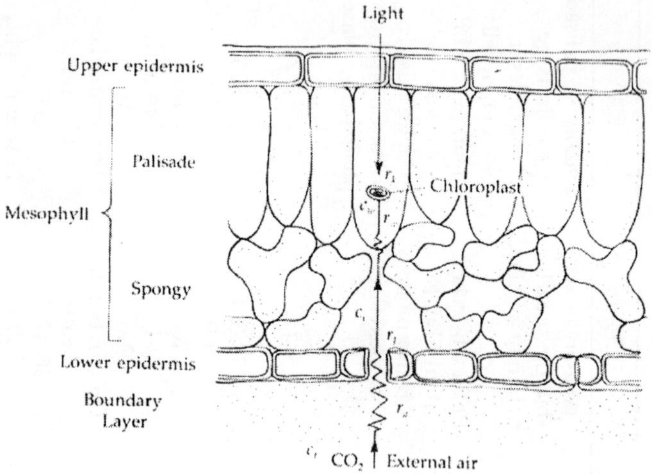

FIGURE 13.5
TRANSPORT PATHWAY, TRANSPORT RESISTANCES, AND CONCENTRATION GRADIENTS IN A PHOTOSYNTHESIZING LEAF

CO_2, C_3 and C_4 plants are reported to have nearly identical photosynthetic rates (Bjorkman *et al.* 1975). However, C_4 plants have a higher fixation rate under conditions where intercellular CO_2 concentrations are low. Low intercellular CO_2 concentrations result when stomatal resistance is high; consequently, C_4 plants have an advantage in arid, warm environments.

13.1.1.2 Light utilization

Vegetation modifies the light environment in different ways depending on the geometry of the plants, the distribution of biomass, spectral properties, and orientation of leaves. The modifications due to these factors are variable seasonally because of the seasonal differences in the vegetation and in the elevation of the sun (the day-length). The highest radiant flux reaches the forest floor between the spring leafless phenoseason and early summer when the canopy is not in full leaf. It is during this period that sufficient light and adequate temperature allows the spring bloom of herbaceous plants on the forest floor. The extinction of light as it travels through a vegetation canopy depends on the total leaf area index (LAI), which is area of leaves projected on a unit area of ground surface. Very little additional sunlight penetrates a canopy of horizontal leaves with a LAI greater than about 3, but when more leaves have a vertical orientation greater light penetration occurs in canopies of higher LAI. Not all leaves within a canopy or even on an individual plant are at the same inclination, and high photosynthesis is possible with large LAI values when leaf orientations are random, but at low LAI leaf orientation has little effect.

It has been long recognized that some plants are adapted to high light environments **(heliophytes)** and some to low light environments **(sciophytes)**. Plants growing in the shade tend to have larger, thinner leaves, with larger, less dense mesophyll cells, longer internodes and less overall pubescence; these adaptations increase utilization efficiency, increase available light, or reduce reflection. In contrast, sun leaves have less area per leaf, are thicker, weigh more per unit leaf area, are more densely distributed on the stem with shorter petioles, and are lighter in colour (less chlorophyll).

The shade leaves require less light to reach their *compensation point* (the CO_2 concentration at which photosynthetic fixation just balances respiratory loss) and they *saturate* (rate of photosynthesis no longer increases with increasing light levels) at lower levels than do sun leaves. The ability of sciophytes to maintain a positive carbon balance in the shade is not necessarily due to an ability to photosynthesize more rapidly than heliophytes in low light, but the key to differential success in shade is a difference in dark respiration rate and therefore a lower light compensation point, allowing them to maintain a positive carbon balance (net photosynthesis) even at very low gross photosynthetic rates. Light saturation levels, in general, are lower in C_3 plants than the C_4 species.

Water Stress: The most immediate effect of water stress (plant-water status when evaporative demands exceed water supply) on the photosynthetic process is an increase in resistance due to stomatal closure, restricting the movement of CO_2 and water. Furthermore, increases in mesophyll resistance to CO_2 uptake occurs as the leaves dry. The fact that C_4 and CAM species have the ability to utilize light while restricting water loss leads to the increased numbers of C_4 and CAM species in hot and dry environments. This is because of their higher water use efficiency, meaning that less water is necessary to fix a molecule of CO_2 in C_4 and CAM plants than in C_3 species.

13.1.1.3 Temperature response

The temperature range over which plants can photosynthesize is surprisingly large. Certain bacteria and cyanobacteria photosynthesize at temperatures as high as 70°C, whereas conifers can photosynthesize extremely slowly at - 6°C or below. In some Antarctic lichens, photosynthesis occurs at -18°C with an optimum near 0°C. In many plants exposed to bright sunlight on a hot summer day, leaf temperatures often reach 35°C or higher with photosynthesis continuing. The most apparent influence of temperature is the limitation of all enzymatically catalyzed reactions. One would expect that reaction rates would increase gradually to optimum temperatures and then decrease sharply at higher temperatures where enzymes denature.

Desert species have higher temperature optima than do arctic or alpine species, and desert annuals that grow during the hot summer months (mostly C_4 species) have higher optima than those that grow there only during winter and spring (mostly C_3 species). This difference is controlled largely by insignificant photorespiration in C_4 plants. Temperature is also a critical factor regulating CAM photosynthesis. Nighttime CO_2 uptake is dependent upon low temperature. For instance, Neales (1973) found that the normal night-day stomatal rhythm of *Agave americana*, a CAM plant, was inverted when night temperature reached 36°C. Kluge (1974) measured no night fixation of CO_2 when plants were subjected to 25°C day and 30°C night temperatures. Nobel (1976) supported these findings for *Agave* deserti in which stomatal resistance increased fivefold when leaf temperatures increased from 5 to 20°C. There is, therefore, strong evidence that success of CAM plants in water limiting environments is dependent on low night temperatures. This conclusion is further supported by the actual geographic distribution of CAM plants; they usually do not live in water limiting areas with high night temperatures.

13.1.1.4 Leaf age

As leaves grow, their ability to photosynthesize increases for a time and then, often even before maturity, begins slowly to decrease. Old, senescent leaves eventually become yellow and are unable to photosynthesize because of chlorophyll breakdown and loss of functional chloroplasts. However, even apparently healthy leaves of conifers that persist several years usually show gradually decreasing photosynthetic rates during successive summers.

13.1.1.5 Carbohydrate translocation

An internal control of photosynthesis is the rate at which photosynthetic products such as sucrose can be translocated from leaves to various parts *sink* organs. It is often found that removal of developing tubers, seeds, or a fruits (strong sinks) inhibits photosynthesis often a few days, especially in adjacent leaves that normally translocate carbohydrates to these organs. Further, species that have high photosynthetic rates also have relatively high translocation rates, consistent with idea that effective transport of photosynthetic products maintains rapid CO_2 fixation. Severe infection of leaves by pathogens often inhibits photosynthesis so much that these leaves become sugar importers rather than exporters; the adjacent leaves then gradually photosynthesize much faster, suggesting that enhanced translocation from them has removed some limitation to their CO_2 fixation.

13.2 Photomorphogenesis

In photomorphogenesis, light in low doses sometimes acts like a trigger to initiate a developmental process that depends upon photosynthetic products for its completion. In other examples of photomorphogenesis, light acts not as a trigger but as a *modulator*, but again the light energy involved is small, and development depends upon photosynthetic products. So photomorphogenetic light initially causes a small change in cells, but this change is somehow amplified greatly so that large morphogenetic events occur. These events vary with the kinds of cells involved, their positions in the plant, and their age (Mohr 1983). Many genes must eventually become activated and others deactivated during photomorphogenesis. As the photomorphogenetic responses are reversibly promoted or inhibited by red (660 nm) and far-red (730 nm) wavelengths of light, they indicate the involvement of the phytochrome pigment in such photoresponses. Phytochrome exists in all photosynthetic organisms (algae to angiosperms) except the photosynthetic bacteria; and it has been detected in cotyledons, coleoptiles, hypocotyls, roots, stems, petioles and leaf blades, vegetative buds, floral tissues, seeds, and developing fruits of higher plants.

Phytochrome (P) has two interconvertible forms: physiologically inactive red absorbing bright blue form (Pr), and physiologically active far-red absorbing green form (Pfr) (Fig. 13.6). The ratio of Pfr to the total amount of phytochrome in both forms is denoted by Φ:

$$\Phi = \frac{P_{fr}}{P_r + P_{fr}} = \frac{P_{fr}}{P_{total}}$$

Red light of 660 nm wavelength converts about 80% of phytochrome into Pfr, so Φ is 0.8. Far-red light (730 nm) removes essentially all Pfr, so Φ approaches zero. Sunlight acts primarily as a red source that forms Pfr than Pr. The Φ value in sunlight is about 0.6. However, when sunlight passes through clouds or through the canopy of leaves in a forest or soil, or during winter months when sun is low, more far-red light (which appears dark-red to human eye) strikes the earth than the red light.

In most species, some of the Pfr gradually disappears in darkness. Two processes account for this. The first is destruction, because after an interval of time in darkness it is no longer possible to regenerate as much Pr in tissues by a far-red exposure; so the total amount of detectable phytochrome is less than before. This destruction probably involves denaturation of the phyotochrome protein. The second process is dark reversion back to Pr, which usually requires a few hours.

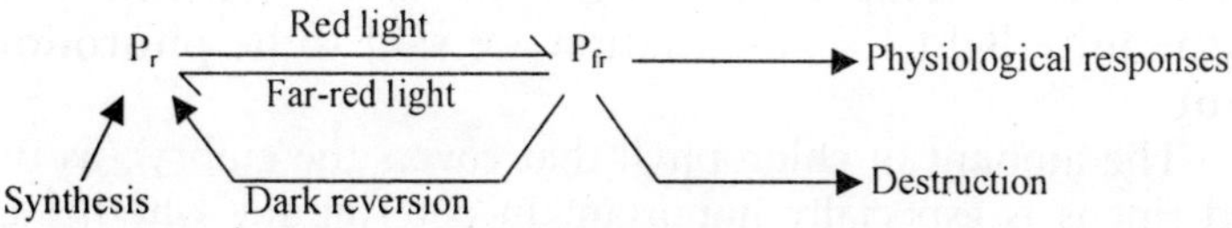

FIGURE 13.6
A SUMMARY OF SOME TRANSFORMATIONS OF PHYTOCHROME

Some of the processes listed in Table 13.2 occur rapidly after appropriate levels of Pr and Pfr are achieved (changes in membrane permeability, changes in cell turgor, enzyme synthesis) whereas others may not be apparent for hours, days, or weeks (flower induction, bulb formation). The initial event is likely a change in membrane permeability to water and solutes (H^+, K^+, Ca^{2+}) which may modify enzyme activity and set in motion a chain of biochemical reactions that eventually are observed later as organ initiation-for instance, flower induction, bulb formation, etc. Another possibility is that a particular Φ value activates specific genes and the eventual synthesis of new enzymes.

TABLE 13.2
SOME PHYTOCHROME-MEDIATED PHOTORESPONSES

1. Elongation (leaf, petiole, stem)
2. Hypocotyl hook unfolding
3. Sex expression
4. Bud dormancy
5. Root development
6. Rhizome formation
7. Bulb formation
8. Leaf abscission
9. Succulency
10. Enlargement of cotyledons
11. Seed germination
12. Flower induction
13. Formation of tracheary elements
14. Changes in rate of cell respiration
15. Synthesis of anthocyanin
16. Increase in protein synthesis
17. Increase in RNA synthesis
18. Auxin catabolism
19. Permeability of cell membranes
20. Photoperiodism
21. Seed respiration
22. Changes in membrane conformation
23. Changes in cell turgor

13.2.1 Seed germination

The word **dormancy** is used as a general description of seeds or buds that fail to grow when exposed to adequate moisture, air, and favourable temperature for growth. In this context, seeds that require light for germination are said to be **photodormant**.

The amount of chlorophyll that covers the embryo as the seed ripens is especially important in determining whether or not seeds of a given species will be photodormant (Cresswell and Grime 1981). In general, embryos that are covered during ripening by material tissues that contain high amounts of chlorophyll require light to germinate, whereas those that are covered by maternal tissues with little or no chlorophyll do not. The apparent reason for this is that chlorophyll absorbs the red wavelengths and prevents Pfr formation in the ripening embryos, so the seeds then require red wavelengths to form Pfr, that promotes germination. The relative lengths of night and day during ripening also affect photodormancy in many species; long days usually favour photodormancy whereas short days favour nondormant seeds (Bewley and Black 1994).

For buried seeds in soil or litter whose germination is promoted by light, germination when they are partly uncovered more nearly assures that the seedlings will be able to photosynthesize, grow and perpetuate that species. A light requirement for buried seeds might distribute germination over several years and thus help perpetuate the species, since only a fraction of the seeds in a soil might be distributed and exposed to light in a given season. An unfavourable growing year might otherwise destroy most of the plants. For seeds whose germination is inhibited by light **(negative photoblastic seeds)**, germination is prevented until they are well covered by litter, when they would more likely have sufficient water to grow. Koller (1969) described two light-inhibited species that inhabit coarse, sandy soils of the Negev desert in Israel, in which germination is prevented unless the seeds are well buried, where moisture is more plentiful than at the soil surface.

Another idea is that phytochrome provides seeds with a clue about whether they are covered by a canopy of other plants or exist in an open area. This possibility has been studied extensively by physiological ecologists. The idea developed from two facts. First, far-red light usually inhibits germination of light-requiring seeds and even seeds that germinate moderately well in darkness. Second, leaves in a canopy transmit considerably more far-red than red light. Most of the blue, red, and some of the green wavelengths are removed by leaves through photosynthesis and reflectance, but most of the far-red passes through to seeds below and converts their active Pfr to inactive Pr . Under such a canopy, no more than 10% of the total phytochrome would exist as Pfr. If seeds ripen on a plant under a canopy transmitting such a high ratio of far-red to red, they are likely to require direct sunlight to form additional Pfr before they will germinate. In a forest, many seeds requiring relatively high amounts of Pfr might never sprout until a fire, death of old trees, or timber removal eliminates the canopy. Evergreen forests seem especially affected in this respect, but deciduous forests are probably also affected in some climates, depending on when their new leaves develop relative to the temperature and light requirements of seeds below. The light-filtering effect of plant canopies is also important in agriculture, because germination of many weed seeds is promoted by sunlight but inhibited by light filtered from crop plants that have developed above them (Frankland and Taylorson 1983).

Ecologically, the light sensitivity of seeds from species that inhabit shaded conditions is different from that of pioneer species from more open areas (Grime 1981). Seeds of species that live in shaded conditions are less inhibited by the far-red light transmitted by plant canopies than are those that invade more open areas.

13.2.2 Photoperiodism

Organisms not only adapt to the physical environment in the sense of tolerating it, but they use the natural periodicities in the physical environment to time their activities and programme their life histories so as to benefit from favourable conditions. Often, this synchronization is concerned with reproduction: it is appropriate and adaptive for young animals to be born at specific times of year, for all members of a given angiosperm species to flower at the same time (ensuring an opportunity for cross-pollination), and for mosses, ferns, conifers, and even some algae to form reproductive structures in a given season. Many other plant responses, such as stem elongation, leaf growth, formation of storage organs, leaf fall, and development of frost resistance, also occur seasonally. Frequently, these seasonal responses are synchronized by *photoperiodism-the length of day and night periods in the 24 h cycle.*

In a nonmountainous region on the equator, sunrise and sunset occur at the same time each day throughout the year, so the relative lengths of the day and night remain constant. Exactly at the poles, the sun remains above the horizon for 6 months each year and below for the other 6 months. Again, the day and night are about equal; each is 6 months long (Table 13.3). As one travels from the equator toward the poles, the days become longer in summer and shorter in winter. This is because the equator is tipped 23.5° to the *plane of the ecliptic* (the earth's orbit around the sun), so that during winter the pole is tipped 23.5° away from the sun and during summer 23.5° toward the sun. Near the times of the summer and winter solstices, when days are longest and shortest, there is little change from day to day; during spring and autumn, the rate of change is much more rapid, as days become longer during spring, shorter during fall.

TABLE 13.3
MAXIMUM LENGTH OF DAY/NIGHT AT DIFFERENT LATITUDES

Latitude	*Longest day/night*	*Latitude*	*Longest day/night*
0°	12 hours	63.4°	20 hours
17°	13 hours	66.5°	24 hours
31°	14 hours	67.4°	1 months
41°	15 hours	69.8°	2 months
49°	16 hours	78.2°	4 months
58.5°	18 hours	90°	6 months

On the basis of their photoperiodic requirement for flowering, plants are classified as *short-day plants (SDPs), long-day plants (LDPs), long-short-day plants, short-long-day plants* and *day-neutral plants*. A SDP is one that flowers on photoperiods shorter than the *critical day-length* which varies from species to species. For example, the critical day-lengths for two SDPs: Maryland Mammoth tobacco and *Xanthium strumarium* are 12 h (12 L/12 D) and 15.5 h (15.5 L/8.5 D), respectively. LDPs, on the ther hand, are induced to flower on photoperiods longer than the critical day-length. For example, the critical day-length for *Hyoscyamus niger* is 11 h (11 L/ 13 D) and it is induced to flower on photoperiods longer than 11 h. To produce flowers, a few species (*Aloe bulbifera, Kalanchoe laxiflora, Cestrum nocturnum*, etc.) require long-days followed by short-days, as occurs in late summer and fall. When maintained either under continuous long-days or continuous short-days, they remain vegetative. The counterparts to these long-short-day plants, the short-long-day plants (*Trifolium repens, Dactylis glomerata, Poa pratensis*, etc.) require short-days followed by long-days, as in spring. There are a few species, at least (*Chenopodium album, Coleus hybrida, Saccharum spontaneum* etc.), that flower only on intermediate day lengths and remain vegetative when days are either too short or too long. They have counterparts that remain vegetative on intermediate-day-length, flowering only on longer or shorter days (*Chenopodium rubrum, Madia elegans, Setaria verticillata,* etc.). Day- neutral plants flower after a period of vegetative growth, regardless of the photoperiod. Some examples of day-neutral plants are *Lycopersicum esculentum, Nicotiana tabacum, Oryza sativa, Gossypium hirsutum, Zea mays,* etc.

The photoperiodic stimulus is perceived by leaves or, in *Pharbitis nil*, by the fully expanded cotyledons. In either case, the organ of perception is leaflike.

The condition a plant must achieve before it will flower in response to photoperiod is called **ripeness to respond** (or **ripeness to flower**). Only a few plants respond to photoperiod when they are small seedlings. The Japanese morning glory (*Pharbitis nil*) responds to short-days in the cotyledonary stage, and some species of *Chenopodium* respond and flower as minute seedlings. Most species, such as cocklebur (*Xanthium strumarium*), must attain a somewhat larger size; cotyledons do not respond. Henbane (*Hyoscyamus niger*) must be 10 to 30 days old before it will respond to long-days.

The number of photoinductive cycles needed to induce flowering differs widely among different plant species. *Salvia occidentalis*, a SDP, requires at least 17 photoinductive cycles to flower, whereas *Plantago lanceolata*, a LDP, needs to receive 25 photoinductive cycles for maximum floral response. In many species, the number of required photoinductive cycles decreases as the plant gets older; that is, the level of ripeness to response increases with age. Once a plant has received the minimum number of photoinductive cycles, it will flower, even if returned to noninductive cycles.

Plants under normal conditions are subjected to a 24 h cycle of light and darkness. The dark period plays an important role in the photoperiodic response, since night interruption by light (*light break*) inhibits flowering of SDPs and promotes flowering of LDPs. *Xanthium* (a SDP) flowers with a single inductive photocycle (9 L/15 D). If the long dark period is interrupted with either white light or red light flowering is inhibited and the plants remain vegetative. Far-red light is without effect; the plant behaves as though exposed to continuous darkness. When a flash of far-red light immediately follows a flash of red light, the inhibitory effect of red light is negated and flower initiation occurs. When red light follows far-red light, the plants remain vegetative as after a night break of white light or red light. On the other hand, *Hyoscyamus* (a LDP with critical day-length 11 h) remains vegetative under a photoperiod of 9 L / 15 D. If the long dark period is interrupted by white light or red light flowering is initiated. Far-red light is without effect and the plants remain vegetative. The red and far-red lights interact with the flowering pattern depending on the wavelength last perceived. Thus, the reversible action of red and far-red light on flower initiation clearly indicates that phytochrome is involved, the same pigment involved in seed germination and other photomorphogenetic responses.

The effect of phytochrome on the flowering of SDPs might be imagined as follows. At the end of the light period the concentration of Pfr is high and the ratio of Pfr to P *(total)* is such that the formation of the flowering stimulus is prevented. During the long dark period Pfr reverts to Pr or is destroyed and the ratio of Pfr to P *(total)* finally drops to a level in which metabolic processes are triggered that lead to the formation of the flowering stimulus. If the dark period is interrupted by red light, Pr is converted to Pfr and Pfr to P *(total)* level is such that the formation of the flower stimulus is prevented. The possible role of phytochrome in LDPs might be as follows. LDPs require a high ratio of Pfr to P *(total)* for the formation of the flowering stimulus. Such a high ratio of Pfr to P *(total)* is attained at the end of a long-day. If the night is too long, Pfr reverts to Pr or is destroyed and the flowering stimulus is prevented from being formed. When the night is interrupted by a flash of red light, Pr is converted to Pfr, thereby raising the ratio of Pfr to P *(total)* to a level that allows the flower stimulus to form. There is ample evidence for the existence of a substance, the flowering stimulus **(florigen)**, that is produced in leaves and translocated to apical and lateral meristems where flower formation is initiated.

Photoperiodism is a widespread phenomenon is nature and many animals also display periods of reproductive activity in response to day-length (Table 13.4). Although likely that the basic mechanisms responsible for these photoperiodic responses differ in plants and animals, there is probably a fundamental similarity in that a time-measuring system (biological clock) is involved.

Organisms possess a physiological mechanism for measuring time which is known as the **biological clock**. The most common and perhaps the basic manifestation is the **circadian** rhythm (*circa*=about; *dies*=day), or the ability to time and repeat functions at about 24 h intervals even in the absence of the conspicuous diurnal clues such as light. In its natural habitat a nocturnal animal such as the flying squirrel (*Glaucomys*) or the white-footed mouse (*Peromyscus*) remains in the nest during the daylight hours and is active at night. The rhythmic activity is exactly 24 h in length and is entrained or "locked into" the daily light-dark cycle generated by the rotation of the earth on its axis. If the animal is brought into the laboratory and placed under constant darkness, the daily rhythm continues, although the onset of activity may shift slightly each day. Other timed events are related to lunar periodicities (which control tides, for example) and seasonal cycles. The mechanism of the biological clock has not been resolved, but two theories are currently proposed: (a) the endogenous timer hypothesis (that is, the "clock" is an internal device that can measure time without environmental clues), and (b) the external timing hypothesis (that is, the internal clock is timed by external signals from the environment). Regardless of the mechanism, the ecological, or selective, advantage of the biological clock is undisputed since it couples environmental and physiological rhythms and enables organisms to anticipate daily, seasonal, and other periodicities in light, temperature, tides, and so forth.

In a cycling environment, it is not difficult to imagine advantages conferred upon organisms by possession of a biological clock. Rhythms or activity in animals, for example, allow a species to occupy a niche not only in space but also in time. A nocturnal animal and diurnal animal might use the same space but at

TABLE 13.4
SOME RESPONSES OF ANIMALS TO PHOTOPERIOD.

Animals	*Long nights and short days*	*Short nights and long days*
1. Red spider mites	Laying of dormant eggs	Laying of nondormant eggs
2. Pulmonate snails	Nonreproductive growth	Laying of eggs
3. Brook trout	Laying of eggs and changing of colour	Nonreproductive growth
4. Junco	Nonreproductive growth and migration (southward)	Mating and migrating (northward)
5. Pheasant	Nonreproductive growth	Laying of eggs
6. Sheep	Mating	Bearing of young (6 month gestation)
7. Horse	Nonreproductive growth and winte fur (pelage) formation	Mating and bearing of young (11 month gestation)

different times. Plant rhythms in metabolism adjust the plant to the light and temperature environment. Rhythms in flower opening must be closely coupled to the time memory of honeybees. The phenomenon of photoperiodism confers upon the organism possessing it the ability to occupy a particular niche in seasonal as contrasted to diurnal time. Several species that flower in response to photoperiodism may do so at different times and in sequence throughout the season, providing a rather constant source of nectar for insect pollinators throughout an entire growing season. Given a time during the season with minimal flowers but high availability of pollinators, there would be a selective advantage to any species able to flower during that time.

13.3 Measurement of Light Intensity

The intensity or brightness of light is usually measured by an instrument called **photometer** or **luxmeter**. This consists of two parts, a photoelectric cell and an amperemeter. The light energy falling on the photocell generates electric current in proportion of the light intensity and it is read in the amperemeter. This meter is so calibrated that the pointer directly gives the light intensity in lux unit. The luxmeter used to measure light intensity under water has a microamperemeter connected through a long electric cord to a photocell. The photocell is sealed in a transparent waterproof case. This helps in taking the light intensity readings outside at the shore or in a boat while the photocell is lowered to different known depths of water in such a way that the photoreceptor disc remains in a horizontal position.

Solar radiation is measured in cal/cm^2/min by **Pyrheliometers** which employ a thin silver or copper plate of 2 cm x 2 cm with a socket fixed in its middle to hold tightly the bulb of a sensitive centigrade thermometer. The upper surface of the plate is painted black.

Secchi-disc consists of a white plate of 20 cm diameter divided into 4 equal parts and the alternate quarters are painted black. At the base of it in the centre a heavy weight is attached so that it remains suspended horizontally in water. The Secchi-disc is gradually lowered down in water till the visibility of the white part is lost, and depth of the Secchi-disc from the surface of the water is measured. Thereafter the disc is gradually raised up and the depth at which it becomes visible again is noted down. The average of these two readings gives the depth of light penetration from where the reflected light is perceived.

14. TEMPERATURE

There is only a very narrow range of temperature within which life processes can occur. Most biological activity proceeds within the range of 0 to 50°C. There is little activity above or below this range. At 0°C and below there is immobilization of water, and liquid water is necessary for life processes. Above 50°C the heat is destructive to vital life processes. The range of temperature variation tends to be less in water than on land, and aquatic organisms generally have narrower limit of tolerance to temperature than equivalent terrestrial life forms.

Heat is the aggregate internal energy of motion of atoms and molecules of a body; temperature shows how hot a living thing is. Heat energy is transferred by means of radiation, convection, or conduction. **Radiation** is the energy propagated as an electromagnetic wave, which would pass through a vacuum with the speed of light (3×10^{10} cm/sec), but would also travel through a medium such as air or water. It is by this process that most of the energy is transferred through space from the sun to the earth. **Convection** is the process of the transfer of heat through movement of a mass or substance from one place to another. Thus, it involves the transfer of heat by means of mass motions of the medium through which the heat is transferred. It is possible only in gases or fluids, for they alone have internal mass motions. **Conduction** is the process of heat transfer through solids by molecular activity. In this process heat is transferred from one part of a body to another or between two objects touching each other.

The solar radiation is transmitted in a broad spectrum that includes not only visible light, but UV radiation at one extreme and heat or infrared radiation at the other extreme. The ozone of the atmosphere, as well as oxygen and nitrogen, is opaque to wavelengths shorter than 0.29 μ, so that most of the UV and shorter wavelengths do not reach the earth. The longer, or infrared, wavelengths are absorbed and re-emitted by water vapour, liquid water, and carbon dioxide. Unless the absorbed radiation is reradiated or in some way transferred, the body absorbing it would continue to increase its energy content. Thus, there is a continual exchange of energy between objects and the environment. Direct sunlight is absorbed by plants, rocks, soil, and the air around them. Infrared radiation (heat) is transferred from plants to the air, from soil to plants, and so forth. This balance between the amount of insolation received from the sun and the outgoing terrestrial radiation is known as the **earth's heat budget** or **solar energy budget** which can be represented by the following equation:

$$S = R + C + G + Ps + LE$$

where S= solar radiation; R= reflected (shortwave, diurnal), or reradiated (longwave, both day and night); C = convecting heating of air; G= heating of solid objects (soil, for example); Ps= photosynthesis; and LE = latent heat of evaporation.

Solar radiation (S) is usually expressed in *langleys per min* which is equivalent to 1 cal/cm^2 (cal or gram-calorie is that quantity of energy that is required to raise the temperature of one gram of water from 14.5 to 15.5°C). At the sea level at high noon, S may be equal to 1.3 ly/min; on a high mountain on a clear day, S may be as high as 1.7 ly/min.

Net radiation (N) is equal to solar radiation minus reflectivity and reradiation, or: N = S - R. This can be diurnally negative, because at night S is zero, and therefore less than R, and there is a net loss of heat from the earth. During the day, S exceeds R, because reflectivity and reradiation are less than solar radiation, and there will be a positive energy balance. During the long winter of the polar regions, N is seasonally negative, and may be so even in more temperate regions. N is always positive in the tropics.

The allocation of energy within the solar budget varies over time, and varies from one community to another. The maximum S incident upon a wet mountain meadow is +1.3 ly/min. Meadow vegetation has an albedo 0.7. **Albedo** refers to the reflectivity of a material summed over all wavelengths for a sunlit surface. S times albedo equals R, and so in this example, R = (0.7) x (1.3

ly/min) = 0.91 ly/min. The rest of the solar radiation is apportioned among G, C, Ps and LE. Mountain meadows are often densely clothed with herbaceous vegetation, and at 100% cover G will be rather low. For various physical reasons, C is also low. Only 1-2% of incident light is usually converted into chemical energy, and photosynthesis here will be lower than usual due to low temperatures. Although Ps is low, LE will be high, simply due to the wetness of this mountain meadow system. A contrasting situation is seen in arid areas. Again, *Smax* is 1.3 ly/min. R will be higher than in mountain meadow. In very arid regions such as Mojave Desert (Barbour *et al.* 1980), plants are very widely distributed, and the open expanses of exposed soil absorb a relatively high fraction of S. Thus, G will be much higher than in the wet mountain meadow system and photosynthesis will be very low, less than 0.1% of S. C will be high, as the denser air absorbs radiant energy from sunlight. There is very little liquid water available, and LE will be correspondingly low.

14.1 Environmental Influences on Temperature

Insolation (exposure to solar radiation) varies temporally and spatially. The primary factors that influence spatial variations in temperature are latitude, altitude, and proximity to water. Topography, cloud cover, vegetation, and slope aspect are contributing factors to temperature variation. The obvious and immediate temporal differences derive from the earth's rotation: the darkened half of the earth is losing heat during the night (reradiating energy), whereas during the day, radiant energy from sunlight usually replaces what was lost, so there is an energy balance.

14.1.1 Latitude

Apart from the obvious diurnal differences in temperature, the primary influence upon temporal variation is latitude. Both latitudinal and altitudinal differences are spatial differences. There are periods when day and night are virtually equal in length. These periods are called **vernal** and **autumnal** (spring and fall, respectively) **equinoxes**. Alternatively, in the Northern

Hemisphere, there are **winter solstice** (longest night) in December and **summer solstice** (longest day) in June. A comparison of solar radiation received on vernal and autumnal equinoxes, and on winter and summer solstices at latitudes ranging from 0 to 90° in Northern Hemisphere is given in Table 14.1.

TABLE 14.1
AVERAGE AMOUNT OF DIRECT RADIATION (CAL/CM²/MIN) IN NORTHERN LATITUDES (LAL 2000)

	Latitudes						
Dates	*0-10°*	*20-30°*	*20-30°*	*30-40°*	*40-50°*	*50-60°*	*60-90°*
A. Received at the upper limit of atmosphere							
Dec 21	0.549	0.465	0.373	0.274	0.173	0.079	0.006
Mar 21	0.619	0.601	0.563	0.509	0.441	0.358	0.211
Jun 21	0.579	0.620	0.664	0.684	0.689	0.683	0.703
Sep 23	0.610	0.592	0.556	0.503	0.435	0.353	0.208
B. Received at the earth's surface if cloudiness and turbidity are considered							
Dec 21	0.164	0.161	0.134	0.082	0.036	0.013	0.001
Mar 21	0.191	0.221	0.206	0.161	0.116	0.094	0.055
Jun 21	0.144	0.170	0.216	0.233	0.183	0.159	0.133
Sep 23	0.170	0.162	0.201	0.183	0.133	0.079	0.028

14.1.2 Altitude

In general, temperature will be lower at higher altitudes. This inverse correlation of altitude with temperature creates a gradient or lapse rate, usually given as about 10°C per 1000 m of elevation. This lapse rate is called **dry adiabatic lapse rate,** which means the cooling of air as it rises without condensation and without cloud formation. There is less atmospheric pressure at the higher elevations, so air there expands and loses energy. Conversely, coming downslope, air is more compressed and therefore warmer. Thus, in addition to latitudinal differences, there will be, at any given latitude, elevational differences in temperature.

14.1.3 Topography

Topography, the height and shape of the land, can play an important part in the distribution of organisms. Temperature decreases with altitude, so that snow lies on high mountains all year round, even at the equator as on Mt. Kilimanjaro in Tanzania. Slopes may be north or south facing, and much better drained than an area of flat land or a hollow. Figure 14.1 summarizes some of the features and effects of topography.

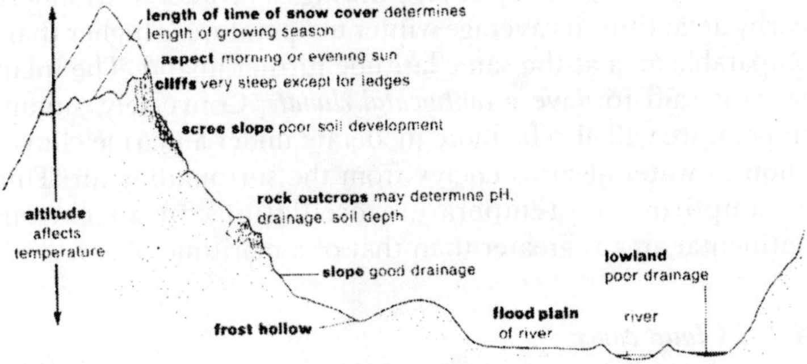

FIGURE 14.1
SOME TOPOGRAPHIC FEATURES AND THEIR ABIOTIC EFFECTS

The decline in air temperature due to increased elevation is not constant, especially in dissected topography. In such a mountainous region, dense and cold air adjacent to the mountain slopes or hill-sides flows down and accumulates in the valleys, while the air aloft is relatively warmer. A surface inversion is produced which results in temperatures increasing with altitude. This type of surface inversion causes killing frosts to occur in the valleys, while temperatures on the hill sides are much above the freezing point.

A further complication of temperature variation in mountainous regions is the width of ravines and valleys. Narrow mountain valleys or canyons receive proportionately less sunlight. The walls therefore reflect less heat into canyon and it will be colder and damper than the surrounding area. By contrast, broad, open mountain valleys build up high temperatures due to the accumulation of heat reflected from the wide sides, and they are consequently hotter and drier than surrounding regions.

14.1.4 Proximity to water

The temperature of a body of water, especially a large lake or ocean, is much more stable than that of a comparable land mass. This is due to the high specific heat of water, its ability to lose energy by evaporation, its high reflectivity, and the contribution of vertical mixing to stability of temperature. Therefore, proximity to a large lake or ocean buffers the climate for adjacent land areas. The water gives up energy during the winter, warming the nearby area; thus, its average winter temperature is higher than a comparable area at the same latitude further inland. The inland region is said to have a *continental climate*. Conversely, summer temperature will also be more moderate under a marine climatic region as water absorbs energy from the surrounding air. Thus, the amplitude of temperature oscillations in an extreme continental area is greater than that of a maritime region.

14.1.5 Cloud cover

The influence of cloud cover is twofold. Water vapour is opaque to certain wavelengths of solar radiation. A place frequently covered by clouds will not be warmed nearly as much as an area under clear skies. There is, however, some reflection of sunlight from clouds to the ground, and some reradiation heat from vegetation, soil surface, and rocks, so that cloud cover may augment direct sunlight and impede heat loss from a land surface. For instance, direct sunlight and sunlight reflected from clouds may simultaneously warm an alpine slope such that radiation values are as high as 2.2 ly/min, which is in excess of the solar constant. The considerations mentioned regarding the moderating influence of liquid water apply also to water vapour. Temperature variations in the tropics vary as little as 2°C on cloudy days, but can vary as much as 9°C on sunny days. Without the moderating influence of cloud cover, deserts experience great diurnal extremes of temperature. Also, frosts are much more likely on clear nights. In general, moisture in the atmosphere has a stabilizing influence on temperature variations.

14.1.6 Plant cover

The sensual impact of a forest is usually one of moist coolness. This is attributable first to the reduction of soil heating due to shading: the soil absorbs energy from the warmer air above it, cooling the forest air. Secondly, because the forest vegetation transpires as photosynthesis proceeds, the forest microclimate will be more humid than adjacent clearings, and the quantity of energy needed to raise the air temperature is increased. The reradiation of heat is impeded by plant cover, so that the nocturnal forest air and soil are warmer than those of nearby clearings. Temperature oscillations are thus damped both day and night by heavy, continuous plant cover. This situation may be changed, or even reversed, if the plants are widely dispersed.

14.1.7 Slope aspect

Due to the inclination of the sun, a south-facing slope in northern temperate latitudes will always experience greater total insolation than will the north-facing slope of the same region. Thus, there will be warmer air and soil, less moisture, and sparser vegetation, in general, on the south-facing slope.

The study of the relationship between climatic factors and periodic phenomena is referred to as *phenology*. A significant climatic factor is temperature. Phenological events concerned with reproductive activity in plants may be retarded by several days on a north-facing slope with respect to plants of the same species on an adjacent south-facing slope. Jackson (1966) found significant correlation between air temperature sums and flowering dates. The increased insolation received by south-facing slopes, as compared to nearby north-facing slopes, results in warmer temperatures earlier in spring and therefore, advanced flowering.

We might relate time range of flowering dates to latitude (northward) and to elevation (upward). For instance, one day of retardation or advance of flowering might be equivalent to more than 27 km in distance northward, or more than 30 m in elevation. Jackson found flowering variation of an average of six days between a north facing slope of a large gorge and the south-facing slope only 46 m distant. Across a small east-west gorge, plants of the same species blooming less than 8 m apart, but on opposite sides, evidenced a delay of 2.8 days on the north-facing slope as compared to the south-facing one. This could be assigned a difference comparable to 80 km (northward) or 85 km (upward). She found a mean range of flowering dates of 7.2 days for all species studied. This may be interpreted as equivalent to geographic distance of 201 km northward and 219 m upward (Hopkins 1938), or as 217 km northward and 168 m upward (Jeffree 1960).

14.2 Temperature-mediated Plant Responses

14.2.1 Energy budget

All objects absorb and transmit or reflect radiation. Energy gains and losses must balance within a leaf, an organism, and the biosphere. Energy is exchanged continuously between a plant and its environment. Incident radiation from a variety of sources (direct sunlight, scattered skylight, reflected sunlight, and infrared wavelengths) is absorbed by a leaf. In turn, a plant allocates energy to maintain a balance, losing heat by convection, reradiation, reflectivity, and evaporation of water. If the air is cooler than a leaf, energy moves away from the leaf. If the air is warmer, the leaf will take on energy.

Cooling results from the evaporation of water (LE). The vaporization of a gram of water requires about 580 cal of energy, a process which cools the leaf much as an evaporative cooler cools a room. There may be a limited water supply and desert plants, for example, cannot depend upon unlimited transpiration for balancing the energy budget. Laminar flow of air removes heat from the leaf (C) if its temperature is warmer than that of the air. The transfer of heat is directly proportional to the square root of the wind velocity and inversely proportional to the square root of the leaf width. In other words, larger the leaf, less effective will be the cooling. At the same air temperature and wind velocity, a small leaf will lose heat more effectively than a larger leaf. Small leaves, dissected leaves, and lobed leaves are characteristic of plants of arid regions.

Day-time temperatures of desert soils and the air immediately above them are often a great deal higher than that of the air at the standard height above the ground for measuring temperature (1.5 m). For instance, the highest standard air temperature of the Mexican Desert in August is 57.8°C, whereas air next to the dark soil surface exceeds 70°C (Gates 1972). As the spring season gives way to early summer, some small desert annuals assume a basket shape, as the rosetted leaves are lifted into the cooler air just a few cm above the soil surface. This adaptation of leaves of desert plants increases chances for heat loss and reduces absorption of heat from the soil (G). Raising leaves may also change the angle of the leaf relative to the sun and thus reduces S.

Deserts plants must deal with strong insolation, often in still air. Insolation (S) may raise leaf temperature above that of the surrounding air. If convection (C) and evaporation (LE) cannot transfer energy as rapidly as it is absorbed, then the leaf will store heat and may then have a temperature of 10°C, or even 15-20°C, above the ambient temperature. Heat exchange is enhanced by winds, which may remove air to within a few mm of the leaf surface. Conversely, heat exchange is hampered in still air, although the layers of warm air above the leaf will tend to rise above cooler air, generating a turbulence which also hastens cooling. Loss of leaf area, and therefore reduction of leaf exposure to insolation (S), is achieved in a variety of ways. Most members of Cactaceae, for instance, have leaves reduced to simple spines. Vertical leaf orientation, an adaptation often exhibited by desert shrubs, is another way to avoid excessive S.

Reflectivity (R) is enhanced by a surface with high albedo. Possession of a white or light coloured leaf surface increases R. Desert holly (*Artiplex hymenyletra*) in the Chenopodiaceae provides a good example of how one desert shrub species is adapted to increased reflectivity by different means and thereby reduce heat load and drought stress. Unlike the cacti, which have virtually no leaves that are photosynthetic, desert holly is an evergreen perennial of the hot desert. Mooney *et al.* (1977) found that the adaptation of this plant is to change the characteristics of the leaves during the year. Those leaves produced during the cool season are nearly twice the size of the summer leaves. Furthermore, young leaves are covered with expanded, hydrated salt bladders which collapse as the summer progresses. The salt solution contained in the bladders crystallizes, giving the leaf a white appearance and increasing the reflectance (R) of the leaf surface. Leaf reflectance (R) (at 550 nm) is inversely correlated with leaf water content. Leaves with 60% leaf moisture exhibit a reflectance above 75% at some wavelengths. Also, at all wavelengths their reflectance is greater than that of leaves with 89% leaf moisture.

Furthermore, the leaves of desert holly are held at a steep angle of 70° from the horizontal, with random orientation to the azimuth. The angle does provide protection by reducing heat load and by allowing greater interception of light during the early morning and late afternoon, which are the times when relative humidity is most favourable. Thus, angle of leaf, reduced leaf size, and high reflectance from the leaf surface enhance the capacity of this desert shrub to remain photosynthetically active even during the hottest summer months. In other words, these adaptations allow the luxury of evergreenness.

14.2.2 Thermoperiodism

Plants are exposed to fluctuating day and night temperatures under natural conditions. Day temperatures are usually higher than night temperatures. The differential may be considerable during the spring and autumn but of lesser magnitude during the summer. Near the equator the temperature differential between day and night is small (except at high elevations) compared to that found in the temperate zone. *Diurnal* (between day and night) difference in temperature is also low in coastal than the inland areas. In general, a phenomenon in which growth and/or development of plants is promoted by alternating day and night temperatures, is called **thermoperiodism**.

Trees that occupy inland, high elevation sites, such as Jeffrey pine (*Pinus jeffreyi*) and red fir (*Abies magnifica*) respond positively to a thermoperiod of 13°C (diurnal difference). These species do set terminal buds, and this restricts upward growth, even though the plant may continue to acquire new biomass. On the other hand, the coast redwood (*Sequoia sempervirens*), considered as the world's tallest trees, does not require a thermoperiod, since it does not live in an area with a large diurnal temperature difference. The coast redwood does not form dormant buds, but continue to increase in height in response to favourable light and temperature regime. Similarly, plants in tropical regions have evolved and developed under a rather restricted range of phototemperatures and photoperiods. A tropical plant, such as sugarcane or coconut palm, cannot flourish outside a rather narrow set of environmental conditions. A temperate zone plant, such as wheat, does poorly in the tropics in absence of appropriate thermoperiod.

14.2.3 Dormancy

Quiescence is the condition of a seed when it is unable to germinate only because external conditions normally required for growth (temperature, moisture, atmosphere) are not present. On the other hand, **dormancy** is the condition of a seed when it fails to germinate even though external conditions are suitable.

Dormancy - a period of inactivity for seeds or plant organs - is a mechanism evolved by plants to escape periods of unfavourable weather, and to allow less competitive species to last through periods of high competition. During times of a short photoperiod and a lack of heat or sunlight or liquid water, growth - whether of an established plant or a new one - is very expensive, if not impossible. The protection afforded by the dormant state, especially in extra-tropical regions, is very significant to the plant, and thus dormancy is a very adaptive feature. The dormancy of apical buds of many woody plants in temperate climates is initiated by short days interacting with cold temperature; this is a photoperiodic response mediated hormonally.

Environmental cues, such as lengthening photoperiod and increasing temperatures, trigger the resumption of activity in more favourable times. The correct interpretation of these cues is vital to the dormant plant or seed. If a dormant beech, for instance, began to leaf out in response to unseasonally warm weather in early November, subsequent frost would kill or injure the new soft tissues, at great cost to the plant. The environmental stimulus that enables the plant to perceive the passage of winter is cold temperature. Degradation of inhibitors and concomitant biosynthesis of growth-stimulating substances take place during the cold period and early spring, and the seed or plant is then prepared to respond to more favourable weather in an appropriate manner.

14.2.4 Stratification

Stratification, the practice of layering the seeds in flats containing moist sand and peat at a low temperature (during the winter), is known to enhance germination. Because seeds in the flats are cooled before they will germinate, a more popular and descriptive term than stratification nowadays is **prechilling**. Many plants, particularly the rosaceous species such as the stone fruits (peach, plum, cherry), many other deciduous trees, several conifers, and several herbaceous **Polygonum** species growing at higher latitudes/altitudes possess seeds which require cold temperatures for germination. That is, the seeds are dormant prior to exposure to cold. Actual germination following release of dormancy may be in response to a variety of environmental cues, including changing photoperiod, increasing heat and water availability, and a changing hormonal regime within the seed.

14.2.5 Vernalization

The flowering of Petkus rye (*Secale cereale*) and of other cereal plants is influenced by exposure to cold temperature during the time of germination. The time required for flowering of Petkus rye may be reduced by half, from 14 weeks to 7 weeks, if the seeds are planted in autumn and exposed to winter cold, rather than being planted in spring. This cold exposure, copied from nature and used commercially, is called *vernalization*, from the Latin word for spring, *vernus*. Seeds of winter strains of certain cereal grasses, if kept near freezing temperatures during germination, will flower the same summer even though not planted until late spring. However, even with vernalization, the appropriate photoperiod is necessary for flowering to be initiated.

14.2.6 Sumorization

The seeds of most desert annuals are subject to excessive heat soon after dissemination, and it probably promotes seed maturation in these ephemerals. Capon and Asdall (1966) found that germination of eight desert annuals native to the Mojave and Sonoran Deserts was enhanced by heat pretreatments. This treatment is called *sumorization* (*sumor* = summer). Maximum germination for these species was reached by the fifth week of storage at 50°C (Table 14.2). Untreated seeds and those stored

TABLE 14.2
PERCENTAGE GERMINATION OF SEEDS OF SOME DESERT ANNUALS IN REPONSE TO TEMPERATURE PRETREATMENT (CAPON AND ASDALL 1966)

Species	*Untreated*	*Storage at 20°C for*		*Storage at 50°C for*					
		4 weeks	*8 weeks*	*1 week*	*2 weeks*	*3 weeks*	*4 weeks*	*5 weeks*	*10 weeks*
**Coreopsis bigelovii*	9	14	30	26	16	24	20	32	2
**Eriophyllum wallacei*	5	16	24	32	48	30	18	36	2
**Euphorbia polycarpa*	1	1	2	2	10	12	4	2	0
**Geraea canescens*	14	14	16	60	74	32	34	48	4
**Salvia columbariae*	6	4	14	80	26	28	14	16	10
***Lepidium lasiocarpum*	0	2	4	53	90	72	61	54	40
***Sisymbrium altissimum*	10	10	12	12	13	92	73	71	48
***Streptanthus arizonicus*	0	3	8	21	23	29	50	50	30
Plantago insularis (not leached)	0	0	0	0	0	0	0	0	0
Plantago insularis (leached)	100	-	-	100	-	-	-	-	-

Mojave Desert spp., ** Sonoron Desert spp.

at 4°C, showed poor germination percentages, especially the three Sonoran species tested. Of these, none showed more than 12% germination without heat pretreatment.

14.3 Measurement of Temperature

The instrument for measuring temperature is thermometer which is a narrow glass tube filled with mercury or alcohol. It works on the principle that mercury expands when heated and contracts when cooled. **Maximum-minimum thermometer** is a special type of U-shaped thermometer in which one arm records the maximum and the other the minimum temperatures (Fig. 14.2). In each thermometer an iron pin is left at the maximum mercury level which indicates the extremes of temperature during the period the instrument is kept for one set of observation. A piece of magnet is used to bring down the pins to existing mercury level before setting the instrument for the next record of diurnal temperature fluctuation.

FIGURE 14.2
A MAXIMUM-MINIMUM THERMOMETER

15. WATER

Water, a substance which may justifiably be called the fluid of life, makes up the greater part of our globe. All life depends on water and may literally be said to swim in it. In many instances, water composes over 90% of an organism and is a participant indirectly or directly in all of its metabolic reactions.

Despite its commonness, water is a most unusual compound, which, due to certain physical and chemical characteristics, is possessed of a number of unique properties. For example, water has a high *specific heat,* which is the reason that living tissues experience considerable heat absorption or loss with only small changes in temperature. Also, water's high *heat of vaporization* (evaporates slowly when heated) causes relatively large amounts of energy to be dissipated under conditions favouring evaporation - a cooling process of considerable magnitude. Another one of the unusual properties of water is that it is less dense as a solid than as a liquid. That is, water as a solid will float on water as a liquid. This property is of obvious advantage to aquatic life in temperate and cold climates where lakes and rivers form ice from the top down, thus allowing life to continue in a liquid medium at lower depths. The adverse consequence of the expansion of water when freezing in living cells is obvious.

Water molecules cling to each other and to many different surfaces; these properties are called *cohesion* and *adhesion*, respectively. Both cohesion and adhesion have major consequences in the rise of water in plants.

The above properties are chiefly due to the configuration of the water molecule. The water molecule is composed of two hydrogen atoms covalently bonded to one side of an oxygen atom, the angle between them being 104.5° (Fig. 15.1). The covalent bonds are very strong and so the water molecule is extremely stable. From Fig. 15.1, one can see that the water molecule is a *polar* molecule; that is, one part of the molecule is *positive* (hydrogen side) and the other part *negative* (oxygen side). Because of the asymmetrical distribution of charges, the water molecules associate with each other (cohesion). The attraction of the positive hydrogen atom of one water molecule for the negative oxygen of another water molecule results in what is called a *hydrogen bond* or *bridge*. Although the hydrogen bond is stronger than the *van der Waals forces* that draw molecules together, it is considerably weaker than the covalent or electrovalent type of bond. There is no limit to the number of water molecules that can become attached to each other in this manner. To use a fanciful example, a lake may be thought of as a loosely held together gigantic molecule, rather than as an accumulation of an astronomical number of discrete water molecules.

The presence of hydrogen bonds is directly responsible for the high *heat of fusion*, high specific heat, and high heat of vaporization of water. That is, the energy required to break hydrogen bonds for the melting of ice, the heating of water, and the evaporation of water is considerably more than the energy needed to overcome van der Waals forces, which are normally found in the grouping of molecules such as methane, ether, benzene, etc.

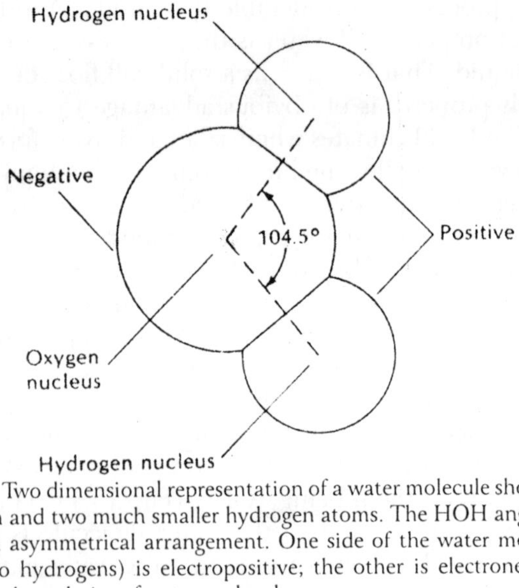

Figure 15.1 Two dimensional representation of a water molecule showing a large oxygen atom and two much smaller hydrogen atoms. The HOH angle of 104.5° results in an asymmetrical arrangement. One side of the water molecule (that with the two hydrogens) is electropositive; the other is electronegative. This accounts for the polarity of water molecule.

As far as life is concerned, the most important property of water is its *solvent action*. Because of its ability to enter into solution with a vast array of compounds, water is sometimes referred to as the "universal solvent". The solvent action of water is due to its ability to form hydrogen bonds and the fact that there is an asymmetrical distribution of charges on the water molecule. For example, compounds such as sugars, alcohols, amino acids, etc., which contain oxygen atoms, hydroxyl groups (-OH), or amino groups ($-NH_2$) are held in water solution by hydrogen bonding. On the other hand, the polar nature of water molecules causes various salts to be held in solution through charge interaction. Therefore, salts can exist in water in the form of ions,

All matter tends to move from areas of high free energy (high capacity to do work) to areas of lower free energy. *Water Potential* (Ψ) is a measure of the free energy of water in comparison to the free energy of pure water. By convention Ψ of pure water at atmospheric pressure and the temperature of the solution being considered are set equal to zero, so the Ψ of an aqueous solution will be some negative number (less than zero). In general, water potential depends on three predominant forces:

$$\Psi = \Psi_m + \Psi_\pi + \Psi_p$$

where, Ψm is a matric potential, Ψ_π osmotic potential, and Ψp the pressure potential. *Matric potential* is a measure of the reduction in free energy caused by the attraction of matrix for water molecules. Macromolecules, soil particles, cell walls, and other surfaces to which water adheres make up the matrix. Such attractions always reduce the free energy of water, and therefore, Ψm is always negative. *Osmotic potential* (Ψ_π) is a measure of the reduction in free energy (always negative) attributable to electrical attraction, reorientation, and increased entropy resulting from solutes in solution. In living systems a selectively permeable membrane must be present since no potential difference could exist without a restriction on the movement of solute molecules (Campbell 1977). *Pressure potential* (Ψ_p) is determined by the positive or negative hydrostatic pressure present in the system. Within living cells Ψp is usually positive; however, it is negative in xylem when transpiration is occurring. Water potential is a negative value except in fully turgid cells where Ψp may balance the negative $\Psi \pi$ and Ψm potentials (Kramer 1969).

15.1 The Hydrologic Cycle

The cyclic nature of water is shown in Fig. 15.2. Condensation of atmospheric water vapour in the form of precipitation, fog, or dew, places water into the biosphere where it is temporarily used or stored in bodies of water; regardless of its immediate fate, processes of evaporation and transpiration eventually cycle the water molecules back to the vapour state.

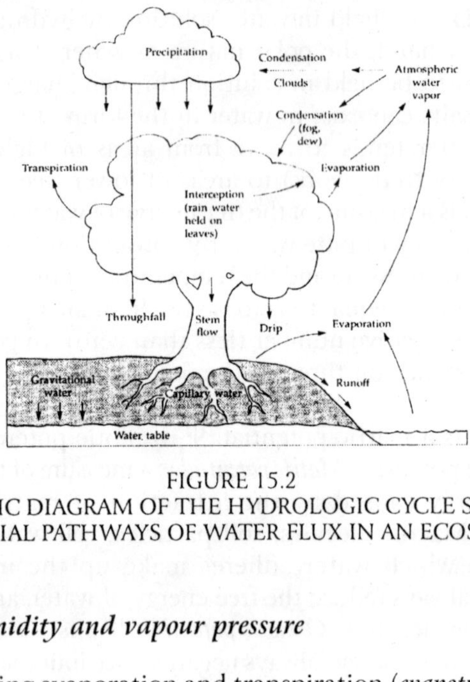

FIGURE 15.2
SCHEMATIC DIAGRAM OF THE HYDROLOGIC CYCLE SHOWING POTENTIAL PATHWAYS OF WATER FLUX IN AN ECOSYSTEM

15.1.1 Humidity and vapour pressure

Understanding evaporation and transpiration (*evapotranspiration*) processes is basic to determining the interchange of energy and water between the plant and the environment. Large amounts of energy are involved in any change in the state of water. For example, it takes most of the solar energy falling on 1 cm^2 on a clear summer day to evaporate a cubic centimeter of water. This rapid use of energy during the evaporation of water from plants (transpiration) causes an important cooling of the leaf surface. Conversely, when water condenses or freezes, a large amount of energy are released. This energy is in the form of heat and is called *latent heat*. When temperatures reach *dew point* (the temperature at which atmospheric water condenses), enough latent heat is released to significantly reduce the rate of cooling. Plants growing near bodies of water can escape frost damage because of this phenomenon (Rosenberg 1974).

Rates of evapotranspiration are indicators of water vapour transfer and energy exchange rates at leaf or soil surfaces. The rate of water loss is determined in part by the water vapour content of the atmosphere.

Humidity represents the amount of water vapour in the air. *Absolute humidity* is the actual amount of water in the air expressed as weight of water per unit of air (g/kg of air, for example). Since the amount of water vapour air can hold (at saturation) varies with temperature and pressure, the humidity of air is expressed as relative humidity. The *relative humidity* of air is the ratio of the quantity of vapour present in any volume of the air to the quantity which would have been present in the same volume had the air been completely saturated. It is also the ratio of the elastic force of vapour at the temperature of the surrounding air. To avoid fractions, it is convenient to multiply this ratio by 100. Instruments used for the purpose of measuring the quantity of aqueous vapour in the atmosphere are called **psychrometers**. In a simple psychrometer, the instrument consists of two thermometers as nearly as possible identical, the one marked dry and the other wet-the bulb of which is wrapped with a piece of cotton wick for the purpose of wetting. The thermometers are placed in a standard meteorological shelter known as the **Stevenson Screen** (Fig. 15.3) to exclude the intensity of the sun's radiant heat. The Stevenson screen consists of a white wooden box raised 1.3 m above the ground on stilts. The roof is double-layered with an intervening air space to exclude much of the direct rays of the sun. The sides of the box are louvered like "venetian blinds" to allow free circulation of the air. One side of the screen is hinged to serve as a door which can be opened and closed to give access to instruments kept inside. The floor of the screen is also louvred. The Stevenson screen normally carries maximum and minimum thermometers, and dry and wet bulb thermometers. Larger ones may also contain a self-recording thermogram and hydrogram.

From the wet bulb evaporation of water takes place on free contact with atmosphere. Evaporation results in cooling of the bulb. In dry atmosphere of low relative humidity there will be greater evaporation and consequently greater cooling or fall in temperature. If there is no depression in the temperature (that is, if both thermometers read the same) it means that the relative humidity is 100%. On the other hand, if the wet bulb thermometer reads less than dry bulb one, as is usually the case, the relative humidity is less than 100 percent, the exact value being determined by consulting the conversion table (Table 15.1).

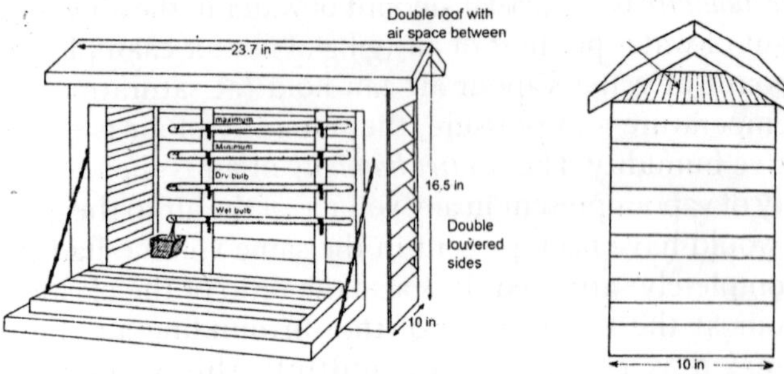

FIGURE 15.3
A STEVENSON SCREEN.

Winds and air masses transport water vapour over various parts of the earth's surface, often over long distances. Falling temperatures cause the water vapour to condense into tiny droplets that form clouds or fog. For precipitation to occur, air must contain **condensation nuclei**; tiny particles on which droplets of water vapour can collect. Volcanic ash, soil dust, pollen, smoke, sea salts, and particulate matter emitted by factories, coal-burning power plants, and vehicles are sources of such particles.

15.1.2 Fog and dew

Dew consists of tiny droplets of water produced by condensation on surface objects rather than on nuclei in the air above the surface. Condensation of dew reaches the greatest levels under clear skies (clouds reduce reradiation), low wind speeds, high vapour pressures (which reduce the amount of cooling necessary), and beneath vegetation with open branching and leaf area spread

TABLE 15.1
CONVERSION OF READINGS OF WET AND DRY BULB THERMOMETER INTO RELATIVE HUMIDITIES (%).

Temperature (°C)	*Depression of wet bulb thermometer (°C)*									
	1.0	*2.0*	*3.0*	*4.0*	*5.0*	*6.0*	*7.0*	*8.0*	*9.0*	*10.0*
0	81	64	46	29	13					
+3	84	69	54	40	25	12				
+6	87	73	60	47	35	23	11			
+9	88	76	65	53	42	32	22	12		
+12	89	78	68	58	48	38	30	21	12	4
+15	90	80	71	62	53	44	36	28	20	12
+18	90	82	73	65	57	49	42	35	27	20
+21	91	83	75	67	60	53	46	39	32	26
+24	92	85	77	70	63	56	49	43	37	31
+27	93	86	79	72	65	59	53	47	41	36
+30	93	86	79	73	67	61	55	50	44	39
+33	93	86	80	74	69	63	57	52	47	42
+36	93	87	81	75	70	64	59	54	50	45
+39	94	87	82	76	71	65	60	56	52	47
+42	94	88	82	77	72	67	62	58	53	49
+45	94	88	83	78	73	68	64	60	54	51
+48	94	89	84	78	74	69	65	61	56	53

over considerable height. Dew formation is also enhanced with cool surfaces, poleward facing slopes, in flats and valleys with cold air draingage, and in areas subject to cold ocean air when conditions are not such that fog will form. No dew forms under a closed forest canopy, presumably because reradiation is reduced; and, even under optimal conditions the theoretical maximum dewfall during a single night is less than 1mm (Slayter 1967; Stone 1963). Dew may be important in reducing water requirements of plants by reducing the leaf-air Ψ gradient even when it never enters the soil.

Frost occurs when the dew point of air falls below the freezing point. When condensation starts with temperature below 0°C, the water vapour in the air passes directly from gaseous to solid state. Sometimes condensation under such conditions results in the formation of ice-crystals commonly called *white frost.*

Fog forms when warm, moist air: (i) passes over cold water or land; (ii) cools as it ascends a mountain mass; or (iii) is cooled by rapid reradiation from soil. Fog only occurs in the lower strata of air as a short of dense "**ground cloud**". Fogs that occur on hills *(hill fogs)* are most common in the morning, even in the tropics, and disperse when the sun rises. In temperate lands, when days are hot and nights are clear and still, fogs may also result from cooling of the land surface by radiation. The lower layers of the air are chilled and water vapour in the atmosphere condenses to form *radiation fog*, or *land fog*. When the cooling surface is over the sea or when a damp air stream is brought into contact with a cold current, *sea fog* is formed. In industrial areas, very thick smoky fog is formed, called *smog*. On the basis of visibility fogs are classified into five categories: (a) *dense fog*, in which buildings or trees situated at a distance of 45 m are not visible; (b) *thick fog*, in which the maximum visibility is 180 m; (c) fog in which visibility is limited to 450 m; (d) *moderate fog*, in which visibility is limited to 900 m; and (e) *thin fog*, in which visibility extends upto 1.8 km.

It is now established that fog is a potentially significant source of moisture, increasing precipitation 1.5 to 3 times in many habitats. Also, dew and fog cause leaching of foliar metabolites (K, Ca, carbohydrates, amino acids, organic acids, etc.) and allelopathic substances from plants.

15.1.3 Precipitation

Water droplets formed around condensation nuclei are called *fog* when in contact with the earth; they are called **clouds** when a layer with no condensation appears between them and the earth. As these water droplets become large enough to respond to gravity, precipitation will fall as **rain.** At temperatures below freezing and when conditions are such that the water arrives at the earth's surface as ice, the precipitation is called **snow.** If the moist air ascends rapidly to the cooler layers of the atmosphere, the water droplets freeze into ice pellets and fall to the earth as **hail** or **hailstones**. As more and more super-cooled water drops accumulate around a hailstone, it increases steadily in size; some of them weigh as much as 1 kg. Very often, the ice-pellets exist as frozen rain-drops, melting and re-freezing on their way down; this forms **sleet**. It is only when the droplets in clouds coalesce into larger drops between 0.2 and 6 mm, that rain falls.

Condensation of atmospheric moisture sufficient to result in precipitation is stimulated by any of three sets of atmospheric conditions. Temperatures generally decrease with altitude so that any condition where warm, moist air rises may cause precipitation. This type of precipitation is classified as either orographic precipitation (*oreos*=mountain) or convectional precipitation. **Orogaphic precipitation** occurs when an air mass is forced upward as it moves across a mountain mass. It is best developed on the *windward slopes* of mountains where the prevailing moisture-laden winds come from the sea. If a volume of gas (water vapour) expands or contracts without exchanging heat with its surroundings, the expansion or contraction is said to be **adiabatic.** As air rises, it expands because pressures are lower at higher elevations. If the expansion is adiabatic, the temperature of the air will decrease, because there will be less heat per unit volume of air. Dry air in the earth's atmosphere cools about 1°C for each 100 m of vertical rise. Thus, if dry air at 30°C on the valley floor is raised by global winds to the mountain top 1500 m above, it will cool to about 15°C unless it is warmed or cooled by the mountainside or sunshine on the way up. This rate of cooling is an expression of the **adiabatic gradient** or **adiabatic lapse rate,** for dry air in the earth's atmosphere. Adiabatic cooling during expansion is the basic reason that higher air is nearly always cooler than lower air. Once the air cools and become completely saturated (relative humidity 100 %), condensation takes place forming clouds and eventually rain. Since it is caused by the relief of the land it is also known as *relief rain*. On descending the *leeward slope*, a decrease in altitude increases both the pressure and temperature, the air is compressed and warmed. Consequently, the relative humidity drops and water holding capacity of air increases greatly. The effect is to increase potential evapotransipration and to decrease precipitation on the lee side. This low precipitation belt is referred to as a **rain shadow** (Fig. 15.4).

Convectional precipitation is most common in regions that are intensely heated, either during the day, as in the tropics, or in the summer, as in temperate interiors. When the earth's surface is heated by conduction, moisture-laden vapour rises because heated air always expands, and becomes lighter. As the air reaches high altitudes it cools and when the saturation point

FIGURE 15.4
OROGRAPHIC PRECIPITATION

is reached torrential downpours occur, often accompanied by thunder and lightning. The summer showers in temperate regions are equally heavy with occasional thunderstorms. Cool season precipitation in temperate areas is usually associated with low pressure centers moving easterly and along cold polar air masses.

Cyclonic or **frontal precipitation** is independent of relief or convection. It is associated entirely with cyclonic activity whether in the temperate regions (depressions) or tropical regions (cyclones). Basically it is due to the convergence (meeting) of two different air masses with different temperatures and other physical properties. As cold air is denser, it tends to remain close to the ground. The warm air is lighter and tends to rise over the cold air. In ascent, pressure decreases, the air expands and cools, condensation takes place and light showers called *cyclonic* (in tropical regions) or *frontal* (in temperate regions) precipitation occur. The heavier and colder air masses eventually pushes up the warmer and lighter air and the sky is clear again.

The spatial and temporal distribution of rainfall is very complex and is closely related to the productivity, distribution, and life forms of the major terrestrial biomes. In general, latitude, temperature, available moisture, atmospheric disturbances, landform barriers, frontal activity, air mass movement, and differential heating are some of the factors involved in causing precipitation. The average annual precipitation over the whole earth is about 80 cm, but this is distributed very unevenly. Equatorial regions and the monsoon areas of Southeast Asia record the highest rainfall. Temperate regions, on the other hand, receive moderate amounts of precipitation. The dry regions of subtropical high pressure belt and Polar Regions receive little rainfall. Rainfall distribution over the year is an extremely important limiting factor for organisms. The situation provided by a 80 cm rainfall evenly distributed is entirely different from that provided by 80 cm rainfall which falls largely during a restricted part of the year. In the latter case, plants and animals must be able to survive long periods of drought. In general, rainfall tends to be unevenly distributed over the seasons in the tropics and subtropics, often with well-defined wet and dry seasons resulting. In temperate climates, rainfall tends to be more evenly distributed throughout the year, with many exceptions.

Snowfall has significant ecological impact in many areas of the world as the prime source of water to replenish underground supplies, as a source of soil moisture, as an insulator in areas of extreme cold, and as physical force important in controlling distribution of vegetation. Snow varies considerably in water content per unit volume but typically 8-12 cm of snow is equivalent to 1 cm of rainfall. Increases in soil water from snow depend on the rate of melt. Sudden increases in temperature may cause rapid snowmelt so that much will be lost as runoff, similar to a very intense rainstorm. Slow melting will increase infiltration and water availability to plants in the immediate vicinity. Temperatures under snow seldom fall much below 0°C, thus protecting plants from extreme cold. Protection from wind also is important where temperatures are low and winds are desiccating. The physiognomic modification called **Krummholz** (flagged appearance of vegetation) is caused by the abrasive and desiccating effects of wind and blowing ice crystals. Patterns of vegetation in subalpine forests are also dependent on snow. Meadows within subalpine forests are often the result of deep snow accumulation and the resulting shorter growing season. Snow avalanches frequently clear large areas of subalpine forest thereby modifying the vegetational mosaic of steep mountain slopes.

15.2 Plant Water Relations

Unlike animals that move, higher plants cannot satisfy their need for water by migrating to favourable water sources. Their roots anchor them to a fixed terrain. The terrestrial higher plant is interposed between the soil, which is a *source* of water, and the atmosphere, which is a *sink*. There is a continual passage of water from soil solution to plant roots, through vascular tissues to leaf cells, and into the external atmosphere (Fig. 15.5). Counteracting the loss of water to the atmosphere is a cuticular layer of waxy material, rather impermeable to water, on the outermost leaf surfaces. Loss of water to the atmosphere is also regulated by stomates in leaf epidermal tissues. Yet even so, the bulk of water absorbed from the soil is lost as water vapour to the atmosphere. Therefore, the terrestrial higher plant is faced at all times during the growth with the need to maintain within itself a quantity of water adequate to support normal functions.

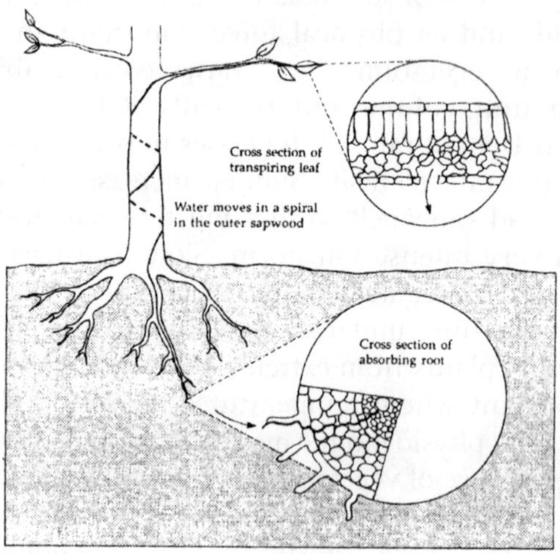

FIGURE 15.5
DIAGRAMMATIC REPRESENTATION OF THE COURSE OF WATER FLOW THROUGH THE SOIL-PLANT-ATMOSPHERE CONTINUUM

For water to move through the plant, it is necessary that $\Psi_{soil} > \Psi_{root} > \Psi_{stem} > \Psi_{leaf} > \Psi_{air}$. Water enters the root primarily through root hairs, not only because they have greatest permeability than older roots, but also because they have a much greater area for absorption. Older roots lack root hairs and become suberized (*suberization* is the formation of suberin in the walls of plant cells, which renders them less permeable to water). From the root hairs, water passes along the course of least resistance through cell walls and between root cells to the endodermis. Here the water and dissolved materials are forced to move through the cell membrane because the *Casparian strip* makes the cell wall impermeable and attaches the membrane to the cell wall. Passage through the endodermis offers considerable resistance, but once the water (and minerals) enter the vascular cylinder and the xylem, resistance is significantly lower. Vascular strands divide into very fine segments upon entering the leaves and are distributed throughout the mesophyll so that almost every cell is within 1 or 2 cells of vascular tissue (Esau 1965), but no vascular tissue is exposed directly to the intercellular spaces adjacent to stomates. From the vascular tissue, water continues to travel in the liquid phase through the cell walls and between mesophyll cells to substomatal cavities where it vaporizes and travels out into the atmosphere. The final loss of water from the plant **(transpiration)** is the most significant force in the movement of water because here the water potential gradient is the steepest. Slatyer (1967) indicated that a common set of water potential measurements along the gradient might be: soil -1 bar, stem -10 bars, leaf -15 bars, atmosphere -1000 bars. The most effective regulation of plant water status is associated with the stomates because of their location in the steepest portion of the gradient. Stomates, because of their ability to open and close, represent the major controlling factor of water flux from the plant to the air. In larger trees movement of water begins in the morning at the very top of the crown and at the tips of the branches, drawing up the column of water extending to the base of the trunk. Then the transpiration stream begins to flow rapidly and soon after sunrise it attains its maximum rate. In the evening flow becomes slower, but until late at night there can be a slow influx to the trunks, so that their water reserves are replenished.

Stomatal transpiration can be formulated as a diffusion process:

$$E_S = \frac{C_l - C_a}{r_a + r_S}$$

that is, the stomatal transpiration Es (in g H_2O /cm^2 /s) is proportional to the difference between the water-vapour content (in g H_2O /cm^3) in substomatal cavities in the leaf (Cl) and in the atmosphere (Ca), and it is limited by the sum of the diffusion resistances-that is, the stomatal resistance (rs) and the boundary -layer resistance (ra). The *boundary-layer resistance* exists external to the leaf, because a leaf is sheathed in a boundary layer of relatively moist air (Fig. 15.6) through which transpired molecules of water vapour move from leaf to atmosphere. The thickness of the boundary layer depends on leaf size and wind speed. Greater wind speeds cause turbulence closer to the leaf surface and reduce the depth of the boundary layer. Smaller leaves have thinner boundary layers because of increased convective exchange between the surface and the free air .Surface texture, leaf shape, and leaf orientation affect air turbulence near the surface and, therefore, the thickness of the boundary layer.

The minimal *stomatal diffusion resistance* (with the stomates wide open) depends upon the size and structural peculiarities of the stomates and upon their arrangement and density. Stomates open and close in response to differences in turgidity stimulated by variations in light, leaf water potential, CO_2 concentrations, and humidity. Turgidity changes are related to osmotic potential variations in the guard cells caused by the movement of K^+ into and out of the cells. When potassium ions enter a guard cell, they cause increased guard cell turgidity and opening of stomates. Water stress will stimulate production of abscissic acid (ABA) which drives the K^+ pump in the other direction, causing stomatal closure. The level of water stress necessary to cause ABA production varies with the drought tolerance of the species. For any species there is a critical Ψleaf at which the stomatal aperture will decrease dramatically, increasing resistance to infinity. Resistance in fully open stomates may range as low as 0.4 s/cm.

The ratio of transpiration loss of water by a plant to its dry matter production during the growing season **(transpiration ratio)** measures the **water use efficiency (WUE)** of the plant species; the larger the ratio, the less efficient is the species in its use of water. Transpiration ratios for most crop plants range from 200 to 500 or more; that is, it takes 200 to 500 g of water or more to bring 1g (dry weight) of a plant to maturity. Clearly higher plants in terrestrial habitats are quite inefficient in their use of water. Yet some plants are more efficient than others. For example, C-4 plants produce two to three times more dry matter per unit of water used than C-3 plants. At high air temperatures

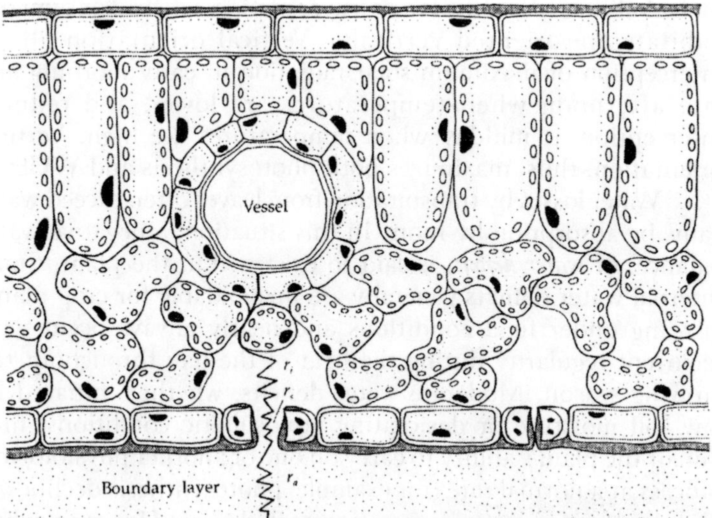

Figure 15.6 Cross section of a leaf showing the resistances associated with water loss; ri is the resistance to vapour movement in the substomatal cavity, rs is the resistance offered by the stomatal pore and guard cells, and ra is the resistance offered by the boundary layer on the leaf surface.

(30°C), larger leaves have lower photosynthetic rates especially when diffusive resistance (resistance to vapour movement in the substomatal cavity) is high. This is a result of overheating because at low resistances, evaporative cooling maintains temperatures which support positive photosynthesis even in leaves 16 cm wide. This evaporative cooling, however, requires large amounts of water and, as a result, the WUE becomes very low. Larger leaves have higher photosynthetic rates than smaller leaves in cool air (10°C) because they are at temperatures above ambient and therefore are closer to the optimum for photosynthesis. Thus, the interactions of temperature, water, and photosynthesis lead to plants with small leaves in warm, arid environments. It also explains the great success of leguminous trees in arid regions. Pinnately compound leaves with small leaflets are common in the desert leguminous trees and are apparently a significant advantage in desert habitats because of reduced heat load and greater WUE. In cooler environments, larger leaves are warmer and, with low diffusive resistance, have high photosynthetic rates. There is a gradient of increasing leaf size from warm arid to cool moist habitats. Also, the larger leaves of plants in arid or semiarid habitats are oriented vertically. Vertical orientation allows interception of maximum solar radiation in early morning and late afternoon when temperatures are lower, and reduces interception at midday when temperatures are high. Vertical orientation, thus, maximizes both photosynthesis and WUE.

Water losses by transpiration from leaves often exceed water gains by absorption by roots. In this situation, a negative water balance, or *water deficit*, is said to exist within the plant. Small internal water deficits generally are unavoidable for crop plants growing under field conditions and nearly always occur with recurrent regularity during the heat of the day throughout the growing season. Moderate water deficits, whether initiated by low soil moisture or desiccating atmospheric conditions, may impair the plant's ability to carry on one or another physiological process at a normal rate (for example, photosynthesis declines as increased water stress causes stomatal closure).However, when stress is great enough, photosynthesis, respiration, protein synthesis, and most other processes involving chemical reactions may be severely reduced because of protein (enzyme) denaturization. Most physiological processes decrease gradually with increasing stress to the point where all turgor is lost, and cease beyond that point. Under moderate stress biochemical processes may proceed but growth may be reduced because of lack of sufficient turgor pressure to cause cell enlargement. Lower overall cell size may increase the drought tolerance of the new tissues (Cutler *et al.* 1977). Newly formed cells may enlarge only when water becomes available in which case, growth much exceeds that expected by the rate of formation of new cells (Oechel *et al.* 1972). Water stress also increases the viscosity of phloem sap as well as reducing mass transfer in the xylem, thus limiting transport of nutrients, photosynthate, and hormones. Plants adjust to water deficits by modifying resistances, by special anatomical characteristics, or by adjusting osmotic concentrations in the cells to overcome aridity caused by low soil water or salinity.

15.2.1 Diffusive resistance and transpiration measurements

The most popular method of measuring leaf resistance is by the use of a **diffusion porometer** (Fig. 15.7) which works on the principle that a leaf loses water vapour as a result of transpiration. A porometer measures the rate of increase in humidity in a small chamber attached to a leaf. Humidity is measured by hygrometers with lithium chloride or aluminum oxide chambers. Thermocouples are incorporated because of the interaction between vapour pressure and temperature. Some models have circulating air to reduce the effect of boundary layer; others have plate humidity sensors which adjust close to the leaf surface to lessen *ra* influences which may be inordinately large in the small, closed porometer chamber.

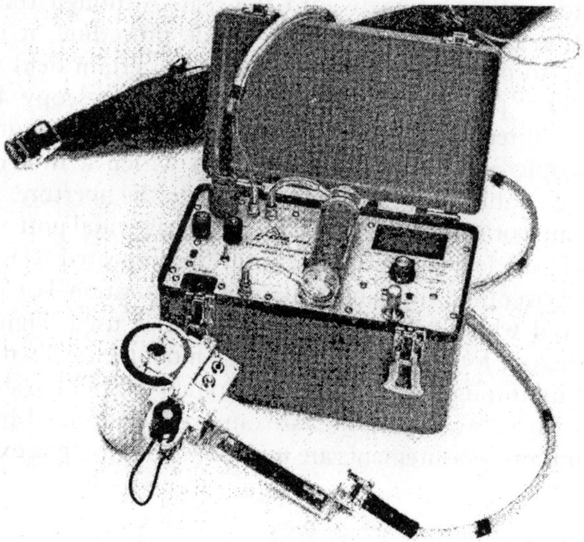

FIGURE 15.7
A DIFFUSIVE RESISTANCE POROMETER, THE LI-COR LI-1600 STEADY STATE POROMETER

Less expensive estimates of leaf resistance can be made using cobalt chloride paper, or calculation from stomatal dimensions using infiltration techniques or direct measurements. In **cobalt chloride method**, pieces of absorbent paper are soaked in a dilute (3 to 5%) solution of cobalt chloride. After drying in an oven, the paper is stored in a desiccator. When dry, the paper is bright blue; its colour changes to pink when exposed to moist air. If a piece of dried paper is transferred quickly from a desiccator to the surface of a leaf and protected from atmospheric moisture with a celluloid or glass cover, a qualitative estimate of the rate of transpiration can be obtained by determining the rate at which the paper changes from blue to pink. In the **infiltration method** a small drop of a liquid of suitable viscosity and surface tension properties-organic solvents, such as xylene, benzene, or kerosene-is applied to a leaf surface. The extent of liquid infiltration through stomatal pores into the leaf and the accompanying "darkening" of the underlying intercellular spaces is proportional to the width of the stomatal pores. The extent of infiltration can be scored either by eye or by the time (seconds) required for a drop of standard size to infiltrate into the leaf. Although the liquid infiltration technique is relatively crude procedure, it is rapid and easy to perform and may be useful in certain field studies, especially if properly calibrated by direct microscopy. Perhaps the most direct method of measuring stomatal aperture is by microscopic examination of leaves in which a **micrometer eyepiece** enables the observer to calculate the aperture.

Transportation rates from detached leaves and potted plants are easily made by **gravimetric analysis**; repeated weighing of the parts over time gives an estimate of transpiration. For attached leaves and whole branches, a **potometer** is used. Figure 15.8 illustrates one form of potometer which monometrically measures water flux through an excised leaf, branch or shoot system of a plant. More accurate measurements are possible when transpiration measurements are incorporated into a **gas exchange system**.

15.3 Plant Adaptations

Depending upon the quantity of water available in their habitats, plants are divided ecologically into three categories: (a) hydrophytes (wet habitat), (b) mesophytes (mesic or moist habitat) and (c) xerophytes (dry habitat).

15.3.1 Aquatic environments and hydrophytes

There are large varieties of habitats whose environment is predominantly characterized by excess of water. They are chiefly divided into two categories: the marine, and the freshwater.

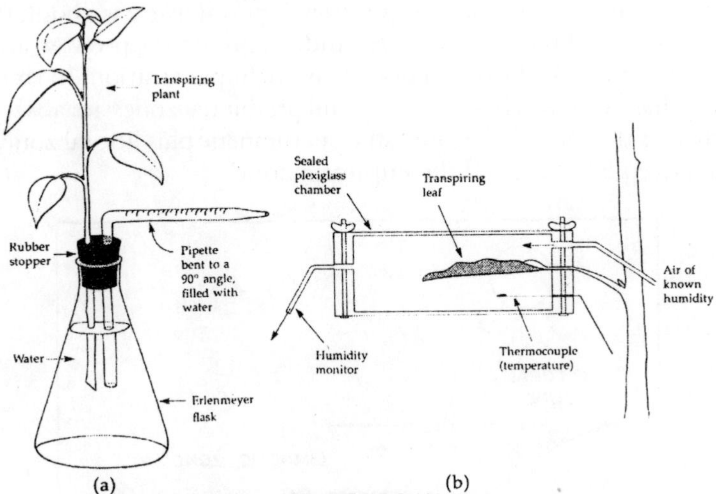

FIGURE 15.8 (A)
POTOMETER FOR THE MEASUREMENT OF TRANSPIRATION OF PLANT SHOOTS OR LEAVES. (B) A GAS EXCHANGE CHAMBER FOR THE MEASUREMENT OF TRANSPIRATION IN CONTINUOUSLY FLOWING AIR

Freshwater habitats may be conveniently considered into two series: (i) standing- water or lentic (*lenis*= calm) habitats (such as pools, ponds, lakes, marshes, swamps); (ii) running-water or lotic (*lotus*= washed) habitats (such as streams, rivulets and rivers). The study of the freshwaters in all their aspects- physical, chemical, geological and biological - is termed **limnology.**

Two major zones are generally evident in streams and rivers: rapids zone and pool zone. **Rapids** is the shallow water zone where velocity of current is great enough to keep the bottom clean of silt and other loose materials, thus providing a firm substrate. **Pool zone** is deeper water where velocity of current is reduced and silt and other loose materials tend to settle to the bottom.

In the ponds and lakes three zones are generally evident:littoral, limnetic and profundal (Fig. 15.9). **Littoral zone** is the shallow-water region with light penetration to the bottom and this zone is generally occupied by rooted plants in natural ponds and lakes. **Limnetic zone** is the open-water zone to the depth of effective light penetration, called the *compensation level* (photosynthesis balances respiration; light intensity is about 1% of full sunlight intensity). **Profundal zone** is deep-water area which is beyond the depth of effective light penetration. In small and shallow ponds both limnetic and produndal zones are absent, whereas the total illuminated stratum (limnetic plus littoral zones) of a deep lake is called the **euphotic zone**.

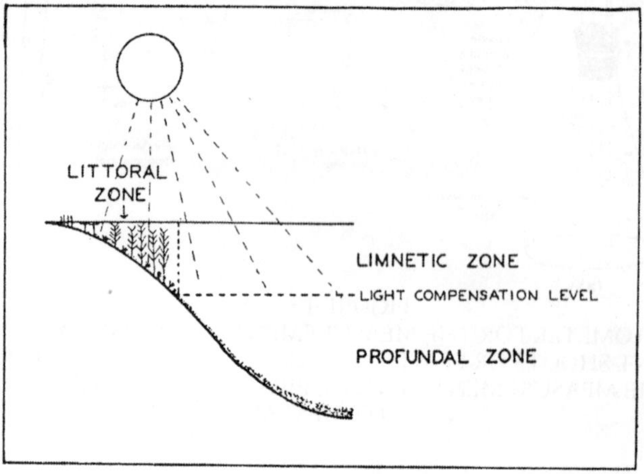

FIGURE 15.9
THE THREE MAJOR ZONES OF A LAKE

In temperate zone lakes, the top waters become warmer than the bottom waters during the summer season. This warmed water expands and therefore, because it is less dense than the cooler water beneath, remains as a surface layer. This stratification prevents mixing of the layers of water, so that although debris from above can sink to the bottom and decay there, there is little transfer of soluble substances such as minerals and dissolved oxygen and carbon dioxide. The upper warm region of such a lake is called **epilimnion**, the bottom cold region **hypolimnion**, and the intermediate zone **thermocline**. With the onset of cooler weather, the temperature of epilimnion drops until it is the same as that of hypolimnion in autumn. It triggers mixing of top and bottom waters; the bottom waters are reoxygenated and the upper ones replenished with nutrients. In winter surface water freezes (0°C) whereas lower water remains comparatively warmer (4°C) in liquid state. In the spring, as ice melts and temperature of epilimnion becomes equal to that of hypolimnion, there is another mixing of top and bottom waters. Thus in temperate lakes mixing of top and bottom waters occurs twice (autumn and spring) a year, and hence such lakes are called **dimictic**. Subtropical lakes having surface temperature that never falls below 4°C, generally show a distinct thermal gradient from top to bottom, but experience only one general circulation period per year, which comes in winter (hence, **monomictic**). High altitude as well as equatorial lakes are more or less continually circulating with only short, if any, stagnation period (hence, **polymictic**). Tropical lakes with high surface temperatures (20 to 30°C) exhibit weak gradients and little seasonal changes in temperature at any depth.

On the basis of productivity and fertility lakes may be either oligotrophic or eutrophic. **Oligotrophic** lakes are deep, with the hypolimnion larger than epilimnion, and have low primary productivity. Littoral plants are scarce and plankton density is low; plankton blooms are rare since nutrients rarely accumulate sufficiently to produce a population eruption of phytoplanktons. Because of low productivity of the top waters, hypolimnion is not subject to severe oxygen depletion (hence a habitat for cold water bottom fishes, such as lake trouts). Almost all high altitude lakes of Nepal are oligotrophic in nature. In contrast, **eutrophic** lakes are shallower, rich in nutrients, and have a greater primary productivity. Littoral vegetation is more abundant, plankton populations are denser, and "blooms" are characteristic. Freshwater bodies near the human settlements in both midland and the terai regions of Nepal are generally eutrophic in nature.

In the marine environment, the shallow shore region is called the **neritic zone**. This region is visited by tides. The portion of the sea away from land is called the **oceanic zone**. The deeper slopes in continuation of the shore up to a depth of 2000 m are called the **bathyal zone**, while still deeper zones (2000-6000 m deep) in the oceanic belt are called the **abyssal zone**. **Oceanography** deals with the study of the sea in all of its aspects - physical, chemical, geological, and biological; whereas **marine ecology** deals with the pattern of relationships between the organisms and the sea environments.

15.3.2 Hydrophytes

All such plants which usually germinate in water and spend all or part of their life cycle in water are called hydrophytes. On the basis of life forms they may be either the phytoplanktons or the macrophytes. **Phytoplanktons**, which are mostly algae, are the floating green plants inhabiting the surface layer of the water bodies, drifting with current. **Macrophytes** are predominantly vascular plants of different categories: (a) marginal emergent, (b) floating - leafed, (c) submerged, and (d) free - floating.

Rooted aquatic marcophytes form concentric zones within the littoral zone, one group replacing the other as the depth of water changes either in space or in time. These zones are (i) **zone of emergent vegetation**: rooted macrophytes of marginal emergent type with the basal part (2 -15 cm) immersed and the principal photosynthetic surface projecting above the water (*Cyperus, Fimbristylis, Polygonum, Typha* etc.) constitute the vegetation of this zone. The emergent plants, together with those on the moist shore, form an important link between water and land environments. They are used for food and shelter for amphibian animals, and they provide a convenient means of entry and exit into the water for aquatic insects which spend part of their lives in water and part on land; (ii) **zone of rooted plants with floating leaves:** the floating - leafed species such as water lilies (*Nymphaea* spp.) and lotus (*Nelumbo nucifera*) are the type of plant species constituting the vegetation of this zone. The horizontal photosynthetic surfaces of these species may more effectively reduce light penetration into water. The undersurfaces of the leaves of these species provide convenient resting places and places for egg deposition by animals; (iii) **zone of submergent vegetation:** rooted macrophytes completely or largely submerged constitute the submergent vegetation. Their leaves tend to be thin and/or finely divided and adapted for the exchange of nutrients with water. Some examples of submerged rooted hydrophytes are *Hydrilla, Potamogeton, Vallisneria,* etc.; (iv) **zone of free-floating hydrophytes:** the free floating hydrophytes occur scattered everywhere in ponds and lakes and keep on changing their position due to water or wind currents. Usually in small ponds such species are aggregated more in the centre where other categories of hydrophytes are few. In form and structure free-floating macrophytes may be (a) large with rosettes of big leaves either floating or rising above the water in the air as in *Trapa* and water-hyacinth, respectively; (b) slightly smaller forms are *Pistia* and *Salvinia*; (c) other variety is of very small surface floating plants with the body reduced to thallus-like structures (*Azolla, Lemna, Spirodela*, etc.).

The general features of aquatic plants are the reduction of protective tissue (epidermis without cuticle), supporting tissue (lack of sclerenchyma), conducting tissue (minimum development of vascular tissue) and absorbing tissue (roots), and the special development of air-chambers (aerenchyma tissue) for the aeration of the internal tissues. The main adaptive features of various types of hydrophytes are:

1. The under-mud parts such as roots, rhizomes or bases of shoots of the hydrophytes are subjected to an upward pull by the above-mud parts on account of their buoyancy. Hence, the under-mud parts of the hydrophytes are generally 4 to 5 times heavier than other parts.

2. In the emergent species, the under-mud parts of stem and root may suffer from paucity of oxygen. Such hydrophytes have stems with swollen bases **(hypertrophy)** containing plenty of interconnected gas spaces *(aerenchyma)* and adventitious roots. While internal transfer of oxygen from shoots to roots is facilitated by aerenchyma, the adventitious roots help in anchorage of the plant body as well as in absorption of the nutrients from the surrounding substrate.

3. The emergent and floating -leafed types of hydrophytes exhibit enhancement in internodal and petiolar growth, respectively to keep their foliage above water to facilitate the movement of air to submerged parts whenever water level in ponds and lakes increases.

4. In many species where some leaves are submerged and some aerial, then they show a different shape and size to suit the situation. **Polymorphism** or different shapes in the leaf of the same individual is called **heterophylly**.

5. In submerged plants, the leaf and shoot organs are provided with cavities or hollow spaces called **lacuna**. The lacunate tissue is filled with air which provides buoyancy to the plant. With the aid of buoyancy the shoot tends to pull up and thus the plant maintains its expanded shape by the aid of air-

filled tissues. In this way the necessity of mechanical tissues such as sclerenchyma or the hard secondary wood, etc. to keep plant erect and in shape does not arise. Leaves are usually thin and filmy. Chloroplast is abundant in epidermal cells because light intensity is low. The stomates are usually absent or if present they are functionless. Many submerged species have dissected leaves which reduce the chance of physical damage from water and increases surface area for a more efficient absorption of materials and increased photosynthetic activity.

6. The leaves of the floating-leafed species are peltate, usually circular in shape and have a strong leathery texture. The upper surface of the leaf is such that water glides off and does not adhere. The leaf margin is strong to resist tearing due to pressure of wind or water current. The petiole is quite flexible so that the leaves may move a little this way or that and adjust themselves in the available space on the water surface. The lower surface in touch with water is usually deficient of stomates. The upper leaf surface has stomates for the easy exchange of gases. Anatomically, the lower side of the leaves is usually lacunate or air-filled. This helps in keeping the leaves afloat. The upper side is full of chlorenchymatous tissue. Despite abundance of air tissues the leaves maintain their shape due to presence of supporting mechanical tissues like sclereids.

7. The free floating species usually grow in waters which are fairly rich in nutrients. The necessity of roots for absorption and anchorage does not arise in many floating species since the entire plant is in contact with water. An erect system like a stem is also unnecessary for such a life form. In most free floating plants (*Azolla, Lemna* etc.) the root and stem are very poorly developed. Leaves are the most characteristic and well developed parts which do most of the functions of a full plant like absorption, maintenance of high buoyancy due to abundant aerenchyma, efficient photosynthesis and vegetative reproduction. However, some of the very fast growing free floating species like *Eichhornia crassipes* have extensive root growth. These roots too have abundant lacunate tissue full of air spaces (Fig. 15.10).

15.3.3 Mesophytes

Mesophytes are plants that normally grow in habitats where water

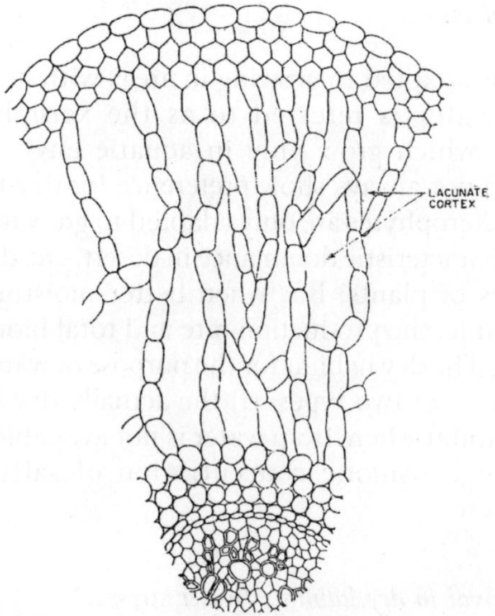

FIGURE 15.10
T.S. OF THE ROOT OF *EICHHORNIA CRASSIPES* SHOWING EXTENSIVE DEVELOPMENT OF AIR FILLED LACUNATE CORTEX

is neither scarce nor abundant. In such habitats the pore space in soil is occupied almost equally by water and air. Such a balanced condition of water and gases is very suitable for plant growth and hence in mesophytic condition we find the best growth of forests or crop plants. Both excessive water and excessive dryness create a large variety of situations adverse to plant growth and only a few species, specially adapted to such environment can grow there. However, in mesophytes, no such adaptation is necessary.

Mesophytes are very extensive on the land surface and most crops like wheat, maize, sugarcane, etc., species in grasslands and of forests are mesophytes. The roots in mesophytic species are quite well developed and young roots have abundant root hairs. There is balanced proportion of mechanical tissues in roots as against their absence in hydrophytes and excessive predominance in xerophytes. The same is true for stems. The leaves are well developed, usually broad and only moderately thick. Stomates are well developed and usually present on the lower surface. The chloroplasts are also abundant and usually dark green in colour.

15.3.4 Xerophytes

A plant that is adapted to growing in areas with low or irregular supplies of water is referred to as the xerophyte. Unlike hydrophytes, which grow only in aquatic environments, the xerophytes do not always show preference for dry habitats for a good growth. Xerophytes are only adapted to grow in dry habitats and assume characteristic dominance in desert, etc. due to absence of other types of plants. But when better moisture conditions become available, the production rate and total biomass increase in xerophytes. The dry habitat for the purpose of water availability to plants may be of two types: (i) the actually dry habitats, and (ii) the wet habitat where from water is not available to plants on account of high osmotic concentration of salts or freezing temperature, etc.

15.3.4.1 Survival in dry habitats (water stress)

The desert **ephemerals** are annual plants that escape the drought by existing only as dormant seeds during the dry season. When enough rain falls to wet the soil to a considerable depth, these seeds germinate and complete life cycle during short periods of mesic conditions before all soil moisture has been exhausted. Typically, the summer ephemerals are C-4 and winter ephemerals C-3 plants. They are eminently well suited to dry regions and thus are xerophytes in the true sense of the word, yet their active and metabolizing protoplasm is never exposed to extremely negative water potentials and is not drought hardy. Hence, the desert ephemerals are the **drought escaper** type of xerophytes. Another type of the drought escapers are the drought-deciduous tree species which avoid water stress by shedding leaves to reduce surface area and water loss during the dry season.

Phreatophytes or the plants restricted to the riparian (riverside) habitats, palms that grow at an oasis, or other plants such as mesquite *(Prosopis glandulosa)* and alfalfa *(Medicago sativa)* that have roots that extend as much as 50 m down, avoid many of the rigours of arid environments by having roots in constant contact with the underground water table. They are **water spenders**.

Succulent species such as the cacti, century plant *(Agave americana),* and various other crassulacean-acid-metabolism (CAM) plants are **water savers**. They resist the drought by storing water in their succulent tissues. Enough water is stored, and its rate of loss is so extremely low (because of an exceptionally thick cuticle and stomatal closure during the day-time) that they can exist for long periods without added moisture. Some of the succulents, specially the cacti, have extensive shallow root systems which help the plant in absorbing surface moisture after a light shower. Absence of broad leaves and abundance of spines further protect the succulents from the ravages of grazing /browsing animals.

Many non-succulent perennial desert plants also have several adaptations that reduce water loss, although they do not actually store water in succulent tissues. Some examples are *Prosopis cineraria*, several species of *Acacia, Calotropis, Nerium, Casuarina equisetifolia* etc. They have very deep root systems and small leaves-which are more efficient energy dissipators and lose less water per unit area than large leaves. Other significant adaptations that apparently reduce transpiration include sunken stomates (Fig. 15.11), shedding of leaves during dry periods, a thick cuticle, and heavy pubescence on leaf surfaces. Although these modifications may indeed reduce the loss of water, they never completely prevent it and are by themselves insufficient protection against extreme drought.

The drought enduring xerophytes lose exceptionally large quantities of water so that their protoplasm is subjected to extremely negative water potentials, yet they are not killed. Such **euxerophytes** (true xerophytes) exhibit tolerance or hardiness rather than mere avoidance. Some of the most spectacular euxerophytes (called **poikilohydric**) are mosses and ferns. Their ability to dry out and then become metabolically active immediately upon rehydration apparently depends upon special features not common to others.

One of the important damaging effects of water stress is the concentration of salts within the cells of plants subjected to water stress. These salts can damage the enzymes that control metabolism and that are thus essential to life. An important adaptation found in many xerophytes that are subjected to water stress is the accumulation of certain organic compounds such as

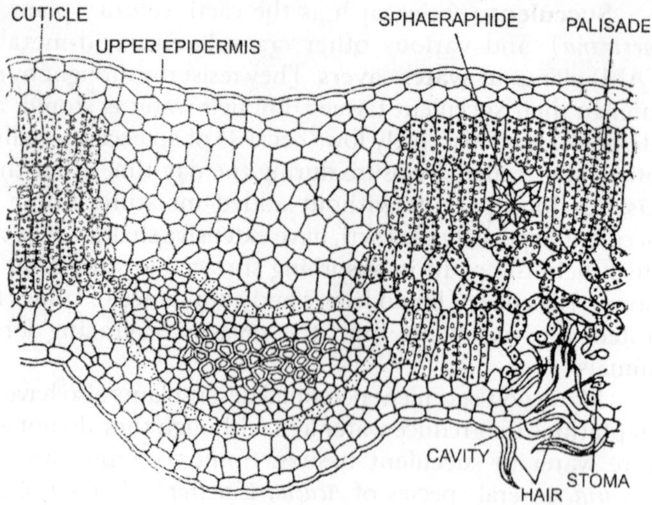

FIGURE 15.11
V.S. OF NERIUM LEAF SHOWING XEROPHYTIC CHARACTERS SUCH AS SPHAERAPHIDES, CAVITIES ON LOWER SURFACES, SUNKEN POSITION OF STOMA AND HAIRS

sucrose, the amino acid proline, and several others that lower the osmotic potential and thus the water potential in cells without limiting enzyme function. Such compounds appear in the cells of many xerophytes as water stress increases, and the resulting drop in osmotic potential is called **osmotic adjustment** or **osmotic regulation**.

15.3.4.2 Survival in saline habitats (salt stress)

Cells of the mesophytes or hydrophytes maintain an osmotic potential equal or slightly more negative than that of the medium (soil or water) where the plants are growing. But when the water medium or soil medium is definitely saline, as in deserts or in soils saturated with brackish waters on the sea coasts or close to the shores of salty waters, the water balance between plants and their habitats becomes disturbed, much to the disadvantage of the former. The osmotic potential of the saline water is very much negative. Evidently this condition tends to extract water from the plant tissues having less negative osmotic potential, rather than help them to absorb water from such a medium. The saline water or soil thus behaves as a physiologically dry medium for such plants. To meet this exigency the plant tissues must maintain a more negative osmotic potential than their saline medium.

Halophytes are plants that live in saline habitats and accumulate relatively large amounts of salts in their organs. The salts are not harmful and in fact are beneficial as long as the quantities are not extreme. Germination, in contrast, is always more successful in a salt-free **milieu**. It is only in the course of development that the halophyte characteristics become fully evident.

In contrast to **glycophytes** (plants sensitive to relatively high salt concentrations), halophytes contain much greater concentrations of chloride and sodium ions. This property enables them to compensate for the high osmotic potential of the soil and to draw water from the ground even if the salt concentration is high there. In the optimal case the salt concentration in the cell sap of halophytes just compensates the concentration in the soil solution. This is possible only if the plants can regulate the salt content of their tissues. They in fact do this by elimination of salt and by increased storage of water (succulence).

(i) **Salt extrusion:** One way to regulate salt balance in mangrove plants is the exudation of salts through salt glands and hairs. These are epidermal structures which actively eliminate salts and thus keep concentration in the leaves within certain limits.

(ii) **Abscission of organs:** Salt may be extruded by shedding of organs such as shoots or leaves which have accumulated salt. Salt evidently induces senility and, as the organ ages, nutrients and stored organics are mobilized to younger parts of the plant, followed by abscission of the organ containing the accumulated salt.

(iii) **Succulence:** Succulence caused by accumulation of water in the tissues serves to dilute salts in cells, thereby reducing solute concentrations. Succulence is widespread among halophytes occurring in damp (wet) saline condition or in arid (dry) habitats. Rapid growth of new tissues also has a diluting effect since salts are distributed to greater volumes of tissues.

15.4 Plant Water Status Measurements

Methods of measuring plant-water status include the pressure chamber, psychrometry, dew point hygrometry, dye method, and relative water content.

In the **pressure chamber method** a living freshly excised, leafy shoot is placed within a chamber able to withstand gas pressures up to 50 or more bars. The cut end of the stem protrudes through a gastight seal (Fig. 15.12). An inert gas (nitrogen) is admitted into the chamber from a compression tank. When the pressure around the leafy shoot is increased sufficiently so that xylem sap appears at the cut surface, equilibrium with respect to the transfer of water between the cells of the shoot and the xylem conduction system is considered reached. The gas pressure (with a negative sign) just sufficient to force xylem fluid to appear at the exposed cut end of a shoot is taken as being equal to the water potential of the leaf cells.

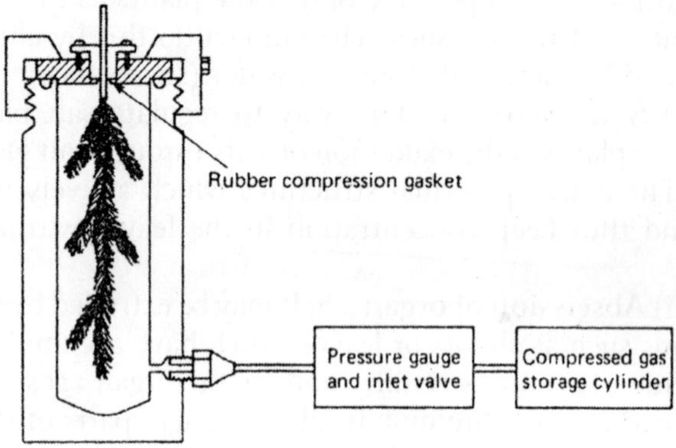

FIGURE 15.11
CROSS SECTION THROUGH A PRESSURE CHAMBER USED TO MEASURE WATER POTENTIAL IN LEAFY SHOOTS (SCHOLANDER 1971)

In a **thermocouple psychrometer** vapour pressure in the air above a tissue sample (a leaf or a piece of a leaf) is measured with a thermocouple. A thermocouple consists of a closed circuit formed by joining together two dissimilar metals at their ends

(Fig. 15.13). A small current will flow in the circuit when one of the junctions is maintained at one temperature and the other at a different temperature. The electromotive force of this current depends on the nature of the metals and is proportional to the difference in temperature between their junctions. The metals usually used in instruments designed for studies of plant water relations are constantan and chromel-P, both of which are alloys. The electromotive force is measured with a sensitive potentiometer. The two junctions of the thermocouple area called the *sensing junction* and the *reference junction*. The sensing junction is wetted with a droplet of pure water and placed in the chamber directly above the tissue sample whose water potential is to be determined. The chamber is immersed in a constant temperature bath. Precise temperature control, within the limits of at least ± 0.005°C, is required to ensure that the water vapour in the air above the tissue sample will remain in equilibrium with water inside the sample during the measurement. Also, the constant temperature bath provides a fixed temperature for the reference junction of the thermocouple. When water evaporates from the wetted sensing junction, it will be cooled. In fact, the rate of evaporation of water from the wetted sensing junction - and hence its temperature - will be determined by the vapour pressure in the chamber. Because of the difference in temperature between the sensing junction and the reference junction, an electromotive force proportional to the difference in the temperature is generated in the thermocouple. Before a thermocouple psychrometer can be used to measure water potential, the electrical output of the instrument must be calibrated against solutions of known water potential. Accurate water potential data for solutions of certain sugars and salts (sucrose, potassium and sodium chloride) are available in the literature (Stadelmann 1966, Michel 1972).

In the **dye method** of measuring tissue water potential several uniform tissue samples (whole leaves from the same branch of a plant) are selected. Each tissue sample is immersed in one of a series of graded (sucrose) "test" solutions of known water potential and each test solution is coloured lightly with a dye, such as methyl orange or methylene blue. A parallel series of control solutions, of the same water potentials as the test solutions, is prepared but not coloured (Fig. 15.14). The dye method depends on changes in density in the test solutions. If water is

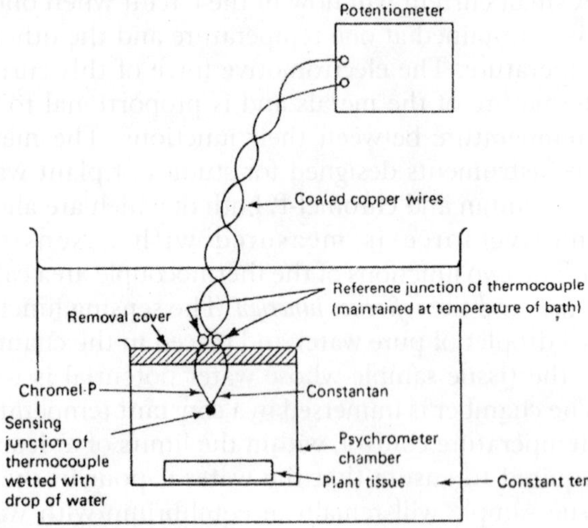

Figure 15.12 The thermocouple psychrometer for measuring water potential in plant tissue is shown schematically. Note the following essential components: (1) a small chamber in which the tissue sample is enclosed; (2) a constant temperature bath; (3) a thermocouple sealed into the removal cover of the psychrometer chamber; (4) a sensitive electrical measuring device capable of measuring very small differences in electromotive force induced in the thermocouple circuit.

transferred into the tissue during the immersion period, the density of the test solution increases. If the water is transferred out of the tissue during the immersion period, the density of the test solution decreases. If the water potential in the tissue sample is the same as that of the test solution, there will be no net water transport between the tissue sample and the test solution during the immersion period. In this case the density of the test solution is unchanged. After an hour or two (that is, when equilibrium with respect to the transfer of water between tissue sample and test solution is attained) the density of each test solution is ascertained as follows. A drop of test solution is placed carefully in the uncoloured control solution of corresponding water potential. The drop - made visible by the presence of the dye - will fall in the control solution if there was an increase in the density of the corresponding test solution. Alternatively, the coloured drop will rise if there was a decrease in the density of the corresponding test solution. If there was no change in density in the test solution, the coloured drop will diffuse away when placed in the corresponding control solution. The water potential in the test solution in which no density change developed during the immersion period is equal to the water potential of the tissue.

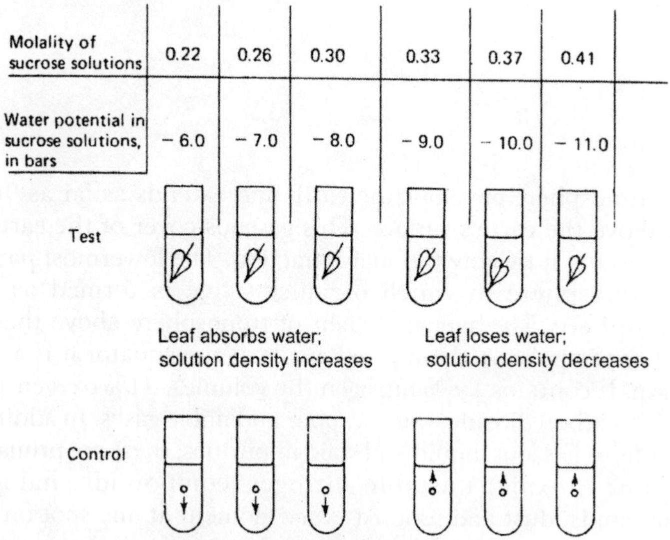

FIGURE 15.13
THE DYE METHOD OF MEASURING TISSUE WATER POTENTIAL DEPENDS ON DENSITY CHANGES THAT TAKE PLACE IN A GRADED SERIES OF COLOURED TEST SOLUTIONS OF KNOWN WATER POTENTIAL WHEN TISSUE SAMPLES ARE IMMERSED IN THEM (KNIPLING 1967)

Relative water content (RWC) provides a simple means by which information concerning the water status of plants can be obtained. RWC is an expression of the current plant water status relative to what it would be if plant were fully hydrated:

$$\text{RWC} = \frac{\text{Fresh weight minus dry weight}}{\text{Fully turgid weight minus dry weight}} \times 100$$

Fully turgid weight is obtained by exposing the tissue to water until it no longer increases in weight. RWC is an accurate measure of the water status but gives no information of the plant water status with respect to soil or atmosphere, and no indication of the plants response to water stress.

16. WEATHER AND CLIMATE

The atmosphere envelops the earth and extends as far as 9600 km above the earth's surface. This gaseous cover of the earth is held around it by gravitational attraction. The lowermost part of the atmosphere in which organisms live is termed as the **troposphere**. The average height of troposphere above the sea level at the poles is about 8 km, while at the equator it is about 16 km. It contains 79 % nitrogen (by volume), 21% oxygen, and 0.03 % carbon dioxide, water vapour and noble gases. In addition it contains gaseous, liquid and solid impurities; these are primarily sulphur dioxide, unstable nitrogen compounds, halogen compounds, dust and soot. At every moment at any spot on the earth, the troposphere has a particular set of physical properties. Examples are: (a) temperature, (b) pressure, (c) humidity, (d) precipitation, (e) sunshine, (f) cloud cover, and (g) wind direction and speed. These short term properties of the troposphere at a particular place and time are **weather**. On the other hand, **climate** is a region's general pattern of atmospheric and weather conditions, including seasonal variations and weather extremes (such as prolonged drought or rain or hurricanes) over a long period. Depending on the terrain and on the density, height and type of vegetation, individual climatic regions of different sizes are formed. Within the large scale "macroclimate", one may distinguish "microclimates" that prevail in specific places such as certain slopes, narrow valleys and stands of vegetation.

16.1 Elements of Weathers and Climates

Weather Bureaux or Meteorological Stations are scattered all over the globe, to obtain data on variables such as atmospheric pressures, precipitation, temperatures, wind speeds and locations of air masses and fronts using some of the most up-to-date devices such as weather balloons, aircrafts, ships, radar, and satellites. These data are fed into computer models to draw weather maps for each of seven levels of the troposphere, ranging from the ground to 16 km up. Computer models use the map data to forecast the weather in each box of the seven-layer grid for the next 12 h. Other computer models project the weather for the next several days by calculating the probabilities that air masses, winds and other factors will move and change in certain ways.

16.1.1 Rainfall

Rainfall including other forms of precipitation (snow, sleet and hail) is always measured by a metal instrument called a **rain gauge** (Fig. 16.1). It consists of a copper cylinder with a metal funnel either 12.5 or 20 cm in diameter, which leads into a smaller copper container or a glass bottle. The hole in the funnel that leads down to the container is very small so that evaporation of the collected rain is minimized. The firmly fastened gauge is kept at least 30 cm above the ground to avoid the splashing. The instrument is sited well away from high buildings, tall trees and other objects. The measurement of the rainfall is done by removing the funnel, emptying the rain in the container into a graduated cylinder with 4 cm diameter. A 2.5 cm of rainfall means the amount of water that would cover the ground to a depth of 2.5 cm, provided none-evaporated, drained off or percolated away. For meteorological recordings, a *rain-day* is reckoned as a period of 24 h with at least 0.25 mm or more rain being recorded. If the amount exceeds 10 mm, it is considered as *wet day*. Only a torrential downpour can account for more than 25 mm of rainfall in a day. In temperate regions, snowfall is carefully melted by warming the funnel and then measured. The daily record of rainfall is added at the end of the month to find the total rainfall for that month. The total for each month is again added at the end of the year to find the *annual rainfall*. The *mean annual rainfall* is obtained from the averages of annual rainfall taken over a period of 35 years (Meher-Homji 1971). For plotting in rainfall maps, places having the same mean annual rainfall are joined by a line called an **isohyet**.

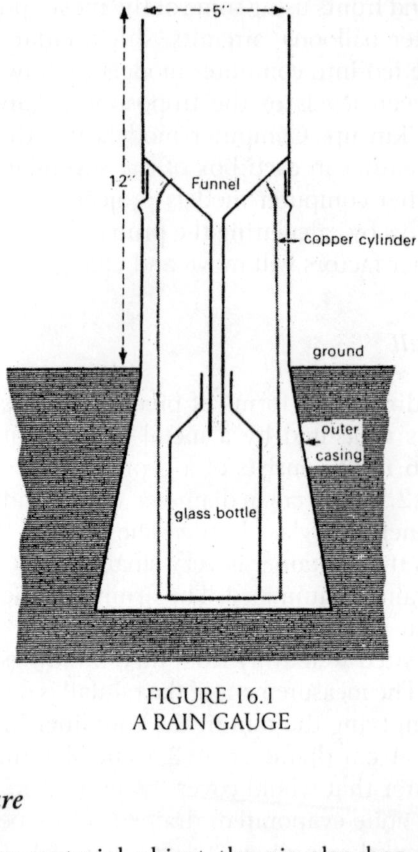

FIGURE 16.1
A RAIN GAUGE

16.1.2 Pressure

Like any other material object the air also has a weight. The pressure of air at a given place is defined as a force exerted in all directions in consequence of the weight of all the air above it. Thus, the mass of a column of air above a given point determines the atmospheric pressure at that point. The atmospheric pressure is measured most accurately with the help of a mercurial barometer, an instrument invented by Galileo and his assistant Torricelli in 1643 (Fig. 16.2). The standard sea-level pressure is 1013.25 mb (millibar) at a temperature of 15°C and at latitude of 45°. This pressure is equivalent to 760 mm. The *millibar* is a force equal to 1000 dynes/cm^2, and a *dyne* is a unit of force approximately equal to the weight of a milligram. On maps places of equal pressure are joined by lines called **isobars**.

Pressure readings vary with a number of factors. A sea-level readings of 760 mm will be halved on mountainous regions of 5.5 km above the sea level. This is because air is less dense above and so the weight (or pressure) is less. The barometer is also sensitive to gravitational forces at different latitudes. The mercury itself also expands with an increase in temperature. Therefore for professional meteorological work which requires very accurate readings, corrections have to be made in respect of altitude, latitude and temperature.

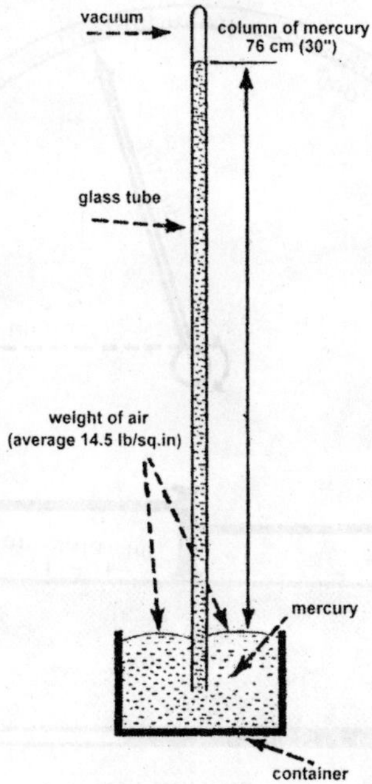

FIGURE 16.2
A MERCURY BAROMETER

Since a mercury barometer that dips in liquid mercury is inconvenient for outdoor measurement, a more portable but less accurate type known as the *aneroid barometer* is used. This comprises a small metal container, with most of the air driven out to form almost a vacuum. As there is practically no pressure at all inside the box, any increase in pressure on the outside of the box will cause the lid to move inwards thus registering high pressure by an indicator on the revolving dial .When there is a decrease in pressure, the lid springs outwards, registering low pressure by the indicator (Fig. 16.3).

In aeroplanes, a modified type of aneroid barometer called an *altimeter* is used. As pressure decreases with altitude at an approximate rate of 25 mm drop in the mercury reading for every 275 m ascent, the altimeter gives the reading in meters for height attained. For a continuous record of pressure changes, the self-recording *barograph* is used.

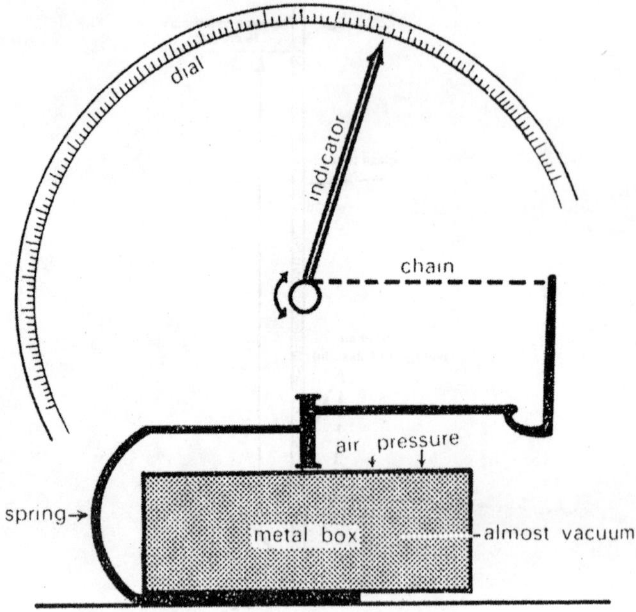

FIGURE 16.3
AN ANEROID BAROMETER

16.1.3 Temperature

The maximum and minimum air temperatures are measured by the maximum and minimum thermometers for the 24 h reading. The maximum thermometer records the highest temperature reached during the day, whereas the minimum thermometer records the lowest temperature which occurs in the middle of the night or early in the morning. Thus, in recording temperature, the maximum temperature is entered in the column for *previous day* and the minimum temperature in the column for *current day* because of their respective period of probable occurrence. The *mean daily temperature* is the average of maximum and minimum temperatures; for example if maximum temperature is 30°C and minimum 20°C, the mean daily temperature is 25°C. On the other, the difference between the maximum and minimum temperatures (30°C-20°C=10°C) of a day gives the *diurnal range* of temperature. The difference between the hottest month (July in the northern hemisphere) and the coldest month (January in the northern hemisphere) gives the *annual range* of temperature. In diagrammatic representations, monthly mean temperatures are shown in simple temperature graphs (Fig. 16.4) or in temperature distribution maps as **isotherms**. For these maps temperatures are reduced to sea level- that is, shown as if the recording station was at sea level. Temperatures decrease at the rate of 10°C per 1000 m of elevation, so for highland stations a higher temperature is shown than was actually recorded.

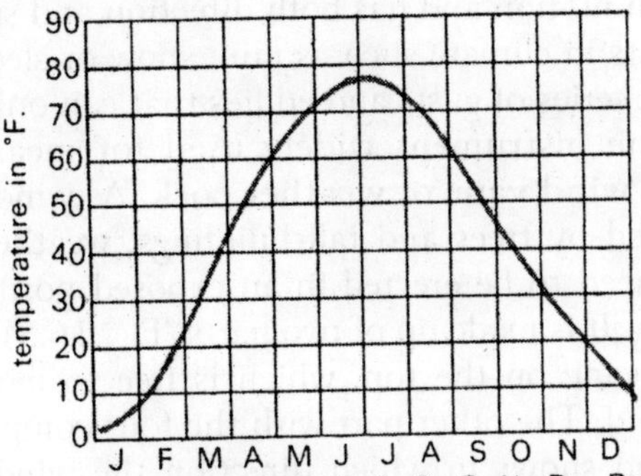

	J	F	M	A	M	J	J	A	S	O	N	D	Range
F°.	8	14	30	47	60	71	77	75	61	48	29	14	69°

Statistical data of temperature of Mukden (Mean monthly temperatures from January to December)
Mean annual temperature is 44°F.

FIGURE 16.4
A TEMPERATURE GRAPH OF MUKDEN, MANCHURIA (42°N, 123°E)

16.1.4 Humidity

Humidity is a measure of the amount of water vapour present in air which varies greatly from place to place at different times of day. The actual amount of water vapour present in the air, which is expressed in grams per cubic meter, is called the *absolute humidity*. But more important from the point of view of weather studies is the *relative humidity*. This is the ratio between the actual amount of water vapour and the total amount the air can hold at a given temperature, expressed as a percentage. Warm air can hold more water vapour than the cold air, so if it contains only half the amount it could carry, the relative humidity is 50%. In the equatorial regions, over 80 % is common in the morning, which means that the air contains four-fifths as much water vapour as it can carry. When the relative humidity reaches 100 %, the air is completely *saturated*. The air temperature is said to be at *dew point*. Further cooling will condense the water vapour into clouds or rain.

16.1.5 Winds

Wind is air in motion and has both direction and speed. Unlike other elements in climate such as rain, snow or sleet, winds are made up of a series of gusts and eddies that can only be felt but not seen. The instrument widely used for measuring wind direction is a **wind vane** or **weather cock**. As wind direction is always blocked by trees and tall buildings, weather cocks and wind vanes need to be erected in an exposed position to get a true direction. It is made up of two parts (Fig. 16.5). One part is an arrow or vane on the top, which is free to move with the prevailing wind. The other part with the four compass points is stationary and shows in which direction the wind is moving. Winds are always named from the direction they blow; an *east wind* is one that blows from east to west and a *south-west wind* is one that blows from the south-west.

The speed of wind is usually measured by an **anemometer** (Fig. 16.6). It consists of four semi-circular cups attached to the ends of horizontal spokes mounted on a vertical spindle. As the concave sides of the cups offer greater resistance to the winds, the horizontal spokes will rotate, moving a central rod which transmits the velocity of the wind in kilometers per hour to an electrically operated dial. In absence of anemometer, a little practice of local wind observations may help in assessing the

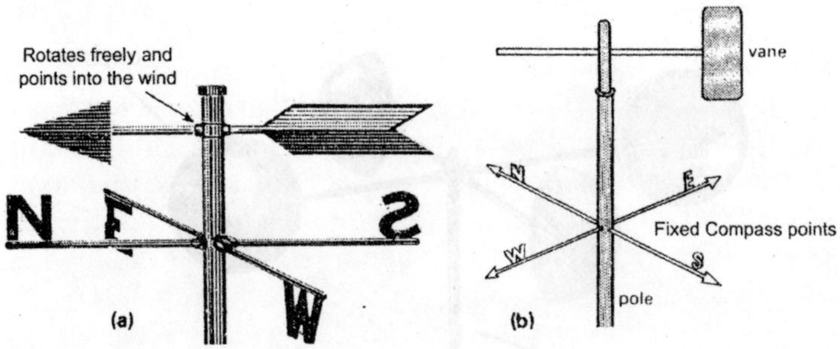

FIGURE 16.5
WIND VANES

speed of winds. The best guide is the **Beaufort wind scale** which was devised by Admiral Beaufort in 1805 for estimating wind speed (Table 16.1).

TABLE 16.1
THE BEAUFORT WIND SCALE

Beaufort scale no.	*Wind description*	*Speed (km/h)*	*Effects (a guide to observation)*
0	Calm	< 1.5	Smoke rises vertically
1	Light air	1.65 - 5	Wind direction shown by smoke-drift but not by wind-vanes
2	Slight breeze	6 - 11	Wind felt on face; leaves rustle; vanes moved by wind
3	Gentle breeze	1 - 19	Leaves and twigs in constant motion; winds extend light flags
4	Moderate breeze	20 - 29	Raises dust and loose paper; small branches moved
5	Fresh breeze	30 - 39	Small trees begins to sway; crested wavelets form on inland water
6	Strong breeze	40 - 50	Large branches in motion
7	Moderate gale	51 - 61	Whole tree in motion; walking inconvenienced
8	Fresh gale	62 - 74	Twigs broken off trees
9	Strong gale	75 - 87	Slight structural damage occurs
10	Whole gale	88 - 100	Considerable structural damage; trees uprooted
11	Storm	101- 120	Widespread damage
12	Hurricane	>120	Widespread devastation (experienced in tropical areas)

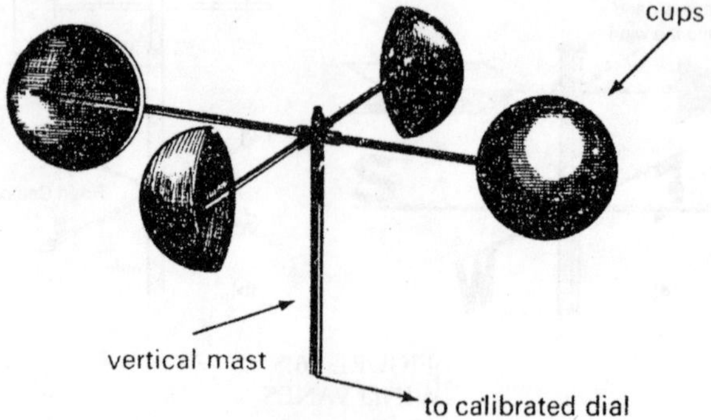

FIGURE 16.6
SIMPLIFIED SKETCH TO ILLUSTRATE THE MAIN FEATURES OF A WIND ANEMOMETER

16.1.6 Sunshine

The sun is the major source of light and heat on the earth. The earth spins on its axis of rotation every 24 h (Fig. 16.7), so that most of the world has 24 h cycles of day and night. Figure 16.7 also shows why the differences between equator and poles occur. As the radiant energy (solar energy) passes through the atmosphere, about 20-40% of it is absorbed by the gases, dust and water vapour in the atmosphere. The energy which reaches the outer atmosphere near the poles (represented by beam B on the diagram) has more atmosphere to travel through before it reaches the earth's surface than does the energy reaching the outer atmosphere near the equator (beam A). More energy is therefore lost in the atmosphere at higher latitudes. Also, because of the curve of the earth's surface, the energy which has passed through the atmosphere at the poles falls obliquely on the earth's surface. It is spread over a larger surface area (D on the diagram) than a beam of the same size shining on the earth near the equator (C). This means that per unit time, the light and heat reaching the earth at the equator is higher than at the poles. Over a year the equator gets twice as much energy as either pole (Woodward 1987).

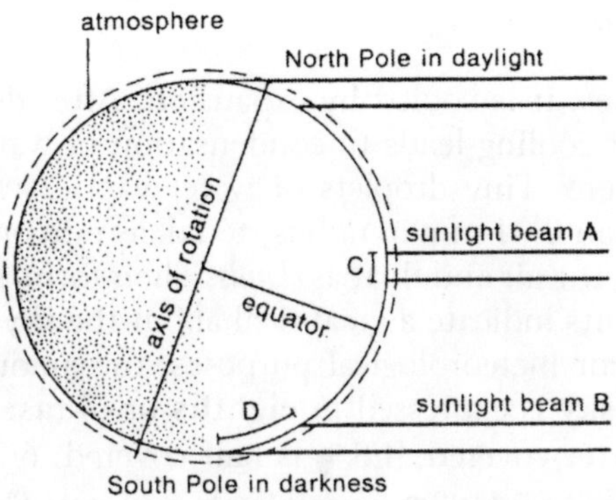

Figure 16.7 A diagram of the earth showing how energy from the sun is distributed differently at the poles and the equator. The energy falling *via* beam A on to a square meter of surface C is greater per unit time than that falling on to the same area of D *via* beam B because of the difference in distance travelled through the atmosphere and the angle of the beam to the surface at the two sites.

Variation in light and temperature with latitude is, however, much more complicated than this indicates. As already mentioned, the earth spins on its axis of rotation every 24 h. This axis, as shown in Fig. 16.7, is tilted at an angle to the sun. This means that, as the earth orbits the sun during each year, the part of the earth directly facing the sun changes. The sun thus appears to be directly overhead at the equator at the two equinoxes, but over the Tropic of Cancer at the northern hemisphere's summer solstice and over the Tropic of Capricorn at the winter solstice. This affects the relative lengths of day and night in any 24 h period. At the equator, day and night are of almost equal length all years, and seasonality is minimal. At high latitudes, inside the Arctic and Antarctic circles, seasonality is extreme. During the summer the sun never sets and total daylight irradiance over 24 h is actually higher than it is in the tropics (where there are only 12 h light). However, during the winter months the sun hardly rises above the horizon and total irradiance is very low.

In the meteorological station, sunshine duration is recorded by a **sun-dial,** 10 cm in diameter, through which the sunrays are focused upon a sensitized card, graduated in hours. A line is made up on the card when it is sufficiently heated, but not when the rays are faint. On the maps places with equal sunshine duration are joined by **isohels**.

16.1.7 Clouds

When air rises, it is cooled by expansion. After dew point has been reached cooling leads to condensation of water vapour in the atmosphere. Tiny droplets of water vapour which are too small (less than 0.01 mm in radius) to fall as rain or snow will be suspended in the air and float as *clouds*. Their form, shape, height and movements indicate a great deal about the sky and weather conditions. For meteorological purposes, the amount of cloud-cover in the sky is expressed in **eighths** or **oktas**; for example, 2/8◔ is quarter covered, 4/8◑ is half covered, 6/8◕ is three - quarters covered, and 8/8● is completely overcast. They are shown on weather maps by discs, shaded in the correct proportions. On maps places with an equal degree of cloudiness are joined by lines known as **isonephs**.

The *classification of clouds* is based on a combination of form, height, and appearance. Four major cloud types and their variations can be recognized.

16.1.7.1 High clouds (heights 6 to 12 km above the ground)

The main types of high clouds are cirrus, cirrocumulus and cirrostratus. The *cirrus* (Fig. 8a) looks fibrous and appears like wisps in the blue sky; it is often called "mares' tails". It indicates fair weather and often gives a brilliant sunset. The *cirrocumulus* (Fig. 8b) appears as white globular masses, forming ripples in a "mackerel sky". The *cirrostratus* (Fig. 8c) resembles a thin white sheet or veil; the sky looks milky and the sun or moon shines through it with a characteristic 'halo'.

16.1.7.2 Medium clouds (height 2 to 6 km above the ground)

The main types are altocumulus and altostratus. The *altocumulus* clouds (Fig. 16.9) are wooly, bumpy clouds arranged in layers and appearing like waves in the blue sky. They normally indicate fine weather. The *altostratus* are denser, greyish clouds with a "watery" look. They have a fibrous or striated structure through which the sunrays shine faintly.

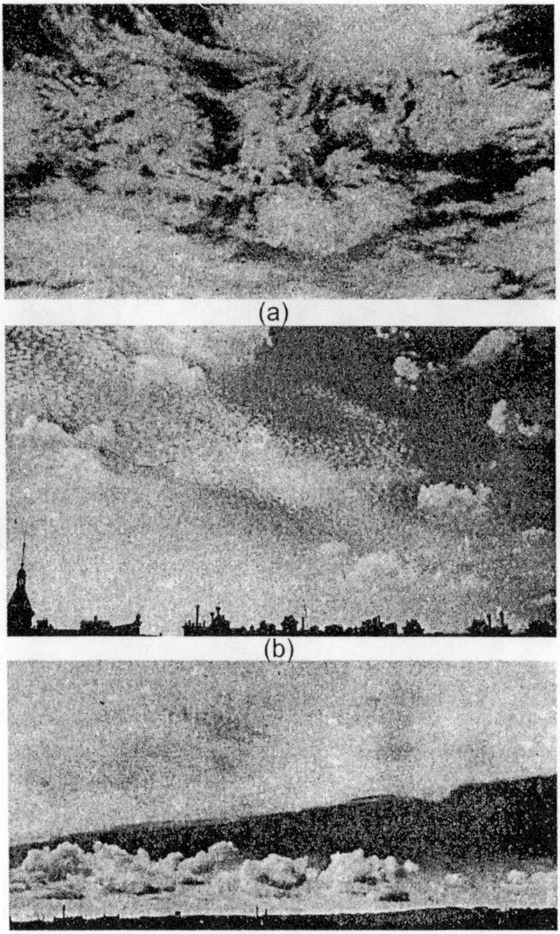

FIGURE 16.8
(A) CIRRUS CLOUD (B) CIRROCUMULUS CLOUD AND (C) CIRROSTRATUS CLOUD

16.1.7.3 Low clouds (heights below 2 km above the ground)

They are mainly stratus or sheet clouds. The main types are stratocumulus, stratus, and nimbostratus. The *stratocumulus* is a rough, bumpy cloud with the waves more pronounced than in altocumulus. There is a great contrast between the bright and shaded parts. The *stratus* is a very low cloud, uniformly grey and thick, which appears like a low ceiling or highland fog. It reduces the visibility of aircrafts, and brings dull weather with light drizzle. The *nimbostratus* is a dark, dull cloud, clearly layered and is also known as a *rain-cloud*. It brings continuous rain, snow or sleet.

FIGURE 16.9
ALTOCUMULUS CLOUD

16.1.7.4 Clouds with great vertical extent

These are mainly cumulus or heap clouds with no definite height (0.5 to 9 km above the ground). The main types are cumulus and cumulonimbus. The *cumulus* (Fig. 16.10a) is a vertical cloud with a rounded top and horizontal base, typical of humid tropical regions, associated with uprising convectional currents. Its great white globular masses look grey against the sun but it is a 'fair weather cloud'. The *cumulonimbus* (Fig. 16.10b) is, in fact, an overgrown cumulus cloud, extending from a base of 0.5 km to over 9 km. Its black and white globular masses take a fantastic range of shapes. Its cauliflower top often spreads out like anvil. This is frequently seen in tropical afternoons. It is also referred to as a 'thunder-cloud' and brings convectional rain, accompanied by lightning and thunder.

Besides clouds, the other elements affecting visibility include haze, mist and fog.

FIGURE 16.10
(A) CUMULUS CLOUD AND (B) CUMULONIMBUS CLOUD

16.2 Pressure And Planetary Winds

Along the equator and within 5 degrees north and south, is the *equatorial low pressure belt,* where there is intense heating, with expanding air and ascending convection currents. This equatorial belt is often termed the **doldrums**, because sailors in the olden days often found themselves becalmed here. It is a zone of *wind convergence.*

About 30°N and S latitudes occur the *sub-tropical high pressure belts* where the air is comparatively dry and the winds are calm and light. It is a region of descending air currents or *wind divergence* and *anticyclones*. It is frequently referred to as the **horse latitudes.**

Around the latitudes 60°N and S are two *temperate low pressure belts* which are also *zones of convergence* with cyclonic activity. The sub-polar low pressure areas are best developed over the oceans, where temperature differences between summer and winters are negligible.

At the North and South Poles 90°N and S where temperatures are permanently low, are the *polar high pressure belts*. Unlike the water masses of the high latitudes in the southern hemisphere, high pressures of the corresponding latitudes in the northern hemisphere are a little complicated by the presence of much land. Some pressure differences between summer and winter can be expected.

16.2.1 The planetary winds

Within the mentioned pattern of permanent pressure belts on the globe, winds tend to blow from the high pressure belts to the low as the planetary winds. Instead of blowing directly from one pressure belt to another, however, the effect of the rotation of earth **(Coriolis force)** tends to deflect the direction of the winds. In the northern hemisphere, winds are deflected to their right, and in the southern hemisphere to their left as shown in the Fig. 16.11. This is known as **Ferrel's law of deflection.** The Coriolis force is absent along the equator but increases progressively towards the poles. For this reason, winds blowing out from the sub-tropical high pressure belt in the northern hemisphere towards the equatorial low become *north-east trade winds* and those in the southern hemisphere become the *south-east trade winds.* These **trade winds** are the most regular of all the planetary winds. They blow with great force and in a constant direction. They were thus helpful to early traders who depended on the wind when sailing the high seas; hence the name ' trade winds '. Since they blow from the cooler sub-tropical latitudes to the warmer tropics, they have great capacity for holding moisture. In their passage across the open oceans, they gather more moisture and bring heavy rainfall to the east coasts of continents within the tropics. As they are off-shore on the west coast, these regions suffer from great aridity and form the trade wind hot deserts of the world, such as the Sahara, Kalahari, Atacama and the Great Australian Deserts.

From the sub-tropical high pressure belts, winds blow towards the temperate low pressure belts as the variable **westerlies.** Under the effect of the Coriolis force, they become the *south-westerlies* in the northern hemisphere and the *north-westerlies* in the southern hemisphere. They are more variable in the northern hemisphere, but they play a valuable role in carrying warm equatorial waters and winds to western coasts of temperate lands. This *warming effect* and other local pressure differences have resulted in a very variable climate in the temperate zones, dominated by the movements of cyclones and anticyclones. In the southern hemisphere where there is a large expanse of ocean, from 40°S to 60°S, westerlies blow with much greater force and regularity throughout the year. They bring much precipitation to the western coasts of continents. The weather is damp and cloudy and the seas are violent and stormy. It is thus usual for seafarers to refer to the westerlies as the *roaring forties*, *furious fifties* and *shrieking* or *stormy sixties,* according to the varying degree of storminess in the latitudes in which they blow.

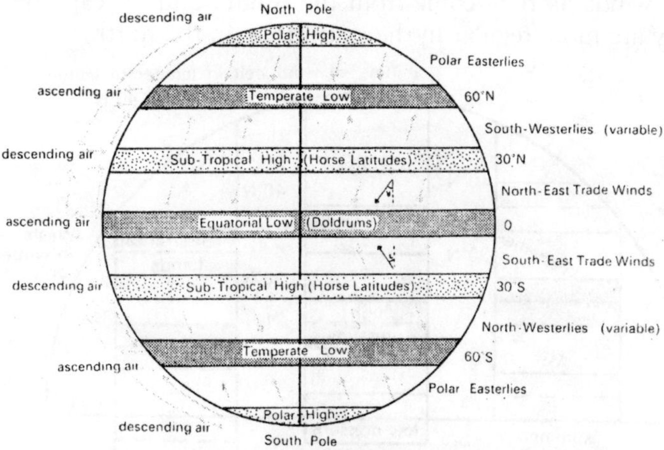

FIGURE 16.11
THE DISTRIBUTION OF WORLD PRESSURE BELTS AND PLANETARY WINDS

Not all the western coasts of the temperate zone receive westerlies throughout the year, however. Some of them like California, Iberia, central Chile, southern Africa and south-western Australia receive westerlies only in winter. This is caused by the *shifting of the wind belts* of such regions which lie approximately between the latitudes 30° and 40°N and S. Due to the earth's inclination, the sun is overhead at midday in different parts of the earth at different seasons. The entire system of pressure and wind belts follows the movement of the midday sun. In the June when the overhead sun is over the tropic of Cancer, all the belts move about 5-10° north of their average position. The 'mediterranean' parts of the southern continents then come under the influence of westerlies and receive rain in June (winter in the southern hemisphere). In the same manner, when the sun is overhead at the tropic of Capricorn in December, all the belts swing 5-10° south of their average position. The 'mediterranean' parts of Europe and California then come under the influence of the westerlies and receive rain in December (winter in the northern hemisphere). This is illustrated in Fig. 16.12.

The *polar easterlies* blow out from the polar high pressure belts towards the temperate low pressure belts. These are extremely cold winds as they come from the tundra and ice-cap regions. They are more regular in the south than in the north.

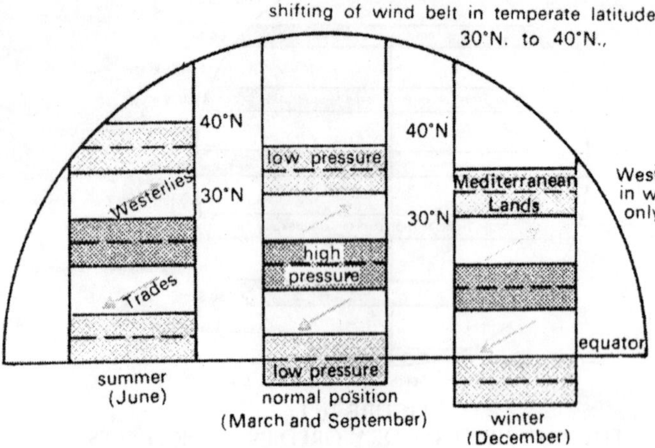

FIGURE 16.12
THE SHIFTING OF THE PRESSURE AND WIND BELTS IN THE NORTHERN HEMISPHERE- SHOWING THEIR POSITIONS IN SUMMER AND WINTER AND AT THE EQUINOXES.

16.2.2 Fohn wind and Chinook wind

Both the Fohn and Chinook winds are *dry winds* experienced on the leeward side of the mountains when descending air becomes compressed with increased pressure. The Fohn wind is experienced in the valleys of the northern Alps, particularly in Switzerland in spring. Chinook winds are experienced on the eastern slopes of the Rockies in USA and Canada in winter.

The air ascending the southern slopes of the Alps expands and cools. Condensation takes place when the air is saturated. Rain and even snow fall on the higher slopes. In descending the northern slope, the wind experiences an increase in pressure and temperature. The air is compressed and warmed. Most of its moisture is lost and wind reaches the valley bottom as a dry, hot wind-the Fohn. It may raise the temperature by 5 to 15°C within an hour. It melts snow and causes avalanches. In North America, it is called Chinook, meaning the 'snow-eater'. But it has its blessings too, it hastens the growth of crops and fruits and thaws the snow-covered pastures. In the Rockies, the Chinook has been known to raise temperature by 20°C within 15 minutes. The occurrence of frequent Chinooks means winter is mild.

16.2.3 Cyclonic activity

Tropical cyclones, typhoons, hurricanes and tornadoes are well-developed low pressure systems into which violent winds blow. In general, typhoons occur in the China Sea, tropical cyclones in the Indian Ocean, hurricanes in the West Indian islands in the Caribbean, and tornadoes in the Guinea lands of West Africa and the southern USA.

Typhoons occur mainly in regions between 6° and 20° north and south of the equator and are most frequent from July to October. In extent, they are smaller than temperate cyclones and have a diameter of only 80 to 320 km, but they have a much steeper pressure gradient. Violent winds with a velocity of over 150 km/h are common. The sky is overcast and torrential downpour is accompanied by thunder and lightning. **Hurricanes** have calm, rainless centers where the pressure is the lowest (about 960 mb) but around this 'eye', the wind strength exceeds 120 km/h. Dense dark clouds gather and violent stormy weather lasts for several hours. **Tornadoes** are small but very violent tropical and subtropical cyclones in which the air is spiraling at a tremendous speed of as much as 800 km/h. A tornado appears as a *dark funnel cloud* 75 to 425 m in diameter. As a tornado passes through a region, it writhes and twists, causing complete devastation within the limits of its passage. There is such a great difference in pressure that houses virtually explode. Tornadoes are most frequent in spring but can occur at almost any time. Fortunately they are not common in many countries and their destructive effects are confined to a small area. Tornadoes are most typical of the USA and occur mainly in the Mississippi basin.

16.2.4 Cyclones and anticyclones

Cyclones are better known as **depressions** and are confined to *temperate latitudes*. The lowest pressure is in the centre and the isobars, as shown in climatic charts, are close together. Depressions vary from 250 to 3000 km in extent. They remain quite stationary or move several hundred kilometers in a day. The approach of a cyclone is characterized by a fall in barometric reading, dull sky, oppressive air and strong winds. Rain or snow falls and weather is generally bad. Winds blow *inwards* into regions of low pressure in the centre, circulating in anticlockwise direction in the northern hemisphere and clockwise in the southern hemisphere (Fig. 16.13a). Precipitation resulting from cyclonic activities is due to the convergence of warm tropical air and cold polar air. *Fronts* are developed and condensation takes place, forming either rain, snow or sleet.

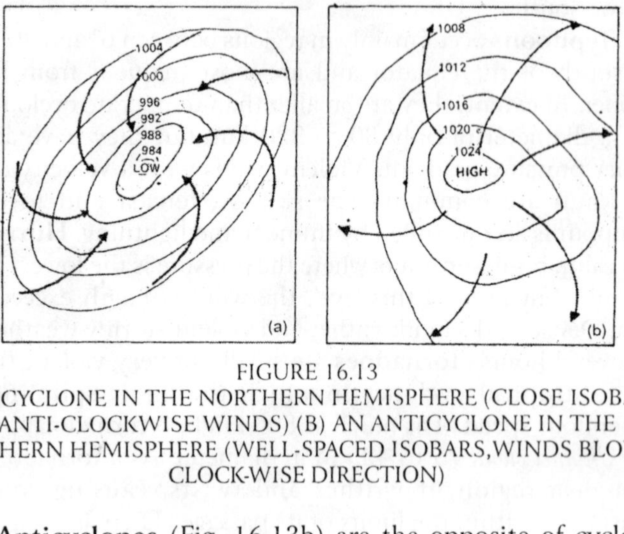

FIGURE 16.13
(A) A CYCLONE IN THE NORTHERN HEMISPHERE (CLOSE ISOBARS, ANTI-CLOCKWISE WINDS) (B) AN ANTICYCLONE IN THE NORTHERN HEMISPHERE (WELL-SPACED ISOBARS,WINDS BLOW IN CLOCK-WISE DIRECTION)

Anticyclones (Fig. 16.13b) are the opposite of cyclones, with *high pressure* in the centre and the isobars far apart. The pressure gradient is gentle and winds are light. Anticyclones normally herald fine weather. Skies are clear, the air is calm and temperatures are high in summer but cold in winter. In winter intense cooling of the lower atmosphere may result in thick fogs. Anticyclonic conditions may last for days or weeks and then fade out quietly. Winds in anticyclones blow outwards and are also subject to deflection, but they blow clockwise in the northern hemisphere and anticlockwise in the southern hemisphere.

16.2.5 Land and sea breezes and monsoons

Land and sea breezes are, in fact, monsoons on a smaller scale. Both are basically caused by *differential heating* of land and sea, the former in a *diurnal* rhythm and the latter in a *seasonal* rhythm. During the day, the land gets heated up much faster than the sea. Warm air rises forming a region of local *low pressure*. The sea remains comparatively cool with a higher pressure so a **sea breeze** blows in from the sea to the land. Its speed and strength is between 8 to 30 km/ h and it is generally stronger in tropical than temperate regions. Its influence does not normally exceed 25 km from the coast. It is most deeply felt when one stands facing the sea in a coastal resort. At night the reverse takes place. As the land cools down much faster than the sea, the cold heavy air produces a region of local *high pressure*. The sea conserves its heat and remains quite warm. Its pressure is comparatively low. A **land breeze** thus blows out from land to sea. Fishermen in the tropics often take advantage of the outgoing land breeze and sail out with it. They return the next morning with the incoming sea breeze, complete with their catch. Land and sea breezes are illustrated in Fig. 16.14. In the same way, **monsoons** are caused. Rapid heating in the hot summer over most parts of India, for example, induces heated air to rise. The south-west monsoon from the surrounding ocean is attracted by the low pressure over the land and blows in bringing torrential rain to the subcontinent. Similarly, in winter when the land is cold, the surrounding seas remain comparatively warm. High pressure is created over Indo-Pakistan and the north-east monsoon blows out from the continent into the Indian Ocean and the Bay of Bengal.

16.3 Circulation Of The Ocean Currents

16.3.1 Circulation of the Atlantic Ocean

The steady trade winds constantly drift two streams of water from east to west. At the 'shoulder' of north-east Brazil, the protruding land mass splits the *south equatorial current* into the *Cayenne current* which flows along the Guiana coast, and the *Brazilian current* which flows southwards along the east coast of Brazil (Fig. 16.15). In the **North Atlantic Ocean,** the Cayenne

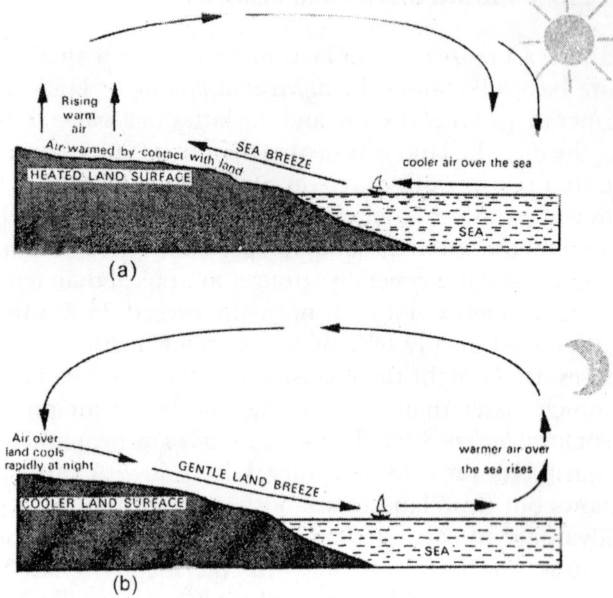

FIGURE 16.14
(A) SEA BREEZE (DAY) AND (B) LAND BREEZE (NIGHT)

current is joined and reinforced by the *north equatorial current* and heads north-westwards as a large mass of equatorial water into the Caribbean Sea. Part of the current enters the Gulf of Mexico and emerges from the Florida Strait between Florida and Cuba as the *Florida current*. The rest of the equatorial water flows northwards east of the Antilles to join the Gulf Stream off the south-eastern USA. The *Gulf Stream Drift* is one of the strongest ocean currents, 55 to 160 km wide, 600 m deep and with a velocity of 5 km/h. The current hugs the coast of America as far as Cape Hatteras (latitude 35°N), where it is deflected eastwards under the combined influence of the *westerlies* and the *rotation* of the earth. It reaches Europe as the *North Atlantic Drift*. The current, flowing at 15 km per day, carries the warm equatorial water for over 1500 km to the coasts of Europe. From the North Atlantic, it fans out in three directions, eastwards to the Britain, northwards to the Arctic and southwards along the Iberian coast, as the cool

Canaries current. Almost two-thirds of the water brought by the Gulf Stream to the Arctic regions is returned annually to the tropical latitudes by dense, cold polar water that creeps southwards in the ocean depths. The Canaries current flowing southwards eventually merges with the north equatorial current, completing the clockwise circuit in the North Atlantic Ocean. Within this ring of currents, an area in the middle of the Atlantic has *no perceptible current.* A large amount of floating sea-weed gathers and the area is called the **Sargasso Sea**.

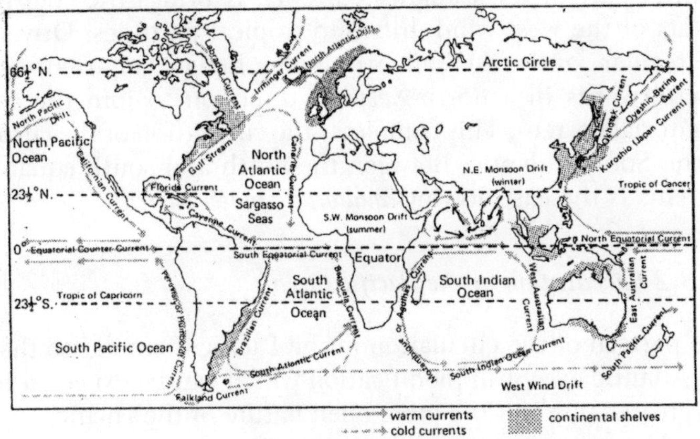

FIGURE 16.15
OCEAN CURRENTS OF THE WORLD

Apart from the clockwise circulation of the currents, there are also currents that enter the North Atlantic from the Arctic regions. These cold waters are blown south by the out-flowing polar winds. The *Irminger current* or *East Greenland current* flows between Iceland and Greenland and cools the North Atlantic drift at the point of convergence. The cold **Labrador current** drifts south-eastwards between west Greenland and Baffin Island to meet the warm Gulf Stream off Newfoundland, as far south as 50°N, where the icebergs carried south by the Labrador Current melt.

The **South Atlantic Ocean** follows same pattern of circulation as the North Atlantic Ocean. The major differences are that the circuit is anti-clockwise and the collection of sea-weed in the still waters of the mid-South Atlantic is not so distinctive. Where the south equatorial current is split at Cape Sao Roque, one branch turns south as the warm **Brazilian Current**. Its deep blue waters are easily distinguishable from the yellow, muddy waters carried hundreds of kilometers out to the sea by the Amazon further north. At about 40°S the influence of the prevailing westerlies and the rotation of the earth propel the current eastwards to merge with the cold *west wind drift* as the *South Atlantic current.* On reaching the west coast of Africa the current is diverted northwards as the cold *Benguela current* (the counter part of the Canaries current). It brings the cold polar waters of the west wind drift into tropical latitudes. Driven by the regular south-east trade winds, the Benguela current surges equatorwards in a north-westerly direction to join the south equatorial current. This completes the circulation of the currents in the South Atlantic. Between the north and south equatorial currents is the east-flowing *equatorial counter current.*

16.3.2 Circulation of the Pacific Ocean

The pattern of the circulation in the Pacific is similar to that of the Atlantic except in modification which can be expected from the greater size and the more open nature of the Pacific.

The north equatorial current flows westwards with a compensating equatorial counter current running in the opposite direction. Due to the greater expanse of the Pacific and the absence of an obstructing land mass the volume of water is very much greater than that of the Atlantic equatorial current. The north-east trade winds blow the north equatorial current off the coasts of Philippines and Formosa into the East China Sea as the **Kuroshio** or *Japan current*. Its warm waters are carried polewards as the **north Pacific drift**, keeping the Alaskan coast ice-free in winter. The cold *Bering current* or *Alaskan current* creeps southwards from the narrow Bering Strait and is joined by the *Okhotsk current* to meet the warm Japan current as the *Oyashio,* off Hokkaido. The cold water eventually sinks beneath the warmer water of the north Pacific drift. Part of it drifts eastwards as the cool **Californian current** along the coasts of the western USA and coalesces with the north equatorial current, to complete the clockwise circulation.

The current system of the **South Pacific** is the same as that of the South Atlantic. The south equatorial current, driven by the south-east trade winds, flows southwards along the coast of Queensland as the **east Australian current**, bringing warm equatorial waters into temperate waters. The current turns eastwards towards New Zealand under the pull of force of the westerlies in the Tasman Sea and merges with part of the cold west wind drift as the *South Pacific current.* Obstructed by the tip of southern Chile, the current turns northwards along the western coast of South America as the cold **Humboldt** or **Peruvian current**. The cold water chills any wind that blows on-shore so that the Chilean and Peruvian coasts are practically rainless. The region is rich in marine phytoplanktons and zooplanktons that attract huge shoals of fish. Consequently, millions of sea-birds gather here to feed on the fish. Their droppings completely whiten the coastal cliffs and islands, forming thick deposits of **guano**, a valuable source of phosphatic fertilizer. The Peruvian current eventually links up with the south equatorial current and completes the cycle of currents in the South Pacific.

16.3.3 Circulation of the Indian Ocean

As in the other oceans as illustrated in Fig. 16.15, the currents of the **South Indian Ocean** form a circuit. The equatorial current, turning southwards past. Madagascar as the **Agulhas** or **Mozambique** current merges with the west wind drift, flowing eastwards and turns equatorwards as the **west Australian current**. In the **North Indian Ocean**, there is a complete reversal of the direction of currents between summer and winter, due to the changes of monsoon winds. In summer from June to October, when the dominant wind is the **South-west monsoon**, the currents are blown from a south-westerly direction as the *south-west monsoon drift.* This is reversed in winter, beginning from December, when the **north-east monsoon** blows the currents from the north-east as the *north-east monsoon drift.* The currents of the North Indian Ocean, demonstrate most convincingly the dominate effects of winds on the circulation of ocean currents.

16.3.4 El Nino and La Nina

Every few years in the Pacific Ocean, normal coastal *upwelling* of cold, nutrient-rich bottom water due to constant trade winds which push surface water away from the land, is affected by changes in climate patterns called the **El Nino-Southern Oscillation (ENSO)**, often called **El Nino.** In an ENSO, the prevailing westerly winds weaken or cease (for reasons that are not clear) and surface water along the south and north American coasts becomes warmer. The normal upwelling of cold nutrient-rich water is suppressed, which reduces primary productivity and causes a sharp decline in the populations of some fish species. A strong ENSO can trigger extreme weather changes over at least two thirds of the globe, especially in lands along the Pacific and Indian Oceans. In 1997 and 1998, the world experienced the strongest ENSO during the past 200 years. It came only 14 years after the second strongest ENSO in1982-1983 and only 2 years after a moderate ENSO between 1992 and 1997. Some models project that if the earth's atmosphere continues to warm, El Nino-like weather may become the norm and cause major ecological and socioeconomic problems (Galtie 1999).

Sometimes an El Nino is followed by its cooling counterpart, **La Nina**, as occurred from July 1998 and 1999. Typically a La Nina means: (i) more Atlantic Ocean hurricanes, (ii) colder winters in Canada and the Northeast, (iii) warmer and drier winters in the southeastern and southwestern United States, (iv) wetter winters in the Pacific Northwest and (v) torrential rains in Southeast Asia. Typically La Nina is worse than El Nino for the united states because: (i) there are more Atlantic coast hurricanes, and (ii) there is an increase in tornadoes spread out over a wider area. According to some climate experts, there are 20 to 30 year climate cycles during which there are more La Ninas and fewer El Ninos.

In brief, one of the things that connect all life on the earth is wind, a vital part of the planet's circulatory system. Masses of air that are warm or cold, wet or dry, and contain air at high or low pressure are constantly moving across land and sea. Weather changes as one air mass replaces or meets another. Cold air is more dense (weighs more per volume) than warm air. Thus, cold air tends to sink through warmer, less dense air. The most dramatic changes in weather occur along a **front**, the boundary between two air masses with different temperatures and densities. A **warm front** is the boundary between an advancing warm air mass and the cooler one it is replacing. Because warm air is less dense than cool air, an advancing warm front, will rise up over a mass of cool air. As the warm front rises, its moisture starts to condense into droplets to form layers of clouds at different altitudes. High, wispy clouds are the first signs of an advancing warm front. Gradually the clouds thicken, descend to a lower altitude, and often release their moisture as rainfall. A moist warm front can bring days of cloudy skies and drizzle. A **cold front** is the leading edge of an advancing mass of cold air. Because cold air is denser than warm air, an advancing cold front stays close to the ground and wedges underneath less dense warmer air. An approaching cold front produces rapidly moving, towering clouds called **thunderheads**. As the overlaying mass of warm air is pushed upward it cools and its water vapour condenses to form large and heavy droplets that fall to the earth's surface as precipitation. As a cold front passes through, we often experience high surface winds and thunderstorms. After the front passes through, we usually experience cooler temperatures and a clear sky. Weather is also affected by changes in atmospheric pressure. An air mass with high pressure called a **high**, contains cool, dense air that descends towards the earth's surface and becomes warmer. Fair weather follows as long as the high-pressure air mass remains over an area. In contrast, a low- pressure air mass, called a **low,** produces cloudy and sometimes stormy weather. This happens because less dense warm air spirals inward towards the centre of the low-pressure air mass. Because of its low pressure and low density, the centre of the **low** rises and its warm air expands and cools. When the temperature drops below the dew point, the moisture in the air condenses and forms cloud. If the droplets in the clouds coalesce into large and heavy drops, then precipitation follows. In addition to normal weather, we sometimes experience weather extremes such as violent storms called: (i) *tornadoes* (which form over land) and (ii) *tropical cyclones* (which form over warm ocean waters and sometimes pass over coastal land). Tropical cyclones are called *hurricanes* in the Atlantic and *typhoons* in the Pacific.

16.4 Bioclimate and Climograph

Climate, the long-term (35 years) interaction of precipitation and temperature, has profound effects not only on the distribution and range of species but, in consequence, on ecosystem. The **bioclimate** is defined as the climate in relation to life, more particular, plant life because most of the animals live in special microclimatic conditions and are subject to migrations. When climatic data are not available for a place, its vegetation type could throw light on the nature of climate. Conversely, when the natural vegetation of a region is destroyed and modified by man as a result of urbanization, cultivation, plantations, introduction of exotic species and artificial landscape creation, the climate of the station could indicate the potential vegetation type with its physiognomy and composition. For example, the permanently humid regions where dry periods are non-existent or of very short duration are those which have tropical rainforest. The periodically humid parts characterized by alternation of wet and dry periods possess semi-deciduous and deciduous forests; the duration and the intensity of the dry season are mild so that changes brought about by its occurrence are limited to deciduousness and greater prominence of grass. In the dry climates, the intensity and duration of drought period brings about a significant degree of xeromorphy in the vegetation; with the result thorn-forests or succulents are prevalent in the landscape.

One of the effective ways of representing the relationship of the temperature and rainfall is the **climograph**, in which temperature is plotted against precipitation. In a climograph, the factor of prime importance is the *regime of rainfall,* that is, the season of occurrence of rains. Four types of regimes are recognized, in general: (i) *irregular* (rainfall without any seasonal rhythm, including trixeric and quadrixeric regime with three or four dry periods in the year), (ii) *Mediterranean* (rains during short days of the year, that is, winter, spring and autumn associated with mean temperature $<15°C$); (iii) *bixeric* (two rainy periods and two dry seasons in the year); and (iv) *tropical* (rains during long days of the year, that is, summer). The next factor is the degree of *aridity-humidity*. Based on the classification of Bagnouls and Gaussen (1957), four types are recognized: (i) *humid*: 0 to 4 months of the year dry; (ii) *sub-humid:* 5 to 8 months of the year dry; (iii) semi-*arid:* 9 to 11 months of the year dry; and (iv) arid: 12 months of the year dry. A month is defined as **dry** when its precipitation (in mm) is less than twice the mean temperature in degree Celsius ($P<2t$). The *rainy season* and the *growing season* from the agronomical and ecological viewpoints are defined as the period in which monthly precipitations in the millimeters are equal to or greater than twice the mean temperature in degree Celsius: P>2t (Bagnouls and Gaussen 1957). The seasons recognized in the Indian subcontinent are the *winter* (the comparatively cooler period of the year; November to February), the *summer* (comparatively hotter period of the year; May to September for the northern stations, but from March to September for the southern stations of the Peninsula). If the mean temperature of March remains below 26°C and that of April below 29°C, these months are considered as the *spring*. The corresponding figures for the mean of the minima are 18°C for March, and 24°C for April. Whenever the March and April temperatures exceed the above values, they are classified as summer months. Finally, the month of October is considered as *autumn*.

16.5 Climate Classification

Climate classification schemes have been devised by many authors. For instance, the climate provinces of Koppen (1918), and the humidity provinces of Thornthwaite (1931), and the life zones of Merriam (1895) have been useful. A new type of life zone, based upon precipitation, moisture, and heat (but only the heat that affects plant growth, *biotemperature*) was developed by Holdridge (1967). Hythergraphs and the climatic diagrams were devised by Smith (1940) and Walter (1973), respectively, to present the climate graphically for a given station. Axelrod and Bailey (1969) proposed an equability system that allows prediction of vegetation type based on mean annual temperature and mean annual temperature range. All of these systems are attempts to make ecological phenomena predictable, an important and essential goal of classification.

16.5.1 Koppen's classification

Wladimir Koppen, a climatologist, developed and published a climate classification in 1918, which he subsequently revised in 1923. Koppen's classification recognizes five principal categories of climate; each category is designated by a capital letter as follows:

A. *Tropical rainy climates* (also called *humid tropical climates;* all months have a mean temperature above 18°C, that is, winterless climates)

B. *Dry climates* (in these climates evaporation exceeds precipitation; there is a constant water deficiency)

C. *Warm temperate rainy climates* (also called *humid mesothermal climates;* these climates have mild winters; the average temperature of the coldest month is below 18°C but above -3°C; the average temperature of the warmest month is over 10°C. In this group of climate both the seasons, winter and summer, are found)

D. *Cold snow forest climates* (also called *humid microthermal climates;* these climates have severe winters; the average temperature of the coldest month is below -3°C and that of the warmest month exceeds 10°C)

E. *Polar climates* (these are summer less climates; the warmest monthly mean is below 10°C.

These climate classes were then subdivided into climate provinces. Koppen's 11 climate provinces and their characteristic vegetation are shown in Table 16.2. Although the system is widely used, it is severely limited in application because it has a narrow base (a single climatic parameter, that of temperature).

16.5.2 Thornthwaite's classification

American climatologist C. Warren Thornthwaite advanced a scheme for North American climates (1931) and later (1933) extended it to the world. Like Koppen's classification, Thornthwaite's is based on natural vegetation but, significantly, takes into account the quantity of moisture lost through evaporation from the soil, and through evapotranspiration from vegetation (Table 16.3).

For the water balance computation of a station, data of mean monthly temperature, monthly precipitation, and latitude are to be known (Table 16.4, Fig. 16.16). The methodology is as follows:

TABLE 16.2
TYPES OF CLIMATES AS CLASSIFIED BY KOPPEN

Main type	*Climate province*	*Symbol*	*Temperature*	*Rainfall*	*Vegetation*
A. Tropical rain	1. Rain forest	Af	Coldest month >18°C Annual variation <3°C	Heavy rain min 6 cm/ month	Dense forest; heavy under-growth
	2. Savanna	Aw	Coldest month >18°C Annual variation <12°C	Dry season with <6 cm/ month	Open grasslands; scattered trees
B. Dry	1. Steppe	BS	Variable winters cold	Rain<50cm/year	Short grass
	2. Desert	BW	High summer temperature	Very dry	Sparse or none
C. Warm, humid (temperate)	1. Mediterranean	Cf	Coldest month between 18°C and -3°C Warmest month >10°C	Ample rain throughout the year	Cotton, wheat and corn
	2. Winter dry	Cw	Coldest month between 18°C and -3°C	Wet summer	Forest
	3. Summer dry	Cs		Dry summer	
D. Cold, humid	1. Cold with no dry period	Df	Mean temperature of summer 10°C, of winter -3°C	Rain even throughout year	Canada
	2. Dry winter	Dw			N.E. Asia only
E. Polar	1. Tundra	ET	Mean temperature of warmest month <10°C	Frozen soil	Moss and small shrubs
	2. Ice cap	EF	Mean temperature of warmest month <0°C		

TABLE 16.3
HUMIDITY PROVINCES OF THORNTHWAITE

Humidity province	*Characteristic vegetation*	*Value of the "moisture index (Im)"*
A. Per humid (wet)	Rain forest	Equal or more than 100
B. Humid	Forest	20 to 99
C. Sub humid	Grassland	0 to 19 (moist subhumid) -19 to -1 (dry subhumid)
D. Semiarid	Steppe	-20 to -39
E. Arid	Desert	Equal or less than -40

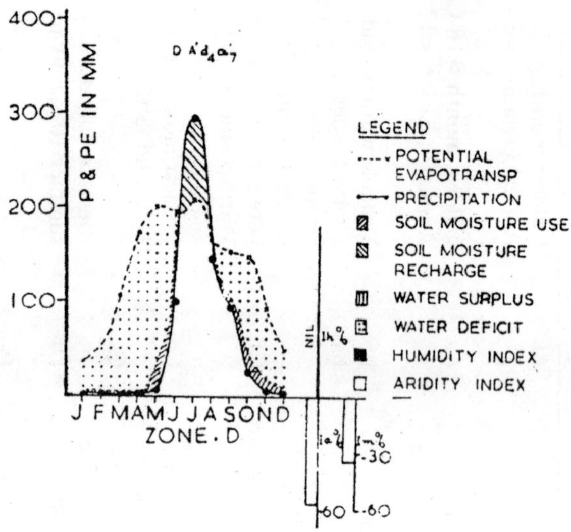

FIGURE 16.16
WATER BALANCE

Heat index (i): Table 16.5 gives the monthly values of "i", corresponding to the mean monthly temperatures ("i" is zero when mean monthly temperature is zero or less than zero). The values of "i" for 12 months of the year are summed up to obtain the value of the *annual heat index.*

TABLE 16.4
COMPUTATION OF WATER BALANCE AS PER THORNTHWAITE'S METHOD (STATION: RAJKOT; LAT N 22°18'; LONG E 70°47'; ALTITUDE 138 MSL)

	Jan.	*Feb.*	*Mar.*	*Apr.*	*May*	*June*	*Jul.*	*Aug.*	*Sep.*	*Oct.*	*Nov.*	*Dec.*	*Annual*
T (°C)	19.4	21.9	26.2	30.0	32.6	32.0	28.7	27.8	27.9	28.1	24.8	26.9	
i	7.79	9.36	12.28	15.07	17.09	16.62	14.09	13.43	13.50	13.68	11.30	12.78	156.89
UPE (mm)	36	54	120	162	175	173	153	146	146	148	84	138	
PE (mm)	35	48	423	171	199	194	177	163	149	146	77	128	1610
P(mm)	1	1	1	3	7	99	293	143	93	25	5	4	675
Δ = P–PE (mm)	–34	–47	–122	–168	–192	–95	+116	–20	–56	–121	–72	–123	
Σ Δ (mm)	–426	–473	–595	–763	–955	–1051	+116	–20	–76	–197	–269	–392	
St (mm)	72	61	41	23	12	9	125	280	232	155	121	80	
ΔSt (mm)	+8	–11	–20	–18	–11	–3	+96	+20	–48	–77	–34	–41	
AE (mm)	9	12	21	21	28	102	177	163	141	102	39	45	850
D(mm)	26	36	102	150	181	93	0	0	8	44	38	82	760
S(mm)	0	0	0	0	0	0	20	0	0	0	0	0	20

Legends: T = Temperature, i = Monthly heat index, UPE = Unadjusted potential evapotranspiration, PE = Potential evapotranspiration, P = Precipitation, Δ = Accumulated potential water loss, St = Storage, Δ St = Changes in soil moisture storage, AE = Actual evapotranspiration, D = Soil moisture deficit, S = Soilmoisture surplus.

TABLE 16.5
MONTHLY VALUE OF HEAT INDEX ("I") CORRESPONDING TO MONTHLY TEMPERATURE °C

T °C	.0	1	.2	.3	.4	.5	.6	.7	.8	.9
0		0.01	0.01	0.01	0.02	0.03	0.04	0.05	0.06	0.07
1	0.09	0.10	0.12	0.13	0.15	0.16	0.18	0.20	0.21	0.23
2	0.25	0.27	0.29	0.31	0.33	0.35	0.37	0.39	0.42	0.44
3	0.46	0.48	0.51	0.53	0.56	0.58	0.61	0.63	0.66	0.69
4	0.71	0.74	0.77	0.80	0.82	0.85	0.88	0.91	0.94	0.97
5	1.00	1.03	1.06	1.09	1.12	1.16	1.19	1.22	1.25	1.29
6	1.32	1.35	1.39	1.42	1.45	1.49	1.52	1.56	1.59	1.63
7	1.66	1.70	1.74	1.77	1.81	1.85	1.89	1.92	1.96	2.00
8	2.04	2.08	2.12	2.15	2.19	2.23	2.27	2.31	2.35	2.39
9	2.44	2.48	2.52	2.56	2.60	2.64	2.69	2.73	2.77	2.81
10	2.86	2.90	2.94	2.99	3.03	3.08	3.12	3.16	3.21	3.25
11	3.30	3.34	3.39	3.44	3.48	3.53	3.58	3.62	3.67	3.72
12	3.76	3.81	3.86	3.91	3.96	4.00	4.05	4.10	4.15	4.20
13	4.25	4.30	4.35	4.40	4.45	4.50	4.55	4.60	4.65	4.70
14	4.75	4.81	4.86	4.91	4.96	5.01	5.07	5.12	5.17	5.22
15	5.28	5.33	5.38	5.44	5.49	5.55	5.60	5.65	5.71	5.76
16	5.82	5.87	5.93	5.98	6.04	6.10	6.15	6.21	6.26	6.32
17	6.38	6.44	6.49	6.55	6.61	6.66	6.73	6.78	6.84	6.90
18	6.95	7.01	7.07	7.13	7.19	7.25	7.31	7.37	7.43	7.49
19	7.51	7.61	7.67	7.73	7.79	7.85	7.91	7.97	8.03	8.10
20	8.16	8.22	8.28	8.34	8.41	8.47	8.53	8.59	8.66	8.72
21	8.78	8.85	8.91	8.97	9.04	9.10	9.17	9.23	9.29	9.36
22	9.42	9.49	9.55	9.62	9.68	9.75	9.82	9.88	9.95	10.01
23	10.08	10.15	10.21	10.28	10.35	10.41	10.48	10.58	10.62	10.68
24	10.75	10.82	10.89	10.95	11.02	11.09	11.16	11.23	11.30	11.37
25	11.44	11.50	11.57	11.64	11.71	11.78	11.85	11.92	11.99	12.06
26	12.13	12.21	12.28	12.35	12.42	12.49	12.56	12.63	12.70	12.78
27	12.85	12.92	12.99	13.07	13.14	13.21	13.28	13.36	13.43	13.50
28	13.58	13.68	13.72	13.80	13.87	13.94	14.02	14.09	14.17	14.24
29	14.32	14.39	14.47	14.54	14.62	14.69	14.77	14.84	14.92	14.99
30	15.07	15.15	15.22	15.30	15.38	15.45	15.53	15.61	15.68	15.76
31	15.84	15.92	15.99	16.07	16.15	16.23	16.30	16.38	16.46	16.54
32	16.62	16.70	16.78	16.85	16.93	17.01	17.09	17.17	17.25	17.35
33	17.41	17.49	17.57	17.65	17.73	17.81	17.89	17.97	18.05	18.12
34	18.22	18.30	18.38	18.46	18.54	18.62	18.70	18.79	18.87	18.95
35	19.03	19.11	19.20	19.28	19.36	19.45	19.56	19.61	19.69	19.70
36	19.80	19.95	20.03	20.11	20.20	20.28	20.31	20.45	20.53	20.62
37	20.70	20.79	20.87	20.96	21.04	21.13	21.21	21.30	21.38	21.49
38	21.56	21.60	21.73	21.81	21.90	21.99	22.07	22.16	22.25	22.33
39	22.42	22.51	22.57	22.68	22.77	22.86	22.95	23.03	23.12	23.21
40	23.30	-	-	-	-	-	-	-	-	-

Unadjusted potential evapotranspiration (UPE): For

temperatures above 26°C, Table 16.6 is used to get the corresponding values of UPE. For temperatures below 26°C, a regression line between the line of convergence through "i" value is drawn. The point where the mean monthly temperature on the 'Y-axis' intercepts the regression line gives the corresponding value of UPE on the 'X-axis'.

TABLE 16.6

TABLE FOR DETERMINING UPE AT TEMPERATURES ABOVE 26°C

T °C	*.0*	*.1*	*.2*	*.3*	*.4*	*.5*	*.6*	*.7*	*.8*	*.9*
26		127.0	129.0	131.0	133.0	135.0	135.9	136.8	137.7	138.6
27	139.6	140.3	141.1	141.4	142.7	143.7	144.4	145.3	146.1	146.9
28	147.8	148.6	149.4	150.2	151.0	151.7	152.4	153.1	153.8	154.5
29	155.4	156.1	156.8	157.5	158.2	158.9	159.6	160.1	160.7	161.3
30	162.1	162.7	163.3	163.9	164.5	165.2	165.8	166.4	167.6	167.6
31	168.0	168.5	169.0	169.5	170.0	170.7	171.2	171.7	172.2	172.7
32	173.5	173.7	173.9	174.3	174.7	175.3	175.7	176.1	176.5	176.9
33	177.1	177.6	178.0	178.4	178.8	179.0	179.3	179.6	179.9	180.2
34	180.5	180.8	181.0	181.4	181.7	181.8	182.0	182.2	182.4	182.6
35	182.9	183.1	183.3	183.3	183.7	183.7	183.8	183.9	184.0	184.1
36	184.3	184.4	184.5	184.6	184.7	184.7	184.7	184.8	184.8	184.9
37	184.9	184.9	184.9	185.0	185.0	185.0	185.0	185.0	185.0	185.0
38	185.7	-	-	-	-	-	-	-	-	-

Potential evapotranspiration (PE): UPE value of each month is multiplied with a correction factor given in Table 16.7 to obtain the corresponding PE value.

Precipitation minus potential evapotranspiration (P-PE= Δ): The values of PE is subtracted from those of precipitation (P) for every month. A negative value indicates the amount by which P (rainfall) fails to supply the potential water need of vegetation covered area, whereas a positive value indicates the amount of excess of water which is available during certain periods in the year.

Accumulated potential water loss (ΣΔ): The negative values of P-PE, representing a potential deficiency of water are summed up month after month, as an aid in the computational steps to follow.

Storage (St): Table 16.8 gives the values of the retention (storage) of soil water based on the values of accumulated potential water loss (ΣΔ). After the soil moisture storage for each

TABLE 16.7
TABLE OF CORRECTION FACTORS BASED ON LATITUDE AND CORRESPONDING DAY-LENGTH IN THE NORTHERN HEMISPHERE TO OBTAIN THE VALUE OF PE FROM UPE

Latitude	*Jan*	*Feb*	*Mar*	*Apr*	*May*	*Jun*	*Jul*	*Aug*	*Sep*	*Oct*	*Nov*	*Dec*
5	1.02	0.93	1.03	1.02	1.06	1.03	1.06	1.05	1.01	1.03	0.99	1.02
6	1.02	0.93	1.03	1.02	1.06	1.04	1.06	1.05	1.01	1.03	0.99	1.01
7	1.01	0.92	1.03	1.02	1.07	1.04	1.07	1.06	1.02	1.03	0.99	1.01
8	1.01	0.92	1.03	1.03	1.07	1.05	1.07	1.06	1.02	1.02	0.98	1.00
9	1.00	0.93	1.03	1.03	1.08	1.05	1.08	1.06	1.02	1.02	0.98	1.00
10	1.00	0.91	1.03	1.03	1.08	1.06	1.08	1.06	1.02	1.02	0.98	0.99
11	0.99	0.91	1.03	1.03	1.09	1.06	1.09	1.07	1.02	1.01	0.97	0.99
12	0.99	0.91	1.03	1.04	1.09	1.07	1.10	1.07	1.02	1.01	0.97	0.98
13	0.98	0.91	1.03	1.04	1.10	1.07	1.10	1.08	1.02	1.01	0.96	0.98
14	0.98	0.91	1.03	1.04	1.10	1.08	1.11	1.08	1.02	1.01	0.96	0.97
15	0.97	0.91	1.03	1.04	1.11	1.08	1.12	1.08	1.02	1.01	0.95	0.97
16	0.97	0.91	1.03	1.04	1.11	1.09	1.12	1.09	1.02	1.00	0.95	0.96
17	0.96	0.91	1.03	1.05	1.12	1.09	1.13	1.09	1.02	1.00	0.94	0.96
18	0.96	0.90	1.03	1.05	1.12	1.10	1.13	1.10	1.02	1.00	0.94	0.95
19	0.95	0.90	1.03	1.05	1.13	1.10	1.14	1.10	1.02	1.00	0.93	0.95
20	0.95	0.90	1.03	1.05	1.13	1.11	1.15	1.11	1.02	1.00	0.93	0.94
21	0.94	0.90	1.03	1.05	1.13	1.11	1.15	1.11	1.02	0.99	0.92	0.94
22	0.94	0.89	1.03	1.06	1.14	1.12	1.15	1.11	1.02	0.99	0.92	0.93
23	0.93	0.89	1.03	1.06	1.14	1.13	1.16	1.12	1.02	0.99	0.92	0.92
24	0.93	0.89	1.03	1.06	1.15	1.14	1.17	1.12	1.02	0.99	0.91	0.92
25	0.93	0.89	1.03	1.06	1.15	1.14	1.17	1.12	1.02	0.99	0.91	0.91
26	0.92	0.88	1.03	1.06	1.15	1.15	1.17	1.12	1.02	0.99	0.91	0.91

Latitude	Jan	Feb	Mar	Apr	May	Jun	Jul	Aug	Sep	Oct	Nov	Dec
27	0.92	0.88	1.03	1.07	1.16	1.15	1.18	1.13	1.02	0.99	0.90	0.90
28	0.91	0.88	1.03	1.07	1.16	1.16	1.18	1.13	1.02	0.98	0.90	0.90
29	0.91	0.87	1.03	1.07	1.17	1.16	1.19	1.13	1.03	0.98	0.90	0.89
30	0.90	0.87	1.03	1.08	1.18	1.17	1.20	1.14	1.03	0.98	0.89	0.88
31	0.89	0.87	1.03	1.08	1.18	1.18	1.20	1.14	1.03	0.98	0.89	0.88
32	0.89	0.86	1.03	1.08	1.19	1.19	1.21	1.15	1.03	0.98	0.88	0.87
33	0.88	0.86	1.03	1.09	1.19	1.20	1.22	1.15	1.03	0.97	0.88	0.86
34	0.88	0.85	1.03	1.09	1.20	1.20	1.22	1.16	1.03	0.97	0.87	0.86
35	0.87	0.85	1.03	1.09	1.21	1.21	1.23	1.16	1.03	0.97	0.86	0.85
36	0.87	0.85	1.03	1.10	1.21	1.22	1.24	1.17	1.03	0.97	0.86	0.84
37	0.86	0.84	1.03	1.10	1.22	1.23	1.25	1.17	1.03	0.97	0.85	0.83
38	0.85	0.84	1.03	1.10	1.23	1.24	1.25	1.17	1.04	0.96	0.84	0.83
39	0.85	0.84	1.03	1.11	1.23	1.24	1.26	1.18	1.04	0.96	0.84	0.82
40	0.84	0.83	1.03	1.11	1.24	1.25	1.27	1.18	1.04	0.96	0.83	0.81

month with negative values of P-PE has been found, the positive values from the line P-PE representing the additions of moisture to the soil must be included. If the accumulative adding brings the value of soil storage over 300 mm, this value is entered until the next negative value of P-PE is reached since the soil cannot hold more water as storage.

Changes in the soil moisture (Δ St): It represents the differences in the amount of the soil moisture storage from one month to the next. When the value of the storage line is 300 mm or more, it is assumed that there is no change in soil storage, although there may be a change in the above surface storage.

Actual evapotransipiration (AE): When precipitation (P) is greater than PE, the soil remains full of water and AE is equal to PE. When P falls below the PE, the soil begins to dry out and AE becomes less than that of potentially possible one. In these months, the AE equals the P plus the amount of water drawn from the soil moisture storage (AE=P+Δ St; disregarding its sign). In short, AE=PE when P is more than or equal to PE; but when P is less than PE, AE=P+ΔSt.

TABLE 16.8

TABLE OF SOIL MOISTURE STORAGE BASED ON THE VALUE OF ACCUMULATED POTENTIAL WATER LOSS (ΣΔ)

	0	*1*	*2*	*3*	*4*	*5*	*6*	*7*	*8*	*9*
0	300	299	298	297	296	295	294	293	292	291
10	290	289	288	287	286	285	284	283	282	281
20	280	279	278	278	277	276	275	274	273	272
30	271	270	269	268	268	267	266	265	264	263
40	262	261	260	260	259	258	257	255	255	254
50	254	253	252	251	250	249	248	248	247	246
60	245	244	244	243	242	241	240	240	239	238
70	237	236	236	235	234	233	232	232	231	230
80	229	228	228	227	225	225	225	224	223	222
90	222	221	220	219	218	218	217	216	215	215
100	214	214	213	212	212	211	210	209	209	208
110	207	207	206	205	204	204	203	202	201	201
120	200	200	199	198	198	197	196	196	195	194
130	194	193	192	192	191	191	190	189	189	188
140	187	187	186	186	185	184	184	183	182	182
150	181	181	180	179	179	178	178	177	176	176
160	175	175	174	173	173	172	172	171	171	170

	0	1	2	3	4	5	6	7	8	9
170	170	169	168	168	167	167	166	166	165	164
180	164	163	163	162	162	161	160	160	159	159
190	158	158	157	156	156	156	155	155	154	154
200	153	153	152	152	151	151	150	150	149	149
210	148	148	148	147	146	146	145	145	144	144
220	143	143	143	142	141	141	140	140	139	139
230	138	138	138	137	136	136	136	135	135	134
240	134	133	133	132	132	132	131	131	130	130
250	130	129	129	128	128	127	127	126	126	126
260	125	125	124	124	124	123	123	122	122	121
270	121	121	120	120	119	119	119	118	118	117
280	117	117	116	116	115	115	115	114	114	114
290	113	113	112	112	112	111	111	110	110	110
300	109	109	109	108	108	108	107	107	106	106
310	106	105	105	104	104	103	103	103	103	103
320	102	102	102	101	101	101	100	100	100	99
330	99	98	98	98	98	97	97	97	96	96
340	96	95	95	95	95	94	93	93	93	93
350	92	92	92	92	91	91	91	90	90	90
360	89	89	89	88	88	88	88	87	87	87
370	86	86	86	86	85	85	85	84	84	84
380	84	83	83	83	82	82	82	82	81	81
390	81	80	80	80	80	80	79	79	79	79
400	78	78	78	77	77	77	77	76	76	76
410	76	75	75	75	74	74	74	74	74	73
420	73	73	72	72	72	72	72	71	71	71
430	71	71	70	70	70	70	70	69	69	68
440	68	68	68	68	67	67	67	67	66	66
450	66	66	66	65	65	65	65	64	64	64
460	64	64	63	63	63	63	63	62	62	62
470	62	62	61	61	61	61	61	60	60	60
480	60	60	59	59	59	59	59	58	58	58
490	58	58	57	57	57	57	57	56	56	56
500	56	56	55	55	55	55	55	54	54	54
510	54	54	54	53	53	53	53	53	52	52
520	52	52	52	52	51	51	51	51	51	50
530	50	50	50	50	50	50	49	49	49	49
540	49	49	48	48	48	48	48	48	47	47
550	47	47	47	47	46	46	46	46	46	46
560	46	45	45	45	45	45	45	44	44	44
570	44	44	44	44	44	43	43	43	43	43
580	43	42	42	42	42	42	42	42	42	41

	0	*1*	*2*	*3*	*4*	*5*	*6*	*7*	*8*	*9*
590	41	41	41	41	41	41	40	40	40	40
600	40	40	40	39	39	39	39	39	39	39
610	38	38	38	38	38	38	38	38	38	37
620	37	37	37	37	37	37	36	36	36	36
630	36	36	36	36	36	36	35	35	35	35
640	35	35	35	34	34	34	34	34	34	34
650	34	34	33	33	33	33	33	33	33	33
660	32	32	32	32	32	32	32	32	32	32
670	32	31	31	31	31	31	31	31	31	31
680	30	30	30	30	30	30	30	30	30	30
690	30	29	29	29	29	29	29	29	29	29
700	28	28	28	28	28	28	28	28	28	28
710	28	27	27	27	27	27	27	27	27	27
720	27	26	26	26	26	26	26	26	26	26
730	26	26	26	26	25	25	25	25	25	25
740	25	25	25	25	25	24	24	24	24	24
750	24	24	24	24	24	24	24	24	23	23
760	23	23	23	23	23	23	23	23	23	23
770	22	22	22	22	22	22	22	22	22	22
780	22	22	22	22	22	21	21	21	21	21
790	21	21	21	21	21	21	21	21	20	20
800	20	20	20	20	20	20	20	20	20	20
810	20	20	20	20	20	19	19	19	19	19
820	19	19	19	19	19	19	19	19	19	18
830	18	18	18	18	18	18	18	18	18	18
840	18	18	18	18	18	18	18	17	17	17
850	17	17	17	17	17	17	17	17	17	17
860	17	17	17	17	16	16	16	16	16	16
870	16	16	16	16	16	16	16	16	16	16
880	16	16	16	16	15	15	15	15	15	15
890	15	15	15	15	15	15	15	15	15	15
900	15	15	14	14	14	14	14	14	14	14
910	14	14	14	14	14	14	14	14	14	14
920	14	14	14	14	13	13	13	13	13	13
930	13	13	13	13	13	13	13	13	13	13
940	13	13	13	13	12	12	12	12	12	12
950	12	12	12	12	12	12	12	12	12	12
960	12	12	12	12	12	12	12	12	12	12
970	12	12	11	11	11	11	11	11	11	11
980	11	11	11	11	11	11	1	11	11	11
990	11	11	11	11	11	11	11	10	10	10
1000	10	10	10	10	10	10	10	10	10	10

1010	10	10	10	10	10	10	10	10	10	10
1020	10	10	10	10	10	10	10	10	10	10
1030	9	9	9	9	9	9	9	9	9	9
1040	9	9	9	9	9	9	9	9	9	9
1050	9	9	9	9	9	9	9	9	8	8
1060	8	8	8	8	8	8	8	8	8	8
1070	8	8	8	8	8	8	8	8	8	8
1080	8	8	8	8	8	8	8	8	8	8
1090	8	8	8	8	8	8	8	8	8	8
1100	7	7	7	7	7	7	7	7	7	7
1120	7	7	7	7	7	7	7	7	7	7
1130	7	7	7	7	7	7	7	7	7	7
1140	6	6	6	6	6	6	6	6	6	6
1190	6	6	6	6	6	6	6	6	6	6
1200	5	5	5	5	5	5	5	5	5	5
1250	4	4	4	4	4	4	4	4	4	4
1350	3	3	3	3	3	3	3	3	3	3
1410	3	3	3	3	3	3	3	3	3	3
1420	2	2	2	2	2	2	2	2	2	2
1490	2	2	2	2	2	2	2	2	2	2
1550	2	2	2	2	2	2	2	2	2	2
1600	1	1	1	1	1	1	1	1	1	1
1700	1	1	1	1	1	1	1	1	1	1
1850	1	1	1	1	1	1	1	1	1	1
1900	0	0	0	0	0	0	0	0	0	0

Moisture deficit (D): The amount by which AE is falling short of PE represents D. In other words, D=PE- AE.

Moisture surplus (S): It is the difference between Δ and Δ St, that is, S= Δ - Δ St.

Moisture index (Im): It is calculated as:

$$I_m = \frac{100S - 60D}{PE} \times 100$$

where S is the annual water surplus, D is annual water (moisture) deficit and PE is the potential evapotransipiration (or water need).

Summer Concentration of Thermal Efficiency (SCTE): It is calculated as:

$$SCTE = \frac{\text{Total of maximum PE for three successive months}}{\text{Total PE for the year}} \times 100$$

In Thornthwaite's classification, the climate of a region is represented by four letters, two standing for the *moisture regime* and the other two for the *thermal regime*. The capital letters in two sets indicate the main moisture or thermal type while the small letters indicate the seasonal variations of moisture effectivity or thermal efficiency (Table 16.9).

TABLE 16.9
CLIMATE FORMULA FOR EACH ADMINISTRATIVE DISTRICT OF NEPAL AS PER THORNTHWAITE'S SCHEME (JHA AND KARN 2001)

District	*PE* (*mm*)	*S* (*mm*)	*D* (*mm*)	*Im* (*%*)	*SCTE* (*%*)	*Climate formula*
1. Accham	1218	356	310	+13.9	35.2	$C_2sA'_1a'_5$
2. Arghakhanchee	1256	0	323	-15.4	33.7	$C_1d\ A'_1a'_5$
3. Baglung	1367	1089	200	+70.9	33.7	$B_3s\ A'_2a'_6$
4. Baitadi	849	455	73	+48.4	41.!	$B_2rB'_2a'_2$
5. Bajhang	817	1164	46	+139.0	42.7	$ArB'_1a'_2$
6. Bajura	816	112	50	+10.0	42.8	$C_2rB'_1a'_2$
7. Banke	1547	121	404	-7.8	34.8	$C_1s\ A'_3a'_5$
8. Bara	1649	658	542	+20.2	33.9	$B_1S_2A'_4a'_6$
9. Bardiya	1611	258	552	-4.5	32.3	$C_1W_2A'_4a'_6$
10. Bhaktapur	1134	286	272	+10.8	36.7	$C_2s\ B'_4a'_4$
11. Bhojpur	1115	250	135	+15.2	33.6	$C_2r\ B'_4a'_6$
12. Chitwan	1554	379	420	+8.2	34.2	$C_2s\ A'_3a'_5$
13. Dadeldhura	1221	293	270	+10.7	35.4	$C_2s\ A'_1a'_5$
14. Dailekh	1217	726	201	+49.7	35.3	$B_2s\ A'_1a'_5$
15. Dang	1643	211	508	-5.7	33.5	$C_1dA'_4a'_6$
16. Darchula	849	483	55	+53.0	41.1	$B_2rB'_2a'_3$
17. Dhading	1658	846	382	+37.2	33.2	$B_1sA'_4a'_6$
18. Dhankuta	1108	52	381	-15.9	35.5	$C_1d\ B'_4a'_5$
19. Dhanusha	1662	376	580	+1.7	33.4	$C_2S_2A'_4a'_6$
20. Dolkha	922	1324	22	+142.2	38.5	$ArB'_3a'_4$
21. Dolpa	768	0	0	0	40.6	$C_2rB'_2a'_3$
22. Doti	1217	0	194	-9.6	35.2	$C_1d\ A'_1a'_5$
23. Gorkha	1282	436	225	+23.5	33.6	$B_1w\ A'_1a'_6$
24. Gulmee	1256	505	302	+25.8	33.7	$B_1sA'_1a'_6$
25. Humla	662	204	126	+19.4	40.9	$C_2wB'_1a'_3$
26. Ilam	1529	88	305	-6.2	31.4	$C_1dA'_3a'_6$
27. Jajarkot	1217	742	241	+49.1	35.3	$B_2S_2A'_1a'_5$
28. Jhapa	1473	1675	314	+100.9	33.4	$AsA'_3a'_6$
29. Jumla	811	0	207	-15.3	40.7	$C_1dB'_1a'_3$
30. Kabhrepalanchok	1120	335	155	+21.6	35.1	$B_1rB'_4a'_5$
31. Kailali	1611	472	506	+10.5	32.3	$C_2wA'_4a'_6$
32. Kalikot	662	346	165	+37.3	40.9	$B_1rB'_1a'_3$
33. Kanchanpur	1611	275	465	-0.2	32.3	$C_1wA'_4a'_6$
34. Kapilvastu	1622	400	543	+4.6	34.0	$C_2S_2A'_4a'_5$
35. Kaski	1367	2701	191	+189.2	33.9	$AwA'_2a'_6$
36. Kathmandu	1188	439	173	+28.2	35.5	$B_1rA'_1a'_5$
37. Khotang	1081	404	175	+27.7	32.9	$B_1sB'_4a'_6$

District	*PE* (*mm*)	*S* (*mm*)	*D* (*mm*)	*Im* (*%*)	*SCTE* (*%*)	*Climate formula*
38. Lalitpur	1134	249	147	+14.2	36.5	$C_2rA'_1a'_4$
39. Lamjung	1343	1717	121	+122.0	33.6	$ArA'_2a'_6$
40. Mahotari	1662	376	580	+1.7	33.3	$C_2S_2A'_4a'_6$
41. Makwanpur	1554	379	420	+8.2	34.2	$C2sA'3a'_5$
42. Manang	713	0	120	-10.1	41.7	$C_1dB'_2a'_2$
43. Morang	1473	292	453	+1.4	33.4	$C_2sA'_3a'_6$
44. Mugu	662	204	126	+19.4	40.9	$C_2wB'_1a'_3$
45. Mustang	774	0	512	-39.7	41.5	$DdB'_2a'_3$
46. Myagdee	1344	497	389	+19.6	33.7	$C_2sA'_2a'_6$
47. Nawalparasi	1644	1242	411	+60.5	33.5	$B_3 sA'_4a'_6$
48. Nuwakot	1385	434	388	+14.5	33.3	$C_2sA'_2a'_6$
49. Okhadhunga	1083	944	86	+82.4	34.4	$B_4rB'_4a'_5$
50. Paanchthar	1109	1299	90	+112.3	35.4	$ArB'_4a'_5$
51. Palpa	1256	906	350	+55.4	33.7	$B_2sA'_1a'_6$
52. Parsa	1649	658	542	+20.2	33.9	$B_1S_2A'_4a'_6$
53. Parwat	1346	879	275	+53.1	33.6	$B_2wA'_2a'_6$
54. Pyuthan	1256	260	305	+6.4	33.6	$C_2sA'_1a'_6$
55. Ramechhap	1083	944	86	+82.9	34.4	$B_4rB'_4a'_5$
56. Rasuwa	1123	74	202	-4.2	35.4	$C_1dB'_4a'_5$
57. Rauthat	1650	478	524	+9.9	34.0	$C_2S_2A'_4a'_5$
58. Rolpa	1275	176	346	-2.5	34.8	$C_1dA'_1a'_5$
59. Rukum	1255	441	254	+22.9	33.8	$B_1S_2A'_1a'_6$
60. Rupendehi	1622	400	543	+4.6	34.0	$C_2S_2A'_4a'_5$
61. Salyan	1256	198	347	-0.8	33.7	$C_1dA'_1a'_6$
62. Sankhuwasabha	1301	158	318	-2.5	32.8	$C_1dA'_2a'_6$
63. Saptari	1473	797	357	+39.6	33.4	$B_1S_2A'_3a'_6$
64 Sarlahi	1662	526	487	+14.1	33.3	$C_2S_2A'_4a'_6$
65. Sinduli	1083	2317	108	+207.9	34.4	$ArB'_4a'_5$
66. Sindhupalchok	1120	2425	39	+214.4	35.5	$ArB'_4a'_5$
67. Sirha	1473	128	443	-9.4	33.4	$C_1S_2A'_3a'_6$
68. Solukhumbhu	541	258	56	+41.5	34.6	$B_2rC'_1a'_5$
69. Sunsari	1473	805	341	+40.8	33.4	$B_2S_2A'_3a'_6$
70. Surkhet	1393	434	215	+21.9	36.0	$B_1wA'_2a'_5$
71. Syangja	1346	473	377	+18.3	33.6	$C_2S_2A'_2a'_6$
72. Tanahun	1282	1283	201	+90.7	33.6	$B_4wA'_1a'_6$
73. Taplejung	1256	821	182	+56.7	33.3	$B_2wA'_1a'_6$
74. Tehrathum	1252	445	225	+24.8	33.1	$B_1wA'_1a'_6$
75. Udaypur	1083	607	121	+49.3	34.4	$B_2rB'_4a'_5$

Legends: PE= potential evapotranspiration, S= soil moisture surplus, D= Soil moisture deficit, Im= moisture index, SCTE=summer concentration of thermal efficiency, A=perhumid (wet), B_4 to B_1=humid, C_2= moist subhumid, C_1=dry subhumid, D=semi-arid, C'_1 to C'_2=microthermal, B'_1 to B'_4=mesothermal, A'_1 to A'_8=megathermal

The climate type categorization on the basis of the moisture regime is done as follows:

Moisture index (Im)	Climate type
100 and above	A (Perhumid)
99 to 80	B_4 (Humid)
79 to 60	B_3 (Humid)
59 to 40	B_2 (Humid)
39 to20	B_1 (Humid)
19 to 0	C_2 (moist sub - humid)
- 1 to -19	C_1 (Dry sub - humid)
-20 to -39	D (semi - arid)
Equal or less than - 40	E (Arid)

Seasonal variation of effective moisture (SVEM) within each climate type is formulated as per the following scheme:

Moist climate *(A, B_4-B_1, C_2)*	*Symbol*	*Dry climate* *(C_1,D,E)*	*Symbol*
Little or no water deficiency throughout the year	r	little or no water surplus throughout the year	d
Moderate summer water deficiency	s	Moderate winter water surplus	s
Moderate winter water deficiency	w	Moderate summer water surplus	w
Large summer water deficiency	S_2	Large winter water surplus	S_2
Large winter water deficiency	W_2	Large summer water surplus	W_2

The climate division based on the thermal efficiency (PE value) is represented by the following notations:

PE (mm)	*Climate type*
427	C'_1 Microthermal
570	C'_2 Microthermal
712	B'_1 Mesothermal
855	B'_2 Mesothermal
997	B'_3 Mesothermal
1140	B'_4 Mesothermal
1282	A'_1 Megathermal
1425	A'_2 Megathermal

1567	A'_3 Megathermal
1710	A'_4 Megathermal
1852	A'_5 Megathermal
1995	A'_6 Megathermal
2148	A'_7 Megathermal
2290	A'_8 Megathermal

On the basis of SCTE value the climate sub-type of a station is recognized as follows:

SCTE	*Climate sub - type*
61.6	b'_1
56.3	b'_2
51.9	b'_3
48.0	b'_4
44.6	a'_1
41.6	a'_2
38.8	a'_3
36.3	a'_4
34.0	a'_5
31.9	a'_6
29.9	a'_7
28.0	a'_8

16.5.3 Holdridge's classification

In Holdridge's system, the depiction of the world life zones (Fig. 16.17) is derived from consideration of biotemperature, latitude, altitude, precipitation, and the ratio of potential evapotransipiration/ /precipitation. **Biotemperature** is the heat that affects plant growth. It is the average for the year of all monthly temperatures which are 0 to 30°C. All temperatures lower than 0°C are considered zero degree for calculation purposes, and those above 30°C are eliminated from consideration. For example, Verkhoyansk in northeastern Siberia has monthly mean temperatures (°C) from January to December as follows: -50.0,-44.0,-32.2, -15.5, 0, +12.2, +13.3, +9.4, +1.7, -15.5, -37.2, and -47.8. Disregarding the magnitude of the negative temperatures and recording them as Os, the sum for the year is +36.6°C. This total, divided by 12, gives a mean annual biotemperature of 3.05°C. At Verkhoyansk the mean annual precipitation is 135 mm. We find the juncture of these two parameters and place Verkhoyansk in the *boreal dry scrub life zone* (Fig. 16.17). By contrast, consider a station where the mean annual biotemperature is 8°C and mean annual precipitation is 350 mm. This falls into a steppe hexagon, the cool, temperate latitudinal region and the montane altitudinal belt. A more precise determination is given by knowing elevation. At 200 m the station is found in a basal life zone and is correctly called *cool temperate Steppe* (Fig. 16.18). However, what of a station with the same parameters but at 1200 m elevation? To calculate the basal belt correctly, we use a lapse rate of 6°C /1000 m: 8°C + (1200 m/ 1000 m) (6°C) = 15.2°C. This means that 8°C at 1200 m is in the same basal belt as 15.2°C at sea level; according to Fig. 16.18, the basal belt for 15.2°C at sea level is warm temperate. The station at 1200 m then falls correctly into the warm temperate montane steppe life zone.

In Holdridge's system, potential evapotransipiration for a given station can be obtained by multiplying the annual biotemperature of the station (in degrees C) by 58.93. Within each altitude zone, life zones are determined by the ratio of potential evapotransipiration to precipitation. If this ratio is less than 0.5, wet forest and wet tundra occur; from 0.5 to 1.0, there is moist tundra; and at ratio above 1.0, the life zones include dry forest, steppe, scrub and dry tundra.

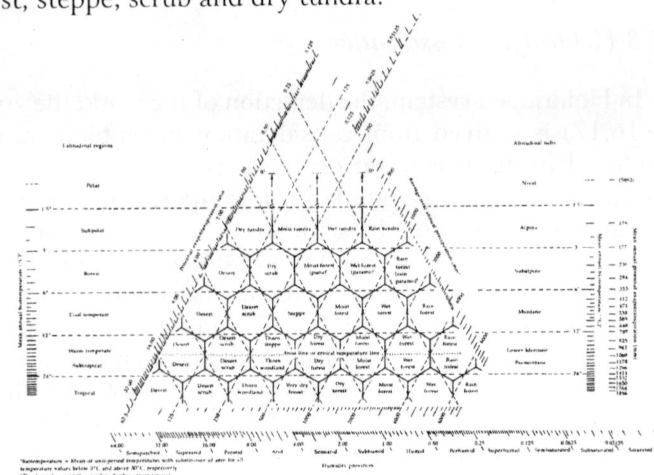

FIGURE 16.17
DIAGRAM FOR THE CLASSIFICATION OF WORLD LIFE ZONES OR PLANT FORMATIONS (HOLDRIDGE 1967)

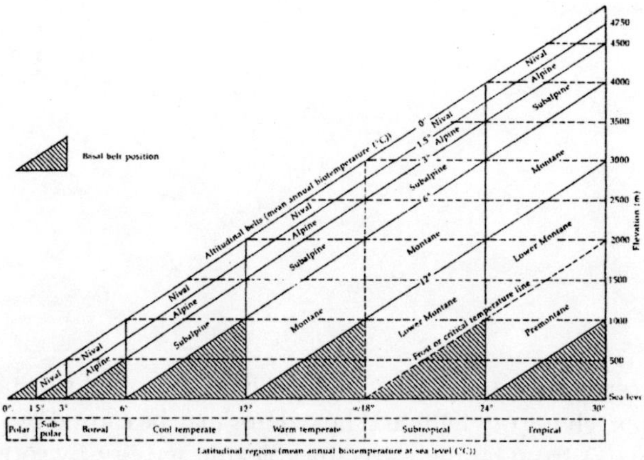

FIGURE 16.18
APPROXIMATE GUIDELINE POSITIONS OF LATITUDINAL REGIONS AND ALTITUDINAL BELTS OF HOLDRIDGE'S LIFE ZONE SYSTEM (BASED ON AN AVERAGE LAPSE RATE OF 6°C/1000 M) (HOLDRIDGE 1967)

17. FIRE

The term "fire" denotes the state of combustion of substance with oxygen, giving out light and heat. Fires may be caused by lightning, by sparks from falling rocks, by volcanic activity, by spontaneous combustion, and by human activity. Of these, only the first and last may develop any periodicity and thus act as a directional and consistent evolutionary force in the community. Climates that consist of hot, dry summers and low humidity are prime situations for the development of thunderstorms with lightning but little or no rain. The vast grasslands of Africa and the USA and the *chaparral* (vegetational areas characterized by shrubs with leaves that are evergreen, hard, thick, leathery, and small) of southwestern USA and the Mediterranean region are naturally occurring situations conducive to lightning-induced fires. Moreover, in the *Old World* (Palaeotropics or Eastern Hemisphere tropics including Tropical Africa, and Tropical Asia comprising of India, Pakistan, Myanmar, Thailand, Indonesia, etc.) and, to a lesser extent, the *New World* (Neotropics or Western Hemisphere tropics including Mexico and a major part of South America), periodic burning by humans has played a part in the evolution of the biota.

The ability of primitive societies to use fire was crucial to their development and success. Every human culture of the recent past is known to have had experience with fire; a site near Nice, France, has yielded evidence that hominids used fire even half a million years ago (Stewart 1963). Although the first sources of fire were assumed to be the results of natural effects (lightning or volcanism), the methods originally used for deliberate firemaking are not clearly known. It is likely, though, that some means of generating heat by friction was first used.

17.1 Types of Fire

The various types of fires are divided into three main groups, based on location and intensity: surface, crown, and ground fires. Fire behaviour and fire intensity are affected by meteorological conditions, by fuel loading, and by fuel moisture and soil moisture.

Surface fires are the least destructive fires, whether in a grassland or in a forest. They are relatively cool, fast moving fires. These fires do not build up high temperatures at the plant and ground levels because they are usually fed by lightweight fuels that are quickly converted to ash. Consequently, the basal portions, root stocks, and tubers are not harmed in grasslands and shrublands. In general, surface fires exert a selective effect on vegetation; they are more limiting to some species than to others and thus favour the development of species with high tolerance to the fire factor. Nitrogen-fixing legumes often thrive after a light burn; and, in regions especially subject to fire, regular light surface fires greatly reduce the danger of severe crown fires by keeping the combustible litter to a minimum.

Crown fires occur in the upper parts of trees and so, by definition, take place only in forest situations. They may result from lightning storms or from intense surface fires by heavy accumulation of litter and debris. This type of fire never occurs in the absence of surface fire. Crown fires are typically very intense, consuming the entire plant community above the ground and often its animal component as well. Because of the severity of destruction, such areas present a barren landscape and typically take years to begin even modest recovery. In the interim the area is subject to erosion following even mild levels of precipitation, thus removing the nutrients required for life and regeneration (Kaufman *et al.* 1993).

Ground fires are termed "retrogressive agents" by Vogl (1974). Although of infrequent occurrence, ground fires can be very destructive, not only of roots, tubers and rhizomes, but of the organic matter in the soil itself. Thus recovery of the plant community may take tens to thousands of years following such a catastrophe. Light, dry soils of peat bogs and grasslands may provide layers of dry organic matter - peat, fern rhizomes, and roots of grasses and other plants - that carry the fire easily. Further, ground fires tend to persist. They are shielded from environmental influences, especially human efforts at extinction, and will continue as long as the fuel remains and there is sufficient energy available to maintain a fire. Thus, when conditions are favourable, a ground fire can serve as a re-ignition source for surface fire (Vogl 1974).

17.2 Fire Effects On Soil

In general, fire has the following effects on soil: soil temperatures are higher, some soil nutrients and organic matter are lost, water holding capacity and soil wettability are changed, and populations of soil microbes are altered.

The effect that fire has on the soil and on soil organisms depends on the intensity of the fire. This intensity, in turn, depends on the heat yield, the availability of fuel, and the rate of spread of the fire. Wakimoto (1977) has symbolized this relationship as:

$$I = H \times W \times R$$

where, I = intensity (kcal/sec/m), H = heat yield (kcal/g), W = available fuel (g/m^2), and R = rate of speed (m/sec). The chemical constitution and various physical characteristics of the fuel influence its heat yield. For example, the pine needle litter has a high surface to volume ratio and is not compacted, so an abundance of oxygen is available. This type of fuel carries a fire readily, but will produce a relatively cool fire. Rates of production and decomposition of litter and fire frequency are the main determiners of fuel availability in the natural environment. The rate of spread of the fire depends on wind speed, nature of the fuel, humidity, and temperature.

A very intense burn will destroy all the organic matter at the soil surface. Some destructive distillations will be affected at levels within the soil column. At temperatures of 200-300°C, 85 % of the organic substances will be destroyed. Nitrogen and potassium loss is a function of temperature. Nitrogen and potassium compounds are subject to volatilization, and will be released and lost by distillation at temperatures of 100-200°C. Although there is little loss of nitrogen up to 200°C, as much as 60 % may be lost at 700°C. Other nutrients such as calcium, sodium, and magnesium will be released and deposited on the soil even at moderate temperatures, and rapidly recycled into the biota (DeBano *et al.* 1977). One consequence of the loss of organic matter by burning is the concomitant loss of the high cation exchange capacity that characterizes organic matter, thus removing a holding substrate for vital nutrients.

Organic matter decreases bulk density, both because it is lighter than a corresponding volume of mineral soil, and because it gives increased aggregate stability to soil. The heat of an intense fire may break down these aggregates, leading to a loss of soil structure, and to lowered water infiltration. Further, there is reduction in the thickness of humus layer due to compaction after a fire. In general, soil pH will be higher after a fire than before mainly due to release of mineral bases in the soluble ash.

The heat energy may be transferred downward in the soil by conduction, convection, and vapour flux. The rate of the transfer is affected most by levels of soil moisture, although there are many other factors involved. Temperatures will not rise above 100°C in a given layer until all the water has been evaporated, and this evaporation requires a great deal of energy, slowing the transfer of heat downward. Even under a hot fire, temperatures do not exceed 200°C 2.5 cm down, and are below 100°C under a light burn 2.5 cm down. In other words, a light surface fire would heat only the top fraction of a cm of mineral soil to near the boiling point. Grassland fires tend to be cooler than forest fires, with soil temperatures in grassland surface fires reaching temperature of 120°C, and the air temperatures rising to 600-700°C. Air temperatures in forest fires may reach greater than 1100°C. In contrast, soil temperatures in grassland fires may rise only 10-15°C above ambient in soils only 1-5 cm down, and only for short periods of time (Daubenmire 1968).

Soil moisture levels are affected by fire, but not in a simple fashion. In certain communities, living moss and humus can take in large amounts of water, effectively preventing rain water from reaching tree roots. The removal of that layer by fire should allow percolation of more water. However, the layer also prevents evaporation, so that water holding capacity is lower and evaporation is higher on burned plots than on unburned plots (Viro 1974). Burned sites also have higher temperatures than unburned sites due to the increased absorption of sunlight by blackened soil surfaces.

The influence of fire on water runoff is temperature

dependent. Following extremely hot fires which completely volatilize hydrophobic substances and thus increase wettability, there may be greater infiltration of water into the soil. However, after less intense fires, runoff is increased, due to lowered infiltration. Increase in the sediment yield is usually observed too, and this may correlate with, for instance, the loss of soil structure, increased nonwettability, or both of these.

In certain communities, hydrophobic substances prevent the movement of water into the soil. This condition is called **nonwettability** and is characteristic of, for instance, ponderosa pine *(Pinus ponderosa)*/ incence cedar *(Calocedrus decurrens)* forest, white fir *(Abies concolar)*/giant sequoia *(Sequioadendron giganteum)* forest, and the chaparral community of California (Barbour *et al.* 1980). Hydrophobic substances which reside in decomposing plant parts accumulate on the soil surface and in the upper part of the soil column in the years between burns. Temperature gradients, established during fires enhance the translocation of these substances into the lower soil layers. The nonwettable layer is thus moved down lower in the soil column, where it decreases water infiltration (Fig. 17.1).

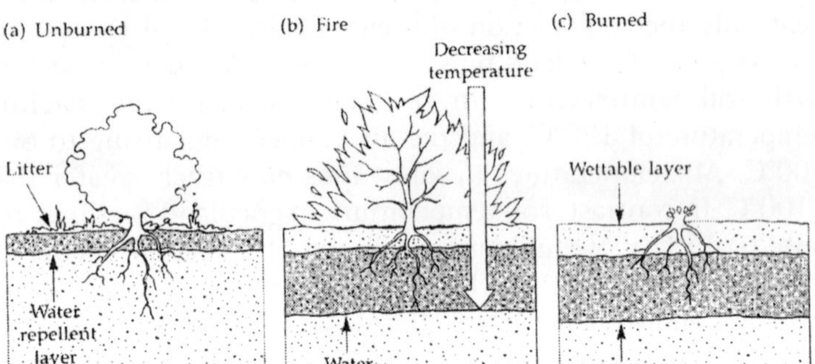

Figure 17.1 Water repellency before, during, and after fire. (a) Before fire, the hydrophobic substances accumulate in the litter and mineral soil immediately beneath it; (b) fire burns the vegetation and litter, causing the hydrophobic substances to move downward among temperature gradients; (c) after fire a water repellent layer is located below and parallel to the soil surface on the burned area (DeBano 1977).

The effect of fire on soil microbiota is usually to reduce fungi, but to increase populations of soil bacteria and actinomycetes. These changes are due not only to the heat of the fire directly, but more importantly to changes in the post-burn environment such as litter removal, changes in soil moisture content, increase in soil pH, etc. Bacterial populations are, in general, favoured over fungi in burned sites, due to the increased pH. Burning enhances the growth of nitrogen-fixing species of the bacterial genera *Azotobacter* and *Clostridium* in the coniferous forest soils. Actinomycetes population growth may be enhanced in the first few post-burn years, especially in the lower levels, because of the enrichment due to soluble minerals being moved downward from surface ash. Nitrogen losses due to volatilization of nitrates, ammonia, and amino acids (as well as through alterations in forms of both organic and inorganic nitrogen), may be recovered, both on grasslands and in the shrublands, by early dominance of the post-burn community by herbaceous legumes. Examples of these nodulated legumes are *Trifolium, Lotus, Lupinus*, etc. Nitrifying bacteria, such as *Nitrosomonas* and *Nitrobacter*, are very sensitive to fire. Their populations are completely destroyed at 140°C, and evidence suggests great mortality even at 100°C. Following a burn, the recovery of such nitrifying bacteria is slow (Dunn and DeBano 1977). As we would expect, any changes in the microbiota are most obvious in the very uppermost soil levels.

17.3 Fire Effects On Vegetation

17.3.1 Forests

Lightning fires have been a common occurrence for millennia on the coastal plain of southeastern United States, and longleaf pine forests (*Pinus taeda, P. serotina, P. clausa, P. elliotii,* and *P. palustris*) there require fire to compete with the scrub hardwoods that otherwise choke them out. Following germination, most pines grow rather uniformly. Longleaf pines reach a height of several centimeters in just a few weeks, but then upward growth ceases. The young seedling sprouts a pompon of long, slender, and drooping needles, which surround the stem apex (Fig. 17.2). This "grass stage" endures for three to seven years, and during that time the plant allocates most of its energy to the production of a sturdy root system, with abundant stored food. Further, the longleaf pines require mineral soil as seedbeds, and most of the seedlings are shade intolerant. In addition, longleaf pines are susceptible to a fungal disease called *brown spot disease*. The causal agent, brown spot fungus *(Septoria acicola)*, matures during the summer months. Sporangia release multitudes of spores to the autumn rains, and they are splashed on low-growing plants, including seedlings of various kinds. Control of the fungus is gained by burning the infected needles before new growth resumes the next spring. In a crown or severe surface fire, the dry needles are burned, as are the competing scrub hardwoods and herbaceous plants. But, the "grass stage" of longleaf pine protects the meristematic regions from heat damage, and the young seedling lives. There is a sensitive period, once the seedling resumes upward growth (terminating the grass stage), until it is about 2 m tall. The mature pines possess the heavy, protective bark necessary to prevent damage to sensitive growing tissue by fire, and, in fact, even young saplings are rather well insulated with corky bark. Fire, then, functions in several ways to promote longleaf pine forests: it clears the mineral soil for seed germination and seedling growth; it reduces competition from hardwoods and herbaceous plants; it eliminates brown spot blight; and, it opens the forest canopy to allow growth of these shade intolerant young trees.

FIGURE 17.2
THE GRASS STAGE OF LONGLEAF PINE. THE LONG, SLENDER NEEDLES PROTECT THE PERENNATING BUD AGAINST FIRE

17.3.2 Grasslands

Both grassland climates and grassland vegetation favour fire. These arid and semiarid regions are usually subject to prolonged drought, and the periods of lowest moisture availability correspond temporally with those of the highest temperatures. Fires caused by lightning are primarily summer fires, and probably the most frequent type of fires in the grasslands. There are many benefits that accrue to those grassland species that survive fire. Silica content of grasses increases, which enhances resistance to decay, and thereby provides more fuel for the next fire. The more immediate advantage is the increased protection afforded to the seeds which remain above the hottest part of the fire on fire resistant seed stalks, thus ensuring survival of the seeds (Vogl 1974). Komarek (1965) pointed out that there are many species of fire-adapted grasses which have very hard stems with a high silica content among the genera *Aristida, Andropogon, Imperata,* and *Setaria.* Tooth adaptations of the fauna of grasslands are determined by the silica content of the flora. Thus, mammals possessing ever-growing teeth, with ridges that are actually sharpened by wear, came to dominate the grassland community. The corollary to the attribute of decay-resistant foliage is the immobilization of nutrients in plant tissue. Apparently, grasslands have become dependent on frequent fire as the agent of soil nutrient renewal, releasing the nutrients from plant tissue and recycling them into new biomass (Mutch 1970). Anderson (1965), writing about Kansas prairie forbs, discussed the benefit of fire for the bluestem (*Andropogon* sp.) range. The indirect benefits of a late spring fire are these: (a) it prevents early annuals from reseeding, (b) it removes the green tops of weedy perennials, and (c) thus it reduces competition. In addition, scarification of certain seeds occur, so that germination is possible. In general, after a fire, one of the following usually occurs, which aids in the survival of grassland plants: (a) the plant resumes life from a protected (perennating) bud at the soil surface or (b) there is increased reproduction from seeds. On the prairie, most of the plants are hemicryptophytes. This means that, at least once a year, the aboveground portion of the plant dies back. Prairies are windy places, and the winds tend to dry the aerial portions of the plants. In addition, the plants have a low growth form, and this allows sunlight penetration which hastens drying. Thus, when the vegetation is dormant, there is often abundant flammable material at the ground level. In addition, the plants have low moisture content, even before drying. We have, then, ideal conditions for a fire: dry, uncompacted, fine fuels, with plenty of available oxygen. These fine fuels will be ashed quickly, and very high temperatures do not build at ground level. Rather, the flames pass very quickly through the vegetation, the highest temperatures occur at the top of the flames, and the heat is dissipated by the prairie wind. Soil surface temperatures in grassland fires reach only about 120°C, and in soil 1-5 cm deep, there is a temperature increase of only 10-15°C above ambient soil temperatures. There is evidence that 60°C is the lethal temperature for the shoots of most land plants. Howver, leaf meristems (perennating buds) of many grasses are 40 mm or so below the soil surface. For example, *Aristida stricta* and *Sporobolus curtissi* have such buried meristems, which are additionally protected by closely packed, persistent leaf sheaths, which tend to exclude oxygen, and therefore do not burn. The living portion of the plant is virtually unharmed by the fire and will resume growth in the favourable time of the year in a nutrient regime actually enhanced by fire (Daubenmire 1968). As we noted previously, the decay-resistant aboveground plant parts contain needed minerals that are recycled by fire. Enhanced reproduction may occur as a direct response to fire: certain grassland species have resistant seeds that require scarification before germination will occur. Legumes, such as species of *Astragalus* and *Trifolium*, are known to produce seeds that require scarification before seed dormancy will be broken. Bermuda grass *(Cynodon dactylon)* exhibits enhanced reproduction from seed following a fire because of increased seed set. However, many grasses do not possess fire-resistant seeds. Table 17.1 summarizes heat tolerances of seeds of various grass species. Note that the survival times are brief, but that tolerances range well into the temperatures attained by surface fires at the soil surface in grasslands (Daubenmire 1968).

17.3.3 Mediterranean vegetation

The Mediterranean climatic regime is found between 30° and 45° north and south latitudes, on the west coast of continents, and in the Mediterranean Sea region. The Mediterranean climate is defined by four characteristics: (a) moderate precipitation concentrated in the cool winter months, and summer drought; (b) in summer, long periods of sunny days and cloudless skies;

TABLE 17.1
HEAT TOLERANCES OF SEEDS OF VARIOUS GRASSES
(DAUBENMIRE 1968)

Taxon	*Maximum temperature with survival*	*Duration of heating*
Avena fatua	104-116°C	5 min
Bromus mollis	116-127°C	5 min
Bromus rigidus	93-104°C	5 min
Stipa pulchra	116-127°C	5 min
Bromus mollis	>200°C	2 min

(c) marine-moderated atmosphere influence year-round; and (d) warm-to-hot summers and mild winters (McCutchan 1977). Within that regime certain areas are occupied by **sclerophyllous** (hard-leaved) plants and by thin, stony soils with little organic matter; these areas also have less that 750 mm of **precipitation** annually. In California (USA), such sites are occupied by a dense scrub called **chaparral** (originally a Basque word for scrub oak, *chabarro*, spelled by the Spanish *chaparro*, and used by them to designate the Mediterranean type scrub of California). In corresponding regions elsewhere are the *matorral* of Chile, the *fynbos* of the Cape of Good Hope, the *phrygana* of Greece, the *garrigue* or *garigue* of dry, calcareous soils of France, known as *macchia* or *macquis* on siliceous soils, and the *tomillares* of Spain (Biswell 1974). In each area the physiognomy of the vegetation is very similar, as are the adaptations to fire. At a higher elevation in the Mediterranean climatic regime chaparral is replaced by coniferous forests as in California.

Mutch (1970) hypothesized that communities maintained by fire possess vegetation that is more flammable than communities not adapted to fire. The woody plants of the chaparral exhibit such flammability, and are characterized as follows: reduced leaves with sclerophyllous adaptations and vertical orientation; intricate branching pattern, resulting in a large surface-to-volume ratio; periodic die-back, resulting in highly flammable twigs; and possession of volatile oils (Philpot 1977). Those species that possess special reproductive mechanisms that enable them to survive fire, or that possess anatomical mechanisms such as thick bark and *epicormic sprouting* (sprouting from the stem), may also have flammable characteristics that enhance fire spread. In general, during the long hot dry season chaparral vegetation becomes dormant, very dry and brittle, and conducive to lightning-induced fire in the autumn. In fact, chaparral is a fire climax community adapted to and maintained by periodic fires which recur every 15 to 100 years (Minnich 1983, Riggan *et al.* 1988) and tend to perpetuate shrub dominance at the expense of trees. If fires occur every 5-10 years, shrubs will be selected against, and grassland will be favoured (Humphrey 1974, Vogl 1974). In general, fire is more destructive of woody plants than of grasses. In the chaparral, most species begin seed production within 3-5 years after a fire, and brushlands rarely burn more frequently than that (Biswell 1974).

Gill (1977) listed the fire adaptive traits for Mediterranean regions as follows: (a) fire-induced flowering; (b) bud protection and sprouting subsequent to fire; (c) *in-soil* seed storage and fire stimulated germination; and (d) *on-the-plant* seed storage and fire stimulated dispersal. Other authors (Lotan 1974, Harvey *et al.* 1977) have commented on the adaptiveness of such traits as thick bark, evanescent branches, rapid growth, and early maturity.

Several of the bulb-forming perennials, which may have been present but not numerous in the chaparral between burns, exhibit enhanced flowering following a fire. Similarly, *Cynodon dactylon* is also known to exhibit enhanced flowering in response to fire. The phenomenon of increased flowering, and concomitantly enhanced seed-set, would appear to be advantageous in two ways: (a) those plants that do produce increased numbers of seeds would be at a competitive advantage over those plants not so adapted; and (b) the seed bed, which receives more light due to canopy removal and has a deposit of soluble mineral ash nutrients, provides an additional opportunity for enhanced growth and reproduction, giving further advantage to those plants that can quickly exploit this resource.

Buds near the base of the shrub may survive a fire and be released from dormancy by the death of shoots above them. Sprouting results as a hormonal response (latent axillary buds are released from inhibition when fire removes the apical meristem) or with the onset of winter rains. The sprouting may be from the stem (epicormic) or may be from a basal burl or *lignotuber*. Protection for the stem bud is provided by the bark.

Buds of lignotubers are protected by the soil as in the genus Arctostaphylos and the ubiquitous chamise *(Adenostoma fasciculatum)*.

There are some chaparral species whose seeds remain dormant in the soil for many years if there is no fire. Table 17.2 shows the response of seeds of deerbrush and mountain whitethorn *(Ceanothus integerrimus, C. cordulatus)* to treatment with hot water. Chamise *(Adenostoma fasciculatum)*, a sprouter which depends on fires every 15-20 years for optimum health and stability, also responds to fire with phenomenal germination of stored seeds: after a burn, as many as 3000 chamise seedlings per square meter may be counted. Seeds of both chamise and mountain whitethorn withstand dry oven temperatures of 138-149°C, the higher temperatures providing the trauma required to rupture the seed coat (Biswell 1974).

TABLE 17.2
GERMINATION OF HEAT-TREATED SEEDS OF DEERBRUSH AND MOUNTAIN WHITETHORN (BISWELL 1974)

Taxon	*Treatment length*	*Germination*
Ceanothus integerrimus	Boiled 20 min	12 %
Ceanothus cordulatus	Boiled 25 min	25 %
Ceanothus cordulatus	Boiled 30 min	0 %

There are coniferous plants that retain viable seeds for many years, or even decades. In the condition called *serotiny* (literally, late to open), the cones of some conifers remain on the tree and remain closed until mechanical breakage severs the vascular connections with the parent plant, or remain closed by resin until the heat of a fire opens them. Serotiny is characteristic of many different plants such as jack pine *(Pinus banksiana)*, some subspecies of lodgepole pine *(P. contorta)*, bishop pine *(P. muricata)*, knobcone pine *(P. attenuata)*, pitch pine *(P. rigida)*, Monterey cyprus *(Cupressus macrocarpa)*, and giant sequoia *(Sequoiadendron giganteum)*. Some subspecies of lodgepole pine require temperatures of 45-50°C (or even higher) to break the resinous bond between the cone scales. Summer soil temperatures may suffice to open low-hanging cones. Knobcone pine, however, requires 200°C for opening the cones. By contrast, giant sequoia cones remain closed due to cone scale turgidity, and open in response to the severence of vascular connections (Harvey *et al.* 1977). Generally, in regions frequently disrupted by fire a high percentage of serotinous-coned trees would be expected (Lotan 1974).

17.4 Prescribed Burning

Stoddard (1936) was one of the first to advocate the use of controlled or "prescribed" burning for increasing both timber and game production, at a time when the "official" or "professional" viewpoint of foresters was that all fire was bad. Biswell (1977) defined prescribed burning as follows: "Prescribed burning is the skillful application of fire to natural fuels under conditions of weather, fuel moisture, etc., that will allow confinement of the fire to a pre-determined area and at the same time will produce the intensity of heat and rate of spread required to accomplish certain planned benefits to one or more objectives of silviculture, wildlife management, grazing, hazard reduction, etc."

In forestry, the main purpose of prescribed burning is to enhance the growth of certain trees, by manipulating the major factors that influence that growth: light, temperature, water, and mineral nutrients. Foresters may desire a change in the relative abundance of species in a forested region, such that desirable species are favoured and competing species are eliminated or at least put at a disadvantage. Facilitation of the renewal of valued species and the provision of optimum conditions for growth are the aims of prescribed burning in the forest. For prescribed burning to be successful, there must be enough fuel to carry a fire and, ideally, the fuel should be a "material that burns readily and uniformly soon after a rain" (Biswell 1958). This allows a surface fire, but, since heavier fuels that might contribute to a crown fire would still be too wet to burn, prevents a holocaust. In general, the loose, well-aerated, high surface-to-volume ratio fuels will dry quickly enough to burn in one to two days after a rain.

Fire is equally important in grassland. Under moist conditions fire favours grass over trees, and under dry conditions fire is often necessary to maintain grassland against the invasion of desert shrubs.

18. SOIL

Three levels can be seen in the section of an exposed hillside diagrammed in Fig. 18.1, which has loose material resting on underlying rocks. The foundation is termed **bedrock.** Above that is the **regolith,** the unconsolidated mantle of weathered rock and soil material, of which the upper 1-2 meters or less of highly weathered, biologically altered material is termed **soil,** derived from the Latin word for floor, *solum.* Soil, along with climate, is a primary agent in plant selection through evolutionary change.

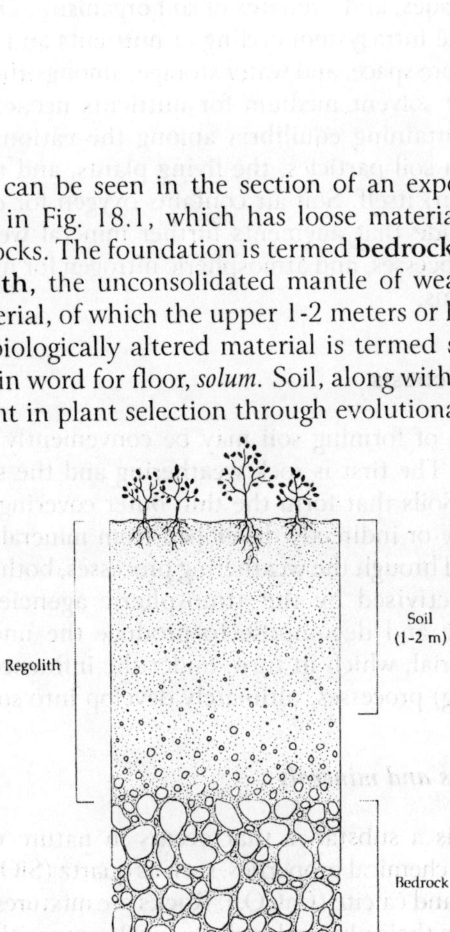

FIGURE 18.1
A VERTICAL SECTION OF SOIL, SHOWING RELATIVE POSITIONS OF BEDROCK, REGOLITH (UNCONSOLIDATED MATERIAL), AND SOIL (HIGHLY WEATHERED PORTION OF REGOLITH)

There are four separate components of soil: mineral grains,

organic matter, water, and air. The mineral substrate provides anchorage, pore space for storage of water and air, and nutrients on an exchange basis. Organic matter refers to the plant and animal residues in various stages of decomposition, as well as to the cells, tissues, and exudates of soil organisms. Organic matter enhances the intrasystem cycling of nutrients and improves soil structure, pore space, and water storage, among other things. Soil water is the solvent medium for nutrients needed by growing plants, maintaining equilibria among the cations and anions adsorbed on soil particles, the living plants, and the soil water (soil solution) itself. Soil air contains oxygen for cellular work, carbon dioxide that augments further mineral weathering and biotropic processes, and atmospheric nitrogen for nitrogen-fixing soil organisms.

18.1 Soil Genesis

The process of forming soil may be conveniently divided into two phases. The first is rock weathering and the second is soil formation. Soils that form the thin outer covering of the earth have directly or indirectly developed from mineral constituents of the rock. Through the weathering processes, both physical and chemical, activised by the atmospheric agencies, the rocks distintegrate and decompose to produce the unconsolidated parent material; which in turn, under the influence of *pedogenic* (soil-forming) processes, ultimately develop into soil.

18.1.1 Rocks and minerals

A **mineral** is a substance that occurs in nature with distinct physical and chemical properties, such as quartz (SiO_2), orthoclase ($KAlSi_3O_8$), and calcite ($CaCO_3$). **Rocks** are mixtures of minerals, and therefore their physical and chemical composition vary with the characteristic of minerals present in them. On the basis of their genesis and structure, rocks are generally grouped into three classes, namely, igneous, sedimentary, and metamorphic.

Igneous rocks are formed by solidification of molten magma from the interior of the earth. They are called **plutonic** when the magma solidifies at great depths; **intrusive** or **dike rocks** when it solidifies at moderate depths; and **extrusive** or **effusive** when solidification takes place on the surface of the earth as a result of volcanic activities. Igneous rock is composed of **primary minerals**, that is minerals which have not been altered chemically since they formed as molten lava solidified (**secondary minerals** are recrystallized products of the chemical breakdown and/or alteration of primary minerals). These include light-coloured primary minerals such as quartz, feldspars, muscovite; dark-coloured primary minerals, namely, hornblende, augite, biotite; coarse-textured primary minerals such as granite, diorite, gabbro, peridotite hornblendite; intermediate - textured primary minerals, namely, rhyolite, andesite, basalt; and fine-textured primary minerals such as felsite obsidian, and basalt glass.

Sedimentary rocks have formed from the consolidation of sediments accumulated through wind or water action at the surface of the earth in past geological age. For example, quartz and sand weathered from granite and deposited on the shore or beach of a prehistoric sea may through geological changes have become cemented into a solid mass called a **sandstone**. Similarly, recemented clays are termed **shale**. Other important sedimentary rocks are listed in Table 18.1 with their dominant minerals. The resistance of a given sedimentary rock to weathering is determined by its particular dominant minerals and by the cementing agent.

Metamorphic rocks are those that have formed by the metamorphism or change in form of igneous and sedimentary rocks subjected to high pressures and temperatures. Some important types of metamorphic rocks are gneiss from granite, slate from shales, marble from limestone, schist from shale, and quartzite from sandstone. Other important metamorphic rocks are listed in Table 18.2.

18.1.2 Weathering processes

Weathering processes (Fig. 18.2) are distinguished into two types, namely mechanical (disintegration) and chemical (decompostition). Disintegration results in a decrease in size of rocks and minerals without appreciably affecting their composition. During decomposition, however, definite chemical changes take place, soluble materials are released, and new minerals are synthesized some of which are resistant end products.

TABLE 18.1
SOME OF THE MORE IMPORTANT SEDIMENTARY AND METAMORPHIC ROCKS AND THE MINERALS COMMONLY DOMINANT IN THEM

Dominant mineral	*Type of Rock*	
	Sedimentry	*Metamorphic*
Calcite ($CaCO_3$)	Limestone	Marble
Dolomite ($CaCO_3.MgCO_3$)	Dolomite	Marble
Quartz (SiO_2)	Sandstone	Quartzite
Clays	Shale	Slate
Variable	Conglomerate*	Gneiss
Variable		Shist

*Small stones of various mineralogical makeup are cemented into conglomerate

Table 18.2 The more common primary and secondary minerals found in soils listed in order of decreasing resistance to weathering under conditions common in humid temperate regions. The primary minerals are found abundantly in igneous and metamorphic rocks, whereas secondary minerals are commonly found in sedimentary rocks.

Primary minerals	*Secondary minerals*	
	Geothite FeOOH	*Most resistant* ↑
	Hematite Fe_2O_3	
	Gibbsite $Al_2O_3.3H_2O$	
Quartz SiO_2		
Muscovite $KAl_3Si_3O_{10}(OH)_2$	Clay minerals / Al-silicates	
Microcline $KAlSi_3O_8$		
Orthoclase $KAlSi_3O_8$		
Biotite $KAl(Mg.Fe)_3Si_3O_{10}(OH)_2$		
Albite $NaAlSi_3O_8$		
Hornblende $Ca_2Al_2Mg_2Fe_3Si_6O_{22}(OH)_2$		
Augite $Ca_2(Al.Fe)_4(Mg.Fe)_4Si_6O_{24}$		
Anorthite $CaAl_2Si_2O_8$		↓
Olivine $(Mg.Fe)_2SiO_4$		Least resistant

Mechanical and chemical processes may be outlined as follows:

(a) **Mechanical (disintegration)**: (a) temperature - differential expansion of minerals, frost action, and exfoliation (b) erosion and deposition - by water, ice, and wind (c) plant and animal influences

(b) **Chemical (decomposition)**: (a) hydrolysis (b) hydration (c) acidification (d) oxidation (e) dissolution.

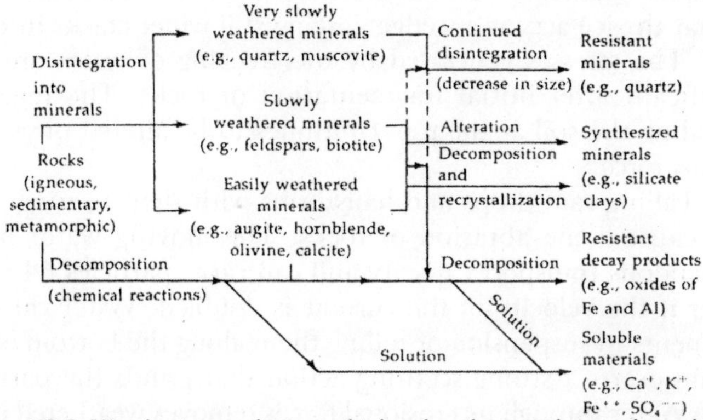

Figure 18.2 Weathering pathways which take place under moderately acid conditions common in humid temperate regions. Major paths of weathering are indicated by the heavier arrows, minor pathways by broken lines. As one would expect, climate modifies the exact relationships. In arid regions physical breakdown (disintegration) would dominate, and soluble ions would not be lost in large quantities. In humid regions decomposition becomes more important, especially under tropical conditions.

18.1.2.1 Mechanical

Mechanical weathering may be attributed to the expansion and contraction of rocks in response to thermal flux. During the day, rocks become heated; at night they often cool much below the temperature of the air. The warming causes some minerals within a given rock to expand more than others. Likewise, when cooled, some minerals contract more than others. With every temperature change, therefore, differential stresses are set up that eventually must produce cracks, thus encouraging mechanical breakdown.

Because heat is conducted slowly, the outer surface of a rock is often warmer or colder than the inner or more protected portions. This differential heating and cooling sets up lateral stresses that, in time, may cause the surface layers to peel away from the parent mass. This phenomenon is referred to as **exfoliation,** which at times is sharply accelerated by the freezing of *included* water. Expansion caused by the freezing of this water is impressive: normal hydrostatic pressure amounts to more than 10,000 kg/cm^2. Ice floats on liquid water, and the water at the air/liquid interface is cooled more rapidly than within the rock, so that the ice acts as a wedge, forcing still wider cracks into the rock. The stresses generated by the freezing of water are also significant after initial fragmentation of rocks. The regolith, including the soil at the top, continues to be altered physically by this force.

Falling rain drops and hailstorms with their beating force also cause some abrasion of rocks. The moving water has a tremendous transport capacity and can carry particles 64 times larger if the velocity of the current is doubled. Water carrying sediments in suspension or rolling them along the bottom of the stream exerts a strong scouring action that grinds the particles finer. Water through its erosional forces removes weathered parts of the rock, thereby exposing fresh surfaces to weathering. Glaciers, during their movement, cause a great deal of cutting and crushing of the bed rocks. Wind, especially when laden with sand particles, causes abrasive action on exposed rocks. Loosely balanced rock boulders sometimes roll down under gravitational force and break into pieces. Physical influences of biological agencies such as plant roots splitting the rocks apart; movement of animals; burrowing by rodents; and cultivation by man, also help in mechanical weathering to some extent.

Under the predominant influence of mechanical weathering, soils usually consist of coarse materials and are often called **skeletal soils**.

As soon as physical disintegration of rock and minerals begins, chemical decomposition starts. This is especially noticeable in warm and humid regions, where chemical and physical processes are intense and tend to accelerate each other. Chemical weathering is accelerated by the presence of water, oxygen, and the organic and inorganic acids that result from the microbial breakdown of plant residues. These agents commonly act in concert to convert primary minerals (feldspars and micas, for example) to secondary minerals (silicate clays, for example) and to soluble forms that carry essential elements that support plant growth.

18.1.2.2 Chemical

(a) Water and its solutions: Water and its dissolved salts and acids are perhaps the most pervasive factors in the weathering of minerals. Through hydrolysis, hydration, and dissolution, water enhances the degradation, alteration, and resynthesis of minerals. A simple example is the action of water on microcline, a potassium-containing feldspar:

$$\underset{\text{(solid)}}{KAlSi_3O_8} + \underset{\text{(liquid)}}{H_2O} \xrightarrow{\text{hydrolysis}} \underset{\text{(solid)}}{HAlSi_3O_8} + \underset{\text{(solution)}}{K^+ + OH^-}$$

$$\underset{\text{(solid)}}{2HAlSi_3O_8} + \underset{\text{(liquid)}}{11H_2O} \xrightarrow{\text{hydrolysis}} \underset{\text{(solid)}}{Al_2O_3} + \underset{\text{(solution)}}{6H_4SiO_4}$$

$$Al_2O_3 + 3H_2O \xrightarrow{\text{hydration}} \underset{\text{(hydrated solid)}}{Al_2O_3.3H_2O}$$

The potassium released is soluble and is subject to adsorption by soil colloids, to uptake by plants, and to removal in the drainage water. Likewise, the silicic acid (H_4SiO_4) is soluble. It can be removed slowly in drainage water, or it can be recombined with other compounds to form secondary minerals such as silicate clays. Hydration is evident through the formation of hydrated aluminum oxide ($Al_2.O_3.3H_2O$). Iron oxides are also commonly hydrated:

$$\underset{\text{(hematite/red)}}{2\,Fe_2O_3 + 3H_2O} \longrightarrow \underset{\text{(limonite/yellow)}}{2Fe_2O_3.3H_2O}$$

Hydration is always accompanied with increase in volume, and hydrated minerals usually become soft and more readily weatherable.

(b) Acid-solution weathering: Weathering is accelerated by the presence of the hydrogen ion in water, such as that provided by carbonic and organic acids. For example, the presence of carbonic acid (H_2CO_3) results in the chemical solution of calcite in limestone:

$$\underset{\text{calcite (solid)}}{CaCO_3} + \underset{\text{(solution)}}{H_2CO_3} \longrightarrow \underset{\text{(solution)}}{Ca^{2+} + 2HCO_3^-}$$

Other much stronger acids, such as nitric acid (HNO_3), sulphuric acid (H_2SO_4), and some organic acids are found in soils. Also present are the hydrogen ions associated with soil clays. Each of these sources of acidity is available for reaction with other soil minerals. An example of the reaction in a moist soil of hydrogen ions with soil minerals is that of an acid clay with anorthite (a feldspar):

$$\underset{\substack{\text{anorthite}\\ \text{(solid)}}}{CaAl_2Si_2O_8} + \underset{\substack{\text{acid clay}\\ \text{(solid)}}}{\boxed{\text{clay}}{}^{H^+}_{H^+}} \longrightarrow \underset{\substack{\text{acid silicate}\\ \text{(solid)}}}{H_2Al_2Si_2O_8} + \underset{\substack{\text{calcium clay}\\ \text{(solid)}}}{Ca^{2+}\boxed{\text{clay}}}$$

The mechanism is simple: the hydrogen ions of the acid clays replace the calcium ions of the fresh minerals that they may closely contact. The unstable acid silicate that results is subjected ultimately to dissolution and subsequent reaction to form silicate clays.

(c) Oxidation: Concomitant with the action of water in weathering is that of oxidation. It is particularly manifest in rocks that contain iron, an element that is easily oxidized. In some minerals, iron is present in the reduced ferrous (Fe^{2+}) form. If ferrous iron is oxidized to the ferric (Fe^{3+}) state, other ionic adjustments must be made because a three-valent ion is replacing a two-valent one. These adjustments result in a less stable mineral, which is then subject to both disintegration and decomposition.

In other cases, ferrous iron may be released from the mineral and is almost simultaneously oxidized to the ferric form. An example of this is the hydration of olivine and the release of ferrous oxide, which may be immediately oxidized to ferric oxide (geothite):

$$\underset{\text{olivine}}{3MgFeSiO_4} + 2H_2O \longrightarrow \underset{\text{serpentine}}{H_4Mg_3Si_2O_9} + SiO_2 + \underset{\text{Ferrous oxide}}{3FeO}$$

$$\underset{\text{ferrous oxide}}{4FeO} + O_2 + H_2O \longrightarrow \underset{\text{geothite}}{4FeOOH}$$

When ions such as Fe^{2+} are removed or are oxidized within the minerals, the rigidity of the mineral structure is weakened and the mechanical breakdown is made easier. This provides a favourable environment for further chemical reactions.

In general, weathering is influenced by three major factors: (i) climatic conditions, (ii) physical properties, and (iii) chemical characteristics of the rocks and minerals.

Climatic conditions, more than any other factor, control the rate and nature of weathering. Under arid conditions where physical forces dominate, the size of the particles has been decreased with relatively little change in composition. The original primary minerals are more prominent, whereas minerals that require water for their synthesis are less so. Physical changes due to temperature fluctuation and wind action are accompanied by only modest chemical action. Consequently, the soils of arid regions are remarkably like the parent materials from which they were formed. In a humid region, however, the forces of weathering are more varied, and most of them are active. Vigorous chemical changes accompany disintegration. New minerals such as silicate clays and oxides of iron and aluminum are more evident. The process is accelerated and intensified by the decay of large quantities of organic matter coming from abundant plant growth. The high year-round temperatures and the luxuriant plant growth in the humid tropics provide optimum conditions for weathering. As a result, in these regions the primary silicate minerals have largely succumbed to intense weathering and only the more highly weathered silicate clays still remain. The more resistant products of chemical weathering, such as the oxides of iron and aluminum, tend to dominate soils of humid tropical regions.

Particle size, hardness, and nature and degree of cementation are among the physical characteristics that influence weathering. Rocks comprised of minerals with large crystals disintegrate more easily than those with fine crystals because of differences in expansion due to temperature changes. Once the rocks have disintegrated into individual minerals, however, the finer crystals are more subject to chemical change than the larger ones. The much larger surface area of finely divided material presents greater opportunity for chemical attack. Hardness and cementation also influence weathering. For example, a dense quartzite or a sandstone cemented firmly by a slowly weathered mineral resists mechanical breakdown and presents a small total surface area for chemical activity. On the other hand, porous rocks such as volcanic ash or coarse limestone, having a larger surface area for chemical attack, are readily broken down into smaller particles and are more easily decomposed.

Chemical and crystalline characteristics also influence the ease of chemical removal or breakdown. Some minerals such as gypsum ($CaSO_4 . 2H_2O$) or calcite ($CaCO_3$) can be solubilized in water saturated with carbon dioxide and are relatively easily removed from parent material. Others, such as the ferromagnesium minerals olivine and biotite, are quite easily weathered. They contain readily oxidizable iron, and their various ions are not very tightly packed in the mineral crystal structures. In contrast, more tightly packed crystal units and lack of oxidizable iron give muscovite considerable resistance to weathering. The stability of soil-forming minerals is also dependent on climatic and biotic conditions. Consequently, no listing of minerals based on their resistance to weathering would be valid for all conditions. However, for humid temperate conditions, the following is a general order of weathering resistance of the sand and silt-sized particles of some common minerals (Barshad 1955): quartz (more resistant) > muscovite and potassium feldspars> sodium and calcium feldspars> biotite, hornblende, and augite> olivine> dolomite and calcite>gypsum.

This order would be changed slightly depending on the climate and other environmental conditions. In any case it accounts for the absence of gypsum, calcite, and dolomite in soils of humid regions and for the predominance of quartz in the coarser fraction of most soils.

18.1.3 Soil parent materials

Inorganic parent materials from which soils are derived may be classified as residual or sedentary, and transported (Fig. 18.3).

When soils develop in place from underlying igneous, sedimentary, or metamorphic bed rocks they are termed to have formed from **residual** materials. Typically, it has experienced long and often intense weathering. In a warm, humid climate, it is likely to be thoroughly oxidized and well leached. Red and yellow colours due to oxides of iron are characteristic when weathering

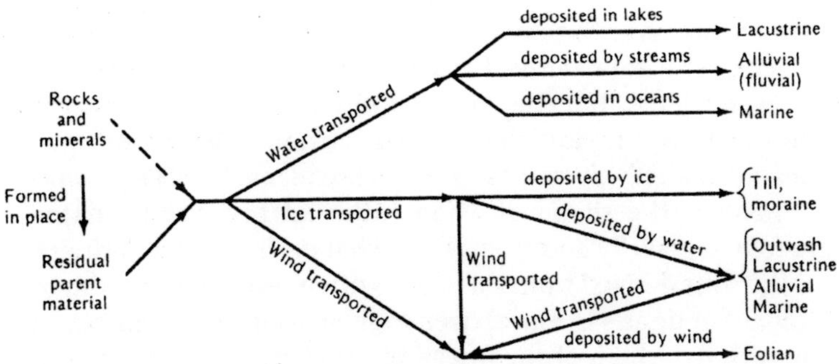

FIGURE 18.3
HOW VARIOUS KINDS OF PARENT MATERIAL ARE FORMED, TRANSPORTED AND DEPOSITED

has been intense as in hot, humid areas. In cooler and especially drier climates, weathering of residual materials is much less drastic and the oxidation and hydration of the iron may be hardly noticeable. Also, the calcium content is often higher and the colours of the weathering mass are subdued.

Soil debris at the foot of a slope that moved there in response to gravity is called **colluvial** material. Colluvial material exists to some extent at the base of all slopes, but it is especially noticeable in mountainous topography where rock slides, slips, and avalanches are common. Parent material developed from colluvial accumulation is dependent on the sources of the material. It is frequently coarse and stony because physical rather than chemical weathering has been dominant. Soils developed from colluvial materials are generally not of great agricultural importance because of their small area, inaccessibility, and unfavourable physical and chemical characteristics. However, some useful timber and grazing lands in very mountainous regions have colluvial materials.

Parent materials that have been transported by water are classified as alluvial, marine and lacustrine. **Alluvial** materials are sediments that have been deposited by flowing water, such as by small streams and large rivers. Sediment carried by the swollen stream is deposited during the flood, with the coarser materials (sand, silt) being laid down near the river channel and finer materials (clay) farther away. If the transported material (that lies along the river) is subject to periodic flooding the deposit is known as a **floodplain**, whereas older deposits that were laid down by the river but are not now subject to flooding are called **terraces**. The soils derived from alluvial deposits are generally rich in nutrients, but they may require drainage and protection from overflow. Streams that leave a narrow valley in an upland area and suddenly descend to a much broader valley below deposit sediment in the shape of a fan **(alluvial fans)**. Fan material generally is gravelly and stony, somewhat porous and well-drained, and the soils derived from this debris often prove very productive. Alluvial fan debris is found over wide areas in mountainous and hilly regions. Much of the finer sediment carried by rivers is not deposited in the floodplain but is discharged into a lake reservoir or ocean into which the stream flows. Some of the suspended material settles near the mouth of the river, forming a delta **(delta deposits)**. A delta often is a continuation of a floodplain (its "front", so to speak). It is clayey in nature and is likely to be swampy as well. Much of the sediment carried away by streams eventually is deposited in oceans, lakes, seas, and gulfs. The coarser fragments are deposited near the shore and finer particles at a distance. Over long periods of time, these underwater sediments build up and can become quite deep. In some areas, changes in the elevation of the earth's crust have resulted in these marine deposits being raised above sea level. The deposits are then subject to weathering and to soil formation, giving rise in some cases to valuable agricultural soils.

The name **drift** is applied to all material of glacial origin, whether deposited by ice or associated waters. The materials deposited directly by ice, called **glacial till**, are heterogeneous mixtures of debris of great diversity, which vary form rocks and boulders to clay. Glacial till is found mostly as irregular deposits called **moraines**, of which there are various kinds. *Terminal moraines* are hilly ridges that were formed when the ice margin was stationary long enough to permit an accumulation of debris. They occupy little land area. As the ice rapidly retreated, a thinner and more level deposition called *ground moraine* was laid down. Ground moraine is by far the most widely distributed of all glacial deposits and is of considerable agricultural significance. As the ice receded, there were periods when the front was relatively stationary again, and hilly *recessional moraines* of limited area were formed. Special small features associated with the moraine are **kames** (conical hills or short ridges of sand and gravel deposited by ice), **eskers** (long, narrow ridges of coarse gravel deposited by ice-walled streams coming from the glacier), and **drumlins** (cigar-shaped hills composed of till and oriented parallel to the direction of the ice movement) (Fig. 18.4). Torrents of water were constantly gushing from the ice lobes of glaciers, especially during the summer. The vast loads of sediment carried by such streams were either dumped immediately or carried to other areas for deposition. As long as the water had ready egress, it flowed away rapidly to deposit its load downstream. An **outwash** plain is formed by streams that are heavily laden with glacial sediment flowing form the ice. Since the sediment is sorted by flowing water, sands and gravels are common. Outwash deposits are found in valleys and on plains where the glacial waters were able to flow away freely. When the ice front came to a standstill where there was no ready escape for the water, ponding began and ultimately very large lakes were formed. The deposits formed in these glacial lakes range from coarse delta materials and beach deposits near the shore, to larger areas of fine silts and clay deposited form the deeper, more still waters at the center of the lake. Areas of inherently fertile soils have developed from these materials as the lakes dried.

Wind-transported parent materials are called **eolian** (or **aeolian**) deposits. During the glacial periods, conditions were ideal for wind erosion. The weather was cold and windy, and vegetation was sparse or nonexistent. In the winters, winds picked up fine alluvial materials, deposited in previous summers by the ice-fed streams, and moved them in areas of present-day depositions. These wind-blown materials, comprised primarily of silt with some fine sand and clay, are called loess (pronounced "luss"). Eolian deposits other than **loess** are sand dunes and volcanic ash. The sand dunes, a menace to agriculture if they are moving, occur in the great Sahara and Arabian deserts as well as in Thar deserts covering the western parts of India and the adjacent parts of Pakistan, whereas eolian loess occur in the Pathwar plateau in west Pakistan.

18.1.4 Soil formation

Soil formation (S) is characterized by Jenny (1941) as dependent on a set of independent variables as follows: climate (cl), organisms

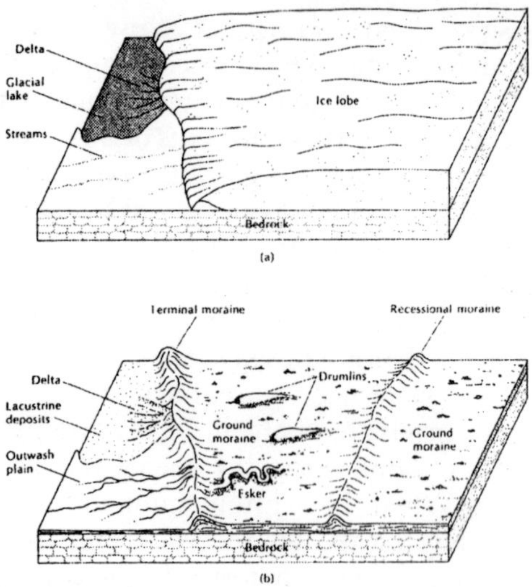

Figure 18.4 Illustration of how several glacial materials were deposited. (a) a glacier ice lobe moving to the left, feeding water and sediments into a glacial lake and streams, and building up glacial till near its front. (b) after the ice retreats, terminal, ground, and recessional moraines are uncovered along with cigar-shaped hills (drumlins), the beds of rivers that flowed under the glacier (eskers), and lacustrine, delta, outwash deposits.

(o), topography (r), rock type or parent material (p), and time (t). Jenny's classic equation is as follows:

$$S = f^{(cl.o.r.p.t...)}$$

Soils are often defined in terms of these factors as "dynamic natural bodies having properties derived from the combined effect of *climate* and *biotic activities*, as modified by *topography*, acting on *parent materials* over periods of *time*".

Climate is perhaps the most influential of the factors because it determines the nature of the weathering that occurs. For example, temperature and precipitation affect the rates of chemical, physical, and biological processes responsible for profile development. For every 10°C rise in temperature, the rates of biochemical reaction double. Furthermore, biochemical changes by soil organisms are sensitive to temperature as well as moisture. Temperature and effective moisture influence the organic matter content of soil. The very modest profile development characteristic of cold areas contrasted with deep-weathered profiles of the humid tropics is further evidence of climatic control.

Climate also influences the natural vegetation. In humid regions, plentiful rainfall provides an environment favourable for the growth of trees. In contrast, grasslands are the dominant native vegetation in semiarid regions, and shrubs and *brush* (rough low-growing bushes) of various kinds in arid areas. Thus, climate also exerts its influence through a second soil-forming factor, the living organisms.

Jenny defined *organisms* as the biotic potential of a site, not just the existing organisms. The biotic potential includes consideration of soil organisms, litter, vegetation type, etc. The soil macrobiota and microbiota interact with the growth medium, altering its capacity to support life. The amount of litter deposited on the soil is determined by the productivity of the community, and the rate of decay or removal of that litter is determined by climate, by the nature of the litter, and by fire frequency. Also, the incorporation of litter derived materials into the soil, the maintenance of soil porosity and permeability, and ultimately soil fertility, all rely at least in part on the activities of soil organism. Vegetation and climate determine what kind of soil if formed. **Podzol soils** develop beneath the coniferous forests in cool, moist climates; podzol soils have a prominent ashy gray upper horizon. **Chernozem** ("black earth") soils develop under grasslands in temperate regions.

The composition of parent material largely determines the chemical make up of its derived soils. For instance, acid soils derive from: (a) aluminum-rich bauxites, (b) silica-rich substrates such as sands, slates, and diatomaceous earth, (c) mine-trailings of lead and zinc deposits, (d) sulphide-rich, hydrothermically altered volcanic rocks and ash, and (e) estuarine deposits. On the other hand, presence of limestone in parent material will delay the development of acidity, a process that moist climates encourage. Soils with poor fertility may develop from ultrabasic **(ultramafic)** rock types rich in silicates of iron and magnesium. For example, serpentine develops from igneous rocks such as peridotite and dunite, which contain silicates of iron and magnesium. Serpentine soils are high in magnesium and low in calcium (Table 18.3). They may also be high in nickel and chromium and deficient in nitrogen and phosphorus; thus, they are not very fertile.

TABLE 18.3
COMPARISON OF CERTAIN SOIL NUTRIENTS IN SERPENTINE AND NONSERPENTINE SOILS (WALKER 1954)

Nutrient	*Nonserpentine soil*	*Serpentine soil*
Replaceable Ca (meq/100g)	1.56	0.21
Replaceable Mg (meq/100g)	0.71	1.89
Typical Ca/Mg ratio	3.0	0.3
Percent N (total, all forms)	0.14	0.01
Percent P (total, all forms)	0.13	0.002
K (meq/100g)	0.83	0.22
Ni (meq/100g)	0.005	0.040
Cr (meq/100g)	0.0001	0.011

Soil may form from transported material or *in situ* parent material (residual soil). *Topography* (which relates to the configuration of the land surface and is described in terms of difference in elevation, slope and so on) influences the stability of a soil body in a variety of ways, including its effect on rainfall patterns and the steepness of slopes. Resistant parent materials may result in steep slopes; parent rocks or materials that erode easily produce wide valleys, rounded hills, and often deep soils. Steep slopes may continually expose new surfaces to weathering and soil formation, as alluvium or **talus** (material moved by gravity) is moved downslope; resident soils, if present at all, are young and shallow. Gentle slopes restrict transport, and often enhance development of deep, strongly developed soils.

Time is a very important factor in soil formation: time for large mineral grains to be disintegrated, decomposed, and changed to secondary minerals; time for finer and finer mineral grains to accumulate; and time for organics and humus to become a significant fraction of the total soil, and also, time for finer grains and soluble anions and cations to be leached away. In general, soils located on alluvial or lacustrine materials, coastal-plains and soils of glaciated regions are poorly developed due to their comparatively recent origin.

It should be emphasized that two or more of the factors influencing soil formation are usually active simultaneously and interdependently; thus, climate and parent material influence vegetation, which, in turn, over a period of time, has helped influence the nature of the parent material available. Likewise, there is a relation between some parent materials and topography. The interaction and interdependence of these factors introduces some complexity in evaluating just how a given soil was formed.

18.1.5 Soil profiles

A vertical section encompassing all the horizons of the soil and extending into its parent material is called a *soil profile*. A common sequence of horizons within a profile is shown in the Fig. 18.5.

The O horizon is the organic layer of a soil profile derived from litter of plant and animal origins. It has two main subdivisions: O_1 and O_2; the former is composed largely of undecomposed organic debris (leaves, etc.), whereas the latter contains partly decomposed organic debris. The O horizon and its subdivisions may be largely or totally lacking in certain desert soils, and may be the only horizons existing in organic soils.

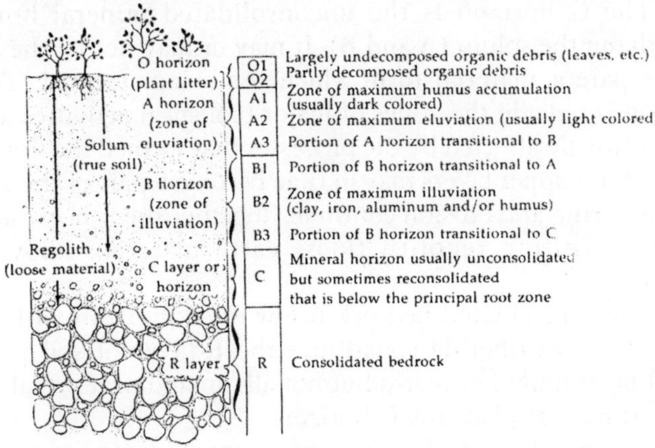

Figure 18.5 A diagrammatic representation of a soil profile (vertical section of the soil) showing soil horizon nomenclature and symbols for minerals soils. Additional subdivisions of the O horizon, not shown, are used in organic soil (Thompson and Troeh 1973).

The A horizon includes the surface soil. It contains a strong mixture of partially decomposed (humified) organic matter, which tends to impart a darker colour than that of the lower horizons. The A horizon is also referred to as the **zone of eluviation** due to maximum leaching or **eluviation** of clay, iron and aluminum oxides, and humus, which leaves a concentration of resistant minerals, such as quartz, in the sand and silt sizes.

The B horizon includes the subsurface part of the soil, which may be a **zone of illuviation**, where the materials lost from A are accumulated. In humid regions, the B horizon is the layer of maximum accumulation of materials such as iron and aluminum oxides and silicate clays. In arid and semi-arid regions, calcium carbonate, calcium sulphate, and other salts may accumulate in the B horizon. Alternatively, the B horizon may be a **zone of alteration** where little material accumulates from A but significant changes in structure or secondary mineral formation take place. A young soil may have A and C horizons, but not B horizon. This lack of B horizon results when there has been no translocation or eluviation of clay from the A horizon downward; such a profile is characteristic of the *chernozem* or "black earth" profiles in the steppe regions. The A and B horizons taken together are referred to as the **solum** or true soil.

The C horizon is the unconsolidated mineral horizon underlying the solum (A and B). It may or may not be the same as the parent material form which the solum formed. The C horizon is outside the zones of major biological activities and is generally little affected by the processes that formed the horizons above it. Its upper layers may in time become a part of the solum as weathering and erosion continue. Together the C horizon and solum constitute **regolith** (loose earth materials above the bedrock).

The consolidated bedrock below the C horizon with little evidence of weathering constitutes the **R-layer** of soil profile. The R layer underlies many, but not all, upland or residual soils, where it may displace the C horizon.

The appearance of a soil profile largely determines the fit of a soil into a soil classification scheme. For example the three orders in 1938 system of soil classification (Baldwin *et al.* 1938) include; (i) **zonal soils,** whose characteristics and profiles are mostly influenced by climate and vegetation (e.g., desert soils, lateritic (red tropical) soils); (ii) **intrazonal soils,** which exhibit unique profiles due to unusual parent material or topography (e.g., saline soils, bog); and (iii) **azonal soils,** which have little or no profile development (e.g., recent alluvial soils, rapidly eroding soils).

18.2 Physical Properties Of Soil

18.2.1 Soil texture

Soil texture is defined as the proportionate composition by weight of sand, silt and clay in a given soil. That is, only mineral components are considered in determining soil texture, even though soil is also composed of organic matter, water and air, with organic materials contributing up to 5% or more of the total.

The United States Department of Agriculture classification of soil particles places the upper limit for *silt* at 0.05 mm in diameter and the lower limit at 0.002 mm. Particles greater than 0.05 mm in diameter are considered *sand,* and particles less than 0.002 mm in diameter are considered *clay* (Brady 1995). Coarse fragments that range from 2 to 75 mm in diameter are termed *gravel* or *pebbles;* those ranging form 75 to 250 mm are called *cobbles* (if round) or *flags* (if flat); and those more than 250 mm across are called *stones* or *boulders.* Although almost any soil contains pebbles and cobbles/flags, such coarse fragments play but a minor role in soil function unless the bulk of soil is composed of these fragments.

Sand grains or gravel, which are often the fragments of rocks and minerals, may be rounded or irregular in shape depending upon the abrasion they have undergone. Silt is essentially microsand particles. They are diverse in shape, and seldom smooth or flat. Quartz (SiO_2) commonly dominates sand, especially the fine sands, as well as the silt separate. In addition, differing quantities of other primary minerals, such as various feldspars (aluminosilicates) and micas (iron and aluminum silicates), usually occur. Gibbsite (hydrons oxide of aluminum), hematite, and goethite (hydrous iron oxides) also are found, usually as coatings on the sand grains. Since sand and silt are dominantly quartz and other primary minerals known for their resistance to weathering, these two fractions have relatively low chemical activity. The primary minerals that may contain nutrient elements in their chemical makeup are generally so insoluble as to make their nutrient-supplying ability insignificant.

Clay particles vary in shape from plate-like to rounded. Material that is without identifiable structure in the fine clay fraction is called **allophone**. Clay particles and bits of organic matter within the soil are referred to as **micelles**, most of which are polyanions. Micelles are negatively charged on their surfaces. The capacity of micelles to provide the primary soil nutrient storage results from (i) the negative charges on the micelles which hold exchangeable cations such as Ca_2^+, Mg_2^+, K^+, H^+, Na^+, NH_4^+ and Al_3^+ that can be released for absorption by plants, and (ii) their very large surface-to-volume ratio. The hypothetical dicing of a mineral grain (Fig. 18.6) allows us to envision the difference in the surface-to-volume ratio of a sand grain with sides 1 mm in length, and the same volume of clay particles with sides 0.001 mm in length. The sand grain has a volume of 1 mm^3 and a surface area of 6 mm^2. If we slice this into bits with sides 0.1 mm in length, there will be 1,000 bits, each with 6 x 10^{-2} mm^2 surface area, or a total of 60 mm^2, which is an increase of ten times more than the surface area of the parent sand grain. Slice again, so that each soil particle has sides 0.01 mm in length, and the surface area increases to 600 mm^2. Slice again, so that clay particles have sides 0.001 mm in length, and surface area now equals 6,000 mm^2, which is 1,000 times that of the parent grain.

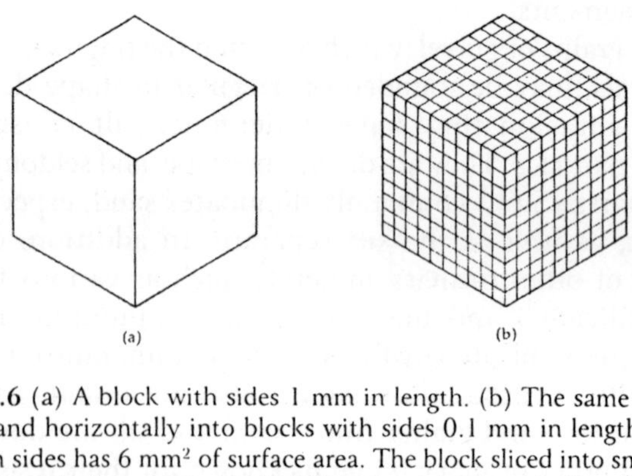

Figure 18.6 (a) A block with sides 1 mm in length. (b) The same block sliced vertically and horizontally into blocks with sides 0.1 mm in length. The block with 1 mm sides has 6 mm^2 of surface area. The block sliced into smaller blocks with 0.1 mm sides has resulted in 1000 blocks with 60 mm^2 of total surface area exposed. Slicing the block into smaller blocks with 0.001 mm sides would result in a surface area of 6000 mm^2, or 1000 times that of the parent grain (Thompson and Troeh 1973).

Water moves into and drains out of a sandy soil with much greater ease than for a finer textured soil, which means that sandy soils are more permeable than fine textured soils. This is because water has strong adhesion and cohesion capacity. Water molecules form a film **(hygroscopic water)** about each clay particle, and easily bridge the gap between particles; the cohesion between water molecules is sufficient to maintain the bridge. Such soils are nutritionally rich, but can be problem agriculturally, due to low permeability and poor aeration. Hygroscopic water also forms a tight film about sand grains, but the surface-to-volume ratio is very low, there are fewer grains than in clays, and the large pore size does not allow efficient cohesion of water molecules, which therefore drain away under the influence of gravity.

In brief, the sand fraction provides *skeletal* (structural) support, aeration and permeability for the soil. The clay fraction provides water holding and nutrient storage capacities. Silt also provides water holding capacity and, through weathering, provides plant nutrients and material to form clays, to replace that lost through leaching.

18.2.1.1 Determination of soil texture

The common field method of determining the textural class of a soil is by its **feel** ("feel" method). In the field, a good estimate can be made of texture by dampening and handling a small amount of the soil. The particles that feel grainy and gritty are sand; those that are silky, like most talcum powder, are silt. If the sample is sticky and if a self-supporting, flexible ribbon can be extruded between the thumb and fingers, there is a high percentage of clay. If it is nonsticky or only slightly sticky, and no ribbon can be extruded, there is a low percentage of clay. The "feel" method is used in soil survey and land classification. Accuracy in such a determination is of great practical value and depends largely on experience.

A more precise determination of soil texture **(particle-size analysis)** can be made in the lab. Particles larger than 2 mm are sieved out of a given sample, which is then treated to oxidize and remove the organic matter (usually with hydrogen peroxide: this also has the effect of rupturing the binding of clays by organic matter). The sample is then dried and weighed. The remaining materials are carefully flushed and shaken through a series or nest of sieves which allow particles 1 mm, 0.5 mm, 0.25 mm, 0.1 mm and 0.05 mm to pass through. The total weight of sand and its distribution can be determined by this process. The fraction passing through the sieves contains silt and clay together, and these can be separated by using a **hydrometer** or **pipette method** of analysis, based on differential rates of settling by silt and clay.

Once the percentage by weight of separate components has been determined, the use of a **soil-texture triangle** (Fig. 18.7) facilitates **textural class** determination. Assume, for instance, that our soil contains 30 % clay, 40 % silt, and 30 % sand. This soil would be called **clay loam**. Make note of two features: the designation is an area, not just a point on the triangle; and, there are four terms commonly applied to soil texture: clay, silt, sand, and loam. **Loam** designates a mixture of sand, silt, and clay that exhibits the properties of each fraction about equally. Thus, our sample exhibits the clay fraction more strongly than does loam, but is still a loamy soil, hence "clay loam". Table 18.4 lists the general terms used by the United States Department of Agriculture to describe soil texture. Accordingly, the **sand** group includes all soils in which the sand separates make up at least 70 % and clay separate 15 % or less of the material by weight. It has two textural classes - **sand** and **loamy sand**. The **silt** group includes soils with at least 80 % silt and 12 % or less clay. It has only one textural class-**silt**. To be designated as **clay**, a soil must contain at least 35 % of the clay separate and, in most cases, not less than 40%. It has three textural classes: **clay**, **sandy clay**, and **silty clay**. Sandy clays may contain more sand than clay. Likewise, the silt content of silty clays usually exceeds that of the clay fraction itself. A loam in which sand is dominant is classified as a **sandy loam**; in the same way, there may occur **silt loams**, **silty clay loams**, **sandy clay loams**, and **clay loams.**

The preceding textural names are used for designating only the mineral particles. When the surface soil is mostly organic matter, the term **muck** or **peat** is used. Muck is the name when the organic surface, approximately 30 cm in depth, is so well decomposed that the plant remains are not recognizable. With less decomposition the soil is called peat.

18.2.2 Soil structure

The term structure relates to the grouping or arrangement of soil particles. It describes the gross, overall combination or

TABLE 18.4
GENERAL TERMS USED BY THE US DEPARTMENT OF AGRICULTURE TO DESCRIBE SOIL TEXTURE IN RELATION TO THE BASIC SOIL TEXTURAL CLASS NAMES

General terms		Basic soil textural class names
Common names	**Texture**	
Sandy soil	Coarse	Sandy Loamy sands
Loamy soils	Moderately coarse	Sandy loam Fine sandy loam
	Medium	Very fine sandy loam Loam Silt loam Silt
	Moderately fine	Clay loam Sandy clay loam Silty clay loam
Clayey soils	Fine	Sandy clay Silty clay Clay

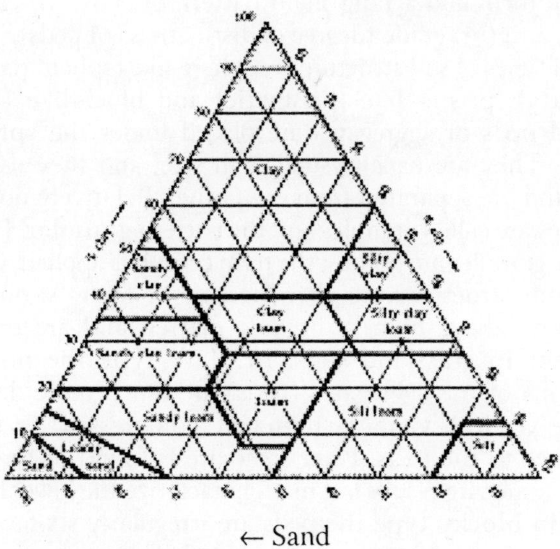

Figure 18.7 Relationship between the particle size distribution of a soil and its class name. In using the diagram the points corresponding to the percentages of silt and clay present in the soil are located on the silt and clay lines, respectively. Lines are then projected inward from % silt parallel to the clay side of the triangle and inward from % clay parallel to the sand side. Note that the third particle size (in this case, sand) is defined by the two projected lines. The name of the compartment in which the lines intersect is the texture class of the soil in question (Thompson and Troeh 1973).

arrangement of the primary soil separates (sand, silt, clay) into secondary units, called **aggregates** or **peds**. A soil profile may be dominated by a single type of structure. More often, however, several types are encountered in the different horizons. Soil conditions and characteristics such as water movement, heat transfer, aeration, and porosity are much influenced by structure.

Aggregate stability, and the shapes and sizes of aggregates, are the most important characteristics of soil structure. **Stability** means the capacity of the peds to retain their structure, that is, to absorb water and not disintegrate. There are various agents which bind these aggregates. Fungal mycelia, microbial exudates, hydrogen and calcium ions, and certain clays promote stability; sodium ions, and expansion and contraction of certain clays promote instability.

Soil structure is best studied in the field under natural conditions and it is described under three categories; (i) type (shape or form and arrangement pattern of peds, (ii) class (size of peds), and (iii) grade (degree of distinctness of peds). The four principal types of soil structure are sphere-like (spheroidal), plate-like (platy), prism- like (prismatic) and block-like (blocky). Rounded peds or aggregates are placed under the **spheroidal** category. They are usually smaller in size, and they usually lie loosely and are separated from each other. Relatively nonporous aggregates are called granules and the pattern **granular**. However, when the granules are porous, the term **crumb** is applied. Granular and crumb structures are characteristic of many surface soils, particularly those high in organic matter, and are especially prominent in grassland soils. In **platy type** the horizontal dimensions of the aggregates (peds) are much more developed than the vertical, giving a flattened, compressed, or lens-like appearance to the peds. Platy structure is found in the surface layers of some virgin soils but may characterize the lower horizons as well. In **blocky type** the peds are irregularly six-faced, with their three dimensions more or less equal. These fragments range from about 1 to 10 cm in thickness. When the edges of the cubes (blocks) are sharp and the rectangular faces distinct, the subtype is designated **blocky**, and when surrounding has occurred, the aggregates are referred to as **subangular blocky**. Block-like structures usually are confined to the sub-soil, and their stage of development and other characteristics have much to do with soil drainage, aeration, and root penetration.

Each primary structural type of soil is differentiated into five **size-classes** depending upon the size of the individual peds. The terms commonly used for the size classes are: very fine or very thin, fine or thin, medium, coarse or thick, very coarse or very thick. The terms thin and thick are used for platy types, while the terms fine and coarse are used for other structural types.

Grade indicates the degree of distinctness of the individual peds. It is determined by noting the stability of the aggregates and the ease with which they separate from other peds. Grade of structure is influenced by the moisture content of the soil. Under moist conditions the soil structure is usually not very distinct. With exposure and drying, the structure becomes much stronger. Four terms commonly used to describe the grade of soil structure are: (a) **structureless:** there are no noticeable peds, such as conditions exhibited by loose sand or cement-like condition of some clay soils; if the appearance is coherent as in compact clay, the term used is **massive,** if non-coherent as in loose sand, **single grain;** (b) **weak:** indistinct formation of peds which are barely durable; (c) **moderate:** moderately well-developed peds which are fairly durable and distinct; and (d) **strong:** very well formed peds which are quite durable and strong.

For naming a soil structure the sequence followed is grade, class, and type; for example, strong coarse angular blocky. Often compound structures are met within the soil under natural conditions. On drying or when gently disturbed, they break into individual small structural peds that constitute the compound structure. For example, large prismatic or columnar types may break into medium blocky structure. Considerable care and experience are needed for proper and accurate designation of the soil structure.

Table 18.5 presents illustrations and descriptions of several common soil aggregates and the classification of soil structures that possess them. In general, granular soils are suitable for plant growth, because both water and air can circulate between the peds. The presence of good structure-granular peds, high degree of aggregate stability- is especially desirable in the topsoil. If the texture of the B horizon is sandy, then structure is of little consequence, since drainage is assured. However, clay soils in B

TABLE 18.5
CLASSIFICATION, DESCRIPTION, AND ILLUSTRATION OF VARIOUS TYPES OF SOIL STRUCTURE (THOMPSON AND TROEH 1973)

Classification	Description	Illustration
Structureless:		
Single grain	Each soil particle independent of all others	
Massive	Entire soil mass clings together, no lines of weakness	
Structured:		
Granular	Primary soil particles grouped into roughly circular peds, such that there is space between them, as they do not fit tightly together. Enhances permeability.	 Granular
Platy	Peds with horizontal dimensions greater than vertical. Common at soil surface, and often associated with lateral movement of water.	 Platy
Blocky	Peds approximately equal in vertical, horizontal dimensions, but fit well together, unlike granular peds. May be angular or subangular.	 Angular blocky Subangular blocky
Prismatic	Peds taller than wide, common in B horizons of well-developed soils. May be columnar, due to eluviation (old age or high sodium content will produce columnar peds of prismatic peds).	 Prismatic
Structure destroyed:		
Puddled	Soils disturbed when wet puddle or run together; structure is destroyed, pores collapse.	

horizon can create a "claypan" or *hard pan,* which has a deleterious effect on plant life.

18.2.3 Density of soil

Density represents weight per unit volume of a substance. Soil density is expressed in two well accepted concepts - particle density and bulk density. The weight per unit of the solid portion of soil is called **particle density.** It depends upon the accumulative densities of the individual inorganic and organic constituents of the soil. Particle density for the normal soils is generally 2.65 g per cubic centimeter. It is higher (2.75 g/cc) if large amounts of heavy minerals such as magnetite, garnet, epidote, zircon, tourmaline, or hornblende, are present in the soil. With increase in organic matter of the soil, the particle density may decrease to 1.1 to 1.4 g/cc.

The oven dry weight of a unit volume of soil inclusive of pore spaces (expressed in gramme per cubic centimeter) is called the **bulk density.** It is estimated by determining the oven dry weight of an undisturbed core of soil of known volume:

$$\text{Bulk density} = \frac{\text{Weight of the oven dry soil}}{\text{Volume of the soil (solids + pores)}}$$

Different types of cylinders are used to collect the soil core in its most natural condition. The bulk density of a soil is always smaller than its particle density. For example, the bulk density of sand is about 1.7 g/cc, whereas that of organic matter like peat is about 0.5 g/cc. Bulk density normally decreases as mineral soils become finer in texture, and clay loams generally have lower bulk densities than sandy soils. It is generally higher in lower profile layers which apparently results from a lower content of organic matter, less aggregation and root penetration, and a compaction caused by the weight of the overlying layers. Since bulk density relates to the combined volumes of the solids and pore spaces, soils with high proportion of pore space to solids have lower bulk densities than those that are more compact and have less pore space.

18.2.4 Porosity of soil

Interstitial space around the soil particles is known as **pore space.** Part of the pore space is filled with water, while the rest is occupied by air. It is calculated as follows:

$$\text{Pore space (\%)} = 100 - \left[\frac{\text{Bulk density}}{\text{Particle density}}\right] \times 100$$

For determining pore space, the soil surface is scraped to a flat plain, and a straight walled pit (10 x 10 x 10 cm) is dug up. The soil removed is collected in a beaker and dried in oven at 105°C to a constant weight. The volume of the pit is determined by filling the pit completely with known volume of sand dropped through a measuring cylinder. The pore space is calculated as:

$$\text{Pore space } (\%) = \frac{2.65 - D}{2.65} \times 100$$

where, D is the volume weight (apparent specific gravity) of soil and is calculated as oven dry weight of soil sample removed from the pit divided by the volume of the pit, and 2.65 is the approximate specific gravity of soil particles.

The total pore space is quite variable among soils. Sandy surface soils show a range of 35-50%, whereas medium- to fine-textured soils vary from 40 to 60% and even more in the presence of organic matter. Porosity normally decreases with depth in the soil, and cropping also tends to lower the total pore space compared to that of uncropped soils.

Two types of individual pore space - **macro** (>0.06 mm diameter) and **micro** (<0.06 mm diameter) - occur in soils. The macropores allow the non-capillary ready movement of air and percolating water and do not hold much water under normal conditions. In contrast, the micropores (capillary) are mostly filled with water in a moist field soil and do not permit much air movement into or out of the soil. The water movement is also slow. Thus, even though a sandy soil has relatively low total porosity, the movement of air and water through such a soil is surprisingly rapid because of the dominance of the macropores. Fine-textured soils, especially those without a stable granular structure, allow relatively slow gas and water movement despite the unusually large volume of total pore space. Here the dominating micropores often stay full of water. Aeration, especially in the subsoil, often is inadequate for satisfactory root development and desirable microbial activity. Obviously, the size of the individual pore spaces rather than their combined volume is the important consideration. The loosening and granulating of fine-textured soils promotes aeration not so much by increasing the total pore space as by raising the proportion of macrospaces.

18.2.5 Soil consistence

It represents, at varying moisture conditions, the degree and kind of cohesion and adhesion of soil material or resistance offered to the forces tending to deform or rupture the soil mass.

Consistence of soil depends on the texture, nature and amount of inorganic and organic colloids, structure, and especially the moisture content of soil. Terms used to describe soil consistence at three moisture levels (wet, moist, and dry) are give in Table 18.6. **Stickiness** (the quality of adhesion to other objects) of wet clay soils (at about field capacity moisture content) are commonly termed **slightly sticky, sticky,** or **very sticky,** and their **plasticity** (the capacity to form moulds) would be indicated as **slightly plastic, plastic,** or **very plastic.** A wet sand, on the other hand, would be **nonsticky** and **nonplastic**. Gradation in stickiness and plasticity would be found with medium-textured soils such as clay loams, silt loams, and some loams. The consistence of moist soils varies from **loose** to **extremely firm.** Sandy soils (and some high in silt) are mostly loose, whereas soils high in sticky clays are usually extremely firm. Moist soils that are easily crushed and yet show coherence when pressed together are termed **friable** or **very friable**, conditions that are prized for ease of manipulation and management. Dry soil with little coherence also will be **loose.** Those with high coherence are **extremely hard** and are so resistant to pressure that clods cannot be broken by hand.

TABLE 18.6
TERMS USED TO DESCRIBE CONSISTENCE OF SOILS

	Wet Soils		*Moist soils*	*Dry soils*
	Stickiness	*Plasticity*		
Increasing coherence	Nonsticky Slightly sticky Sticky Very sticky	Nonplastic Slightly plastic Plastic Very plastic	Loose Very friable Friable Firm Very firm Extremely firm	Loose Soft Slightly Hard hard Very hard Extremely hard

Cementation refers to brittle, hard consistence caused by some cementing agents such as calcium carbonate, silica, or oxides or salts of iron and aluminum; and, the degree of cementation of soil materials, independent of soil moisture, is also considered in identifying soil consistence. Terms such as *weakly cemented, strongly* *cemented*, and *indurated* are used to define categories of cementation.

Contraction and expansion in volume of soil with changing moisture content are more marked in clayey soils. Shrinkage on drying in fine clayey soils and highly organic soils produces prominent cracks. With increase in moisture, the water is imbibed, especially between the lattice structure of clay particles and leads to swelling.

18.2.6 Soil colour

The colour of the soil varies widely amongst various kinds of soils as well as within different horizons of a soil profile. The variation in soil colour are due mainly to the organic matter content which generally imparts black to dark grey tinges; iron compounds which are responsible for red, brown, and yellow tinges; and silica, lime and other salts which give light, white, and grey tinges. The surface horizon of soil is commonly darkest, due mainly to the presence of organic matter. Red colour is associated with unhydrated ferric oxides, whereas yellow colour indicates some degree of hydration. Red colour is generally suggestive of relatively old and intensely weathered, well-drained soils such as lateritic soils in tropics. Brown is the most common soil colour and is due to a mixture of the organic matter and iron oxides. Some of the bluish and greenish colours are due to the presence of ferrous compounds created by reducing conditions in imperfectly drained soils such as marsh and high-water-table bottom lands. Colour variegation or mottling in soils indicates alternating good and poor aeration, a condition not conducive to proper plant growth. Also, the colour of soil when wet is normally darker than when it is dry; and the dark coloured soils absorb more heat than light coloured soils.

The soil colours are best determined by comparison with standard colour charts. The **Munsell soil colour chart** is commonly used for this purpose. The colour of the soil is a result of the light reflected from the soil, and it depends upon the combination of three simple variables of colour; namely, hue, value and chroma. **Hue** is the dominant spectral colour. **Value** is the brilliance or total quantity of light. **Chroma** is the relative purity or saturation of the dominant spectral colour. The munsell colour notations are systematic numerical and letter designations of each of these threee variables. For example, the numerical notations 2.5 YR 5/6 constitutes 2.5YR as hue, 5 as value, and 6 as chroma. The equivalent soil colour is "red".

18.2.7 Soil temperature

The temperature of a soil greatly affects the physical, biological, and chemical processes occurring in that soil. In cold soils, chemical and biological rates are slow. In cold soils, chemical and biological rates are slow. Biological decomposition can come to a near standstill, thereby limiting the rate at which nutrients such as nitrogen, phosphorus, sulphur, and calcium are made available. Also, absorption and transport of water and nutrient ions by higher plants are adversely affected by low temperature. Plant growth slackens below 18°C soil temperature. The microbial oxidation of ammonium ions to nitrate ions, which occurs most readily at temperatures 27-32°C, is negligible when the soil temperature is lowered to about 10°C. Alternate freezing and thawing subject the soil aggregates to pressures as ice crystals form and expand and alter the physical structure of soil. Freezing and thawing of the upper layers of soil can also result in **heaving** of perennial forage crops such as alfalfa. This action, which is most severe on bare, imperfectly drained soils, can kill alfalfa plants, as well as those of some clovers, and trefoil. Changes in soil temperature have the same effect on shallow house foundations and roads with fine material as a base. Such structures show the effects of heaving in the spring.

The amount of heat radiation from the sun that enters the soil is controlled by climate, colour of the soil, altitude and aspect of the land, and the vegetative cover present on the soil. Soil moisture is the most vital controlling factor in soil temperature. *Specific heat* (the amount of heat required to raise the temperature of one gram of a substance by 1°C) of pure water is 1 cal/g; and, that of dry soil is about 0.2 cal/g. Hence, a dry soil is more easily heated than a wet one. The temperature of the upper few centimeters of wet soil is commonly 3-6°C lower than that of a moist or dry soil. This is partially due to evaporation and partially to specific heat. The evaporation of water from soil surfaces requires a large amount of energy, 540 kcal (kilocalories) for every kilogram of water vapourized. This energy must be provided by solar radiation or it must come from the surrounding soil. In either case, evaporation has the potential of cooling the soil.

The heat radiation from the sun slowly penetrates the profile, largely by conduction. The rate of conduction in soil is influenced by a number of factors, the most important of which is the moisture content of the soil layers. Heat passes from soil to water about 150 times more easily than form soil to air. As the water content increases in the soil, the air content decreases, and the transfer resistance is lowered remarkably.

With a clear sky, the air temperature in temperate regions rises from lowest in the morning to a maximum at about 2 p.m. The surface soil, however, does not reach its maximum until later in the afternoon because of the usual lag. This retardation is greater and the temperature change is less as the depth increases. Thus, the deep subsoils are less apt to freeze in the winter months, thereby protecting the roots of perennial crop plants and trees.

Man can influence the soil temperature through drainage management of soils and by soil cover. Good structured soils warm up sooner than those with poor structure because they become dry more quickly. Draining a soil also helps to make the soil warmer.

Soil temperatures are influenced by soil cover and especially by organic residues or other types of **mulch** placed on the soil surface. In periods of hot weather, mulches keep the surface soil cooler than where no cover is used; in contrast, during cold spells in winter, they moderate rapid temperature declines. Mulches tend to buffer extremes in soil temperatures. In warm regions, mulching may provide the double benefit of increased rooting in the more fertile topsoil plus decreased evaporation of water from the soil surface.

18.2.8 Soil air

For the respiration of the roots of plants as well as microorganisms in the soil, oxygen must be supplied and CO_2 removed. Through the soil aeration, there is an exchange of these two gases between the soil and the atmosphere. In a well-aerated soil, this exchange is sufficiently rapid to prevent the deficiency of O_2 or the toxicity of excess CO_2. For most land plants, the supply of O_2 in the soil air must be kept above 10%, and that of CO_2 and other potentially toxic gases such as methane must not be allowed to build up excessively.

Under field conditions, poor soil aeration occurs under two conditions: (i) when the moisture content is so high that there is little or no room for gases as in waterlogged condition, and (ii) when the exchange of gases with the atmosphere is so slow that desirable levels of soil gases cannot be maintained. The latter may occur even when sufficient total air space is available. The more rapid the usage of O_2 and the corresponding release of CO_2, the greater is the need for the exchange of gases between the soil and the atmosphere above. This exchange is facilitated by two mechanisms: **mass flow** and **diffusion**. Mass flow (bulk flow) of air, which is due to pressure differences between the atmosphere and soil air, involves movement of air molecules *en masse.* It is enhanced by fluctuations in soil moisture content. As water moves into the soil during a rain or from irrigation, air must be forced out. Likewise, when soil water is lost by evaporation from the soil surface or is taken up by plants, air is drawn into the soil. Nevertheless, mass flow is less important than diffusion in determining the total exchange of air that occurs between soil and atmosphere.

Most of the gaseous interchange in soils occurs by diffusion. Through this process, each gas moves in a direction determined by its own partial pressure. The *partial pressure* of a gas in a mixture is simply the pressure this gas would exert if it alone were present in the volume occupied by the mixture. Thus, if the pressure of air is 1 atmosphere (approximately 1 bar), the partial pressure of O_2, which makes up about 21% of the air by volume, is approximately 0.21 bar.

Diffusion allows extensive gas movement from one area to another even though there is no overall pressure gradient for the total mixture of gases. There is, however, a concentration gradient for each individual gas, which may be expressed as *partial pressure gradient.* As a consequence, even though the total soil-air pressure and that of atmosphere may be the same, a higher concentration of O_2 in the atmosphere will result in a net movement of this particular gas into the soil. Carbon dioxide and water vapour normally move in the opposite direction since the partial pressures of these two gases are generally higher in the soil air than in the atmosphere.

18.2.8.1 Factors affecting soil aeration

The most important factors influencing the aeration of well-drained soils are those that determine the volume of the soil's macropores which affect the total air space as well as gaseous exchange and biochemical reactions. Soil texture, bulk density, aggregate stability, and organic matter content are among the soil properties that help determine macropore content and, in turn, soil aeration.

The concentrations of both O_2 and CO_2 are definitely affected by microbial decomposition of organic residues. Incorporation of large quantities of manure, crop residues, or sewage sludge may alter the soil air composition appreciably. Respiration by higher plant roots and by soil organisms around the roots is also a significant process. Also, subsoils are usually more deficient in O_2 than are top soils because the total pore space as well as the macropore space is generally much lower in the deeper horizons.

18.2.8.2 Effects of soil aeration

Soil aeration influences many soil reactions and, in turn, soil properties. The most obvious of these reactions are associated with the microbial breakdown of organic residues. Poor aeration slows down the rate of decay as evidenced by the relatively high organic matter level of poorly drained soils, especially in swampy areas.

The nature as well as the rate of organic decay is determined by the O_2 content of the soil; in its presence aerobic organisms are active and oxidation reactions such as the following occur:

$$C_6H_{12}O_6 + 6O_2 \longrightarrow 6CO_2 + 6H_2O$$

whereas, in the absence of O_2, anaerobic organisms take over, and much slower breakdown occurs through reactions such as the following:

$$C_6H_{12}O_6 \longrightarrow 3CO_2 + \underset{\text{(methane)}}{3CH_4}$$

Less complete decomposition than that described in the two preceding equations may yield certain organic acids and ethylene gas (C_2H_4), which can be toxic to higher plants.

The level of O_2 largely determines the forms of several inorganic elements (Table 18.7). The oxidized states of nitrogen and sulphur are readily utilizable by higher plants. In general, the oxidized forms of other elements are also more desirable for most crops grown on acid soils of humid regions. This is because the reduced forms of iron and managanese are sometimes present at toxic levels in moist acid soils. In the drier areas, the opposite is generally true, and reduced forms of elements such as iron and manganese are preferred. In the neutral to alkaline soils of these drier areas, oxidized forms of iron and manganese are tied up in highly insoluble compounds, resulting in deficiencies of these elements.

TABLE 18.7
OXIDIZED AND REDUCED FORMS OF SOME IMPORTANT ELEMENTS

Element	*Normal form in well-oxidized soils*	*Reduced form found in waterlogged soils*
Carbon	CO_2	CH_4
Nitrogen	NO_3^-	N_2, NH_4^+
Sulphur	SO_4^{2-}	H_2S, S^{2-}
Iron	Fe^{3+}(ferric oxides)	Fe^{2+} (ferrous oxides)
Manganese	Mn^{4+}(manganic oxides)	Mn^{2+} (manganous oxides)

Higher plants are adversely affected in at least three ways by conditions of poor aeration: (i) the growth of the plant, particularly the roots, is curtailed; (ii) the absorbtion of nutrients and water is decreased; and (iii) formation of certain inorganic compounds toxic to plant growth is favoured.

18.2.8.3 Oxidative-reduction (redox) potential

An indication of the oxidation and reduction states of systems (including those in soils) is given by the oxidation-reduction potential or redox potential (Eh). It provides a measure of the tendency of a system to reduce or oxidize chemicals and is usually measured in volts (V) or millivolts. If Eh is positive and high, strong oxidizing conditions exist. If it is low and even negative, elements are found in reduced forms (Table 18.8). In a well-drained soil, the Eh is in the 0.4-0.7 V range. As aeration is reduced, the Eh declines to a level of about 0.3- 0.35 V.

TABLE 18.8
OXIDIZED AND REDUCED FORMS OF CERTAIN ELEMENTS IN SOILS AND THE REDOX POTENTIALS (Eh) AT WHICH CHANGE IN FORMS COMMONLY OCCURS (PATRICK AND REDDY 1978)

Oxidized form	*Reduced form*	*E_h at which change of form occurs*
O_2	H_2O	0.38 to 0.32
NO_3^-	N_2	0.28 to 0.22
Mn^{4+}	Mn^{2+}	0.28 to 0.22
Fe^{3+}	Fe^{2+}	0.18 to 0.15
SO_4^{2-}	S^{2-}	-0.12 to -0.18
CO_2	CH_4	-0.2 to -0.28

18.2.9 Soil water

Water used by plants comes from water held by the soil, and availability of soil water for plant growth is ultimately determined by soil water potential (ψsoil), which is the difference between the free energy of soil water and that of pure water (ψ). The ψ of pure water has been assigned a value of 0 (zero) bars. A **bar** is a pressure unit directly relatable to energy per unit mass (1 bar = 0.1 MPa = 0.987 atmospheres = 10^6 ergs/ g).The chemical energy of water in the soil is lower than pure water and therefore Ψ_{soil} values are expressed as negative numbers.

The Ψ_{soil} may be broken into several components which influence its chemical energy:

$$\Psi_{soil} = \Psi_m + \Psi_\pi + \Psi_g$$

where, Ψ_m is matric potential, Ψ_π the osmotic potential, and Ψ_g denotes the gravitational potential.

A material with surfaces that bind water is called a **matrix**, and the **matric potential** of soil water is the result of two phenomena, adhesion (or adsorption) and capillarity (Fig. 18.8). The soil solids tightly adsorb water, whereas capillary forces are responsible for water's being held in the capillary pores. The net effect of these phenomena is to reduce the free energy of soil water as compared to that of unadsorbed pure water. Consequently, ψm are always negative. The ψm exerts its effect not only on soil moisture retention but on soil water movement as well. Differences between the ψm of two adjoining zones of a soil encourage the movement of water. Water moves from a moist zone (high free energy) to a dry zone (low free energy). Although this movement may be slow, it is extremely important in supplying water to plant roots.

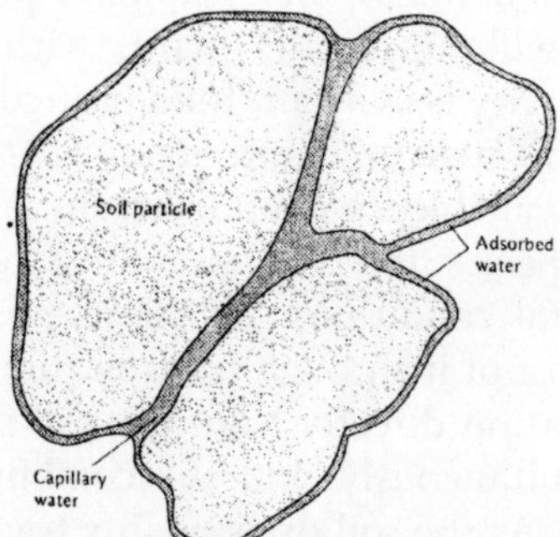

FIGURE 18.8
TWO "FORMS" OF WATER THAT TOGETHER GIVE RISE TO MATRIC POTENTIAL (ADSORBED AND CAPILLARY). THE SOIL-SOLIDS TIGHTLY ADSORB WATER, WHEREAS CAPILLARY FORCES ARE RESPONSIBLE FOR WATER'S BEING HELD IN THE CAPILLARY PORES

The **osmotic potential** (Ψ_π) is attributable to the presence of solutes in the soil - in other words, to the soil solution. The solutes may be inorganic salts or organic compounds. They reduce the free energy of water, primarily because the solute ions or molecules attract the water molecules. Unlike the ψm, the Ψ_π has little effect on the mass movement of water in soils. Its major effect is on the uptake of water by plant roots. In soils high in soluble salts, Ψ_π may be greater in the soil solution than in the plant root cells. This leads to constraints in the uptake of water by the plants.

The force of gravity acts on soil water the same as it does on any other body, the attraction being toward the earth's center. Water entering the soil will fill most pore spaces and will drain downward through pores in response to the pull of gravity (the

gravitation potential ψ_g). While the ψm and ψ_π are always negative because they reduce the free energy of water, the ψ_g above the reference point (elevation) is always positive.

Gravitational potential is only significant in saturated soils. Water moving through soil in response to gravitational forces is *gravitational water*. Once gravitational water has moved from the soil or is dispersed in the soil, the soil is left at field capacity (FC). **Field capacity** is the amount of water that can be retained by a soil after gravitational water has been removed. At this time, water has moved out of the macropores, and its place has been taken up by air. The micropores (capillary pores) are still filled with water and will supply the plants with needed water. In general, field capacity is constant for a particular soil, depending on soil texture, soil structure, and organic content.

The water available for plant use from soils at field capacity depends on the species, the physiological state of the individual, and environmental conditions. As plants absorb water from a soil, they lose most of it through transpiration. Some water also is lost by evaporation directly from the soil surface. Those two losses occur simultaneously, and the combined loss is termed *evapotranspiration*. As the soil dries, plants begin wilt to conserve moisture during the daytime. At first the plants will regain their vigour at night, but ultimately they will remain wilted night and day. This lower limit of water availability is designated as the **permanent wilting point (PWP)**. PWP is reached when ψsoil is equal to or below the minimum osmotic potential of the plant so that water cannot be removed from the soil and the plant wilts, and remains wilted even if placed in a cool, dark, humid chamber. Addition of soil water may allow recovery. The amount of water between FC and PWP is called **available water** or **capillary water**, since it is retained against the pull of gravity by capillary action and matric forces. Plants are able to utilize capillary water to varying extents but cannot remove all water from the soil. Below the PWP, water is held so firmly to the soil matrix that it is removable only under conditions surpassing biological tolerance. The most important components of water potential in nonsaturated soils are matric potential and, in some case such as saline soils, osmotic potential.

The relationship between FC, soil water content, and ψsoil depends to a large degree on soil texture (Fig.18.9). Clay soils hold large volumes of water at relatively lower water potentials than do coarse textured soils. This is because of the greatest total pore space and greater particle-water contact space in the fine textured soils.

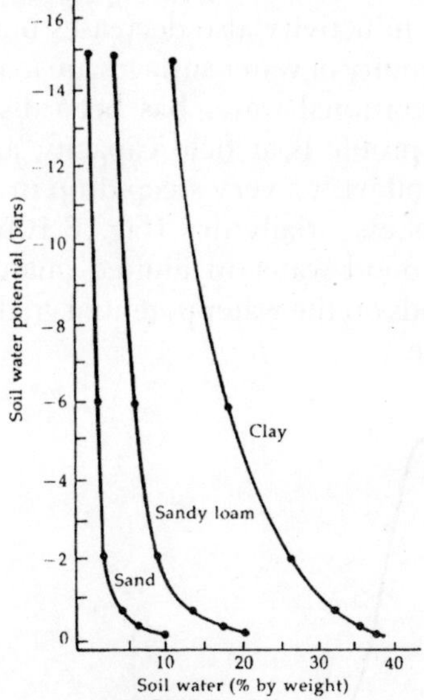

FIGURE 18.9

TYPICAL WATER CHARACTERISTICS CURVES FOR SAND, SANDY LOAM, AND CLAY SOILS (1 BAR = 0.1 MP_a) (SLATYER 1967)

18.2.9.1 Movement of soil water

Availability of water within a soil depends, in part, on how quickly and how far water will move upward from the water table toward an absorbing root, or into the soil during a rain storm. **Infiltration** and movement of soil water depends on factors which determine soil water potential. Water will infiltrate into a dry soil faster than a wet soil if they have comparable texture and structure, because the gradient in water potential is greater in the dry soil. The rate of flow of water through a saturated soil (v) is:

$$V = -K\frac{\Delta\Psi_{soil}}{D}$$

where K is the **hydraulic conductivity** (the capacity of a soil to transport water); ψsoil is the water potential of the soil; and D is the depth in the soil. As ψsoil drops, there is a significant drop in hydraulic conductivity (Fig.18.10a), especially in sandy soils where the continuity of the water is broken. In finer textured soils hydraulic conductivity also decreases but to a lesser extent because the continuity of water surfaces are maintained at a lower ψsoil. After gravitational water has been dispersed, the upper zone of the soil profile is at field capacity, and then there is a narrow zone of soil with a very steep drop in water content; the soil below that is essentially dry (Fig.18.10b). Water may rise from an underground water in limited quantities. The rate of movement depends on the water potential gradient and the depth to the water table.

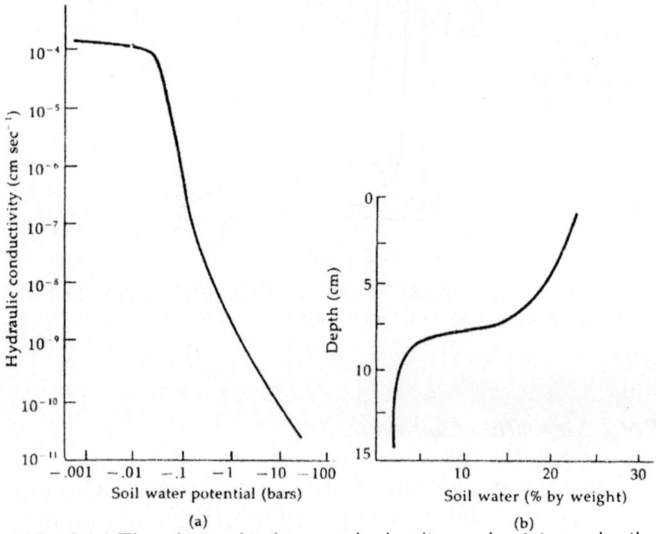

Figure 18.10 (a) The relationship between hydraulic conductivity and soil water potential in a loam soil (1 bar = 0.1 MPa). (b) The water content profile of an initially dry loam soil at one day following irrigation (Slatyer 1967).

Lateral movement of water through the soil to zones of absorption by roots depends on the water potential gradient and the hydraulic conductivity of the soil. Root water potential must be lower than ψsoil for absorption to occur. This water potential difference can be quite low when ψsoil is -5bars or higher, whereas at -15 bars, a significantly greater drop in potential is necessary for absorption. This is a result of the decrease in hydraulic conductivity as ψsoil drops. The distance through which water will move towards the dry zone surrounding a root depends on the plant's tolerance for stress and the soil properties which influence hydraulic conductivity.

18.3 Soil Chemistry

One of the most important aspects of soil chemistry is the state and the relationship between chemical elements in soil systems: as soluble ions in the soil solution, as adsorbed ions on charged particles, and as constituents of mineral and organic particles. Soil chemistry deals primarily with reactive materials, which are the chemicals that are of significance to living things. Reactive materials tend toward equilibria in a changeable environment. There is virtually no correlation between the absolute and relative amounts of an element present in the soil and the significance of that element to life. Oxygen (49.5%), silican (25.8%), and aluminum (7.5%) are the most abundant, the second most abundant, and the third most abundant elements, respectively, in the earth's crust. Yet, they are of little consequence to life and are relatively inaccessible due to the bound nature of their chemical compounds. By contrast, the most important elements in soil chemistry (Table 18.9) are, all together, less than 16% of the earth's crust materials. Soil pH, cation exchange capacity, and the nature of the interface between the solid phase and the liquid phase are all important aspects of soil chemistry.

18.3.1 Soil colloids

If a spoonful of ordinary clay soil is placed in a glass of water and agitated thoroughly a murky liquid, which appears uniformly brown in colour, results. If we allow this mixture to stand, it starts to clear rapidly, the larger particles of soil settling out first, followed by smaller and smaller particles. However, after a considerable length of time, it becomes evident that not all of the soil is going to settle out. Some of it remains indefinitely in suspension. The stable heterogeneous mixture that results is called a **colloidal suspension** - in this case composed of very minute

TABLE 18.9
ELEMENTS IMPORTANT IN SOIL CHEMISTRY AND THEIR CHEMICAL SYMBOLS AND PRINCIPAL FORMS (THOMPSON AND TROEH 1973)

Element	*Principal Ions*
Aluminum	Al^{3+}
Boron	$B_4O_7^{2-}$
Calcium	Ca^{2+}
Carbon	CO_3^{2-}, HCO_3-
Chlorine	Cl^-
Cobalt	CO^{2+}
Copper	Cu^{2+}
Hydrogen	H^+, OH^-
Iron	Fe^{2+}, Fe^{3+}
Magnesium	Mg^{2+}
Manganese	Mn^{2+}, MnO_4^-
Molybdenum	MoO_4^{2-}
Nitrogen	NH_4+, NO_2-, NO_3-
Oxygen	with other elements
Phosphorus	$H_2PO_4^-$, HPO_4^{2-}
Potassium	K^+
Sodium	Na^+
Sulphur	SO_4^{2-}
Zinc	Zn^{2+}

soil particles (clay micelles) suspended in water. The suspended phase is termed the **dispersed phase**, and the medium in which the dispersion takes place is called the **dispersion medium**.

Particles of colloidal dimension range in size from 1 to 200 millimicrons (mμ.) A millimicron is one-thousandth of a micron (μ) or one-millionth of a millimeter. They are too small to be seen with the light microscope, but are large enough to scatter light. Because of this ability to scatter light, the presence of colloidal particles may be detected with the ultramicroscope.

Because of their small size, all soil colloids expose a large external surface per unit mass. The external surface area of 1g of colloidal clay is at least 1000 times that of 1g of coarse sand. Some colloids, especially certain silicate clays, have extensive internal surfaces as well which occur between plate-like crystal units that make up each particle. The total surface area of soil colloids ranges from 10 m^2/g for clays with only external surfaces to more than 800 m^2/g for clays with extensive internal surfaces.

Soil colloidal surfaces, both external and internal, characteristically carry negative charges, although some mineral colloids in very acid soils have a net electropositive charge. The colloidal particles, referred to as *micelles (microcells),* attracts hundreds of thousands of positively charged cations, such as H+, Al^{3+}, Ca^{2+}, and Mg^{2+}. This gives rise to an ionic double layer (Fig. 18.11). The colloidal particle constitutes the *inner* ionic layer, being essentially a huge anion, with both external and internal layers that are negative in charge. The *outer* layer is made up of a swarm of rather loosely held (adsorbed) cations attracted to the negatively charged surfaces. In addition to the adsorbed cations, a large number of water molecules are also associated with colloidal particles. Some are attracted to the adsorbed cations, each of which is hydrated. Others are held in the internal surfaces of the colloidal particles.

18.3.1.1 Types of soil colloids

There are four major types of colloids present in soils: layer silicate clays, iron and aluminum oxide clays, allophane and associated amorphous clays, and humus.

The most important property of the **layer silicate clays** is their crystalline, layer-like structures. Each particle is made up of a series of layers much like the pages of a book. The layers comprise primarily horizontally oriented sheets of silicon, aluminum, magnesium, and/or iron atoms surrounded and held together by oxygen and hydroxyl groups. The formula of one of these clays, kaolinite ($Si_2Al_2O_5(OH)_4$), illustrates their general chemical make up. Silicate clay minerals are the most prominent inorganic colloids in soils of temperate areas and are prominent in soils of the tropics as well.

Iron and aluminum oxide clays dominate in the highly weathered soils of the tropics and subtropics. Examples are gibbsite ($Al_2O_3.3H_2O$) and goethite ($Fe_2O_3.3H_2O$). These are amorphous and not as sticky and plastic as the layer silicate clays. At high pH values, the micelles carry a small negative charge and are surrounded by a corresponding equivalent of cations. In very acid soils, however, some Fe, Al oxides are positively charged and attract negatively charged anions instead of cations.

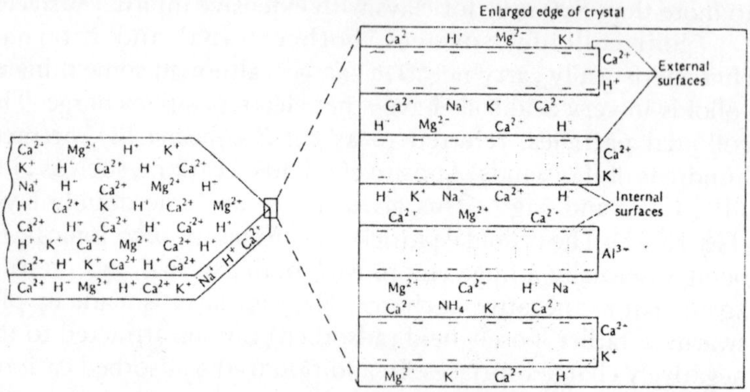

Figure 18.11 Diagrammatic representation of a silicate clay crystal (micelle) with its sheet-like structure, its innumerable negative charges, and its swarm of adsorbed cations. The enlarged schematic view of the edge of the crystal illustrates the negatively charged internal surfaces of this particular particle, to which cations and water are attracted. Note that each crystal unit has definite mineralogical structure.

Perhaps the most significant amorphous silicate colloid is **allophane.** This mineral is a somewhat poorly defined aluminum silicate with a general composition approximating $Al_2O_3.2SiO_2H_2O$. It is most prevalent in soils developed from volcanic ash. Allophane has a high capacity to adsorb cations but also adsorbs anions.

The colloidal organization of **humus** has some similarities to that of clay. A highly charged micelle is surrounded by a swarm of cations. The humus colloids are not crystalline, however. They are composed basically of carbon, hydrogen and oxygen rather than of silicon, aluminum, iron, oxygen and hydroxyl groups. The organic colloidal particles vary in size, but they may be as small as the silicate clay particles. The negative charges of humus are associated with partially dissociated enolic (-OH), carboxyl (-COOH), and phenolic (-⬡-OH) groups; these groups in turn, are associated with central units of varying size and complexity. As in the case for the Fe, Al oxides, the negative charge associated with humus is dependent on soil pH. Under very acid conditions, the negative charge is not very high, lower than that of some of the silicate clays. With a rise in pH, however, the hydrogen ions dissociate first from the carboxyl groups and then the enolic and phenolic groups. This leaves a greatly increased negative charge on the colloid. Under neutral to alkaline conditions, the electronegativity of humus per unit weight greatly exceeds that of the silicate clays. In these high pH soils, the adsorbed hydrogen is replaced by calcium, magnesium and other cations.

18.3.1.2 Structure of layer silicate clays

The use of X-ray electron microscopy and other techniques has demonstrated that silicate clay particles are crystalline and that each particle comprises individual layers or sheets. The mineralogical organization of these layers varies from one type of clay to another and markedly affects the properties of the mineral.

The most important silicate clays are known as **phyllosilicates** (*phyllon*=leaf) because of their leaf-like or platelet structure. As shown in Fig. 18.12, they are comprised of two kinds of horizontal sheets, one dominated by silicon, the other by aluminum and/or magnesium.

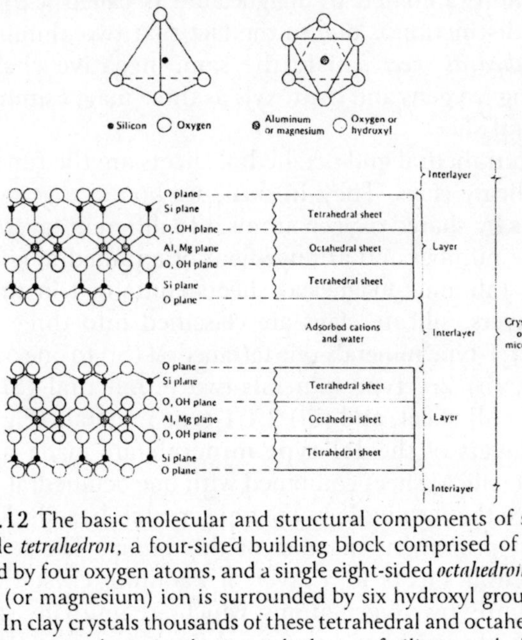

Figure 18.12 The basic molecular and structural components of silicate clays. (a) A single *tetrahedron*, a four-sided building block comprised of a silicon ion surrounded by four oxygen atoms, and a single eight-sided *octahedron*, in which an aluminum (or magnesium) ion is surrounded by six hydroxyl groups or oxygen atoms. (b) In clay crystals thousands of these tetrahedral and octahedral building blocks are connected to give horizontal planes of silicon and aluminum (or magnesium) ions. These planes alternate with planes of oxygen atoms and hydroxyl

groups (heavy circles). The silicon plane and associated oxygen/hydroxyl planes make up a *tetrahedral sheet*. Similarly, the aluminum/magnesium plane and associated oxygen /hydroxyl planes comprise the *octahedral sheet*. Different combinations of tetrahedral and octahedral sheets are termed layers. In some silicate clays these layers are separated by *interlayers* in which water and adsorbed cations are found. Many layers are found in each *crystal* or *micelle* (microcell).

The basic building block for the silica-dominated sheet is a unit composed of one silicon atom surrounded by four oxygen atoms. It is called the **silica tetrahedron** because of its four-sided configuration. An interlocking array of a series of these silica tetrahedra tied together horizontally by shared oxygen anions gives a **tetrahedral sheet.**

Aluminum and/or magnesium ions are the key cations in the second type of sheet. An aluminum (or magnesium) ion surrounded by six oxygen atoms or hydroxyl groups gives an eight-sided building block termed **octahedron**. Numerous octahedra linked together horizontally comprise the octahedral sheet. An aluminum-dominated sheet is known as a **dioctahedral sheet**, whereas one dominated by magnesium is called a *trioctahedral sheet.* The distinction is due to the fact that two aluminum ions in a *dioctahedral sheet* satisfy the same negative charge from surrounding oxygens and hydroxyls as three magnesium ions in a trioctahedral sheet.

The tetrahedral and octahedral sheets are the fundamental units of silicate clays. They, in turn, are bound together within the crystals by shared oxygen atoms into different layers. On the basis of the number and arrangement of tetrahedral (silica) and octahedral (alumina-magnesia) sheets contained in the crystal units or layers, silicate clays are classified into three different groups: (i) 1:1-type minerals-one tetrahedral (Si) to one octahedral (Al) sheet, (ii) 2:1 type minerals-two tetrahedral (Si) to one octahedral (Al) sheet, and (iii) 2:1:1-type minerals (Fig. 18.13).

The layers of the **1:1-type minerals** are made up of one tetrahedral (silica) sheet combined with one octahedral (alumia) sheet- hence the terminology 1:1-type crystal. In soils, **kaolinite** is the most prominent member of this group. The tetrahedral and octahedral sheets in a layer of kaolinite crystal are held together tightly by oxygen atoms, which are mutually shared by the silicon and aluminum cations in their respective sheets. These layers, in turn, are held together by hydrogen bonding.

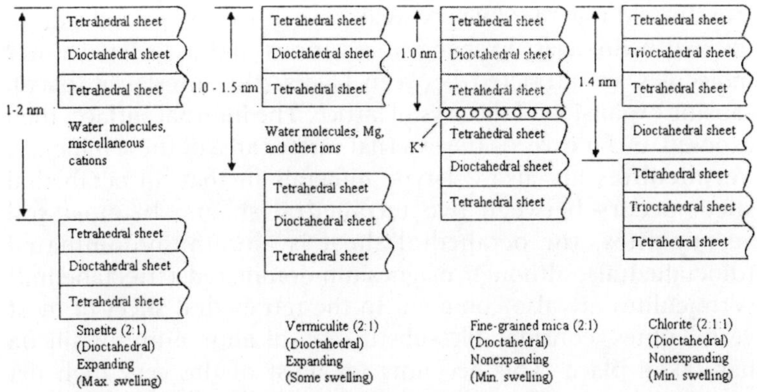

Figure 18.13 Schematic drawing illustrating the organization of tetrahedral and octahedral sheets in three 2:1-type minerals (smectite, vermiculite, and fine-grained mica) and one 2:1:1-type, chlorite. The aluminum-dominated dioctahedral minerals are illustrated for the 2:1 types because such minerals are most commonly found in soils. Trioctahedral chlorite in which magnesium ions dominate is the most common form of this mineral in soils. Note maximum interlayer expansion in smectite, with reduced expansion in vermiculite due to moderate binding power of numerous Mg^{2+} ions. Fine-grained mica and chlorite do not expand because K^+ ions (fine-grained mica) and an extra trioctahedral sheet (chlorite) tightly bind the 2:1 layers together. The interlayer spacings are shown in nanometers (1 nm = 10^{-9} m).

Consequently, the structure is fixed, and no expansion ordinarily occurs between layers when the clay is wetted. Cations and water do not enter between the structural layers of a 1:1-type mineral particle. The effective surface of kaolinite is thus restricted to its outer faces or to its external surface area; this accounts for its low capacity to adsorb cations.

The crystal units (layers) of **2:1-type minerals** are characterized by an octahedral sheet sandwiched between two tetrahedral sheets. Three general groups have this basic crystal structure. Two of them, **smectite** and **vermiculite**, include expanding-type minerals, while the third, **fine-grained micas (illite)**, is relatively nonexpanding. The smectite group is noted for interlayer expansion, which occurs by swelling when the minerals are wetted, the water entering the interlayer space and forcing the layers apart. **Montmorillonite** is the most prominent member of this group in soils. The flake-like crystals of smectites are composed of 2:1 -type layers and these structural layers are loosely held together by very weak oxygen-to-oxygen and cation-to-oxygen linkages. Exchangeable cations and associated water molecules are attracted between layers (the interlayer space), causing expansion of the crystal lattice. The internal surface, then exposed, by far exceeds the external surface area of these minerals. **Vermiculites** are also 2:1-type minerals in that an octahedral sheet occurs between two tetrahedral sheets. In most soil vermiculites, the octahedral sheet is aluminum-dominated (dioctahedral), although magnesium-dominated (trioctahedral) vermiculites are also common. In the tetrahedral sheet of most vermiculites, considerable substitution of aluminum for silicon has taken place. This accounts for most of the very high net negative charge associated with these minerals. Water molecules, along with magnesium and other ions, are strongly adsorbed in the interlayer space of vermiculites. However, they act primarily as bridges holding the units together rather than as wedges driving them apart. The degree of swelling is, therefore, considerably less for vermiculites than for smectites. For this reason, vermiculites are considered *limited-expansion clay minerals,* expanding more than kaolinite but much less than the smectites. The cation exchange capacity of vermiculites usually exceeds that of all other silicate clays, including montmorillonite and other smectites, because of the very high negative charge in the tetrahedral sheet. Vermiculite crystals are larger than those of the smectites but much smaller than those of kaolinite. **Micas** are the type minerals in the group of *nonexpanding minerals.* **Muscovite** and **biotite** are examples of unweathered micas found in the sand and silt separates. Weathered minerals similar in structure to these micas are found in the clay fraction of soils. They are called **fine-grained micas**, Like smectites, fine grained micas have a 2:1 type crystal. However, the particles are much longer than those of the smectites. Also, the major source of charge is in the tetrahedral sheet where about 20% of the silicon sites are occupied by the aluminum atoms. This results in a high net negative charge in the tetrahedral sheet, even higher than those found in vermiculites. To satisfy this charge, potassium ions are strongly attracted in the interlayer space and are just the right size to fit snugly into certain spaces in the adjoining tetrahedral sheet (Fig. 18.13). The potassium thereby acts as a binding agent, preventing expansion of the crystal. Hence, fine-grained micas are *quite non-expansive.*

The **2:1:1 type silicate group minerals** is represented by **chlorites,** which are common in a variety of soils. Chlorites are basically iron-magnesium silicates with some aluminum present. In a typical chlorite clay crystal, 2:1 layers, such as are found in vermiculites, alternate with a magnesium-dominated trioctahedral sheet giving a 2:1:1 ratio (Fig. 18.13). Magnesium also dominates the trioctahedral sheet in the 2:1 layer of chlorites. Thus, the crystal unit contains two silica tetrahedral sheets and two magnesium-dominated trioctahedral sheets, giving rise to the term 2:1:1 or 2:2-type structure. The negative charge of chlorites is about the same as that of fine-grained micas and considerably less than that of the smectites or vermiculites. Also, there is no water absorption between the chlorite crystal units, which accounts for the non-expansive nature of this mineral.

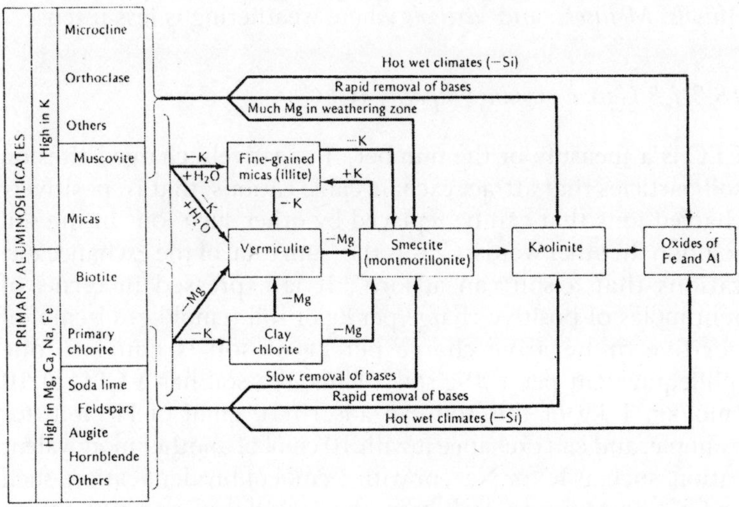

Figure 18.14 General conditions for the formation of the various layer silicate clays and oxides of iron and aluminum. Fine-grained micas, chlorite, and vermiculite are formed through rather mild weathering of primary aluminosilicate minerals, whereas kaolinite and oxides of iron and aluminum are products of much more intense weathering. Conditions of intermediate weathering intensity encourage the formation of smectite. In each case silicate clay genesis is accompanied by the removal of soluble elements such as K, Na, Ca, and Mg.

There are a variety of processes by which individual clays are formed. Fine-grained micas, chlorite, and vermiculite are formed through rather mild weathering of primary aluminosilicate minerals, whereas kaolinite and oxides of iron and aluminum are products of much more intense weathering (Fig. 18.14). Conditions of intermediate weathering intensity encourages the formation of smectites. In each case silicate clay genesis is accompanied by the removal of soluble elements such as potassium, sodium, calcium, and magnesium.

The clay of any particular soil is generally made up of a mixture of different colloidal minerals, and in a given soil, the mixture may vary from horizon to horizon. The well-drained and well-weathered *Oxisols* of humid and subhumid tropics tend to be dominated by oxides of iron and aluminium. Kaolinite is commonly the dominant silicate mineral in *Ultisols* (whose weathering intensity is next only to Oxisols). The smectite, vermiculite, and fine-grained mica groups are more prominent in *Alfisols, Mollisols,* and *Vertisols* where weathering is less intense.

18.3.1.3 Cation exchange capacity (CEC)

CEC is a measure of the number of negatively charged sites on soil particles that attract exchangeable cations, that is, positively charged ions that can be replaced by other such ions in the soil solution. In other words, CEC is the "sum total of the exchangeable cations that a soil can adsorb". It is expressed in terms of centimoles of positive charge per kg of soil (cmol/kg) (1 cmol of positive or negative charge per kg of soil is equal to one milliequivalent per 100 g soil). Thus, if a soil has a CEC of 10 cmol/kg, 1 kg of this soil can adsorb 10 cmol of H^+ ion, for example, and can exchange it with 10 cmol of another monovalent cation, such as K^+ or Na^+, or with 5 cmol of bivalent cation, such as Ca^{2+} or Mg^{2+}. In each case, the 10 cmol of negative charge associated with 1kg of soil is attracting 10 cmol of positive charges, whether they come from H^+, K^+, Na^+, NH_4^+, Ca^{2+}, Mg^{2+}, Al^{3+}, or any other cation.

Three factors strongly influence the CEC of a soil: its clay content, the kinds of clay minerals or amorphous colloids that it contains, and the humus content. The negative charge, which provides for nutrient storage on clay mineral particles, is a result of, for instance, substitution of ions within clay mineral structures or ionization of hydrogen ions from OH groups, which also produces negatively charged surface sites. If an Mg^{2+} or an Fe^{2+}

replaces an Al^{3+} within the octahedral sheet of a layer of silicate clay, there will be a net negative charge which must be balanced on or near that layer. A silicate clay particle with an expanding lattice (such as montmorillinite) will provide not only external surface sites but also internal exchange sites between adjacent layers. Within silicate clays with a non-expanding lattice (for example, kaolinite) there are no interlayer exchangeable cations. Highway engineers value kaolinite clays because they have little tendency to swell and shrink. However, kaolinite clays provide low nutrient storage capacity for plant growth for the agriculturists.

Two major factors will determine the relative proportion of the different cations adsorbed by clays. First, these ions are not all held with equal tightness by the soil colloids. The order of strength of adsorption, when the ions are present in equivalent quantities, is $Al^{3+} > Ca^{2+} > Mg^{2+} > K^+ = NH_4^+ > Na^+$. Second, the relative concentration of the cations in the soil solution will help determine the degree to which adsorption occurs. Thus, in the soil solution of very acid soils, the concentration of both H^+ and Al^{3+} are high, and these ions dominate the adsorbed cations. At neutral pH and above, however, the concentrations in the soil solution of both H^+ and Al^{3+} are very low, and, consequently, the adsorption of these ions is minimal. In neutral to moderately alkaline soils, Ca^{2+} and Mg^{2+} dominate. In some poorly drained soils of arid regions, salts high in sodium accumulate, and the adsorption of Na^+ becomes much more prominent. The dominance of Na^+ ions results in a dispersed condition of the soil colloids, eliminates the macropores, and makes the soils largely impervious to water penetration. Most plants will not grow under these conditions.

Let us consider the interaction of vegetation, soil texture, rainfall regime, and soil microorganisms, in reference to cation exchange in a particular ecosystem, the tropical rainforest (TRF). Soils of the TRF are usually highly weathered and *laterized*: there is little or no horizon development in the reddish-brown loam, which is comprised of aluminum and ferric sesquioxides ($Al_2O_3.Fe_2O_3$). Lateritic soils (Oxisols) are very acidic and nutrient poor. The litter layer is extremely thin because mobilization of nutrients occurs at a very rapid pace in the warm, moist rainforest. Even wood is quickly recycled by subterranean termites.

The luxuriant growth of vegetation (52.5 t/ha/yr gross productivity), however, belies the extremely poor nutrient content of most TRF soils. Walter (1973) stated that the entire nutrient reserve required by the forest is contained in the phytomass. What mechanism allows for litter recycling so that mobile nutrients are not leached away by the inevitable rains? Work by Went and Stark (1968) showed that feeding roots of TRF trees are directly connected to litter. These rootlets exploit the hyphae of mycorrhizal fungi to obtain nutrients from the forest litter. As long as the TRF is not disturbed, the lack of soil storage compartment is unimportant. The nearly complete recycling of materials allows for stability for hundreds or even thousands of years. Yet, deforestation through burning, clearing of the land for cultivation, or logging, removes the only significant storage facility, the *phytomass*, and quickly allows the mobilized ions to move out of the system. The luxuriant growth of virgin forest is never (in human terms) renewed.

18.3.2 Soil solution

The readily available ions are present in the soil solution, and maintain an equilibrium with adsorbed ions. These, in turn, maintain an equilibrium with absorbed mineral ions (Fig. 18.15). Plant removal of nutrients from the soil solution lowers the concentration of those nutrients in the rhizosphere; ions move toward the rhizosphere down a chemical gradient. This shift due to solubility constants of the ions involved, enhances the release of adsorbed ions, and so forth.

For example, plant uptake of phosphorus- roots removing P from the soil solution- establishes a concentration gradient in the soil solution such that the movement of P is toward the root. Under optimum conditions, daily uptake of P in grassland may be as much as 50 times the quantity of P in the solution pool, the pool itself being replenished from the *labile pool* (that is, from adsorbed ions). When conditions of temperature and water are exactly right, the labile pool has the potential of replacing the solution pool 250 times daily. With continued uptake, the concentration of P in the cytoplasm of root cells may exceed concentrations (by up to two orders of magnitude) in ambient soil solution. It is well known that mycorrhizae greatly enhance uptake of P. This enhancement is attributed to the tremendously enlarged contact zone of root and soil, mediated through hyphal fungi.

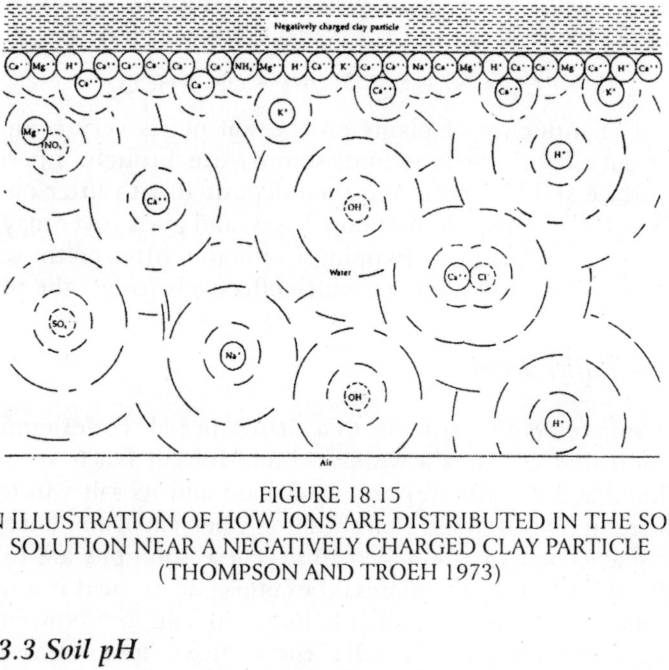

FIGURE 18.15
AN ILLUSTRATION OF HOW IONS ARE DISTRIBUTED IN THE SOIL SOLUTION NEAR A NEGATIVELY CHARGED CLAY PARTICLE (THOMPSON AND TROEH 1973)

18.3.3 Soil pH

Suitability of soil as a medium for the growth of plants and desirable microorganisms depends upon soil reaction-that is, whether the soil solution is acidic, alkaline, or neutral. Typical soils exhibit pH values that range from 4 to 8, although certain soils may have values higher or lower. For instance, saline soils commonly exhibit pH values ranging from 7.3 to 8.5. A pH of less than 6.6 is considered an acid soil.

Soil pH is largely controlled by soil colloids and their associated exchangeable cations. Soil acidity is common in all regions where precipitation is high enough to leach away appreciable quantities of exchangeable base-forming cations (Ca^{2+}, Mg^{2+}, K^+ and Na^+) from the surface layers of soils. In contrast, alkalinity (a characteristic of arid and semiarid regions) occurs when there is a comparatively high degree of saturation with base-forming cations. The presence of calcium, magnesium, and sodium carbonates also can result in a preponderance of hydroxyl ions over hydrogen ions in the soil solution. When salts of strong bases and weak acids, such as Na_2CO_3, K_2CO_3, and $MgCO_3$ dissolve, they undergo hydrolysis and produce OH^- ions which greatly enhance alkalinity. For Na_2CO_3, the reaction is:

$$2Na^+ + CO_3^{2-} + HOH \rightleftharpoons 2Na^+ + OH^- + HCO_3^-$$

The influence of plants on the soil pH is very complex. Nutrients (both cations and anions) are brought up from subsurface soil by plants and then deposited with litter on the surface. Grasses tend to use more bases, and grass cover may act to keep the soil pH from dropping. Coniferous litter, on the other hand, provides acid leachate, which effectively lowers the pH.

18.3.4 Buffer action

Buffering refers to resistance to a change in pH. For example, a solution that contains a weak acid and its salt (such as acetic acid and sodium acetate) or a weak base and its salt will resist changes in hydrogen ion concentration when small amounts of a strong acid or base are added to it. These solutions are called buffer solutions. In the aforesaid example, acetic acid is a weak acid and, therefore, only slightly ionize in solution. Now, if we add a small amount of NaOH, the hydroxyl ions released in solution are neutralized by the free hydrogen ions in the buffer solution. This causes some more acetic acid to ionize, thus restoring the original hydrogen ion concentration. As more NaOH is added, more acetic acid ionizes until all of the acetic acid in solution is ionized. At this point, any further addition of NaOH will cause an abrupt rise in pH. If a small amount of HCl is added to the acetic acid- sodium acetate buffer, the hydrogen ions that are released unite with free acetate ions to form undissociated acetic acid. Therefore, there is no change in hydrogen ion concentration. As more of the HCl is added, more of the free acetate ions (sodium acetate exists in the solution as free sodium and acetate ions) are converted to acetic acid until finally all of the acetate ions have been converted. When this happens any further addition of HCl will cause an abrupt drop in pH.

Soils also exhibit buffering action. The colloidal complex is similar to a mixture of weak acids and their salts, and it functions as a buffer exactly in the same way as does an acetic acid-sodium acetate solution. The basis of buffer capacity in soil lies in the reserve hydrogen of the exchange complex and its associated metallic cations. Other factors being equal, the higher the exchange capacity of the soil, the greater will be its buffer capacity. Thus an acid soil with high amounts of clay and/or organic matter will require larger amounts of lime for affecting a certain change in pH than an acid soil with smaller amounts of clay and/or organic matter.

18.3.5 Soil acidity

Soils are acidic for one or more of the following reasons: (i) leaching due to heavy rainfall, (ii) origin of soil from acid parent material, (iii) use of acid-forming fertilizers, and (iv) microbial action. The magnitude of leaching is the main factor involved in determining whether or not the soil formed will be acidic. The rainfall carries lime and other bases downward beyond the reach of plant roots. Humid regions are the most susceptible to acid soil formation unless there is a large accumulation of $CaCO_3$ in the soil. Thus, countries having an annual rainfall of 1000 mm or more have a high percentage of acid soils. Rainwater containing dissolved CO_2 from air or from soil is particularly effective in dissolving and leaching calcium from the soil. Some soils have developed from parent materials which are acidic, such as granite. Here too, the leaching effect is predominant in creating acidity.

The use of ammonium sulphate and ammonium nitrate fertilizers have long been known to increase soil acidity. Ammonium ions from ammonium sulphate when added to the soil replace calcium (the dominant cation) from the exchange complex and the calcium sulphate formed is liable to be lost in leaching.

Several kinds of microorganisms are active in the soil. These are responsible for many processes such as decomposition of organic residues and nitrification. As a result of microbial activity, acids are constantly being formed. These, on liberation, seek a base either from free $CaCO_3$ or from the exchange complex. If the exchange complex is low in base saturation, these acids are not neutralized and cause the soil solution to be acidic.

In general, two adsorbed cations-hydrogen and aluminum-are largely responsible for soil acidity. The mechanisms by which these two cations exert their influence depends on the degree of soil acidity and on the source and nature of the soil colloids. Under very acid soil conditions ($pH<5$), there is a larger number of adsorbed aluminum ions than adsorbed hydrogen ions. The preponderance of the adsorbed aluminum ions arises from the fact that aluminum hydroxide in clay micelles reacts with hydrogen ions as follows:

$$Al(OH)_3 + 3H^+ \rightarrow Al^{3+} + 3H_2O$$

Aluminum ions released by this reaction are strongly adsorbed by soil colloidal micelles and displace other cations. In moderately acid soils (pH value between 5.0 and 6.5) aluminum hydroxyl ions produce hydrogen ions by the following hydrolysis reactions:

$$Al(OH)^{2+} + H_2O \rightarrow Al(OH)_2^+ + H^+$$

$$Al(OH)^{2+} + H_2O \rightarrow Al(OH)_3 + H^+$$

Soils that are neutral to alkaline are not dominated by either hydrogen or aluminum ions. The permanent charge exchange sites are now occupied mostly be exchangeable Ca^{2+}, Mg^{2+} and other base-forming cations. Both the hydrogen and hydroxyl ions have been largely replaced. Most of the aluminum hydroxyl ions have been converted to gibbsite by reactions such as:

$$Al(OH)_2^+ + OH^- \rightarrow \underset{\text{Gibbsite (insoluble)}}{Al(OH)_3}$$

More of the pH- dependent charges have become available for cation exchange, and the H^+ ions released therefrom move into the soil solution and react with OH^- ions to form water. The place of hydrogen on the exchange complex is taken by Ca^{2+}, Mg^{2+} and other base forming cations.

Soil acidity is commonly decreased by adding carbonates, oxides, or hydroxides of calcium and magnesium, compounds that are referred to as **agricultural limes**. More than 90 % of agricultural lime is calcium carbonate, some is calcium and magnesium carbonate, and a much smaller quantity is calcium oxide, calcium hydroxide or wood ash. The common liming materials are : (i) calcic limestone ($CaCO_3$), which is ground limestone, (ii) dolomite limestone ($CaMg(CO_3)_2$), from ground limestone high is magnesium, (iii) quicklime (CaO), which is burned limestone, (iv) hydrated (slaked) lime ($Ca(OH)_2$), coming from quicklime which has changed to the hydroxide form on reaction with water, (v) coral shell lime, (vi) chalk ($CaCO_3$), resulting from soft limestone, (vii) blast furnace slag ($CaSiO_3$ and Ca_2SiO_4), a byproduct of the iron industry. Some slags contain phosphorus and a mixture of CaO and $Ca(OH)_2$; this product is called **basic slag** and is valued mostly for its phosphorus content, (viii) miscellaneous sources, such as wood ash, ground oyster shell, press mud and byproduct lime resulting from paper mills, sugar factories, tanneries, and water softening plants, and by-product $CaCO_3$ from fertilizer factories using the gypsum process. All of the liming materials mentioned have value for supplying calcium or calcium and magnesium, raising the pH, and making aluminum, manganese and iron less toxic. For example, when an acid soil is limed with calcium carbonate, calcium ions displace many of the adsorbed aluminum ions and hydrogen ions. Desorbed aluminum ions are converted to aluminum hydroxide ($Al^{3+} + 3H_2O \rightarrow Al(OH)_3 + 3H^+$). Desorbed hydrogen ions as well as hydrogen ions produced by above reactions react with carbonate ions (from calcium carbonate). Because the resulting bicarbonate ions are weakly dissociated, hydrogen ions are effectively removed. Thus, the pH of the soil increases.

Although most plants grow best in soils with a neutral or slightly acid reaction, a few species, such as cranberry, blueberry, azalea, heather, and rhododendron thrive in acid soils. The effects of soil acidity are usually more noticeable in meadows and pastures than on arable land or short-term leys, for very many species of plants are growing together in competition with each other on these uncultivated soils. Plants which tolerate acidity a little better than their neighbours will spread at their expense and come to dominate the flora of acid soils, even though they do not necessarily thrive best in acid soils in the absence of competition. Soil acidity can have a further important effect, for pastures on soils which are too acid for the larger species of earthworm to flourish will have a surface mat of dead vegetation, since there is no longer any mechanism by which the dead herbage can be mixed with the soil. Liming the soil, to allow earthworms to dominate the soil fauna, will alter the composition of the herbage, will encourage grass species that are more palatable to domestic stock, and will cause the soil to lose its slightly **mor-like** (raw humus/unincorporated organic materials) character and become more **mull-like** (humus blended into the upper mineral layers). The use of basic slag on hill pastures in the United Kingdom, by being both a liming material and a phosphatic fertilizer, cause coarse grasses, such as species of *Nardus* and *Agrostis* to be replaced by more nutritious species, such as ryegrass and the Poas (Russell 1973).

18.3.5.1 Saline and sodic soils

In soils of arid and semiarid regions, lack of extensive leaching leaves the level of base-forming cations quite high. As a result, the pH is commonly 7 or above, positively charged aluminum and aluminum hydroxyl ions are absent, and hydrogen ions are at extremely low levels. Adsorbed Ca^{2+} and Mg^{2+} ions dominate the exchange complex, although concentrations of Na^{+} and K^{+} ions are generally higher than in acid soils. Organic matter is lower than in soils of humid regions. A well-developed profile in low rainfall areas usually carries at some point (usually in the C horizon) a calcium carbonate accumulation greater than that of its parent material. The lower the rainfall, the nearer the surface this layer will be. As a result, these soils may have alkaline subsoils and alkaline or neutral surface layers.

Laboratory measurement of the **electrical conductivity** (the ability of the salt in the solution to conduct electricity) of the soil solution extracted from a saturated sample of soil gives an indication of the salt levels. The salinity is commonly expressed in terms of the electrical conductivity (EC), which is measured in decisiemens per meter (ds/m) (1 ds/m=1 millimhos per centimeter or mmho/cm).

The sodium status of the saline and sodic soils is characterized by the **exchangeable sodium percentage (ESP)** complemented by the **sodium adsorption ratio (SAR)**. The ESP identifies the degree to which the exchange complex is saturated with sodium:

$$\text{ESP} = \frac{\text{exchangeable sodium (cmol / kg)}}{\text{cation exchange capacity (cmol / kg)}}$$

ESP levels of 15 yield pH values of 8.5 and above. Higher levels may bring the pH to at least 10. The SAR gives information on the comparative concentrations of Na^+, Ca^{2+}, Mg^{2+} in soil solution. It is calculated as:

$$SAR = \frac{Na^+}{\sqrt{\frac{1}{2}(Ca^{2+} + Mg^{2+})}}$$

where, Na^+, Ca^{2+}, and Mg^{2+} are the concentrations (in cmol/kg) of the sodium, calcium and magnesium ions in the soil solution. SAR of a soil extract takes into consideration that the adverse effect of sodium is moderated by the presence of calcium and magnesium ions. The SAR also is used to characterize the irrigation water added to these soils.

Using EC, ESP and SAR characteristics and soil pH, salt affected soils are classified as saline, saline-sodic, and sodic (Table 18.10).

TABLE 18.10
PROPERTIES OF NORMAL, ACIDIC, SALINE, SODIC, AND SALINE-SODIC SOILS

	pH	*EC (ds/m)*	*SAR**
Normal	6.5-7.2	<4	<13-15
Acid	<6.5	<4	<13-15
Saline	<8.5	>4	<13-15
Saline-sodic	<8.5	>4	>13-15
Sodic	>8.5	<4	>13-15

*Exchangeable sodium percentage (ESP) and SAR are closely related in most soils.

Saline soils contain a concentration of neutral soluble salts sufficient to interfere seriously with the growth of most plants. The EC of a saturated extract of soil solution is more than 4 ds/m. The ESP is less 15, and the pH is less than 8.5 because the salts are neutral, dominated by the chlorides, sulphates, and carbonates of calcium, magnesium, and sodium (the base-forming cations). Saline soils have been called **white alkali** soils because a surface incrustation, if present, is white in colour. It is formed especially in the dry season when the net movement of soil moisture is upward; salts dissolved in the soil water move to the surface where they are left as a crust when water evaporates.

Although the Na^+ ion concentration may be somewhat higher than those of Ca^{2+} and Mg^{2+}, the SAR is less than 13. Those soils can be reclaimed by merely leaching out the salts, an important consideration in their management. Of course, the leaching water must have a low SAR to prevent an increase in sodium adsorption, and, in turn, in soil pH.

The **saline-sodic** soils contain appreciable quantities of neutral soluble salts and enough sodium ions to seriously affect most plants. The ESP is greater than 15, and the EC of a saturated extract is more than 4 ds/m. The pH is commonly 8.5 or less because of the presence of neutral salts. The SAR is at least 13 in these soils. Leaching of saline-sodic soils, unlike the result with saline soils, will markedly raise the pH, unless calcium or magnesium salt concentration are high in the soil or irrigation water. Increase of the pH occurs because once the neutral soluble salts are removed, the exchangeable sodium readily hydrolyzes and thereby sharply increases the OH^- ion concentration of the soil solution:

$$\underset{\text{(soil solid)}}{Na^+ \boxed{\text{Micelle}}} + \underset{\text{(soil solution)}}{H_2O} \rightleftharpoons \underset{\text{(soil solid)}}{H^+ \boxed{\text{Micelle}}} + \underset{\text{(soil solution)}}{Na^+ + OH^-}$$

If the neutral salts are removed, the pH rises above 8.5 and the mineral colloids become dispersed, resulting in a tight, impervious soil structure. At the same time, sodium toxicity to plants is increased.

Sodic soils do not contain any great amount of soluble salts. The detrimental effect of these soils on plants is due not only to the toxicity of Na^+, HCO_3^-, and OH^- ions but also to reduced water infiltration and aeration. Sodic soils have ESP more than 15, SAR more than 13, pH above 8.5 often rising to 10 or higher, and EC less than 4 ds/m. Owing to the deflocculating influence of the sodium, sodic soils usually are in an unsatisfactory physical condition. Because of the extreme alkalinity resulting from the high sodium content, the surface of sodic soils is usually discoloured by the dispersed humus carried upward by the capillary water and deposited when water evaporates. Hence, the name **black alkali** has been used to describe these soils. Sometimes located in same areas called **slick spots**, sodic soils may be surrounded by soils that are relatively productive.

Three kinds of general management practices have been used to maintain or improve the productivity of saline and sodic soils. The first is eradication or removal of the salts; the second is a conversion of some salts to less injurious forms; and the third is designated control. Saline soils in which the soluble salts contain appreciable amounts of calcium and magnesium do not develop into sodic soils by the action of leaching water. The most common methods used to free such soils of excess salts are installation of drainage systems and leaching and flushing. A combination of the two, flooding after drainage ditches have been installed, is the method most thorough and satisfactory. When this method is used in irrigated regions, heavy and repeated applications of water can be made. The salts that dissolve are leached from the solum and drained away. In sodic soils, the exchangeable sodium is so great as to make the soil almost impervious to water. But even if water could move downward in sodic soils, the water alone would not leach out the excess exchangeable sodium. The sodium must be replaced by another cation and leached out of the reach of plant roots. The use of gypsum ($CaSO_4.2H_2O$) on sodic soils is commonly recommended for the purpose of exchanging Ca^{2+} for Na^+ on clay micelles and removing sodium bicarbonates from the soil solution:

$$2NaHCO_3 + CaSO_4 \rightarrow CaSO_3 + Na_2SO_4 + CO_2$$

$$Na_2CO_3 + CaSO_4 \rightleftharpoons CaCO_3 + Na_2SO_4 \text{ (leachable)}$$

$$\begin{matrix}Na^+\\ Na^+\end{matrix}\boxed{\text{micelle}} + CaSO_4 \rightleftharpoons Ca^{2+}\boxed{\text{micelle}} + Na_2SO_4 \text{ (leachable)}$$

Several tonnes of gypsum per hectare are usually necessary. The gypsum should be thoroughly mixed into the soil, and the soil must be kept moist to hasten the reaction. The treatment must be supplemented later by a through leaching of the soil with irrigation water to leach out the sodium sulphate. Elemental sulphur and sulphuric acid can be used to advantage on salty lands especially where sodium carbonate abounds. The sulphur, upon biological oxidation yields sulphuric acid, which not only changes the sodium carbonate to the less harmful sodium sulphate but also decreases the alkalinity. The reactions of sulphuric acid with the compounds containing sodium may be shown as follows:

$$Na_2CO_3 + H_2SO_4 \rightleftharpoons CO_2\uparrow + H_2O + Na_2SO_4 \text{ (leachable)}$$

$$\begin{matrix} Na^+ \\ Na^+ \end{matrix} \boxed{\text{micelle}} + H_2SO_4 \rightleftharpoons \begin{matrix} H^+ \\ H^+ \end{matrix} \boxed{\text{micelle}} + Na_2SO_4 \text{ (leachable)}$$

Not only is the sodium carbonate changed to sodium sulphate, a mild neutral salt, but the carbonate radical is removed from the system. The retardation of evaporation is an important feature in the control of salty soils. Slowing evaporation will not only save moisture but also will retard the translocation upward of soluble salts into the root zone. Farm manure is very useful in such an attempt.

18.3.6 Nutrient availability

The general relationship between soil pH and plant nutrient is shown in Fig. 18.16. The range of availability of *essential elements* (*macroelements* required in relatively large amounts for plant growth) - nitrogen, phosphorus, potassium, calcium, magnesium, and sulphur- is best at pH range of 6.5 to 7.5. On the other hand, the availability of *minor* or *trace elements* (*microelements* required by plants in only extremely small quantities) - iron, manganese, chlorine, boron, copper, and zinc- is more in the acid range than in the neutral or alkaline range. Nevertheless, since the requirement of plants of these minor elements is small, the quantities available at pH 6.5 to 7.5 are usually enough for satisfactory plant growth. Boron deficiency frequently occurs when too much lime is added to an acid soil. On the other hand, molybdenum is not often deficient in acid soils but becomes more available as the soil is limed. All in all, a soil in the intermediate pH range (6.5-7.5) presents the most satisfactory biological environment. Nutrient conditions are favourable without being extreme, and phosphorus availability is at a maximum. Among the soil microbes, bacteria and actinomycetes function best at intermediate and high pH values, whereas activities of fungi tend to predominate in acid soils.

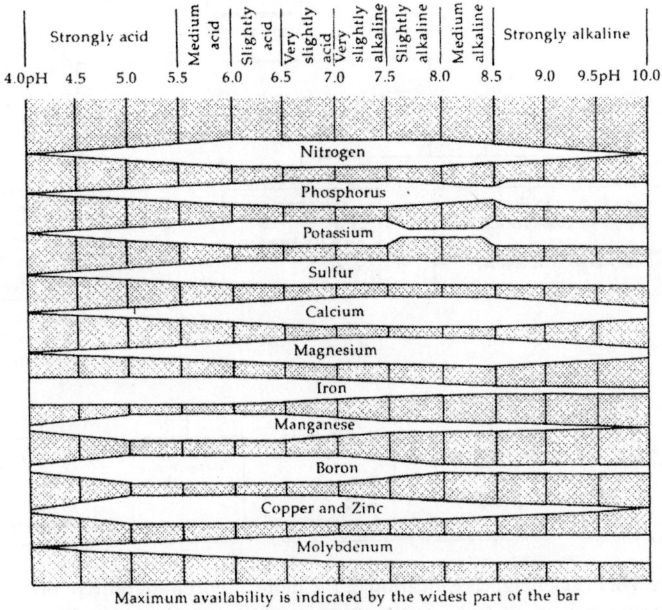

FIGURE 18.16
A SCHEMATIC ILLUSTRATION OF THE RELATIONSHIP BETWEEN PH AND PLANT NUTRIENT AVAILABILITY (THOMPSON AND TROEH 1973)

18.4 Soil Organisms

Soil organisms are vital to the cycle of life on the earth. They incorporate plant and animal residues into the soil. They create humus-the organic constituent essential for good physical and chemical soil conditions. They release essential plant nutrients in inorganic forms during decomposition of organic matter that can be absorbed by plant roots or leached from the soil. In addition to these, soil organisms, particularly earthworms, are responsible for mechanically incorporating organic residues into the soil, and they have open channels through which water and air can flow. The important groups of soil organisms are depicted in Fig. 18.17.

The principal soil microorganisms include bacteria, fungi, actinomycetes, algae, rotifers, nematodes and protozoa. Bacteria are unicellular organism which range in size from 1 μm to about

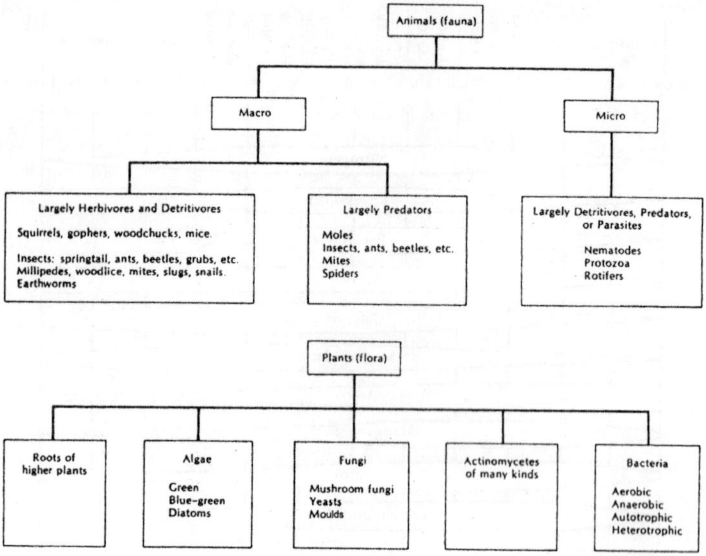

Figure 18.17: General classification of the more important groups of organisms commonly present in soils. Some animals subsist on plants (herbivores and detrivores), some consume other animals (predators), and some live off plants or other animals but do not consume them (parasites).

500 μm, but usually 1-10 μm in diameter. Further, bacteria have various shapes, occurring as *cocci* (spherical), *bacilli* (rod-shaped), and *spirilla* (helical). In the soil, bacteria exist as mats, clumps, and filaments called *colonies*. Soil bacteria are either autotrophs or heterotrophs. The autotrophs obtain energy from the oxidation of inorganic substances such as ammonium, sulphur, and iron *(chemoautotrophs)* and most of their carbon from CO_2. They are not very numerous but play vital roles in controlling nutrient availability to higher plants. For example, decay of animal and plant proteins gives rise to ammonium compounds which the autotrophic bacteria *Nitrosomonas* and *Nitrococcus* oxidize to nitrites; subsequently, autotrophic *Nitrobacter* bacteria oxidize nitrites to nitrates. Because nitrate is the form preferentially absorbed by most plants, this oxidation has implications for plant nutrition. Other autotrophic bacteria are responsible for sulphur oxidation, which yields sulphate ions that can be readily taken up by plants. Similarly, bacterial oxidation and reduction of inorganic ions such as iron and manganese influence the

availability of these elements to plants. Most soil bacteria are heterotrophic- that is, both their energy and their carbon come directly from organic matter. Heterotrophic bacteria, along with fungi and actinomycetes, account for the general breakdown of organic matter in soils. A critical process that depends largely on bacteria is *nitrogen fixation*-the biochemical combination of atmospheric nitrogen with hydrogen to form ammonia which is then incorporated into nitrogen compounds usable by plants. The process can be carried out by aerobic *(Azotobacter)* and anaerobic *(Clostridium)* bacteria in soils independent of plants, but the amount of nitrogen fixed is much greater if the bacteria are intimately associated with plant roots (the bacterium *Rhizobium* living in root nodules of certain leguminous plants, for example).

Actinomycetes are filamentous, often profusely branched, unicellular *Gram-positive* bacteria. Reproduction is either by total fragmentation of hyphae and/or by production of spores in specialized areas of the mycelium. Actinomycetes are typically aerobic organisms and develop best in moist, well aerated soils at pH values between 6.0 and 7.5. They are especially numerous in soils high in humus, such as old meadows or pastures, where the acidity is not too great. The aroma of freshly ploughed land so noticeable at certain times of the year is probably owing to actinomycetes as well as to certain moulds. Actinomycetes undoubtedly are of great importance in the decomposition of soil organic matter and the liberation of its nutrients.

Several hundred species of algae have been isolated from soils, the most prominent being the same throughout the world. They develop most readily in damp soils exposed to the sun, hence usually have their maximum development in spring and autumn when the soil is damp, the sun not too hot and other vegetation sparse. They develop most freely on fertile soils well supplied with nutrients, and tend to be less numerous on light, infertile, acid soils. The green algae are the dominant group in acid soils, but as the soils become more neutral the blue-green algae and the diatoms become equally important, and on fertile soils the blue-green may be the dominant group. Blue-green algae are especially numerous in paddy fields, and when such lands are flooded, appreciable amounts of atmospheric nitrogen are fixed or changed to a combined form by these algae. Blue-green algae, growing within the leaves of aquatic fern *Azolla*, are able to fix nitrogen in quantities significant for paddy production.

The main groups of fungi inhabiting soils are moulds and mushrooms. Moulds are filamentous fungi occurring in most soils (acidic, neutral, or alkaline soils), and the low pH tolerance of moulds is especially important in decomposing the organic residues in acid forest soils, where bacteria and actinomycetes are less active. The most common genera of moulds found in soils are *Penicillium, Mucor, Rhizopus, Fusarium* and *Aspergillus.* Mushroom fungi are found in forests and grasslands where there are ample moisture and organic residues. Some of these mushrooms are edible, and their production is domesticated. The plant body consists of two parts-vegetative (mycelium) and reproductive (basidiocarp). The aboveground reproductive part (commonly called the *fruiting body*) is only a small part of the total organism in comparison to the extensive mass of hyphae found in the underlying soil or organic residue. Although mushrooms are not as widely distributed as the moulds, these fungi are of significance locally, especially in the breakdown of woody tissue. In general, fungi are the most versatile and perhaps the most persistent of any soil organisms in their ability to decompose organic residues. Cellulose, starch, gums, and lignin, as well as the more easily affected proteins and sugars, succumb to fungal attack. In affecting the processes of humus formation and aggregate stabilization, moulds are more important than bacteria since they continue to decompose complex organic residues even after bacteria and actinomycetes have essentially ceased to function. An economically important, symbiotic association between numerous fungi and the roots of higher plants is called *myccorhizae*, a term meaning "fungus root". The association is of great practical significance because it markedly increases the availability to plants of several essential nutrients including phosphorus, zinc, copper, calcium, magnesium, manganese, and iron, especially from infertile soils. On the other hand, the symbiotic association provides the fungi with sugars and other organic exudates for use as food.

Of the abundant microscopic animal life in soils, three groups are of some importance: nematodes, protozoa, and rotifers. Protozoa are probably the simplest form of animal life and are the most varied and numerous of microanimals in soils. They include amoeba, ciliates, and flagellates. Protozoa generally thrive best in moist well-drained soils and are most numerous in surface horizons. Some protozoa are predators on soil bacteria and other microflora. Overall, however, the protozoa generally are not sufficiently abundant in soils to be a major factor in organic matter decay and nutrient release. Nematodes-commonly called threadworms or eelworms- are found in almost all soils, often in surprisingly large numbers. In size, they are almost wholly microscopic, seldom being large enough to be seen with the naked eye. Nematodes are either saprophytes living on decaying organic matter or predators on the nematodes, bacteria, algae, and protozoa. But some nematodes, especially those of the genus *Heterodera*, can infest the roots of practically all plant species. Rotifers thrive under moist conditions, especially in swampy land where their number may be great. Just how important rotifers are in soils is unknown, but their activities are likely confined largely to peat bogs and wet areas of mineral soils.

The important soil invertebrates are earthworms, termites, and ants. Earthworms, especially in the upper 15-35 cm of soil, ingest organic matter as well as soil. As these materials pass through the earthworm's body, they are subjected to digestive enzymes as well as to grinding action within the animal. The weight of the material passing through their bodies (casts) each day may equal the weight of the earthworm. Compared to the soil itself, the casts are definitely higher in bacteria, organic matter, and available plant nutrients (Table 18.11). Earthworms prefer a well-aerated but moist habitat. Therefore they are found mostly in medium textured upland soils where the moisture capacity is high rather than in droughty sands or poorly drained lowland. Earthworms must have organic matter as a source of food. Consequently, they thrive where farm manure or plant residues have been added to the soil. A few species are reasonably tolerant to low pH, but most earthworms thrive best where the soil is not too acid.

Termites, or "white ants" are most prominent in savannas and forests of tropical and subtropical areas (both humid and arid). In building their mud nests or mounds termites transport soil from lower layers to and above the surface soil, thereby bringing about extensive mixing of soil materials and the plant

TABLE 18.11
COMPARATIVE CHARACTERISTICS OF EARTHWORM CASTS AND SOILS (DE VLEESCHAUWER AND LAL 1981)

Characteristic	*Earthworm Casts*	*Soils*
Silt and clay (%)	38.8	22.2
Bulk density (megagram/m^3)	1.11	1.28
Structural stability*	849	65
Cation exchange capacity (cmol/kg)	13.8	3.5
Exchangeable Ca^{2+} (cmol/kg)	8.9	2.0
Exchangeable K^+ (cmol /kg)	0.6	0.2
Soluble P (ppm)	17.8	6.1
Total N (%)	0.33	0.12

*Numbers of raindrops required to destroy structural aggregates

residues they use as food. The effect of termites on soil productivity is generally less beneficial than that of earthworm because the digestive processes of termites, aided by microorganisms in their gut, are generally more efficient than those of earthworms. Ants have less widespread influence on soil properties than earthworm and termites; however, some ants produce conspicuous mounds while others have underground nests, and there is considerable turnover in the soil associated with the mounds and nests because subsoil is brought to the surface. In general, earthworms, termites and ants assist in the aeration and drainage of the soil through their channels and burrows; they macerate and grind up the plant litter they eat, excreting it in the form more readily available to the attack of the soil microorganisms; and they distribute this macerated plant debris together with some of the microflora throughout the volume of soil in which they are working.

Roots of higher plants grow and die in the soil and in doing so supply the soil fauna and microflora with food and energy. In addition, the living roots physically modify the soils as they push through cracks and make new openings of their own. Tiny initial channels are increased in size as the roots swell and grow. By removing moisture from the soil plant roots bring about further physical stresses that stimulate soil aggregation. The roots also support myriad microorganisms with their exudates, which further stabilize these aggregates. As they later decompose, the roots provide material for the creation of humus not only in top but also to greater soil depths.

18.5 Soil Organic Matter

Organic matter supplies essential plant nutrients, and also functions as a source of food for soil microbes and invertebrates. It commonly accounts for as much as one third or more of the cation exchange capacity of surface soils and is responsible for the stability of soil aggregates.

The original source of the soil organic matter is aboveground (tops) and belowground (roots) parts of plants which annually supply large quantities of organic residues. As these organic materials are decomposed and digested by soil organisms, they become part of the underlying soil by infiltration or by actual physical incorporation. Animals usually are considered secondary sources of organic matter.

The average moisture content of plant residues is 75%, and the dry matter is mostly carbon and oxygen, with less than 10% each of hydrogen and inorganic elements (ash). The actual organic compounds in plant tissue are many and varied; however, they can be grouped into a small number of classes, as shown in Fig. 18.18. The carbohydrates, which range in complexity from simple sugars and starches to cellulose, are usually the most prominent of the organic compounds found in plants. Lignins, which are complex compounds with ring-type structures, are found in older plant tissue and especially woody tissues. Fats and oils, which are somewhat more complex than carbohydrates and less so than lignins, are found primarily in seeds. Proteins contain-in addition to carbon, oxygen, and hydrogen-nitrogen and smaller amounts of other elements such as sulphur, manganese, copper, and iron.

All organic compounds usually begin to decompose simultaneously when fresh plant tissue is added to a soil. In terms of ease of decomposition, the organic compounds may be arranged as: sugars, starches, and simple proteins> crude proteins> hemicelluloses> cellulose >fats, waxes, etc > lignins. The sugars and simple proteins decompose most readily; at the other extreme, lignins are the most resistant to breakdown. When organic tissue is added to soil, the bulk of the material (sugars, starches, celluloses, for example) undergoes enzymatic oxidation with CO_2, water, energy, and heat as the major products:

$$\text{-(C,4H)} + 2O_2 \xrightarrow{\text{enzymatic oxidation}} CO_2 + 2H_2O + \text{energy}$$

(carbon- and hydrogen – containing compounds)

The plant proteins also succumb to microbial decay, yielding not only CO_2 and water but also amino acids such as glycine (CH_2 NH_2 COOH) and cysteine ($CH_2HSCHNH_2COOH$). In turn, these nitrogen and sulphur compounds are further broken down, eventually yielding simple inorganic ions such as NH_4^+, NO_3^- and SO_4^{2-}. Similar breakdown of other organic compounds release inorganic phosphates as well as cations such as Ca^{2+}, Mg^{2+}, K^+, and Na^+. Although nitrates and sulphates are subject to leaching, the phosphates tend to be retained in insoluble calcium, iron and aluminum compounds. The cations released to the soil solution are subject to adsorption by negatively charged soil colloids. They can later be released for plant uptake, and some will be removed be leaching waters. The overall process of release of inorganic ions by decomposition of organic matter is called **mineralization**.

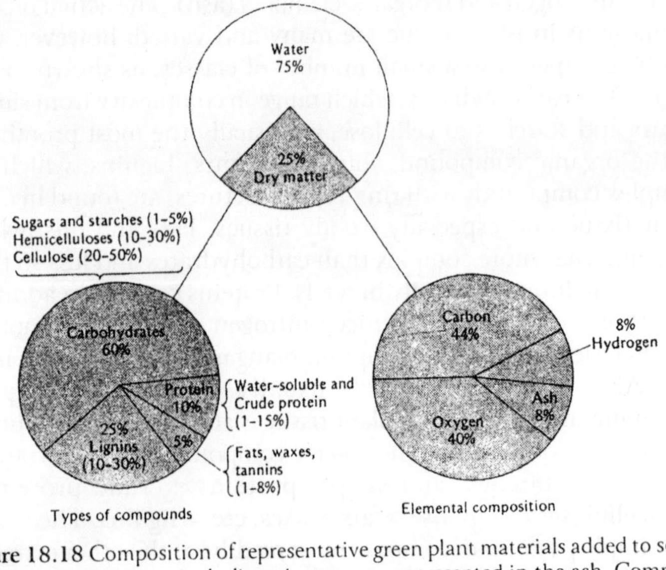

Figure 18.18 Composition of representative green plant materials added to soils. All inorganic elements, including nitrogen, are represented in the ash. Common ranges in the percentage of compounds present are shown in parentheses (Waksman 1948).

General purpose decay bacteria, fungi, and actinomycetes are fully active during decomposition of organic matter. The microbial tissue may at times account for as much as one third of the organic fraction of the soil. At the height of microbial activity, even the original soil organic matter is subject to some breakdown. As soon as the easily decomposed food is exhausted, microorganisms begin to decline in number. When the microbial cells die, their bodies are devoured by living microbes, with the continued evolution of CO_2 and water. As further reduction in available food occurs, microbial activity continues to decline, and the general-purpose decay organisms again sink back into comparative quiescence. The organic matter now remaining is mostly a dark, heterogeneous colloidal mass of resistant and newly synthesized (by microorganisms) compounds, usually referred to as **humus**.

There are two general groups of compounds that collectively make up humus, the *nonhumic* group and the *humic* group. The nonhumic materials are less complex and less resistant to microbial attack than those of the humic materials. They are made up of easily identifiable organic compounds: polysaccharides such as cellulose and its decomposition products; polyphenols such as tannins; polyquinones; proteins and their decomposition products; a host of hydrocarbons, organic acids, alcohols, esters, and aldelydes; and polyuronides, which are not found in plants but have been synthesized by the soil microbes. The humic materials are amorphous, dark in colour, and not so easily identified as the nonhumic materials. Many humic materials contain reactive groups-carboxyl, amine, phenolic hydroxyl group-combined with chains and rings, forming rather large and complex molecules. They are very resistant to microbial attack. Figure 18.19 shows the approximate proportion of organic carbon from plant residues that are found in humic and non humic compounds one year after the residues were added to the soil.

The humus particles have the dimensions of clay particles and the two mix together intimately in the soil. Like the clay micelles, the humic colloids are also characterized by the presence of surface charges, but these arise from exposed carboxyl (-COOH) and hydroxyl (-OH) groups from which the hydrogen may dissociate and be replaced by other cations. At high pH values the cation exchange capacity of humus on a mass basis (150-300

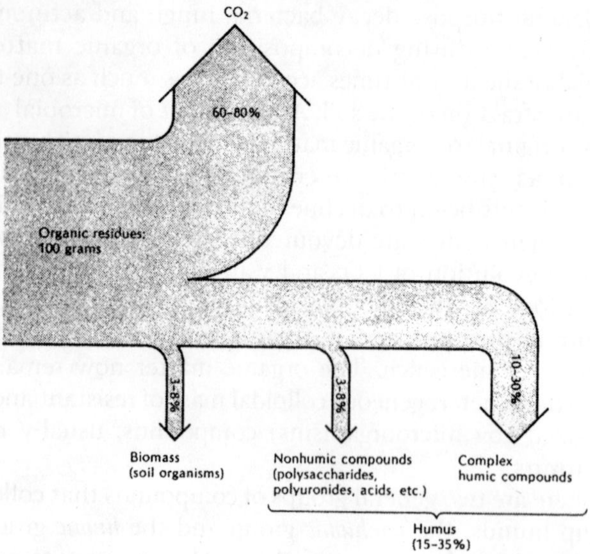

Figure 18.19 Diagram to illustrate the disposition of 100 g of organic residues 1 year after they were incorporated into the soil. More than two thirds of the residues have been oxidized to CO_2 and less than one third remains in the soil, some in the bodies of soil organisms but a larger component as humus.

cmol/kg) far exceeds that of most silicate clays. The water holding capacity of humus on a mass basis is 4-5 times that of the silicate clays; humus has a very favourable effect on aggregate formation and stability; and, the significance of slowly decomposing humus is that it provides a secure supply of continually available nutrients for plant growth.

18.5.1 Determination of soil organic matter

Walkley and Black's (1934) method is a quicker and easier method for determining soil organic matter. Five gram soil sample is transferred to a dry 500 ml Erlenmeyer flask to which 10 ml of normal potassium dichromate solution and 20 ml of concentrated suphuric acid are added successively. The flask is shaken for 2 to 3 minutes and allowed to stand on a sheet of asbestos for 30 minutes. Thereafter, 200 ml distilled water, 10 ml of 85% phosphoric acid, and 1 ml of diphenylamine indicator are added into the flask. The mixture thus obtained is titrated against normal ferrous sulphate solution filled in burette, until the colour of the mixture becomes green. Thereafter, 0.5 ml of normal potassium dichromate is further added to the flask, and the solution is titrated against normal ferrous sulphate solution until the last traces of blue colour disappears. The percentage of soil carbon (C) is calculated as follows:

$$C\ (\%) = \frac{V_1 - V_2}{W} \times 0.003 \times 100$$

where, V_1 = volume of normal potassium dichromate; V_2 = volume of normal ferrous sulphate solution in ml; and, W= weight of the soil in gram. Percentage of organic matter is obtained by multiplying the percentage of organic carbon by the factor 1.724.

18.5.1.1 Carbon: Nitrogen (C/N) ratio

The fresh plant residues contain large amounts of carbon and comparatively small amounts of total nitrogen. When fresh plant residues are added to the soil, the heterotrophic flora-bacteria, fungi, and actinomycetes-become active and multiply rapidly. Under these conditions, nitrate nitrogen essentially disappears from soil because of the persistent microbial demand for this element to build microbe tissues. This results in wide C/N ratio which may be 40:1 or more; indicating that little or no mineral nitrogen (NH_4^+ or NO_3^-) is available to higher plants. The time interval of nitrate depression may be only a week or so, or it may last throughout the growing season depending upon the rate of decay of organic residues. For example, the greater the quantity of decomposable residues applied, the longer *nitrification* (nitrate formation) will be blocked. In contrast, organic residues with lower C/N ratio interfere less with nitrification and yield nitrogen more quickly. In general, mature residues, whether legume or nonlegume, generally have much higher C/N ratio than do younger materials. When the decomposition of fresh organic residues has proceeded to the extent that the C/N ratio is reduced to 20:1 or so, available nitrogen in soil tends to increase.

In general, low C/N ratios (10:1 or smaller) in soil organic matter generally indicate an advance stage of decomposition and resistance to further microbial decomposition. A wide range of

C/N (35:1 or more) indicates little decomposition, susceptibility to further and rapid decomposition and slow nitrification. The C/N ratio tends to be lower in soils of arid regions than those of humid regions when annual temperatures are about the same. The ratio is also lower in warmer regions than in cooler ones if the rainfalls are about equal. Also, C/N ratio is lower in subsoils than for corresponding surface layers.

18.6 Soil Taxonomy

As soil formation proceeds, there is great variability in the soil profile characteristics from one location to another. Differences are found in properties such as soil depth, clay and organic matter content, and the degree of horizon development. The naming of soils allows one to impart a great deal of information by invoking the name of a certain soil. A name is applied to an individual kind of soil that has a definable existence on the landscape. A unit of soil, large enough such that its profile and horizons can be studied and its properties defined, is called a **pedon**. This is the smallest practical volume that can be properly referred to as soil. A pedon may range from $1 m^2$ to $10 m^2$ in area, depending on the continuity or cyclic discontinuity on the soil's horizons within short linear distances (1 to 7 m). Pedons of soils with continuous horizons have the smallest area ($1 m^2$). Groups of similar adjacent pedons, termed **polypedons**, form the basis for soil survey and soil classification.

For names to be useful, a classification system must be scrupulously followed, and names must be assigned on the basis of certain specified properties. Characteristics that are used to define a soil are termed **differenting characteristics** or **differentiae**. Soil scientists strive for a classification scheme that reflects a natural relatedness, and therefore the scheme of naming is called **soil taxonomy**.

18.6.1 Differentiating properties of soils

The properties of a polypedon should serve as differentiae. There are six defined diagnostic surface horizons or **epipedons**. An epipedon is a horizon, or horizons, that forms in the upper part of the soil profile. A subsurface horizon (often B, but perhaps part of A) is also of diagnostic value and the names, descriptions and genesis or mode of formation of these horizons are also important in soil classification. The diagnostic epipedons are as follows:

(a) **Mollic epipedon**: from the Latin **mollis**, meaning soft; it is normally granular or easily workable, has appreciable organic matter and high base saturation (>50%).
(b) **Anthropic epipedon**: formed by human manipulation over a long time period under semiarid and arid conditions.
(c) **Umbric epipedon**: from the Latin **umbra**, meaning shade, thus dark soil; it has appreciable organic matter and a low base saturation (<50%).
(d) **Histic epipedon**: from the Greek **histos**, meaning tissue, thus a thin layer of muck or peat.
(e) **Plaggen epipedon**: from the German **plaggen**, meaning sod, thus a surface layer of 50 cm or more thick, produced by long-continued manuring.
(f) **Ochric epipedon**: from the Greek **ochros**, meaning pale; a soil too dry, too thin, too high in chroma, etc., to qualify as a mollic or umbric epipedon.

The diagnostic subsurface horizons include argillic, agric, natric, spodic, placic, cambic, oxic, duripan, and the like (Table 18.12).

The United States Department of Agriculture recognizes six categories of classification in soil taxonomy: (i) order (the broadest category), (ii) suborder, (iii) great group, (iv) subgroup, (v) family, and (vi) series (the most specific category). These categories may be compared with those used for the classification of plants: phylum (the broadest category), class, subclass, order, family, genus, and species. A major feature of the system is the nomenclature used to identify different soil classes. The names of the classification units are combinations of syllables, most of which are derived from Latin or Greek. Since each part of a soil name conveys a concept of soil character or genesis, the name automatically describes the general kind of soil being classified. For example, soils of the order **Aridisols** (from the Latin **aridus**,

TABLE 18.12
MAJOR FEATURES OF THE DIAGNOSTIC SUBSURFACE HORIZONS IN MINERAL SOILS

Subsurface horizons	*Major features*
Argillic	Silicate clay accumulation
Natric	Argillic, high in sodium, columnar or prismatic structure
Spodic	Organic matter, Fe and Al oxides accumulation
Cambic	Changed or altered by physical movement or by chemical reactions
Agric	Organic and clay accumulation just below plough layer resulting from cultivation
Oxic	Highly weathered, primarily mixture of Fe, Al oxides and nonsticky-type silicate clays
Duripan	Hard pan, strongly cemented by silica
Fragipan	Brittle pan, usually loamy textured, weakly cemented
Albic	Light coloured, clay and Fe and Al oxides mostly removed
Calcic	Accumulation of $CaCO_3$ or $CaCO_3.MgCO_3$
Gypsic	Accumulation of gypsum
Salic	Accumulation of salts
Kandic	Accumulation of low activity clays
Petrocalcic	Cemented calcic horizon
Petrogypsic	Cemented gypsic horizon
Placic	Thin pan cemented with iron alone or with manganese or organic matter
Sombric	Organic matter accumulation
Sulphuric	Highly acid with Jarosite mottles

dry, and *solum,* soil) are characteristically dry soils. Those of the order *Inceptisols* (from the Latin *inceptum,* beginning) are soils only with beginnings of profile development.

The names of suborders automatically identify the order of which they are a part. For example, soils of the suborder *Aquolls* are the wetter soils (from the Latin *aqua,* water) of the *Mollisols* order. Likewise, the name of the great group identifies the suborder and order of which it is a part. *Argiaquolls* are *Aquolls* with clay or *argillic* (Latin *argilla,* white clay) horizons.

The nomenclature as it relates to the different categories in the classification system might be illustrated as follows:

Mollisols .... order
Aquolls ....suborder

Argiaquolls .... great group
Typic Argiaquolls .... sub group

Note that the three letter *oll* identify each of the lower categories as being in the *Mollisols* order. Likewise, the suborder name *Aquolls* is included as part of the great group and subgroup name. If one is given only the subgroup name, the great group, suborder, and order to which the soil belongs are automatically known.

Family names in general identify subsets of the subgroup that are similar in texture, mineral composition, and soil temperature at a depth of 50 cm. Thus the name *Typic Argiaquolls, fine, mixed, mesic,* identifies a family in the *Typic Argiaquolls* subgroup with a fine texture, mixed clay mineral content, and mesic (8-15°C) soil temperature.

Soil series names have local significance since they normally identify the particular locale in which the soil is first found. Thus, series such as Fort Collins, Cecil, Miami, Norfolk, and Ontario identify soils first described near the city or locality named.

18.6.2 Soils orders

There are eleven soil orders, and all these orders have names with a common end, *sol* (from the Latin *solum*, soil). All these soil orders along with suborders are listed in Table 18.13.

Entisols (little if any profile development) are weakly developed mineral soils without natural genetic (subsurface) horizons or with only the beginnings of such horizons. Most have an ochric epipedon and a few have man-made anthropic or agric epipedons. The extremes of highly productive soils on recent alluvium and infertile soils on barren sands as well as shallow soils on bedrock are included. The agricultural productivity of the Entisols varies greatly depending on their location and properties. With adequate fertilization and a controlled water supply, some Entisols are quite productive; in fact, Entisols developed on alluvial floodplains are among the world's most productive soils. However, restrictions on the depth, clay content, or water balance of most Entisols limit the intensive use of large areas of these soils.

TABLE 18.13
SOIL ORDERS AND SUBORDERS IN SOIL TAXONOMY. ENDING OF THE SUBORDER NAMES IDENTIFIES THE ORDER IN WHICH THE SOILS ARE FOUND

Order	*Suborder*	*Order*	*Suborder*	*Order*	*Suborder*
Entisols	Aquents	Alfisols	Aqualfs	Aridosols	Argids
	Arents		Boralfs		Orthids
	Fluvents		Udalfs	Spodosols	Aquods
	Orthents		Ustalfs		Ferrods
	Psamments		Xeralfs		Humods
Inceptisols	Andepts	Ultisols	Aquults		Orthods
	Aquepts		Humults	Histosols	Fibrists
	Ochrepts		Udults		Hemists
	Plaggepts		Ustults		Saprists
	Tropepts		Xerults		Folists
	Umbrepts	Oxisols	Aquox	Andisols	Aquands
Mollisols	Albolls		Perox		Cryands
	Borolls		Torrox		Torrands
	Rendolls		Udox		Udands
	Udolls		Ustox		Ustands
	Ustolls	Vertisols	Torrerts		Xerands
	Xerolls		Uderts		Vitrands
			Usterts		
			Xererts		

Inceptisols (few diagnostic features) have ochric or umbric epipedons and/ or cambic subsurface horizons. The horizons of Inceptisol are thought to form quickly and result mostly from the alteration of parent materials. Hence, subsurface horizons associated with more advanced weathering and/ or accumulation of silicate clays, iron and aluminum oxides, and organic matter are absent. Profile development of Inceptisols is more advanced than that of the Entisols order but less so than for the other orders. Many agriculturally useful soils are included along with others for which productivity is limited by factors such as slow drainage of water. Inceptisols formed under continually warm conditions **(tropepts)** are found in southwestern India, whereas wet Inceptisols **(aquepts)** are found in areas along with the Ganges and Amazon rivers.

Mollisols (dark soils of grasslands, mollic epipedon base rich) include some of the world's most important agricultural soils. They are characterized by a mollic epipedon, or surface horizon, which is thick, dark, and dominated by base-forming cations such as Ca^{2+} and Mg^{2+} (high base saturation). They may have an argillic (clay), natric, albic, or cambic horizon, but not an oxic or spodic one. The surface horizon generally has granular or crumb structures, largely resulting from the organic matter present, that are not hard when the soils are dry. This justifies the use of a name that implies softness. Most of the Mollisols have developed under grass vegetation. They are called **Udolls** (from the Latin *udus,* humid) if soil moisture is not limiting; **Ustolls** (from the Latin *ustus,* burnt) if soil is intermittently dry during the summer; and **Xerolls** (from the Greek *xeros,* dry), the driest of the Mollisols. The high native fertility of Mollisols (especially the Udolls) makes them the world's more productive soils.

Alfisols (argillic or natric horizon, medium-high bases) are moist mineral soils having no mollic epipedon or oxic or spodic horizons. They have grey to brown surface horizons (commonly an ochric epipedon), medium- to high-base status, and contain an illuvial horizon in which silicate clays have accumulated. The cation exchange capacity of the clay horizon is more than 35% saturated with base-forming contains (Ca^{2+}, Mg^{2+}, K^{+}, Na^{+}), This horizon is termed *argillic* if only silicate clays are present and *natric* if, in addition to the clay, it is more than 15% saturated with sodium and has prismatic or columnar structure. Alfisols may be *Udalfs* (from the Latin *udus,* humid), *Xeralfs* (from the Greek *xeros,* dry), *Boralfs* (from the Greek *boreas,* northern), *Ustalfs* (from the Latin *ustus,* burnt) or *Aqualfs* (from the Latin *aqua,* water). The Alfisols appear to be more strongly weathered than the Inceptisols but less than so than the Spodosols. They are formed in cool to hot humid areas but also are found in the semiarid tropics. Most often Alfisols are developed under native deciduous forests, although in some cases grass is the native vegetation. In general, Alfisols are productive soils for crop production.

Ultisols (argillic horizon, low bases) are soils with an argillic (clay) horizon and a low-base status (less than 35% of the exchange capacity satisfied with base-forming metallic cations). Their mean annual temperatures at 50 cm depth are above 8°C. They are more highly weathered and acidic than Alfisols but less acid than Spodosols. Most Ultisols have developed under moist conditions in warm tropical climates. Except for the wetter members of the order, their subsurface horizons are commonly red or yellow in colour, evidence of accumulation of oxides of iron. Ultisols are formed on old land surfaces, normally under forest vegetation, although savanna or even swamp vegetation is also common. Depending upon the prevailing conditions, Ultisols are typified as *udults* (moist, well-drained), *humults* (high in organic matter) and *xerults* (Mediterranean-type climates). In general, Ultisols are located mostly in regions of long growing seasons and of ample moisture for good crop production, however, they are naturally not as fertile as Alfisols or Mollisols.

Oxisols (oxic horizon, highly weathered) are the most highly weathered soils. They have ochric or umbric epipedons, but their most important diagnostic feature is a deep oxic subsurface horizon. This horizon is generally very high in clay-size particles dominated by hydrous oxides of iron and aluminum. Weathering and intense leaching have removed a large part of silica from the silicate minerals in this horizon. Oxisols occupy old land surfaces and occur mostly in the tropics. Oxisols having short dry season or none are called *orthox*, whereas those Oxisols having hot dry summers are called *ustox*.

Vertisols (dark, swelling clays) are characterized by a high content (>30%) of sticky or swelling-and-shrinking-type clays to a depth of 1 m, which in dry seasons causes the soils to develop deep, wide cracks. A significant amount of material from the upper part of the profile may slough off into the cracks, giving rise to a partial "inversion" of the soil. This accounts for the term *invert*, which is used to characterize this order in a general way. If moist condition prevents cracks from persisting for more than three months the Vertisols are called *uderts* (from the Latin *udus*, humid). On the other hand, if cracks persist for more than three months of the year they are called *userts* (from the Latin *ustus*, burnt); and those with hot, dry summers, *xererts* (Greek *xeros*, dry). Vertisols are found mostly in subhumid to semiarid environments where the average soil temperatures are higher than 8°C. Their very fine texture and marked shrinking and swelling characteristics make them less suitable than other soils for crop production and for building foundations and highway bases.

Aridisols (dry soils) are dry soils characterized by an ochric epipedon generally light in colour and low in organic matter. They may have a horizon of accumulation of calcium carbonate (calcic), gypsum (gypsic), soluble salts (salic) or sodium (natric). Except where there is groundwater or irrigation, the soil layers are moist only for short periods during the year. These short moist periods may be sufficient for drought-adapted desert shrubs and annual plants, but not for conventional crop production. If ground water is present near the soil surface, soluble salts often accumulate to levels that most crop plants can not tolerate. Aridisols with a horizon of clay accumulation are called *argids* (from the Latin *argilla*, white clay), and those without argillic or natric horizons, the *orthids*.

Spodosols (spodic horizon, forests, low bases) have a spodic horizon, a subsurface horizon with an accumulation of organic matter and of oxides of aluminum with or without iron oxides. The illuvial horizon, which is not very thick, usually occurs under an eluvial horizon, normally an albic horizon, which is light in colour and, therefore has been described as "wood-ash". Spodosols occur only in moist to wet areas and are most common in cold temperate areas. Forest species low in metallic ion contents, such as pine trees, seem to encourage the development of Spodosols. As the litter from these low base species decomposes, strong acidity develops. Percolating water leaches acids down the profile, and the upper horizons succumb to this intense acid leaching. Most minerals except quartz are removed. In the lower horizons, oxides of aluminum and iron as well as organic matter precipitate, thus yielding the Spodosol profiles. The common Spodosols described above are called *orthods*, whereas Spodosols seasonally saturated with water are called *aquods*.

Histosols (organic soils) are characterized by high organic carbon content (above 20% organic matter or 12% organic carbon). Histosols are typically formed in water-saturated environment such as that found in peat bog. They can form in any climate from the equator to permafrost areas so long as water is available. In uncultivated areas, the organic matter retains much of the original plant tissue form. Upon drainage and cultivation of the area, the original plant tissue tends to disappear as decomposition takes place.

Andisols order specifically includes soil developed in volcanic ash and other ejecta. Since these materials have been deposited in relatively recent time, they have not been highly weathered. The upper layers of Andisols are characteristically dark in colour and have low bulk densities.

Different soil types as recognized by UNESCO, their form and distribution from equator to poles, and the associated vegetation types are listed in Table 18.14.

TABLE 18.14
MAJOR CLASSIFICATIONS (UNESCO SYSTEM) FOR THE SOILS OF THE WORLD AND THE VEGETATION TYPES MOST COMMONLY ASSOCIATED WITH THEM

Soil type	*Soil form*	*Vegetation type*
Laterite (latosol)	Weathered red	Wet tropical forest
Margalite (tropical black)	Base rich black	Seasonal tropical forest
Desert soils	Hot dry	Desert
Chernozem	Dry black (with calcite)	Grasslands (steppe)
Chestnut soils	Dry organic black or red- brown	Grasslands (prairie)
Grey-brown podzol (brown earth)	Grey-brown mull	Deciduous forest
Podzol	Grey banded mor	Boreal conifer forest
Peats	Organic peat	Bog
Tundra soil or podzol (if well drained)	Organic on mineral base (with perma-frost)	Tundra

19. ENVIRONMENTAL IMPACT ASSESSMENT (EIA)

19.1 Introduction and Terminology

EIA (Fig.19.1) is a planning tool that can be applied to any developmental project to ameliorate potentially negative environmental impacts by specifying appropriate mitigation measures. EIA can be viewed as both a planning tool and as a mechanism for decision making and is generally used to accomplish the followings:

- identify potential (both beneficial and adverse) environmental impacts
- analyze the significance of environmental impacts
- determine whether the adverse impacts can be mitigated
- recommend preventive and mitigation measures
- identify alternatives to the proposed project
- recommend whether the proposed project should be implemented

Development projects are instrumental to Nepal's economic growth and to fulfill the basic needs of its growing population. Yet there have been cases of such projects degrading the very basis for sustained development by exploiting natural resources without adequate concern for the maintenance of environmental quality. As a result, development projects that have been formulated to overcome poverty, illiteracy, and malnutrition have also created environmental problems such as disease, flooding, erosion, landslides, and deforestation, negating the objectives of the development project. Economic growth which proceeds without attention to the proper management and replenishment of environmental resource base cannot be sustained. Development and conservation must go together.

The need for EIA was stressed in Nepal's seventh five year plan (1985- 1990), reemphasized in the eighth five year plan (1990-1995) and made mandatory since the ninth five year plan (1995-2000), for all major development projects related to sectors including tourism, water resources, transportation, urbanization, agriculture, forestry, industry, etc.

19.1.1 Terminology

(a) **Adverse/Beneficial impacts:** Adverse impacts are harmful to human interests over either the short or long term. Beneficial impacts improve the economic or quality of life for a population and improve resource base.

(b) **Auditing:** An activity to check on past and current environmental performance.

(c) **Baseline studies:** Work done to collect and interpret information on the status/trends of the environment likely to be affected by a development action.

(d) **Compensation:** A form of mitigation which involves action such as financial payment to people affected by a development action or the protection/creation of a habitat or similar feature which is similar to the one to be adversely affected.

(e) **Cost - benefit analysis:** Technique used in comparing the ratio of costs to benefits for a development activity. Comparison is made through conversion of costs/ benefits to monetary units, for example, the US $. It has not yet been successfully used to incorporate all environmental, social and health impacts of concern.

(f) **Development actions:** Often used to refer to the range of individual tasks to be performed during the construction and operation of a facility. Examples of such tasks are: vegetation removal, construction of fences, construction of access road, transport of raw material, etc. Each of these specific tasks can cause environmental impacts.

(g) **Development proposal:** Consists of a wide range of

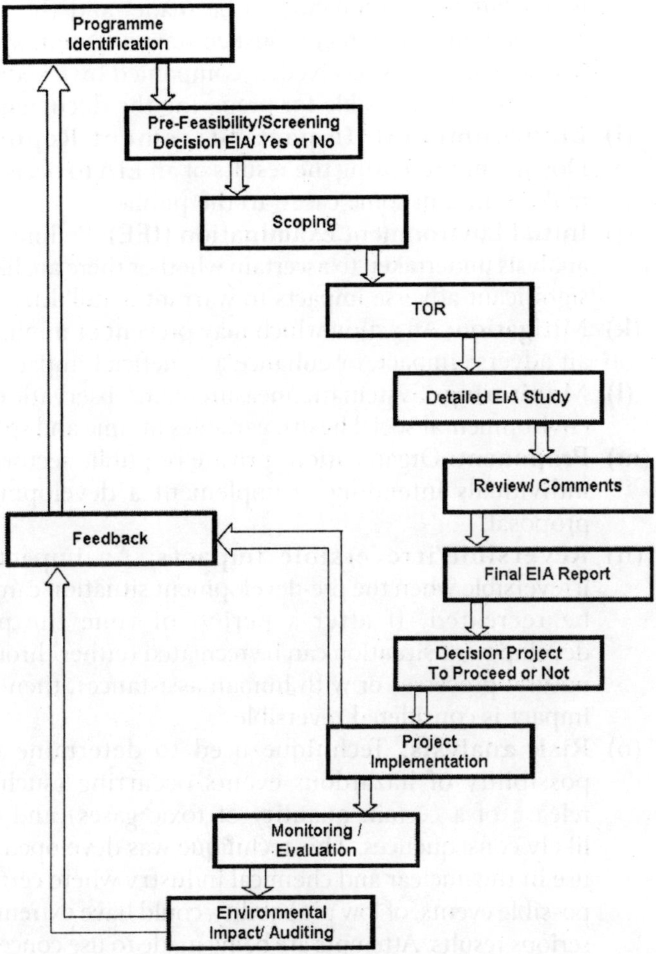

FIGURE 19.1
OVERALL PROCESS OF ENVIRONMENTAL IMPACT ASSESSMENT

human activities which involve a transformation on the natural biophysical environment to increase the goods and services available to society.

(h) Draft environmental statement: A draft statement is produced and circulated to relevant government agencies and made publicly available. A period of time is allocated for comments to be made, and the final environmental statement must contain a section which lists all comments received, accompanied by responses from those responsible for preparing the document.

(i) **Environmental Impact Statement/Reports:** Document presenting the results of an EIA to decision-makers and, in some cases, to the public.

(j) **Initial Environment Examination (IEE):** Preliminary analysis undertaken to ascertain whether there are likely significant adverse impacts to warrant a 'full' EIA.

(k) **Mitigation:** An action which may prevent or minimize an adverse impact, or enhance a beneficial impact.

(l) **Monitoring:** Systematic measurement/observation of environmental/social health variables in time and space.

(m) **Proponent:** Organization (private or public sector) or individuals intending to implement a development proposal.

(n) **Reversible/irreversible impacts:** An impact is irreversible when the pre-development situation cannot be recreated. If after a period of time the pre-development situation can be recreated (either through natural processes or with human assistance), then the impact is considered reversible.

(o) **Risk analysis:** Technique used to determine the possibility of hazardous events occurring (such as release of a certain quantity of toxic gases) and the likely consequences. This technique was developed for use in the nuclear and chemical industry where certain possible events, of low probability, could have extremely serious results. Attempts are being made to use concepts from probable risk analysis to characterize environmental impacts, the occurrence and nature of which are not easy to predict with any degree of accuracy.

(p) **Scoping:** Activity to identify possible significant impacts to be investigated in EIA work.

(q) **Screening:** Activity undertaken to decide if an EIA statement/report is required for a specific development action. Screening is undertaken before EIA work begins.

(r) **Terms of Reference (TOR):** A document specifying

the requirements of the EIA implementation, consultations to be held, data to be produced and form/content of the EIA statement/report. It is often produced as an output of the scoping exercise.

19.2 Screening and Initial Environmental Examination (IEE)

Many projects are considered by government each year. EIA needs only to be applied to those actions which may significantly affect the environment. It is therefore important to establish mechanisms for identifying projects requiring EIA; this process of selection is referred to as "screening".

Development projects have biophysical as well as social and economic impacts. The scale of the project is the key for determining whether an environmental assessment is warranted. The scale of a project reflects project size, the number of people potentially affected, and the extent to which natural systems are disrupted. Sufficient understanding of these factors is necessary for the initial screening decision.

In brief, the screening process places project proposals within the following three categories:

(i) Project clearly requiring an EIA;
(ii) Project not-requiring an EIA;
(iii) Project for which the need for an EIA is not clear.

For the initial screening decision, development projects can be divided into three broad categories or schedules based on *threshold criteria* (that is, environmental parameters which, if exceeded, would require an EIA).

Schedule 1: Proposals likely to have significant but easily identifiable adverse impacts on the environment and for which mitigation measures are able to be readily prescribed, require an IEE report. These include proposals relating to:

1. Use of pesticides over 1 ton but less than 10 tons
2. Livestock development schemes
3. Agricultural development schemes
4. Fisheries
5. Water supply schemes serving over 2000 and less then 10,000 people

6. Housing development schemes
7. Rural electrification schemes
8. River training works
9. Afforestation and Deforestation schemes
10. Watershed management schemes
11. Protected area management plans
12. Population control schemes
13. Subsidy, incentive and credit schemes
14. Introduction of exotic species
15. Construction of district and municipal roads
16. Small scale irrigation schemes
17. Electricity generation schemes (<5 MW)
18. Waste management, drainage and disposal system serving over 2,000 but less than 10,000 people.

In addition, the following industrial proposals are required to have an initial environmental examination report prepared:

(i) **Chemical industry** with a production not exceeding 10 t/day.

(ii) **Petrochemical works** including storage and processing works, all sizes

(iii) **Non-ferrous metal industries** with primary smelting having production less than 10 t/day, except aluminum and copper.

(iv) **Non-metallic Industries**

- (a) Cement with production not exceeding 30 t/day
- (b) Lime with production not exceeding 100 t/day (rotary kiln), or <50 t/day (vertical kiln)
- (c) Brick and tile manufacture with production over 3 million/yr but not exceeding 10 million/yr
- (d) Extraction of minerals and mining covering an area less than 2 ha
- (e) Extraction of aggregates or marble or boulders with a capacity less than 100 cu m/day

(v) **Iron and steel industries**

- (a) Iron ore as a raw material with production less than 100 t/day
- (b) Scrap iron as raw material with production less than 10 t/day

(vi) **Pulp and paper industries** with production less than 5 t/day

(vii) **Leather and textile industries**

(a) Leather tanning with capacity of <100 pieces of hides per day
(b) Textile industries with dyeing facilities having a capacity less than 10 million m/yr
(c) Carpet manufacturing having capacity of less than 500 sq m/day

(viii) **Food industries**

(a) Abattoirs including knackeries having capacity not exceeding 100 head of animal/day
(b) Canning and bottling works having work space not exceeding 200 sq m
(c) Brewery and distillery works which include infusion, boiling and fermentation having production capacity not exceeding 2500 l/day.

Schedule-2: Proposals which are likely to have a significant adverse impact on the environment require an environmental impact assessment report to be prepared. These include proposals relating to:

1. Use of fertilizers, more than 100 t
2. Use of pesticides, more than 10 t
3. Construction of highways and feeder roads
4. Multipurpose reservoir development
5. Electricity generating schemes of greater than 5 MW capacity
6. Medium and large scale irrigation schemes
7. Airport construction
8. Waste management, drainage and disposal systems serving more than 10,000 people
9. Resettlement programmes
10. Water supply schemes serving more than 10,000 people.

With respect to industrial proposals, an EIA report is required for:

(i) **Chemical industries** where production capacity of each product or combined product is greater than 100 t/day.

(ii) **Petrochemical works** including storage and processing works, all sizes

(iii) **Non-ferrous metals industries** (primary smelting)

(a) Aluminum- all sizes
(b) Copper- all sizes
(c) Others- producing 50 t/day or more.

(iv) **Non-metallic industries**

(a) Cement - with production over 40 t/h
(b) Lime - with production over 100 t/day (rotary kiln) or 50 t/day and more (vertical kiln)
(c) Brick and tile manufacture with production over 10 million/yr
(d) Extraction of minerals and mining covering an area over 5 ha
(e) Extraction of aggregates or marbles or boulders with a capacity of over 100 cu m/day
(f) Asbestos industries of all sizes
(g) Radio - active related industries of all sizes.

(v) **Iron and steel**

(a) Iron ore as a raw material-having a production > 100 t/day.
(b) Scrap iron as raw material-having a production more than 200 t/day.

(vi) **Pulp and paper**

(a) Wood based raw material - having a production more than 50 t/day
(b) Raw material other than wood - having a production more than 100 t/day.

(vii) **Establishment of industrial estates** for medium and large scale industries covering an area of 50 ha or more.

(viii) **Leather and textile**

(a) Leather tanning with a capacity of over 100 pieces/ day

(b) Textile industry with dyeing facilities having a capacity of more than 10 million m/year.
(c) Carpet manufacturing with capacity of over 500 sq m/day.

(ix) Food industry

(a) Abattoirs including knackeries with capacity of 100 heads of animal/day
(b) Canning and bottling works having work space of 200 sq m
(c) Brewery and distillery works which include boiling and fermentation with production capacity of >25000 l/day.

(x) Coal industry works of all sizes
(xi) Hazardous waste industry of all sizes or any capacity of the following nature:

(a) Construction of incineration plant
(b) Construction of recovery plant (off site)
(c) Construction of secure land fill facility
(d) Construction of storage facility (off site)
(e) Construction of treatment facilities.

Schedule 3: Proposals for projects which are proposed to be located within or near environmentally sensitive areas are required to have an EIA report prepared unless exempted by the Act. These areas include the following categories:

1. Areas of unique historic, cultural, archaeological, scientific or geological interest
2. Wetlands
3. Ecologically fragile areas
4. National parks, wildlife sanctuaries and protected areas
5. Wilderness areas, containing rare or endangered species of animals or plants and their habitat
6. Semi - arid, alpine and tundra areas
7. Flood or other hazard zones
8. Residential, school, and hospital areas
9. Major sources of public water supply
10. Water bodies (static or lotic).

The proposed project in initially screened through project schedules. If the project concerned falls into project schedule - 1, then the application of an IEE is sufficient, if into schedule - 2, then an EIA is necessary. In cases where the proposed project falls in schedule - 3, an EIA is required unless exempted by the Act.

19.2.1 Initial Environmental Examination (IEE)

Projects where the requirement for an EIA needs to be ascertained should be subjected to an IEE. It is carried out to determine whether potentially adverse environmental effects are significant or whether mitigation measures can be adopted to reduce or eliminate these adverse effects.

To prepare an IEE, it is necessary to initially make a checklist that identifies and briefly describes the project activities to be implemented and natural resources to be affected. Typical activities include siting of the project, resource demand, waste production and regulation, policies and guidelines. For any particular project only a few of these may be significant and therefore the first step is to narrow the list of activities likely to be significant for the particular project.

The interaction matrix shown in Table 19.1 illustrates the process of identifying significant impacts. The horizontal column in the matrix describes the proposed project activities, and the vertical column lists the environmental parameters that might be affected. The response of each environmental parameter to each project action is prepared by an interaction cell and should be checked by a rating number ranging from 1 to 3, in which 1 represents no significant impact, 2 moderate impact and 3 major impact. In general, screening and IEE should be conducted during pre-feasibility stage of a project.

Agencies Responsible for the screening process:

- National Planning Commission Secretariat
- Line Ministries and Departments
- Project proponents.

19.3 Scoping and Preparation of TOR

19.3.1 Scoping

The next important step following the screening procedure is determining what should be the *coverage* or *scope* of the EIA study for a project proposal identified as having potentially significant environmental impacts. This activity is referred to as "scoping"; and it helps in developing and selecting alternatives to the proposed action and in identifying the issues to be considered in an EIA. It is also a procedure designed to establish the terms of reference (TOR) for an EIA.

Scoping is required to ensure the following:

(a) Identify concerns and issues for consideration in an EIA
(b) Ensure a relevant EIA
(c) Enable those responsible for an EIA study to properly brief the study team on the alternatives and on impacts to be considered at different levels of analysis
(d) Determine the assessment methods to be used
(e) Provide an opportunity for public involvement in determining the factors to be assessed
(f) Facilitate early agreement on contentious issues
(g) Save time and money
(h) Establish TOR for EIA study.

The approach and techniques used in scoping are commonly a part of any decision - making process. Issues are identified, evaluated, and organized for presentation to decision makers. The basic components of a scoping exercise are as follows:

(a) Making a plan for public involvement
(b) Assembling relevant existing information
(c) Distribution of information to affected interests
(d) Identifying major issues of public concern
(f) Establishing priorities for environmental assessment
(g) Developing a strategy for addressing priority issues.

A major purpose of scoping at an early stage in project planning is to clearly define all the communities and agencies which should be allowed to influence decisions relating to the

TABLE 19.1
CHECK LIST OF ENVIRONMENTAL PARAMETERS FOR IEE: ITEMS IN INTERACTION MATRIX CHECKED AS (1) INDICATE

IDENTIFICATION OF ACTIVITIES			CHECK- LIST OF ENVIRONMENTAL																								
			SITES SELECTION AND PREPARATION													CONSTRUCTION											
			ACCESS ROADS	SITE SURVEYING	SOIL TESTING	HYDROLOGICAL TESTING	ENVIRONMENTAL SURVEY	SITE CLEARING	BURNING	EXCAVATION	DRAINAGE ALTERATION	STREAM CROSSING	EQUIPMENT	WASTE DISPOSAL AND RECOVERY	PRODUCT STORAGE	ACCESS ROADS	SITE CLEARING (DEFORESTATION)	EXCAVATION	BLASTING AND DRILLING	DEMOLITION	BUILDING RELOCATION	CUT AND FILL	ANCILLARY TRANSMISSION LINES AND PIPELINES	EROSION	DRAINAGE ALTERATION	STREAM CROSSING	EQUIPMENT MOVEMENTS
AREAS OF POTENTIAL ENVIRONMENTAL EFFECTS	PHYSICAL	WATER																									
		NOISE																									
		LAND																									
		ATMOSPHERE																									
	BIOLOGICAL	SPECIES AND POPULATION																									
		HABITATS AND COMMUNITIES																									
	AESTHETIC	LAND																									
		ATMOSPHERE																									
		WATER																									
		FLORA AND FAUNA																									
		MAN MADE OBJECTS																									
		COMPOSITION																									
	SOCIAL	INDIVIDUAL ENVIRONMENTAL INTERESTS																									
		INDIVIDUAL WELL BEING																									
		SOCIAL INTERACTIONS																									

INSIGNIFICANT IMPACT; IF CHECKED BY (2) AN IEE IS ADEQUATE; IF CHECKED BY (3) A FULL SCALE EIA IS NECESSARY

PARAMETERS FOR IEE																													FUTURE AND RELATED ACTIVITIES			
									OPERATION AND MAINTENANCE																							
LABOUR FORCE	WASTE DISPOSAL	PRODUCT DISPOSAL	PRODUCT STORAGE	ABANDONMENT	RECLAMATION	REFORESTATION	FERTILISATION	ANCILLARY TRANSMISSION LINES AND PIPELINES	FOREST CLEARING	EXCAVATION	SPOIL AND OVERBURDEN	BLASTING AND DRILLING	DREDGING	EQUIPMENT OPERATION	OPERATIONAL FAILURES	ENERGY REQUIREMENTS	ENERGY GENERATION	AUTOMOBILE AIRCRAFT VESSEL MOVEMENT	PEDESTRIAN MOVEMENT	UTILITIES	WASTE DISPOSAL AND RECOVERY	PRODUCT STORAGE	SPILLS AND LEAKS	EXPLOSIONS	DE-ICING SNOW REMOVAL AND DISPOSAL	PEST CONTROL	DUST CONTROL	ABANDONMENT	URBANISATION	INDUSTRIAL DEVELOPMENT	TRANSPORTATION	ENERGY REQUIREMENTS

TABLE 19.2
RELEVANCE MATRIX FOR ENVIRONMENTAL STUDY (ARUN III HYDRODAM)

● CERTAIN EFFECT IS EXPECTED, ○ EFFECT IS POSSIBLE

ACTIONS — PREPARATION PHASE

	IMPACT AREAS	ENGINEERING STUDIES (FIELD ACTIVIES, SEISMIC PROSPECTION, ABORING SURVEY)	PAYMENT OF COMPENSATION	REAL COMPENSATION	RESETTLING	LAND REFORM AND REFORM OF LAND USE	DEVELOPMENT OF AGRICULTURAL LAND AND FOREST	IMPROVEMENT AND CONSTRUCTION OF PUBLIC BUILDING AND INSTALLATION	LAND USE OF ENTIRE PLANT	QUARRIES (STONE AND GRAVEL)	TIMBER CUTTING	STORAGE AND DEPOSIT OF CONSTRUCTION MATERIAL	IMPROVEMENT OR MODIFICATION OF EXISTING ROAD	ROAD BED	BRIDGE	PROTECTING MEASURES	PACKING	TRANSPORTATION BY ROAD, AEROPLANE, HELICOPTER	DIVERSTION TUNNEL	COFFER DAM
NON-BIOTIC AREA	1. AIR (1.1. PRODUCTION, 1.2 NOISE INTENSITY)	●								○		○	○	●	●	●	●	●	●	●
	2. WATER (2.1. SURFACE WATER, 2.2 UNDERGROUND WATER)	●								?	?	?	?						?	?
	3. SOIL ROCK (3.1. CULTIVATED SOIL, 3.2 NATURAL SOIL, 3.3 RAW SOIL AND ROCK)	●								?	?	?	?							
	4. UNDERGROUND	○																		
	5. SURFACE (LANDSCAPE)						?		?	?	?	?		?	?	?	?			
	6. CLIMATE (6.1. SUNSHINE, 6.2 MACRO CLIMATE)																			
BIOTIC AREA	7. FLORA ANA FAUNA OF THE SOIL	○							?	?				?						
	8. SURFACE FAUNA TERRESTRIAL ECOSYSTEM	○							?	?				?						
	9. PLANT FORMATION OF TERRESTRIAL ECOSYSTEM (9.1. FOREST, 9.2 GRASS LAND)								?	?	?			?						
	10. AQUATIC ECOSYETEM (10.1. RIVER AND BROOKS, 10.2 PONDAGE)	○							?										?	
SOCIAL AREA	11. POPULATION STRUCTURE (11.1. AGE, 11.2 PROFESSION)	○	○	○	●		?													
	12. POPULATION MOVEMENT	○	○	●	○		?		?									?		
	13. HEALTH (13.1. PHYSICAL HEALTH, 13.2 PHYSICAL HEALTH)																			
	14. SECURITY (14.1. MATERIAL SECURITY, 14.2 EXISTENAL SECURITY, 14.3 SOCIAL SECURITY)																			
	15. AESTHETIC					?	?		?	?	?			?	?	?	?		?	?
	16. FREEDOM OF CHOICE AND DECISION		●	●	●															
	17. COMMUNICATION	○																?		
CULTURAL AREA	18. INSTITUTIONS (18.1. SCHOOLS, 18.2 CULTURAL SITES)							?												
	19. SCENES OF VALUES		○	○	○															
	20. TRADITIONS AND CUSTOMS (20.1. FESTIVALS, 20.2 TRADITIONAL LAND USE, 20.3. ART & CRAFT																			
	21. CONTENT OF CULTURAL LANDSCAPE (21.1. NATURAL FEATURES TO BE PROTECTED)																			
ECONOMIC & POLITICAL	22. PROFESSIONAL STRUCTURE																			
	23. AGRICULTURE AND FORESTRY																			
	24. TRADE						?											?		
	25. INDUSTRY																			
	26. TOURISM																			
	27. PUBLIC SOURCE		○	○	○	○	○													
	28. INFRASTRUCTURE (28.1. ROADS, 28.2 PUBLIC BODY INSTALLATION, 28.3. SUPPLY SERVICE)					●		?	?				?							
	29. MANAGEMENT OF PUBLIC FEATURES		●	○	○															
	30. INTEREST OF INDIGENOUS FOUNDATIONS (30.1. DEVELOPMENT, 30.2 ADMINISTRATION)	○	○	○	○	○	○	○	○									?		
	31. EXTERNAL (31.1. ECONOMIC, 31.2 OTHER PUBLIC, 30.3. PRIVATE)																	?		

I PHASE																																				OPERATION PHASE							
POWERHOUSE	TAIL RACE	SWITCH YARD	ERECTION OF TOWERS AND INSTALLATION OF HV LINE	BLASTING AND CLEANING OF LOOSE ROCK	ERECTION OF PROTECTIVE CONSTRUCTION AGAINST ROCK FALLING	PROTECTIVE FORCE FOR LANDSLIDE	REFORESTATION	LANDSCAPING	TEMPORARY ACCOMODATION	STORAGE OF CONSTRUCTION MATERIAL	SUPPLY AND REUSE LOGISTICS	CONSTRUCTION WORKERS AND AXILLARY PERSONNEL TECHNICIAN	TECHNICIAN	PROFESSIONAL PERSONNEL	SERVICE AND ADMINISTRATIVE PERSONNEL	CONTACT WITH FOREIGN WORKERS	CONTACT WITH STRANGERS	CONTACT WITH SCENE AND TECHNOLOGY	BONDAGE	BUILDINGS AND CONSTRUCTION	TRANSMISSION AND LINE	ROAD	ENERGY PRODUCTION	PERMANENT JOB	CONTACT FOR MAINTENANCE AND REPAIR	TAXES (REVENUE)	NEW SUPPLY OF SERVICES AND CONSUMER GOODS	OUTDOOR ACTIVITIES AND SPORTS	FISHING	HIRING AND CAMPING	LATENT RISK	CONNECTING ACROSS THE VALLEY	SEASONAL VARIATION OF WATER LEVEL	EMERGENCY DRAINAGE	RETENTION OF GRAVEL AND SILT	DRODGING OF PONDAGE	PRODUCTION OF FISH	CHANGE OF EROSION PROCESS	LOSS OF WATER	TRANSPORT OF GRAVEL AND SILT	MAINTENANCE OF BREAKS (PRESSURE DRAF AND POWER LINE)	TRANSPORT OF GOODS AND PERSONS	PROTECTIVE CONSTRUCTION
?	?	?	?	?				?														?																					
?	?	?						?		?		?	?	?	?				?															?									
		?	?		?	?	?	?	?	?									?																			?					
?	?																																										
?	?	?	?	?	?	?	?	?	?										?																?			?			?		
																			?			?																					
							?	?											?																								
							?	?											?																								
							?	?											?			?																					
																			?														?		?			?	?	?			
												?	?	?	?									?	?																	?	
											?	?	?	?	?							?		?	?																	?	
												?	?	?	?				?					?			?	?		?							?						
												?	?	?	?		?							?			?																
	?	?	?	?	?	?	?	?	?																													?					
																?								?	?																	?	
											?	?	?	?	?	?	?	?						?						?		?										?	
																										?																	
												?	?	?	?	?								?			?	?														?	
																?																										?	
																?																										?	
												?	?	?	?	?				?	?	?		?	?				?								?				?	?	?
																					?	?					?		?						?		?		?			?	
											?	?	?	?	?							?		?	?		?		?			?					?					?	
																						?	?				?		?													?	
																						?					?		?	?		?					?					?	
											?	?	?	?	?							?	?	?	?																	?	
											?	?	?	?	?						?	?	?	?		?						?										?	
												?	?	?	?					?		?	?	?		?											?						
											?	?	?	?	?	?	?	?	?	?		?	?	?		?	?		?		?			?			?				?	?	
																			?	?		?	?	?			?				?						?					?	

proposal. The plan should identify whom to talk to and when and how to undertake the communication exercise. Consent must be obtained from the authorities and government agencies concerned. The project proponent, relevant experts, local people affected, as well as special interest groups should be considered for inclusion in the list of persons to be covered by the communication plan. Methods for involving affected interests and for collecting information include:

(a) Securing written submissions from government agencies and the public
(b) Holding community meetings and public hearings
(c) Conducting preliminary field study/site observations
(d) Conducting workshops/seminars
(e) Establishing an intersectoral task force.

For formulating appropriate mitigation measures, relevant information on the nature of the project, its potential environmental impacts and alternatives should be processed and assembled into an information package with the key issues arranged in order of priority. For those issues which need further information in order to be resolved, *terms of reference (TOR)* should be prepared in order to define guidelines for future study. For major projects, it is always advisable to issue a general public notice inviting public comment and to hold public meetings at the project site as well as at central level, to facilitate consultation and interaction in preparing the information package.

19.3.2 Preparation of TOR for EIA studies

If the screening exercise indicates the need for an EIA for a particular project, a TOR is prepared to provide specific guidelines for the full-scale EIA study. It assists in:

- Systematizing the working procedure
- Delineating the specific activities to be implemented
- Fitting the EIA study into the context of existing policies, rules and administrative procedures
- Accomplishing the work within a specified time frame
- Giving emphasis to the most important elements for study

- Providing technical guidance in delineating the specific environmental aspects for study.

19.3.2. Format for TOR

(i) Introduction

(a) Purpose of terms of reference

(b) Responsible party for preparing the EIA report.

(ii) EIA guidelines General guidelines to be followed should be provided, together with the procedure to be used in preparing the EIA report.

(iii) Background information

(a) General background studies and relevant reports should be described

(b) Specific background studies and reports related to the project and location should be listed or described

(c) Relevant policy and legislation should be mentioned

(d) The relevance to local planning perspectives should be discussed.

(iv) Specific EIA guidelines

(a) The environmental components which need detailed or further study should be listed. The indicators of each listed component to be measured or assessed might be specified.

(b) Action required to minimise possible adverse impact (relocation, increased taxes, compensation etc.).

(c) Cost - benefit analysis of project in terms of its environmental implications.

(d) EIA report should propose a monitoring programme for assessing the actual impacts during the projects' operation and beyond.

(e) The following tasks should be mentioned in the EIA study proposal:

- **Work tasks:** Each specific task to be done

should be sufficiently described.

- **Study schedule:** The proposed plan for carrying out the EIA study should be indicated.
- **Review sessions:** Periodic reviews of the work might be needed during study.
- **Study team:** Expertise and specialists needed for the proposed project study team should be described.

(f) Conclusion on whether the project should be implemented or whether it poses an unacceptable risk.

(v) Time constraints:

(a) The time required for the EIA report to be completed should be specified (that is, within 3 to 12 months).

(b) A description of EIA tasks and completion dates should be stated.

(vi) Budget

The estimated budget allocation for the EIA study and report should be given.

(vii) EIA report format

(viii) Other relevant information

(ix) Reference list

(x) Annexes, Maps, Photographs, Figures, Tables, Chart Drawing, Checklists.

While the TOR prepared at the initial stage provides direction and guidelines for further study, it should always be flexible and open to modification as the EIA work progresses.

19.4 EIA Report

The EIA report should be concise and limited to addressing significant environmental issues. The detail and sophistication of analysis should be commensurate with the potential impact. The report should be prepared with the responsibilities of the target users, such as project proponent, designer and decision makers in mind.

19.4.1 EIA Report format (main body)

The EIA study report should include the following sections:

(i) **Executive summary**: The report should include a precise and non-technical description of significant results and recommended actions, summarized in less than 5 pages.

(ii) **Project description**: This section should outline the activities of the proposed project, including its technical, geographic, ecological, economic, social and physical context. A schedule of activities should be provided for each phase of the project cycle, along with the objectives and strategies for each activity.

(iii) **Baseline information**: Depending upon the nature of the project, the EIA report should also present baseline information pertaining to the geophysical, biological and socio-economic situation of the area under study including any changes anticipated prior to project implementation. Data gaps and limitation should also be mentioned.

(iv) **Identification of environmental impact**: The EIA report should identify likely impacts, adverse as well as beneficial, envisaged as a consequence of the proposed project activities at different stages of the project cycle (Fig. 19.2). Impacts should be predicted and quantified as far as possible in terms of spatial and temporal contexts.

(v) **Alternative analysis**: Design, site technology and operational alternatives should be compared systematically in terms of their potential environmental impact; capital and recurring costs; suitability to local conditions; and institutional training and monitoring requirements. To the extent possible, the environmental costs and benefits should be quantified and economic values attached for each of the alternatives.

(vi) **Mitigation measures**: The EIA report should propose pragmatic mitigation measures for all activities that are likely to have an adverse impact on the environment. As mitigation or abatement measures can not be expected to totally eliminate adverse impacts, it is recommended that compensatory measures be proposed as well. It is essential that the cost effectiveness of mitigation measures be analyzed against viable alternatives.

(vii) **Review of policy and legal framework**: Based on the nature and magnitude of the project, a review should be undertaken of the policy and legislative framework relating to the project, with an examination of the weaknesses and strengths of their legal aspects. The report should suggest necessary amendments of such policies and legislation if these are likely to hinder environmental conservation and development.

(viii) **Monitoring**: The EIA report should clearly specify the nature of monitoring required, stipulating who should undertake this activity, how much it would cost and what other inputs are necessary. Time schedules should be specified for monitoring.

(ix) **Auditing**: The EIA report should include environmental audit design and justification.

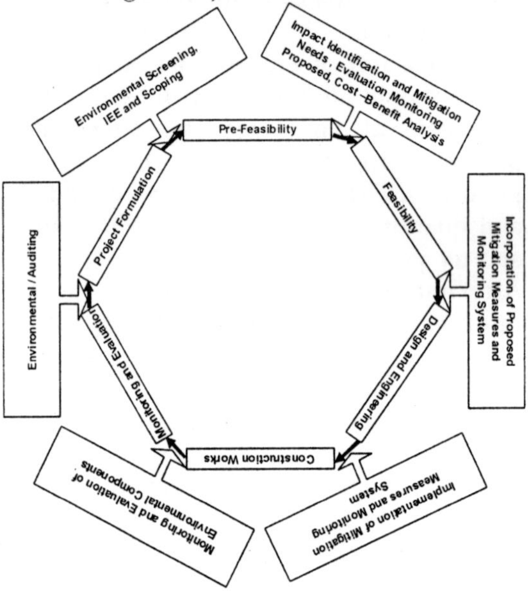

FIGURE 19.2
PROJECT CYCLE AND ITS IMPLEMENTATION STAGES

19.4.2 EIA report annexes

The EIA document should be accompanied by a comprehensive series of annexes relevant to the project operation and project area in question. The annexes should include all significant materials used during the preparation of the EIA report. The annexes should either be presented as a continuation of the main report or within a companion volume. The annexes should include the following:

- Related maps such as topographical, geological, land use, land capability, district maps
- Aerial photographs of the project and surrounding areas
- Questionnaire or checklist employed for field investigations
- Figures or drawings related to the EIA such as flow charts, diagrams, etc.
- Climatological, meteorological and hydrological data, if any, in the form of a time series table
- Data about the existing flora and fauna in the project area
- Geological and risk assessment data, if any
- Information on air and water quality and noise level, before and after the project, if available
- Relevant matrix or flow diagram for EIA
- Audio-visual aids, such as photographs, slides, transparencies, recordings, video films (if used)
- Agricultural data, such as cropping patterns, livestock, soil, recommended doses of chemical fertilizer, etc.

The following details should be provided in the annexes:

- Study team personnel and organizations involved in EIA preparation
- Individuals and organizations contacted during the study, including addresses and telephone numbers
- References listing the written materials used in study preparation
- Record of inter-agency interaction/ forum/ meetings including lists of invitees and actual participants, with a brief summary of the discussions and observations.

19.4.3 EIA Report References

EIA draft report should list references in the text of the main report in chronological order such as:

- Ross,W.A.1987. Evaluating environmental impact statements. *Journal of Environmental Management 25:* 133 - 147.
- Nepal Electricity Authority (NEA) 1990. *Socio-economic and environmental impact of Arun III hydro-project.* Mimeographed report. pp. 1- 250.
- Jha, S. 1990. *Conservation for development in Nepal.* National Book Organization, New Delhi.

19.4.4 Agency responsible for conducting the EIA

The project proponent should prepare the EIA report. Government agencies or private sector organizations should hire consultants for this purpose.

19.5 Identification of Environmental Impact

19.5.1 Objectives

Environmental impact implies an alternation of environmental conditions or the creation of a new set of adverse or beneficial environmental consequences caused or induced by the action under consideration. Impact identification starts at the early stage of scoping when available data on both the project and surrounding environment are studied. As the EIA study progresses more data become available, especially on environment and socio-economic conditions. The preliminary identification of impacts from scoping may be confirmed or "new" impacts may be identified as requiring investigation.

19.5.2 Types of impacts

The following categories of impacts should be considered:

19.5.2.1 Socio-economic impacts

A study of socio-economic impacts would examine project action that alters the existing social and economical conditions of communities within or around the project location. It may prove either adverse or beneficial. For example, an expanded irrigation facility designed to enhance agricultural production would be beneficial, while the project might also result in waterlogging that would produce adverse consequences.

19.5.2.2 Biological and physico-chemical impacts

Impacts in this category relate to effects on biophysical resources such as vegetation, wildlife, crops, and aquatic life. Impacts affecting soil and land forms, or creation of a propensity for soil erosion, floods and sedimentation, would be considered as physical impacts. Chemical impacts relate to project activities that cause a chemical change in air/water/soil quality. Smoke emitted from a brick factory, for example, may change the SO_2 content of air, while untreated effluent discharged directly into a river by a paper factory may change the chemical characteristics of the river.

19.5.2.3 Cultural impacts

Project impacts on cultural heritage should be considered including historic sites, religious shrines or areas, or traditional practices that may be affected.

The types of impact should be categorized as direct, indirect and/or cumulative. *Direct impact* refers to an alteration of the existing environmental condition as a direct consequence of project activity. Construction of an irrigation diversion weir on a stream bank, for example, may have a direct impact on the aquatic ecosystem and stream valley vegetation and also directly cause erosion. *Indirect impact* results when one component of the environment produces repercussions for other related components. While an environmental impact produced by a single activity may not be significant, a series of similar types of impacts created by more than one project, or by combined effects of several impacts from the same project, may be *cumulative*.

19.5.3 Methods of impact identification and comparison

Some simple methods used for impact identification and comparison are: (a) checklist method, (b) matrix method, and (c) network method.

19.5.3.1 Checklist

This method starts with the preparation of a list of environmental, social and economic factors which may be affected by a proposed development project. One of the most effective and widely used methods combines a checklist with 'thresholds of concern' (TOC). A checklist of environmental resources is usually compared against a TOC value. Any impact that exceeds the TOC value is considered to be 'significant' for decision-makers (Tables 19.2 and 19.3).

19.5.3.2 Matrix method

The simplest method is to combine a checklist of development actions and a checklist of environmental components within a two dimensional matrix as a means to identify potential impacts (Tables 19.2 and 19.3). This is referred to as an *interaction matrix.*

19.5.3.3 Network method

The network method is an expansion of the matrix method, and introduces a cause-condition-effect relationship.

19.5.4 Impact prediction

The next step after impact identification is impact prediction in terms of impact magnitude, extent and the duration. Prediction should be based on the available environmental baseline data that are likely to be affected by the implementation of certain developmental activities. These anticipated changes or impacts should be described in quantitative or qualitative terms.

An environmental impact prediction should, at minimum, perform the following;

- Determine the initial reference or baseline state
- Estimate the future state without the proposed action
- Estimate the future state with the proposed action.

Impact prediction takes into account:

19.5.4.1 Magnitude of impact

This is defined by the severity of each potential impact. It also indicates whether the impact is irreversible or if reversible the potential rate of recovery. The magnitude of impact is considered high if a major adverse impact is not mitigated. A major adverse impact would affect the potential subsistence/recreational/ commercial use of biophysical resources, with the result that the value of resources would be reduced far below the publicly acceptable level. Moderate to minor unmitigated impact of a similar nature will result in resources being still usable but at some inconvenience to the public.

19.5.4.2 Extent of impact

The spatial extent or the zone of influence of the impact should always be determined. An impact can be site specific or limited to the project area; a locally occurring impact within the watershed of the proposed project; a regional impact that may extend beyond the watershed; and national, affecting resources on a national scale.

19.5.4.3 Duration of impact

Environmental impacts have a temporal dimension that needs to be considered in an EIA. Impacts arising at different phases of the project cycle may need to be considered. An impact that generally lasts for only 3 years after project initiation may be classified *short term*. An impact which continues for 10 years or more but less than 20 years may be defined as *medium term,* and impacts that last beyond 20 years are considered *long term.*

19.5.4.4 Impact ranking methodology

A numerical scale should normally be developed in order to provide a quantitative assessment of the various types of predicted impact. The following ranking methodology has been used in many studies:

Magnitude		*Extent*		*Duration*	
High/major	60	Regional	60	Long-term	60
Moderate	20	Local	20	Medium-term	10
Minor	10	Site specific	10	Short-term	05

A sum of the impact value for magnitude, extent, and duration provides a maximum value of 180 and a minimum value of 25.

19.6 Mitigation Measures and Review of Draft EIA Report

19.6.1 Mitigation measures

Mitigation measures are actions which reduce, avoid or offset the potential adverse environmental consequences of development activities. The objective of mitigation measures is to maximize project benefits and minimize undesirable impacts. Although a wide range of mitigation measures may be proposed, the following are relevant to most development projects.

19.6.1.1 Consideration of alternatives

The considering of alternatives to a proposed project is one of the key functions of an EIA, which involves an examination of alternative ways of achieving the objectives of a proposed project. The aim is to arrive at a development option which maximizes the benefits while minimizing unwanted impacts. Important aspects of development proposals for which alternatives need to be considered include: scale, technology, location, fuel, raw materials, design, time schedule, and economic factors.

A comparison of adverse impacts against advantages should also include considering the alternative of dropping the project altogether. However, adoption of any alternative should not substantially alter the objectives of a proposed project if it is determined that the project should proceed.

19.6.1.2 Introduction of compensatory measures

Compensatory measures are actions which compensate for adverse impacts that are unavoidable. Possible compensatory measures include:

- Restoration of damaged resources
- Rehabilitation of displaced settlements
- Compensation for affected persons.

19.6.1.3 Introduction of corrective measures

Corrective measures may be adopted to reduce adverse impacts to acceptable levels. The following are examples of the types of corrective measures that can be used :

- Installation of pollution control devices
- Construction of wastewater treatment plant
- Construction of fish ladder (in dams/weirs).

19.6.1.4 Introduction of preventive measures

Some potential adverse impacts may be reduced or eliminated before occurrence by introducing preventive measures. Examples of preventive measures include:

- Implementation of health education programme
- Initiation of a public awareness programme.

19.6.1.5 Implementation of mitigation measures

Implementation of mitigation measures requiring funding should be estimated and included in the EIA report. All proposed

TABLE 19.3
ENVIRONMENTAL IMPACT MATRIX OF ARUN III PROJECT

Resource/ Project Activity	*Impact*	*Magni-tude*	*Extent*	*Duration*	*Mitigation*	*Responsibility*
Microclimate						
1. Reservoir operation	Changes in temperature and humidity	L	L	L	None required	
2. Vegetation/ forest removal	Changes in temperature and	L	L	S	Protection of vegetation on south facing slopes;	HMG
					Reduce to the extent possible soil dump sites on forested areas;	JV/Contractor
					Implement reforestation programmes;	GON
					Replace adit road with shorter roads;	DOR/SWK
					Combine adit and powerhouse construction camps;	JV/Contractor
					Evaluate alternates to dam site access road	DOR/SWK
					Implement ground cover/erosion protection/ revegetation measures	JV/Contractor
Air Quality						
1. Vehicle emission	Increase of air pollution possible health	L	L	S	Reduce travel distances by placing construction camps close to work areas;	JV/Contractor
					Replace adit road with shorter road;	DOR/SWK
					Eliminate spoil dumps and associated travel to exetent possible;	JV/Contractor
					Engines to comply with international standards for exhaust gases;	Contractor

					Maintenance of engines and exhaust gas check; Dust control with water in dry season	Contractor Contractor
2. Raod dust	Increase of air borne dust during dry season	L	L	S	Reduce distances between camps and work areas; Black top heavy use areas; Breathing protection	JV/Contractor Contractor Contractor
3. Dust/emissions from concrete silos, crushing plants and quarries	Increase of air borne dust	L	L	S	Install dust collectors, sprays and emission control equipment	Contractor
Birds and mammals						
1. Construction of project facilities	Loss of habitat; interruption of migratory routes for mammals; local disturbance of breeding sites	L	L	L	Implementation of reforestation programme; Reduction of spoils dumps to the extent possible; Replacement adit access road with shorter road; Combine adit and powerhouse camp sites	HMG JV/Contractor DOR/SWK JV
Demographics						
1. Project construction	Concentration of approx. 3,500 project work force; displacement of approximately 144 households	H	L	S	Site camps near workplace and at a distance from indigenous settlement; Rehabilitation programme; Rural development and training programme compensation	JV HMG/Project HMG/Project
2. Project	Permanent NEA staff of					

Operation	approx. 203					
3. Permanent NEA Camps	Permanent NEA staff of approx. 203; development of larger economic centre resulting in increased population.	M	L	L	None required None feasible	
Socio-economic and cultural						
1. Project construction	Loss of arable land	H	L	L	Intensify production on remaining land; Reclaim arable land after construction;	HMG/Project JV/Contractor
	Loss of forest resources	L	L	L	Control deforestation; Forestry conservation and development programme;	HMG HMG/Project
	Displacement of approx. 144 families; interruption of normal farming activities	H	L	L	Minimize project land requirement; Compensation and rehabilitation programme; Intensify production on remaining land; Provide skill training for employment on construction	JV/Contractor HMG/Project HMG/Project contractor
	Requirement for additional food for local people, workforce and immigrants	M	L	S	Encourage production of crops and livestock workforce and other outlets Contractor to provide logistics for his work force	HNG/Project JV/Contractor
	Requirement for additional fuel supplies	M	L	S	Contractor to provide kerosene and electricity to workforce; limit use of local fuelwood; Prohibit unauthorized felling of trees and use of open fire;	Contractor HMG/Project JV/Contractor

					Forestry development programme	HMG/Project
	Disruption of traditional, social and cultural patterns by large scale immigration	M	L	S	Training and strengthening of villages;	HMG/Project
					Limited work contracts for non-indigenous labour	JV/Contractor
Topography, soils, landuse and vegetation						
1. Road construction	Changes in drainage system; loss of vegetation cover increased;run-off; slopes; creation of fill areas; loss of cultivated land	M	L	S	Permanent and temporary road construction Schaffner (1987) or other suitable methods;	JV/Contractor
					Replace adit access road with shorter road;	DOR/SWK
					Reduce nos of swithchbacks in favour of longer	JV/Contractor
					Reduce to extent possible spoit dumps and associated roads;	JV/Contractor
					Combine adit and powerhouse campus;	JV/Contractor
					Implement slope protection measures;	JV/Contractor
					Reclaim and revegetate all temporary roads;	JV/Contractor
					Salvage soils for use in reclamation	JV/Contractor
2. Quarry and borrow pits	Changes in drainage system; loss of vegetation cover; cutting of slopes; creation of steep highwall sections; loss of cultivated land	H	L	S	Fill borrow pits with muck;	JV/Contractor
					Backfill to stable contrours;	JV/Contractor
					Provide top soil;	JV/Contractor
					Excavate by having benches wide enough for later cultivations;	JV/Contractor
					Provide for controlled drainage;	JV/Contractor
					Reforest and revegetate disturbed slopes;	JV/Contractor

No.	Component	Impact				Mitigation	Responsibility
						Make temporary facilities subject to aproval	JV/Contractor
3.	Spoil dumps	Loss of cultivated land; creation of fill areas; construction of temporary access roads	H	L	S	Eliminate by using selected spoil for concrete, for road maintenance, as backfill for quarry and borrow pits;	JV/Contractor
						Make temporary facilities subject to approval	JV
4.	Temporary Construction Camps	Temporary loss of cultivated lands; Changes in drainage system; loss of vegetation cover; loss of productive forest lands; construction of access roads; waste control	L	L	L	Select sites. Alternatives or a dditional sites proposed by contractor to be subject of bid evaluation and approval;	JV/Contractor
						Combine adit and powerhouse construction camps;	JV/Contractor
						Impelement erosion protection measures;	JV/Contractor
						Reduce distance betwee camps and work areas;	JV/Contractor
						Salvage soils prior to grading;	JV/Contractor
						Replace soils and reclaim as cultivated land;	JV/Contractor
						Install drainage control systems for long term use;	JV/Contractor
						Revegetate all disturbed slopes with soil stabilizing plant species;	JV/Contractor
						Make temporary facilities subject to approval;	JV
						Reduce space requirement by specifying at least two storey buildings	JV
5.	Permanent NEA Camps	Loss of cultivated land; displacement of local residents; changes in drainage system; increase in erosion potential; creation of terraces; cutting of slopes	M	L	L	Salvage soils prior to grading;	JV/Contractor
						Minimize height or benches;	JV/Contractor
						Install drainage control systems;	JV/Contractor
						Revegetate slopes;	JV/Contractor
						Reduce area needs by eliminating free standing single family structure;	JV
						Develop two storey buildings where posible to	

	reduce building sizes; Compensation payment and additional measures	HMG

Legends

Magnitude	*Extent*	*Duration*		
H = High	L = Local	L = Long term (OVer 20 years)	JV	Joint Venture Arun III
M = Medium	R = Regional	M = Medium (Over 10 years)	DOR	Department of Raods
L = Low	N = National	S = Short term (Below 10 years)	SWK	Scott Wilson Kirkpatrick (Road Consultant)
	I = International		NEA	Nepal Electricity Authority
			GON	Government of Nepal

mitigation measures should be integrated in the project design so that these measures may automatically form part of the construction and operational phases of the project.

19.6.2 Review of draft EIA report

A draft EIA report should be released for public review and comments. The draft EIA report and comments received should be reviewed by an independent body, and not by the project proponent alone.

19.6.2.1 Objectives

The main review objectives should be to critically examine whether the draft EIA report meets the following criteria:

- The EIA report responds fully to the TOR presented at the beginning of the study
- The EIA report is in concurrence with National EIA Guidelines
- The report addresses the key environmental issues which need to be answered prior to making a decision
- The study identifies all significant adverse environmental impacts likely to arise through project implementation, with mitigation measures for each impact
- Reasonable alternatives have been suggested to the proposed action.

19.6.2.2 Review parameters

The following parameters should be carefully checked while reviewing the EIA report to examine whether they have been adequately treated:

Impact identification:

- Does the project have an impact on any environmentally sensitive area?

- Is there a clear statement of significant beneficial/ adverse impacts?
- Have the risks been evaluated?
- Has attention been paid to off-site effects, including trans-boundary effects, and to the possible time-lag before effects are manifested?

Mitigation measures:

- What mitigation measures are proposed and what alternative sites have been considered?
- What lessons from previous similar projects have been incorporated into the EIA?
- Have concerned populations and groups been effectively involved?
- Is adequate consideration given to provision of compensation for loss or damage to property, or for resettlement?
- Are institutional arrangements adequate to implement recommended mitigation measures?
- Does the EIA report specify who will be responsible for the monitoring and the standards enforcement programme?

Agencies responsible for EIA review:

- Project proponent (Department/ Ministries)
- NPC/ Environment Division
- Related line ministries / departments
- NGOs and representatives from affected population.

The project proponent should receive comments and recommendations from reviewers and then request the EIA study team to revise the report.

19.7 Impact Monitoring

A serious shortcoming of most EIAs is the absence of baseline data and impact monitoring during the construction, completion and operation phases of large developmental projects. Without such data and programmes, it is impossible to test impact predictions and the success of mitigative measures.

Environmental monitoring is one of the most important components of an EIA. It is essential for:

- Ensuring that impacts do not exceed legal standards
- Checking the implementation of mitigation measures in the manner described in the EIA report
- Providing early warning of potential environmental damage.

19.7.1 *Principles of monitoring*

If the EIA monitoring process is to generate meaningful information and improve implementation of mitigation measures, it must accomplish the following:

- Carefully determine the indicators to be used in monitoring activities
- Collect meaningful and relevant information
- Apply measurable criteria in relation to chosen indicators
- Pass objective judgements on the information collected
- Draw tangible conclusions based on the processing of information and objective judgements
- Make rational decision based on the conclusion drawn
- Recommend improved mitigation measures to be undertaken by the implementing authorities.

19.7.2 *Types of monitoring*

Various types of monitoring activities are briefly described below:

19.7.2.1 Baseline monitoring

A survey should be conducted of basic environmental parameters in the area surrounding the proposed project before construction begins, so that subsequent monitoring can assess changes in those parameters over time against the baseline.

19.7.2.2 Impact monitoring

The ecological and socio-economic (including public health), parameters within the project area must be measured during the project construction and operational phases in order to detect environmental changes which may have occurred as a result of project implementation.

19.7.2.3 Compliance monitoring

This form of monitoring employs a periodic sampling method, or continuous recording of specific environmental quality indicators or pollution levels to ensure project compliance with recommended environmental protection standards.

Monitoring should be regular and performed over a long duration. Interruptions in monitoring may result in insufficient data to draw accurate conclusions concerning project impact.

The main aim of EIA monitoring is to provide the information required to ensure that project implementation has the least possible negative environmental impact on the people and landscape ecology.

19.7.3 Level or intensity of monitoring

It is not possible to monitor all the aspects covered by the impact identification process. A selection, or "scoping out" should therefore be made of the most important and critical parameters that could influence the project and its surrounding environment. The level or intensity is to be determined on the basis of the potential severity of the environmental impact being monitored.

19.7.4 Institutional aspects

There needs to be a firm institutional commitment by the agencies responsible to the monitoring process, particularly in regard to the following:

- Willingness on the part of the institutions involved and organizational personnel to support the monitoring process with the necessary level of resources and authority

- Maintaining continuity in the monitoring programme
- Technical capabilities of the personnel involved must be developed
- Integrity or honesty of the process must be maintained
- Decisions must be taken based on a thorough review of results
- Monitoring information must be made available to all agencies concerned
- Necessary institutional reforms need to be made within the planning and implementation agencies and among the hierarchy of the planning organizations in the country.

EIA monitoring responsibility should be given to EIA monitoring agencies created in planning organizations such as NPC, Ministry of Population and Environment, regional level organizations and district level planning organizations of the government.

The costs involved in EIA monitoring should be borne by government since it is responsible for ensuring the welfare of the people.

19.8 EIA Evaluation

An EIA evaluation is primarily employed to assess the performance of environmental protection programmes undertaken by responsible public or private sector agencies. EIA evaluation is conducted following the implementation of a project, particularly after data generated by EIA monitoring activities have been analyzed. The evaluation division of the ministry concerned is responsible for the overall evaluation.

During evaluation, indicators derived from important project parameters are checked on a regular basis. These parameters include the following:

- The environmental issues raised in the EIA report
- Areas of actual environmental impact
- Areas of unanticipated environmental impacts
- Adequacy of mitigation measures suggested
- Adequacy of resource allocation for implementing mitigation measures

- Involvement of appropriate authorities
- Cost/benefit of environmental mitigations
- Adequacy of project implementation monitoring
- Adequacy of monitoring data for evaluating predictions made in the EIA report
- Recommendations made to improve mitigation, including new remedial measures if necessary.

19.9 Environmental impact Auditing

Auditing, which is usually performed only once for each project, refers to the examination and assessment of a certain type of performance. In the case of an EIA, an audit assesses actual environmental impact, the accuracy of prediction, the effectiveness of environmental impact mitigation and enhancement measures, and the functioning of monitoring mechanisms. The audit should be carried out by the agency which approves the project, with assistance from other relevant organizations, if required. The auditing should be scheduled immediately after project completion or when considered necessary.

Audits are not required in all cases. However, at the project approval stage, both project proponent and authorizing agency should consider whether the application of a particular audit technique is likely to result in new information or an improvement in management practices. Particular attention should be given to the projected cost-effectiveness of any proposed audit and to technical difficulties likely to be encountered.

In general, the following types of audit are recommended for different aspects of the EIA process:

(a) **Decision point audit:** It examines the effectiveness of EIA as a decision-making tool
(b) **Implementation audit:** It ensures that consent conditions have been met
(c) **Performance audit:** It examines the responses of agencies concerned with project management
(d) **Project impact audit:** It examines environmental changes arising from project implementation
(e) **Predictive technique audit:** It examines the accuracy and utility of predictive techniques by comparing actual

against predicted environmental effects

(f) **EIA procedure audit:** It examines critically the methods and approach adopted during the EIA study.

19.10 Popular Participation

The project proponents, implementors and authorizing agencies should be responsible for encouraging community involvement, from the planning stage onwards, in all economic developmental projects that could result in adverse environmental impacts. The public, proponents and implementors should jointly undertake information collection and assessment prior to decisions being made to proceed with a project, and also following the project implementation.

Public involvement should be guaranteed during the following phases of project planning:

- Project identification, feasibility and scoping
- IEE
- Detailed EIA study
- Monitoring, evaluation and audit.

The following agencies, groups and individuals should participate in the EIA process:

- Local beneficiaries, target groups, user groups, affected groups, special interest groups (women)
- Relevant government and private sector agencies
- Local leaders and academic groups
- Relevant NGOs
- Recognized experts.

Methods to involve people at different levels include:

- Community meetings and information sharing
- Inter-personal contact
- Dialogue with user groups and local leaders
- Questionnaires/survey/interview
- Local community review of draft EIA report and monitoring/evaluation results
- Panels comprising representatives of concerned organizations and local people.

REFERENCES

Allee, W.C., A.E. Emerson, O. Park, T. Park and K.P. Schmidt. 1949. *Principles of animal ecology*. W.B. Saunders, Philadelphia.

Ambasht, R.S. 1969. *A text book of plant ecology*. Students' Friends and Co., Varanasi.

Anderson, K.L. 1965. Fire ecology- some Kansas prairie forbs. *Proc. Tall Timbers Fire Ecol. Conf.*, no. 4. Tall Timbers Research Station. Tallahassee, Fl. pp.153-159.

Anonymous. 1982. Pandas eat a lot, but digest little. *New Scientist* 2 September.pp. 620.

Avers, C.J. and R.H. Goodwin. 1956. Studies on roots.IV. Effects of coumarin and scopoletin on the standard root growth pattern of *Pheleum pretense*. *American Journal of Botany 43*: 612-620.

Axelrod, D.I and H.P. Bailey. 1969. Paleotemperature analysis of Tertiary floras. *Paleogeography, Paleoclimatology, Paleoecology 6*: 163–195.

Ayala, F.J. 1968. Genotype, environment and population numbers. *Science 162*: 1453-1459.

Bagnouls, F. and H. Gaussen. 1957. Les climates biologiques et leur classification. *Annals de Geography 355*: 193–220.

Baldwin, M., C.E. Kellogg and J. Thorp. 1938. Classification of soils on the basis of their characteristics. In *Soils and man: 1938 yearbook of agriculture*. US Dept. of Agriculture, Washington D.C., pp. 979-1001.

Barbour, M.G. 1970. Seedling ecology of *Cakile maritima* along the California coast. *Bulletin of the Torrey Botanical Club 97*: 280-289.

Barbour, M.G., J.H. Burk and W.D. Pitts. 1980. *Terrestrial plant ecology*. The Benjamin /Cummings Publishing Co., Inc., California.

Barnes, R.S.K. and R.N. Hughes. 1988. *An introduction to marine ecology*. 2nd edn. Blackwell Scientific Publications, Oxford.

Barret, J. and C.M. Yonge. 1958. *Collins pocket guide to the sea shore*. Collins, London.

Barshad, I. 1955. Soil development. *In*: F.E. Bear (ed.), *Chemistry of the soil*. Reinhold, New York.

Bazzaz, F.A. 1990. The response of natural ecosystems to the rising global CO_2 levels. *Annual Review of Ecology and Systematics 21*:167-196.

Beatley, J.C. 1967. Survival of winter annuals in the northern Mojave Desert. *Ecology 48*: 745-750.

Bell, R.J.V. 1971. A grazing system in the Serengeti. *Scientific American 225* (1): 86-93.

Bergamini, D. 1965. *The land and wildlife of Australia*. Time-Life International.

Bewley, J.D. and M. Black. 1994. *Seed: physiology of development and germination*. 2nd edn. Plenum Press, New York and London.

Bhatt, D.D. 1977. *Natural history and economic botany of Nepal*. Orient Longman Ltd., New Delhi.

Bhowmik, P.C. and J.D. Doll. 1984. Allelopathic effects of annual weed residues on growth and nutrient uptake of corn and soybeans. *Agronomy Journal 76*: 383-388.

Bidwell, R.G.S. 1974. *Plant physiology*. Macmillan, New York.

Billings, W.D. 1952. The environmental complex in relation to plant growth and distribution. *Quarterly Review of Biology 27*: 251-265.

Billings, W.D. 1970. *Plants, man, and the ecosystem*. 2nd edn. Wadsworth, Belmont, CA.

Biswell, H.H. 1958. Prescribed burning in Georgia and California compared. *Journal of Range Management 11*: 293-298.

Biswell, H.H. 1974. Effects of fire on chaparral. *In:* T.T. Kozlowski and C.E. Ahlgren (eds.), *Fire and ecosystems*. Academic Press, New York.

Biswell, H.H. 1977. *Giant sequoia fire ecology*. University Extension Course X 417.4. University of California, Davis, CA.

Bjorkman, O., H.A. Mooney and J. Ehleringer. 1975. Photosynthetic characteristics of plants from habitats with contrasting thermal regimes: comparisons of photosynthetic responses of intact plants. *Carnegie*

Institution of Washington Yearbook 74: 743-748.

Black, C. C. 1973. Photosynthetic carbon fixation in relation to net CO_2 uptake. *Annual Review of Plant Physiology 24*:253-286.

Blackman, F.F. 1905. Optima and limiting factors. *Annals of Botany 14* (74): 281-295.

Bliss, L.C. 1988. *In*: M.G. Barbour and W.D. Billings (eds.), *North American terrestrial vegetation*. Cambridge University Press, Cambridge. pp. 1-32.

Booth, W.E. 1941. Revegetation of abandoned fields in Kansas and Oklahoma. *American Journal of Botany 28*: 415-422.

Bowers, W.S., T. Ohta, J.S. Cleere and P.A. Marsella. 1976. Discovery of insect anti-juvenile hormones in plants. *Science 193*: 542-547.

Bradshaw, A.D. 1965. Evolutionary significance of phenotypic plasticity in plants. *Advances in Genetics 13*: 115-155.

Brady, N.C. 1995. *The nature and properties of soils*. 10th ed. Prentice-Hall of India, New Delhi.

Braun–Blanquet, J. 1932. *Plant sociology: the study of plant communities*. McGraw–Hill, New York.

Burkholder, P.R. 1952. Cooperation and conflict among primitive organisms. *American Scientist 40*: 601-631.

Calow, P. 1977. Conversion efficiencies in heterotrophic organisms. *Biological Reviews of the Cambridge Philosophical Society 52*: 385-409.

Campbell, G.S. 1977. *An introduction to environmental biophysics*. Springer–Verlag, New York.

Capon, B. and W. Van Asdall. 1966. Heat pretreatment as a means of increasing germination of desert annual seeds. *Ecology* 48: 305-306.

Cates, R.G. and G.H. Orians. 1975. Successional status and the palatability of plants to generalized herbivores. *Ecology 56*: 410-418.

CBS (Central Bureau of Statistics). 2002. *Preliminary results of population census 2001*. HMG/Nepal, National Planning Commission Secretariat, Kathmandu.

Chapin, F.S. 1980. The mineral nutrition of wild plants. *Annual Review of Ecology and Systematics 11*: 233-260.

Chapman, J.L. and M.J. Reiss. 1992. *Ecology: principles and applications*. Cambridge University Press, Cambridge.

Chaudhary, R.P. 1998. *Biodiversity in Nepal (status and conservation)*. Tecpress Books, Bangkok and S. Devi, Saharanpur, Uttar Pradesh, India.

Chou, C.H. and S.J. Chou. 1979. Autointoxication mechanism of *Oryza sativa*. II. Effects of cultural treatments on the chemical nature of paddy soil and on rice productivity. *Journal of Chemical Ecology 5*: 839-859.

Clements, F.E. 1916. *Plant succession: an analysis of the development of vegetation*. Carnegie Institution of Washingten Publ. 242, Carnegie Institution of Washington, Washington, D.C.

Clements, F.E. 1936. Nature and structure of the climax. *Journal of Ecology 24*: 252-284.

Cloudsley-Thompson, J.L. and M.J. Chadwick. 1964. *Life in deserts*. Dufour, Philadelphia PA.

Cody, M.L. 1974. Optimization in ecology. *Science 183*: 1156-1164.

Cody, M.L. 1986. Roots in plant ecology. *Trends in Ecology and Evolution 1*: 76-78

Cohen, M.N. 1980. Speculations on the evolution of density measurement and population regulation in *Homo sapiens*. *In*: M.N. Cohen, R.S. Malpass and H.G. Klein (eds.), *Biosocial mechanisms of population regulation*. Yale University Press. pp.275-303.

Colton, C.E. and F.A. Einhellig. 1980. Allelopathy mechanisms of velvetleaf (*Abutilon theophrasti* Medic.) on soybean. *American Journal of Botany 67*: 1407-1413.

Cottam, W.P., J.M. Tucker and R. Drobnick. 1959. Some clues to great basin postpluvian climates provided by oak distributions. *Ecology 40*: 361-377.

Cowles, H.C. 1911. The causes of vegetative cycles. *Botanical Gazette 51*: 161-183.

Creber, G.T. 1977. Tree-rings: a natural data-storage system. *Biological Reviews of the Cambridge Philosophical Society 52*: 349-383.

Cresswell, E.G. and J.P. Grime. 1981. Induction of a light requirement during seed development and its ecological consequences. *Nature* 291: 583-585.

Cronquist, A. 1988. *The evolution and classification of flowering plant*. 2nd edn. The New York Botanical Garden, New York.

Currey, D.B. 1965. An ancient bristlecone pine stand in eastern

Nevada. *Ecology 46*: 564-566.

Cutler, J.M, D.W. Rains and R.S Loomis. 1977. The importance of cell size in the water relations of plants. *Physiologia Plantarum 40*: 255-260.

Dahlman, R .C. and C. L. Kucera. 1965. Root productivity and turnover in native prairie. *Ecology 46*: 84-89.

Dansereau, P. 1951. Description and recording of vegetation upon a structural basis. *Ecology* 25: 172-229.

Darwin, C.R. 1859. *On the origin of species by means of natural selection, or the preservation of favoured races in the struggle for life*. John Murray, London.

Daubenmire, R.F. 1952. Forest vegetation of northern Idaho and adjacent Washington and its bearing on concepts of vegetation classification. *Ecological Monongraphs 22*: 301-330.

Daubenmire, R.F. 1968. Ecology of fire in grasslands. *Advances in Ecological Research 5*: 209-266.

Daubenmire, R.F. 1968. *Plant communities*. Harper and Row, New York.

De Bano, L.F., P.H. Dunn and C.E. Conrad. 1977. Fire's effect on physical and chemical properties of chaparral soils. *Proc. Symp. Env. Cons. Fire and Fuel Managmt. in Medit. Ecosyst*. U.S. Dept. of Agriculture Forest Service, Washington D.C. pp.65-74.

De Vleeschauwer, D. and R. Lal 1981. Properties of worm casts under secondary tropical forest regrowth. *Soil Science* 132: 175-181.

De Wit, C.D. 1960. *On competition*. Versl. Landouwkl. Onderzoek. no. 66.8 Wageningen, Center for Agricultural Publications and Documentation, The Netherlands.

De Wit, C.D. 1961. Space relationships within populations of one or more species. *In*: F.L. Milthorpe (ed.), *Mechanisms in biological competition.* Cambridge University Press, New York. pp. 314-429.

Deevey, E.S.1947. Life tables for natural populations of animals. *Quarterly Review of Biology 22*: 283-314.

Deevey, E.S.1960. The human population. *Scientific American 203*: 194-204.

Deevey, E.S. 1970. Mineral cycles. *Scientific American 223* (3): 148-160.

Devi, S.R., F. Pellissier and M.N.V. Prasad. 1997. Allelochemicals. *In*: M.N.V. Prasad (ed.), *Plant ecophysiology*. John Wiley and Sons, Inc., New York. pp.253-303.

Dice, L.R. 1952. *Natural communities*. University of Michigan Press, Michigan.

Dix, R.L. 1957. Sugar maple in forest succession at Washington D.C. *Ecology 38*: 663-665.

Dobremez, J.F. 1992. *Mise an point d'une methode cartographique d'etude des montagnes tropical. Le Nepal, ecologies et phytogeographie*. University of Grenoble, France (D. Sc Thesis).

Drury, W.H. 1956. Bog flats and physiographic processes in the upper Kuskokwin river region, Alaska. *Gray Herbarium Contributions 178*: 1-30.

Drury, W.H. and I.C.T. Nisbet. 1973. Succession. *Journal of the Arnold Arboretum 54*: 31-368.

Dunn, P.H. and L.F. DeBano. 1977. Fire's effect on physical and chemical properties of chaparral soils. *Proc. Symp. Env. Cons. Fire and Fuel Managmt. in Medit. Ecosyst*. U.S. Dept. of Agriculture Forest Service, Washington D.C. pp.75-84.

East Consultant. 1990. *Environmental impact study of drinking water supply in Kathmandu and Lalitpur from outside the valley, Kathmandu*. Report, East Consultant, Kathmandu.

Edwards, J.S. 1988. Life in the allobiosphere. *Trends in Ecology and Evolution 3*: 111-114.

Egler, F.E. 1954. Vegetation science concepts. I. Initial floristic composition, a factor in old-field vegetation development. *Vegetatio 4*: 412-417.

Einhellig, F.A. 1989. Current status and future challenges in allelopathy. *In*: *Report on the workshop on chemical interactions between organisms*, Santiago, Chile, Nov. 12-17, 1989. International Foundation for Science, Stockholm. pp.93-106.

Ellenberg, H. 1958. Bodenreaktion (einschlieplich Kalfrage). *In*: W. Ruhland (ed.) *Handbuch der pflanzenphysiologie*, vol. 4. Springer-Verlag, Berlin. pp. 638-708.

Elton, C. 1927. *Animal ecology*. Sidgwick and Jackson, London.

Epstein, E. 1972. *Mineral nutrition of plants:principles and perspectives*. John Wiley & Sons, Inc., New York.

Esau, K. 1965. *Plants anatomy*. Wiley, New York.

Faegri, K. and L. van der Pijl. 1971. *The principles of pollination ecology*. Pergamon press, New York.

Feeny, P.P. and H. Bostock. 1968. Seasonal changes in the tannin content of oak leaves. *Phytochemistry 7*: 871-880.

Fisher, R.F., R.A. Woods and M.R. Glavicic. 1978. Allelopathic effects of golden rod and aster on young sugar maple. *Canadian Journal of Forest Resources 8*: 1-9.

Forbes, S.A. 1884. The lake as a microcosm. *Bulletin of the Peoria Science Association*: 77-87.

Foster, N.W. and I.K. Morrison. 1976. Distribution and cycling of nutrients in a natural *Pinus banksiana* ecosystem. *Ecology 57*: 110-120.

Frankland, B. and R. Taylorson. 1983. Light control of seed germination. *In*: W. Shropshire Jr. and H. Mohr (eds.), *Photomorphogenesis, encyclopedia of plant physiology,* New Series, vol. 16. Springer-Verlag, Berlin. pp. 428-456.

Friday, A. and D.S. Ingram (eds.). 1985. *The Cambridge encyclopedia of life sciences*. Cambridge University Press, Cambridge.

Galtie, A. 1999. Is El Nino now a man–made phenomenon? *The ecologist 29* (2): 64–67.

Gates, D.M. 1965. Energy, plants and ecology. *Ecology 46*: 1-13.

Gates, D.M. 1972. *Man and his environment: Climate*. Harper and Row, New York.

Gause, G. F. 1934. *The struggle for existence*. Williams and Wilkins, Baltimore.

Geiger, R . 1965. *Atmosphare der Erde; Karte*
1: Jahrliche Sonnenstrahlung ; Karte
5: Jahrliche Niederschlag; Karte
6: Jahrliche Wirkliche Verdunstung . Darmstadt, Perthes.

Glass, A.D.M. 1975. Inhibition of phosphate uptake in barley roots by hydroxyl benzoic acids. *Phytochemistry* 14: 2127-2130.

Glass, A.D.M. 1976. The allelopathic potential of phenolic acids associated with the rhizosphere of *Pteridium aquilinum*. *Canadian Journal of Botany 54*: 2440-2444.

Gill, A.M. 1977. Plant traits adaptive to fires in Mediterranean land ecosystems. *Proc. Symp. Env. Cons. Fire and Fuel Managmt. in Medit. Ecosyst*. U.S. Dept. of Agriculture Forest Service, Washington D.C. pp.17-26.

Gimingham, C.H. 1972. *Ecology of heathlands*. Chapman and Hall, London.

Gleason, H.A. 1962. The individualistic concept of the plant association. *Bulletin of the Torrey Botanical Club 53*: 1-20.

Goldsworthy, A. 1976. *Photorespiration*. Carolina Biology Readers, no. 80, Carolina Biological Suppliers, Burlington.

Good, R. 1953. *The geography of flowering plants.* 2nd edn. Longmans, Green and Co., Inc., London.

Gooday, G.W. 1988. The potential of the microbial cell and its interaction with other cells. *In*: J.M. Lynch and J.E. Hobbie (eds.), *Microorganisms in action: concepts and applications in microbial ecology.* Blackwell Scientific Publications, Oxford. pp.7-32.

Gosz, J. R., G.E. Likens and F. H. Bormann. 1973. Nutrient release from decomposing leaf and branch litter in the Hubbard Brook Forest, New Hampshire. *Ecological Monographs 43*: 173-191.

Graunt, J. 1662. *Natural and political observations mentioned in a following index, and made upon the bills of mortality*. Roycraft, London.

Gregor, J.W. 1946. Ecotypic differentiation. *The New Phytologist 25*: 254-270.

Grime, J.P. 1977. Evidence for the existence of three primary strategies in plants and its relevance to ecological and evolutionary theory. *American Naturalist 111*: 1169-1194.

Grime, J.P.1979. *Plant strategies and vegetation processes.* Wiley, New York.

Grime, J.P. 1979. Plant strategies in shade. *In*: H. Smith (ed.), *Plants and the daylight spectrum*. Academic Press, Inc., New York. pp. 159-186.

Grummer, G. 1955. *Die gegenseitige Beeinflussing hoherer Pflanzen-Allelopathie*. Gustav Fischer, Jena, Germany.

Grubb, P.J. 1986. The ecology of establishment. *In*: A.D. Bradshaw, D.A. Goode and E. Thorpe (eds.), *Ecology and design in landscape.* Symposium of the British Ecological Society 24. pp. 83-97.

Guttermann, Y.1972. Delayed seed dispersal and rapid germination as survival mechanisms of the desert plant *Blepharis persica* (Burm.) Kuntze. *Oecologia 10*: 145-149.

Haeckel, E.H.P.A. 1866. *Generelle Morphologie der Organismen*. Reimer, Berlin.

Hall, A.B., U. Blum and R.G. Fites. 1982. Stress modification of allelopathy of *Helianthus annuus* L. debris on seed germination. *American Journal of Botany 69*: 776-783.

Hamilton, W.A. 1988. Microbial energetics and metabolism. *In*: J.M. Lynch and J.E. Hobbie (eds.), *Microorganism in action: concepts and applications in microbial ecology*. Blackwell Scientific Publications, Oxford. pp.75-100.

Harper, J.L. 1964. The individual in the population. *Journal of Ecology 52* (supplement): 149-158.

Harper, J.L and J. Ogden. 1970. The reproductive strategy of higher plants.I. The concept of strategy with special reference to *Senecio vulgaris*. *Journal of Ecology 58*: 681-698.

Harper, J.L. and J. White 1974. The demography of plants. *Annual Review of Ecology and Systematics 5*: 419-463.

Harris, G.A. 1967. Some competitive relationships between *Agropyron spicatum* and *Bromus tectorum*. *Ecological Monographs 37*: 89-111.

Hartmann, H.T., D.E. Kester, F.T. Davies (Jr.), and R.L. Geneve, 1997. *Plant propagation: principles and practices*. 6 th edn. Prentice-Hall, Inc., New Jersey, USA.

Harvey, H.T., H.S. Shellhammer and R.E. Stecker. 1977. *Giant sequoia ecology: fire and reproduction*. U.S. Dept. of the interior, National Park Service Sci. Mono 12. U.S. Dept. of the interior, National Park Service, Washington D.C.

Harvey, H. W :1950. On the production of living matter in the sea off Plymouth. *Journal of Marine Biology 29*: 97-137.

Hassan, F.A. 1980. The growth and regulation of human population in prehistoric times. *In*: Cohen, M.N., R.S. Malpass and H.G. Klein (eds.), *Biosocial mechanisms of population regulation*. Yale University Press. pp.305-319.

Hickman, J.C.1975. Environmental unpredictability and plastic energy allocation strategies in the annual *Polygonum cascadense*. *Journal of Ecology 63*: 689-701.

Hickman, J.C.1977. Energy allocation and niche differentiation in four coexisting species of *Polygonum* in western North America. *Journal of Ecology 65*: 317-326.

HMG/Nepal (His Majesty's Government of Nepal). 2002. *Nepal biodiversity strategy*. Ministry of Forest and Soil Conservation, Kathmandu.

Holdridge, L.R. 1967. *Life zone ecology*. Tropical science centre, Costa Rica.

Hopkins, A.D. 1938. *Bioclimatics: a science of life and climatic relations*. U.S. Department of Agriculture, Washington D.C.

Horn, H.S. 1971. *The adaptive geometry of trees*. Princeton University Press, Princeton, N.J.

Horn, H.S. 1974. The ecology of secondary succession. *Annual Review of Ecology and Systematics 5*: 25-37.

Horn, H.S. 1975. Forest succession. *Scientific American 232 (5)*: 90-98.

Howarth, R.W., A. Giblin, J. Gale, B.J. Peterson and G.W. Luther III. 1983. Reduced sulphur compounds in the pore waters of a New England salt marsh. *Environment, Biochemistry and Geology Bulletin* (Stockholm) *35*: 135-152.

Humboldt, A.von. 1807-1834. *Voyage aux regions equinoxiales du nouveau continet*.30 vols. Paris.

Humphrey, R.R. 1974. Fire in the deserts and desert grassland of North America. *In*: T.T. Kozlowski and C.E. Ahlgren (eds.), *Fire and ecosystems*, Chapter 10. Academic Press, New York.

Hunt, W.H. 1977. A simulation model for decomposition in grasslands. *Ecology* 58: 469-484.

Hutchings, M.J. and J.P. Barkham. 1976. An investigation of shoot interactions in *Mercurialis perennis* L., a rhizomatous perennial herb. *Journal of Ecology 64*: 723-743.

Hutchinson, G.E. 1958. Concluding remarks. *Cold Spring Harbour Symposium on Quantitative Biology 22*: 415-427.

Jackson, J.B.C. 1968. Bivalves: spatial and size-frequency distributions of two intertidal species. *Science 161*: 479-480.

Jackson, M.T. 1966. Effects of microclimate on spring flowering phenology. *Ecology 47*: 407-415.

Janzen, D.H. 1970. Herbivores and the number of tree species in tropical forests. *American Naturalist 104*: 501- 528.

Jeffree, E.P. 1960. Some of the long term means from the phenological reports (1891-1948) of the Royal Meteorological Society. *Quarterly Journal of the Royal Meteorological Society 86*: 95-103.

Jenny, H. 1941. *The factors of soil formation*. Mc Graw-Hill, New York.

Jha, S. 1990. *Conservation for development in Nepal*. National Book Organisation, New Delhi.

Jha, S. 2003. *Ecological study of some selected grasses and forbs found in Morang district of Nepal.* Tribhuvan University, Kathmandu (Ph.D. Thesis).

Jha, S. 2004. The wetland ecology. *Our Nature 2*: 47-52.

Jha, S. and A. Karn. 2001. Climatic analogues for the administrative districts of Nepal. *Journal of Natural History Museum (Tribhuvan University) 20*: 69–82.

Jordon, C.F. 1985. *Nutrient cycling in tropical forest ecosystems.* John Wiley, Chichester.

Jordan, D.F and J. R. Kline. 1972. Mineral cycling: some basic concepts and their application in a tropical rain forest. *Annual Review of Ecology and Systematics 3*: 33-50.

Kauffman, J.B., R.L. Sanford, D.L. Cummings, I.H. Salcedo and E.V.S.B. Sampaio. 1993. Biomass and nutrient dynamics associated with slash fires in neotropical dry forests. *Ecology 74*: 140-151.

Kluge, M. 1974. Metabolism of carbohydrates and organic acids. *In*: H. Ellenberg (ed.), *Progress in botany*, vol. 36, Springer-Verlag, Berlin. pp. 90-98.

Kenward, R.E. 1978. Hawks and doves: factors affecting success and selection in goshawk attacks on woodpigeons. *Journal of Animal Ecology 47*: 449-460.

Kershaw, K.A. 1973. *Quantitative and dynamic plant ecology.* 2nd edn. American Elsevier, New York.

Kira, T. 1975. Primary production in forests. *In*: J. P. Cooper (ed.), *Photosynthesis and productivity in different environments.* Cambridge University Press, Cambridge.

Kittredge, J. 1944. Estimation of the amount of foliage on trees and stands. *Journal of Forestry 42*: 905-912.

Klemm, F.B. 1990. *The fluid earth: physical science and technology of the marine environment.* Curriculum Research and Development Group, College of Education, University of Hawaii, Honolulu.

Knipling, E.B. 1967. Measurement of leaf water potential by the dye method. *Ecology 48*: 1038-1041.

Koepp, D.E., L.M. Rohrbaugh, E.L. Rice and S.H. Wender. 1970. The effect of x-radiation on the concentration of scopolin and cuffeoxylquinic acids in tobacco. *Radiation Botany 10*: 261-265.

Koepp, D.E., L.M. Southwick and J.E. Bittell. 1976. The relationship of tissue chlorogenic acid concentration and leaching of phenolics from sunflower grown under varying phosphate nutrient conditions. *Canadian Journal of Botany 54*: 593-599.

Koller, D. 1969. The physiology of dormancy and survival of plants in desert environments. *In*: H.W. Woolhouse (ed.), *Dormancy and survival*. Symposia for the Society of Experimental Biology, no. 23, Academic Press Inc., New York. pp. 449-469.

Komarek, E.V. 1965. Fire ecology – grasslands and man. *Proc. Tall Timbers Fire Ecol. Conf.* no. 4. Tall Timbers Research Station, Tallahassee, Fl. pp. 169-220.

Koppen, W.1918. Klassifikation der Klimate nach Temperature, Niederschlat, und jahres Lauf. *Petermann's Mitteilungen 64*: 193–203.

Kramer, P.J. 1969. *Plant and soil water relationships: a modern synthesis*. McGraw Hill Book Co., New York.

Kruckeberg, A.R. 1954. The ecology of serpentine soils. III. Plant species in relation to serpentine soils. *Ecology 35*: 267-274.

Kuchler, A.W. 1967. *Vegetation mapping*. Ronald Press, New York.

Lal, D.S. 2000. *Climatology*. Sharda Pustak Bhawan, Allahabad.

Larcher, W. 1975. *Physiological plant ecology*. Springer-Verlag, Berlin–Heidelbeg–New York.

Larcher, W. and H. Bauer. 1981. Ecological significance of resistance to low temperature. *In*: O.L Lange, P.S. Nobel, C.S. Osmond and H. Ziegler (eds.), *Encyclopedia of plant physiology. New series Vol. 12A, Physiological plant ecology*. Springer-Verlag, Berlin-Heidelberg–New York. pp. 403-437.

Langlet, O. 1959. A cline or not a cline– a question of scots pine. *Sylvae Genetica 8*: 13-22.

Leopold, A.C. and P.E. Kriedemann. 1975. *Plant growth and development*. 2nd edn. McGraw-Hill, New York.

Liebig, J. 1840. *Chemistry in its agriculture and physiology*. Taylor and Walton, London.

Lieth, H. 1968. The measurement of calorific values of biological material and determination of ecological efficiency. *In*: F.E. Eckardt (ed.), *Functioning of terrestrial ecosystem at primary level,* Proceedings Copenhagen symposium, UNESCO.

Lieth, H. 1975. The measurement of calorific values. *In*: H. Lieth and R.H. Whittaker (eds.), *Primary productivity of Biosphere,*. Springer–Verlag. pp.119-130.

Likens, G.E., F.H. Bormann, R.S. Pierce, J.S. Eaton and N.M Johnson. 1977. *Biochemistry of a forested ecosystem*. Springer-Verlag, New York.

Longman, K.A. and J. Jenik. 1974. *Tropical forest and its environment*. Longman, London.

Lotan, J.E. 1974. Cone serotiny-fire relationships in lodgepole pine. *Proc. Tall Timbers Fire Ecol. Conf.*, no. 4. Tall Timbers Research Station. Tallahassee, Fl. pp.267-278.

Lotka, A.J. 1925. *Elements of physical biology*. Williams and Wilkins, Baltimore.

Loucks, O.L. 1970. Evolution of diversity, efficiency, and community stability. *American Zoologist 10*: 17-25.

Lucas, Y., F.J. Luizao, A. Chauvel, J. Rouiller and D. Nahon. 1993. The relation between biological activity of the rainforest and mineral composition of soils. *Science 260*: 521-523.

Luther, G.W., T.M. Church, J.R. Scudlark and M. Cosman. 1986. Inorganic and organic sulphur cycling in salt-marsh pore waters. *Science 232*: 746-749.

MacArthur, R.H. 1958. Population ecology of some warblers of northeastern coniferous forests. *Ecology 39*: 599-619.

MacArthur, R.H. and E.R. Pianka. 1966. On optimal use of a patchy environment. *American Naturalist 100*: 603-609.

Mahadevan, S., Z. Futehally and D. Lahiry. 1988. *Biology*. NCERT, New Delhi.

Majupuria, T.C. and R. K. Majupuria. 1998. *Wildlife, national parks and reserves of Nepal*. S. Devi, Saharanpur, Uttar Pradesh, India and Tecpress Books, Bangkok, Thailand.

Malthus, T.R. 1798. *An essay on the principle of population*. Macmillan, New York.

Marx, D.H. and N.C. Schenck. 1983. Potential of mycorrhizal symbiosis in agricultural and forest productivity. *In*: T. Kommédahl and P.H. William (eds.), *Challenging problems in plant health*. American Phytopathological Society, Minnesota. pp. 334-347.

McCutchan, M.H. 1977. Climatic features as a fire determinant. *Proc. Symp. Env. Cons. Fire and Fuel Managmt. in Medit. Ecosyst*. U.S. Dept. of Agriculture Forest Service, Washington D.C. pp.1-11.

Meher–Homji, V.M. 1971. Bio-climatic variability with special reference to India. *Tropical Ecology 12* (2): 155–176.

Merriam, C.H. 1894. Laws of temperature control of the geographical distribution of animals and plants. *National Geographic 6*: 229-238.

Mertens, J.A.L. 1969. The influence of brood size on the energy metabolism and water loss of nestling great tits *Parus major major*. *Ibis 111*: 11-16.

Mertz, W. 1981. The essential trace elements. *Science 213*: 1332-1338.

Michel, B.E. 1972. Solute potentials of sucrose solutions. *Plant Physiology 50*: 196-198.

Miller, W.J. and M.N. Neathery. 1977. Newly recognized trace mineral elements and their role in animal nutrition. *BioScience 27*: 674-679.

Minnich, R.A. 1983. Fire mosaics in southern California and northern Baja California. *Science 219*: 1287-1294.

Mitsch, W.J. and J.G. Grosselink. 1986. *Wetlands*. Van Nostrand Reinhold Co., New York.

Mohr, H. 1983. Pattern specification and realization in photomorphogenesis. *In*: W. Shropshire Jr. and H. Mohr (eds.), *Photomorphogenesis: Encyclopedia of Plant Physiology*, New Series, vol. 16A. Springer-Verlag, Berlin. pp. 338-357.

Molisch, H. 1937. *Der einfluss einer pflanze auf die andere-allelopathie*. Gustav Fischer, Jena, Germany.

Monk, C.D. 1966. An ecological significance of evergreenness. *Ecology 47*: 504-505.

Moorby, J. and R.T. Besford. 1983. Mineral nutition and growth. *In*: Lauchli and R.L. Bieleski (eds.), *Encyclopedia of plant physiology, New series vol. 15B Inorganic plant nutrition*. Springer Verlag, Berlin. pp. 481-527.

Mooney, H.A., J. Ehleringer and O. Bjorkman. 1977. The energy balance of leaves of the evergreen desert shrub *Atripex hymenelytra*. *Oecologie 29*: 301-310.

Moore, R.M. (ed.) 1970. *Australian grasslands*. Australian National University Press, Canberra, Australia.

Mueller-Dombois, D. and H. Ellenberg. 1974. *Aims and methods of vegetation ecology*. John Wiley and Sons, Inc., New York.

Muller, C.H. 1969. Allelopathy as a factor in ecological process. *Vegetatio 18*: 248-357.

Muller, C.H. and R. del Moral. 1966. Soil toxicity induced by terpenes from *Salvia leucophylla*. *Bulletin of the Torrey Botanical Club 93*: 130-137.

Mutch, R.W. 1970. Wildland fires and ecosystems – a hypothesis. *Ecology 51*: 1046-1051.

Neales, T.F. 1973. The effect of night temperatures on CO_2 assimilation, transpiration, and water use efficiency in *Agave americana* L. *Australian Journal of Biological Science* 26: 705-714.

Neill, S.R. and J.M. Cullen.1974. Experiments on whether schooling by their prey affects the hunting behaviour of cephalopods and fish predators. *Journal of Zoology* (*London*) *172*: 549-569.

Nice, M.M. 1941. The role of territory in bird life. *American Midland Naturalist 26*: 441-487.

Nobel, P.S. 1976. Water relations and photosynthesis of a desert CAM plant, *Agave deserti*. *Plant Physiology 58*: 576-582.

NPC (National Planning Commission), HMG Nepal, in collaboration with IUCN- The World Conservation Union. 1992. *National environmental impact assessment guidelines*. Kathmandu.

O'Brien, S.J., D.E. Wildt, D. Goldman, C.R. Merril and M. Bush. 1983. The cheetah is depauperate in genetic variation. *Science 221*: 459-462.

O'Brien, S.J., M.E. Roelke, L. Marker, A Newman, C.A. Winkler, D. Meltzer, L. Colly, J.F. Evermann, M. Bush and D.E. Wildt. 1985. Genetic basis for species vulnerability in the cheetah. *Science 227:* 1428-1434.

Odum, E.P. 1963. Limits of remote ecosystems containing man. *American Biology Teacher 25*: 429-443.

Odum, E.P. 1971. *Fundamentals of ecology*. W.B. Saunders, Philadelphia.

Odum, H.T. 1957. Trophic structure and productivity of Silver Springs, Florida. *Ecological Monographs 27*: 55-112.

Oechel, W.C., B.R. Strain and W.R. Odening. 1972. Tissue water potential, photosynthesis, ^{14}C-labeled photosynthetic utilization and growth in the desert shrub *Larrea divaricata*. *Ecological Monographs 42*: 127-141.

Oerke, E.C., H.W. Dehne, F. Schonbeck and A. Weber. 1994. *Crop production and crop protection : Estimated losses in major Food and Cash Crops*. Elsevier.

Olson, J. H. 1963. Energy storage and balance of producers and decomposers in ecological systems. *Ecology 44*: 322-331.

Oosting, H.J. 1942. An ecological analysis of the plant communities of Piedmont, North Carolina. *American Midland Naturalist 28*: 1-26.

Oosting, H.J. 1958. *The study of plant communities*. W.H. Freeman and Co., San Francisco.

Oosting, H.J. and W.D. Billings. 1951. A comparison of virgin spruce-fir forest in the northern and southern Appalachian system. *Ecology 32*: 84-103.

Paczoski, J. 1921. *Osnowy fitosocjologji* Izd. Stud. Comitet Tech., Cherson.

Patrick, W.H. and C.N. Reddy 1978. Chemical changes in rice soils. In: *Soils and rice*. The International Rice Research Institute, Los Banos. pp.361-379.

Pearson, G.A. 1942. Herbaceous vegetation a factor in natural regeneration of ponderosa pine in the southwest. *Ecological Monographs 12*: 315-338.

Perino, J.V. and P.G. Risser. 1972. Some aspects of structure and function in Oklahoma old-field succession. *Bulletin of the Torrey Botanical Club 99*: 233-239.

Peter, G.A. 1978. Blue-green algae and algal associations. *BioScience 28*: 580-585.

Philpot, C.W. 1977. Vegetative features as determinants of fire frequency and intensity. *Proc. Symp. Env. Cons. Fire and Fuel Managmt. in Medit. Ecosyst*. U.S. Dept. of Agriculture Forest Service, Washington D.C. pp.12-16.

Pianka, E.R. 1970. On r-and K-selection. *American Naturalist 104*: 592-597.

Polunin, O. 1950. Plant hunting in the Nepal Himalayas. *Geography Magazine 23*: 132-147.

Putman, R.J. 1994. *Community ecology*. Chapman and Hall, London.

Rabotnov, T.A.1969. On coenopopulations of perennial herbaceous plants in natural coenoses. *Vegetatio 19*: 87-95.

Rajbhandari, K.R. 2002. Flora of Nepal: 200 years march. *In*: S. Noshiro and K.R. Rajbhandari (eds.), *Himalayan botany in the twentieth and twenty–first centuries*. The Society of Himalayan Botany, Tokyo. pp.76-93.

Rau, M.A. 1975. *High altitude flowering plants of west Himalaya*. Botanical Survey of India, Howrah.

Raunkiaer, C. 1934. *The life forms of plants and statistical plant geography*. Clarendon Press, Oxford.

Ricklefs, R.E. 1980. *Ecology*. 2nd edn. Nelson, Sunbury-on-Thames, Middlesex.

Rhoades, D.H. and R.G. Cates.1976.Toward a general theory of plant antiherbivore chemistry. *Recent Advances in Phytochemistry 10*: 168-213.

Rice, E.L. 1984. *Allelopathy*. 2nd edn. Academic Press, Inc., Orlando, USA.

Richards, P.W. 1973. The tropical rainforest. *Scientific American 229* (6): 59-67.

Riggin, P.J., S. Goode, P.M. Jacks and R.N. Lockwood. 1988. Interaction of fire and community development in chaparral of southern California. *Ecological Monographs 58*: 155-176.

Root, R.B. 1967. The niche exploitation pattern of the blue-gray gnatcatcher. *Ecological Monographs 37*: 317-350.

Rosenberg, N.J. 1974. *Micro-climate: the biological environment*. Wiley, New York.

Russell, E.W. 1973. *Soil conditions and plant growth*. 10th edn. The English Language Book Society and Longman, London.

Ruttner, F. 1963. *Fundamentals of limnology*. 3rd edn. University of Toronto Press, Toronto.

Salisbury, F.B. and C.W. Ross. 1986. *Plant Physiology*.CBS Publisher & Distributors, Delhi.

Schleich, H.H. and W. Kastle (eds.). 2002. Amphibians and reptiles of Nepal. Koenigtein, Koeltz Scientific Book, Germany.

Schimper, A.F.W. 1898. *Pflanzengeographie auf physiologischer Grundlage*. Fischer, Jena, Germany.

Scholander, P.F., H.T. Hammel, E.A. Hemmingsen and E.D. Bradstreet.1964. Hydrostatic pressure and osmotic potential in leaves of mangroves and some other plants. *National Academy of Sciences 52*: 119-125.

Schouw, J.F. 1822. *Grundtraek til en almindelig plantegeographie*. Glydendal, Copenhagen, Denmark.

Scott, D.A. (ed.). 1989. *A directory of asian wetlands*. IUCN, Gland and Cambridge.

Slatyer, R.O. 1967. *Plant-water relationships*. Academic Press, New York.

Shakya, P.R., M.K. Adhikari, K.R. Rajbhandari, R.P. Chuadhary and K.K. Shrestha. 1997. *Flora of Nepal*, Country paper presented at International Seminar-cum-Workshop on Flora of Nepal, April 15-16, Organizers Central Department of Botany, T.U., and Department of Plant Resources (HMG/Nepal), Kathmandu.

Shannon, C.E. and W. Weaver. 1949. *The mathematical theory of communication*. University of Illinois Press, Urbana, IL.

Sharma, C.K. 1999. Physiography. *In*: T.C. Majupuria and R. Kumar (eds.), *Nepal-nature's paradise*. M. Devi, Gwalior. pp. 4-8.

Shelford, V.E. 1913. *Animal communities in temperate America*. University of Chicago Press, Chicago.

Shrestha, J. 2001. Taxonomic revision of fishes of Nepal. *In*: P.K. Jha, S.R. Baral, S.B. Karmacharya, H.D. Lekhak, P. Lacoul and C.B. Baniya (eds.), *Environment and agriculture: Biodiversity, agriculture and pollution in South Asia*. Ecological Society (ECOS), Kathmandu, Nepal. pp. 171-180.

Shrestha, T.B. and R.M. Joshi, 1996. *Rare, endemic and endangered plants of Nepal*. WWF Nepal program, Kathmandu.

Shrestha, T.K. 2001. *Birds of Nepal-field ecology, natural history and conservation*. Kathmandu, Nepal.

Simpson, E.H. 1949. Measurement of diversity. *Nature* 163: 688.

Singh, J.S. and P.S. Yadava .1974. Seasonal variation in composition, plant biomass and net primary productivity of a tropical grassland at Kurukshetra, India. *Ecological Monographs 44:* 351-376.

Singh J.S. and S.P. Singh. 1992. *Forests of Himalaya*. Gyanodaya Prakashan, Nainital.

Smith, A.D. 1940. A discussion of the application of a climatological diagram, the hythergraph, to the distribution of natural vegetation types. *Ecology* 21: 184–191.

Sorenson, T. 1948. A method of establishing groups of equal amplitude in plant society based on similarity of species content. *K. Danske, Vidensk. Selsk*. 5: 1-34.

Spomer, G.G. 1973. The concepts of "interaction" and "operational environment" in environmental analyses. *Ecology 54* (1): 200-204.

Stadelmann, E.J. 1966. Evaluation of turgidity, plasmolysis, and deplasmolysis of plant cells. *In*: D.M. Prescott (ed.). *Methods in cell physiology*, vol 2. Academic Press, New York. pp.143-216.

Stainton, J.D.A. 1972. *Forests of Nepal*. John Murray, London.

Stewart, O.C. 1963. Barriers to understanding the influence of fire by aborigines on vegetation. *Proc. Tall Timbers Fire Ecol. Conf.*, no. 4. Tall Timbers Research Station. Tallahassee, Fl. pp.117-126.

Stoddard, H.L. 1936. Relation of burning to timber and wildlife. *Proc. First National Academy Wildlife Conference 1*: 1-4.

Stone, E.C. 1963. The ecological importance of dew. *Quarterly Review of Biology 38*: 328-341.

Tappeiner, J.C. and A.A. Alm. 1975. Undrgrowth vegetation effects on the nutrient content of litterfall and soils in red pine and birch stands in northern Minnesota. *Ecology 56*: 1193-1200.

Tansley, A.G. 1935. The use and abuse of vegetational concepts and terms. *Ecology 16*: 284-307.

Thompson, L.M. and F.R. Troeh. 1973. *Soils and soil fertility*. Mc Graw-Hill, New York.

Thomson, J.W. 1974. Lichenolgoy in North America 1947-1972. *Annals of the Missouri Botanical Garden 61*: 45-55.

Thornthwaite, C.W. 1931. The climates of North America: according to a new classification. *Geographical Review* 21: 633–655.

Tso, T.C., M.J. Kasperbauer and T.P. Sorokin. 1970. Effect of photoperiod and end-of-day light quality on alkaloids and phenolic compounds of tobacco. *Plant Physiology 45*: 330-333.

Turesson, G. 1922. The genotypic response of the plant species to habitat. *Hereditas 3*: 211-350.

Turesson, G. 1925. The plant species in relation to habitat and climate. *Hereditas 6*: 147-236.

Turesson, G. 1930. The selective effect of climate upon the plant species. *Hereditas 14*: 99-152.

UNESCO. 1973. *International classification and mapping of vegetation*. UNESCO, Paris.

Upreti, B.N. 1991. National parks and protected areas. In: *Background papers to the National Conservation Stretegy for Nepal*, Vol II, National Planning Commission, HMG/Nepal and the World Conservation Union (IUCN), Kathmandu. pp. 477-509.

Vasek, F.C, H.B. Johnson and D.H. Eslinger. 1975. Effects of pipeline construction on creosote bush scrub vegetation of the Mojave Desert. *Madrono 23*: 1-13.

Viro, P.J. 1974. Effects of forest fire on soil. *In*: T.T. Kozlowski and C.E. Ahlgren (eds.), *Fire and ecosystems*, Chapter 2. Academic Press, New York.

Vogl, R.J. 1974. Effects of fire on grasslands. *In*: T.T. Kozlowski and C.E. Ahlgren (eds.), *Fire and ecosystems*, Chapter 5. Academic Press, New York.

Volterra,V. 1926. Fluctuations in the abundance of a species considered mathematically. *Nature 118*: 558-560.

Wakimoto, R.H. 1977. Chaparral growth and fuel assessment in southern California. *Proc. Symp. Env. Cons. Fire and Fuel Managmt. in Medit. Ecosyst*. U.S. Dept. of Agriculture Forest Service, Washington D.C. pp. 412-418.

Walker, R.B. 1954. The ecology of serpentine soils. II. Factors affecting plant growth on serpentine soils. *Ecology* 35: 259-266.

Walker, W.D. 1970. Direction and rate in some British postglacial hydroseres. *In*: D. Walker and R.West (eds.), *The vegetational history of the British Isles*. Cambridge University Press, Cambridge. pp. 117-139.

Walkley, A.J. and I.A. Black 1934. An examination of method for determining organic carbon and nitrogen in soil. *Journal of Agricultural Science* 25: 589-609.

Walter, H. 1973. *Vegetation of the earth in relation to climate and the eco-physiological conditions*. Springer-Verlag, New York.

Walter, H. 1979. *Vegetation of the earth and ecological systems of the geobiosphere*. 2nd edn. Translated from the third, revised German edition by J. Wieser. Springer-Verlag, New York.

Waksman, S.A. 1948. *Humus*. Williams and Wilkins, Baltimore.

Warming, J.E.B. 1909. *Oecology of plants*. Oxford University Press, London.

Watanabe, R., W.J. McIlirath, J. Skok, W. Chorney and S.H. Wender. 1961. Accumulation of scopoletin glucoside in boron-deficient tobacco leaves. *Archives of Biochemistry and Biophysics 94*: 241-243.

WCMC (World Conservation Monitoring Centre), 1992. *Global*

biodiversity-status of the earth's living resources (B. Groombridge ed.). Chapman and Hall, London.

Went, F.W. and N. Stark. 1968. Mycorrhiza. *Bioscience 18*: 1035-1039.

Whittaker, R.H. 1969. New concepts of kingdoms of organisms. *Science 163*: 150-160.

Whittaker, R.H. 1975. *Communities and ecosystems.* 2nd edn. Macmillan, New York.

Whittaker, R.H. and G.M. Woodwell. 1972. Evolution of natural communities. *In*: J.A. Wiens (ed.), *Ecosystem structure and function*, Annual Biology Colloquia, Volume 31. Oregon State University, Corvallis, OR.

Wiegert, R.G. and J.T. McGinnis.1975. Annual production and disappearance of detritus on three South Carolina old fields. *Ecology 56*: 129-140.

Wilson, E.O. 1999. *The diversity of life.* W.W. Norton and Co., Inc., New York.

Woodward, F.I. 1987. *Climate and plant distribution.* Cambridge University Press, Cambridge.

Yost, S.E. 1987. The effect of shade on petiole length in the *Viola fimbriatula-sagittata* complex. *Brittonia 39*: 180-187.

Zucker, M. 1963. The influence of light on synthesis of protein and of chlorogenic acid in potato tuber tissue. *Plant Physiology 38*: 575-580.

